AF906132

Ensuring Secure Connectivity Through AI-Powered Wireless Systems

Valentina E. Balas
https://orcid.org/0000-0003-0885-1283
Aurel Vlaicu University of Arad, Romania

Hari Mohan Pandey
https://orcid.org/0000-0002-9128-068X
Bournemouth University, UK

Shilpa Mehta
https://orcid.org/0000-0002-4691-3185
Independent Researcher, New Zealand

Vineeta Singh
https://orcid.org/0000-0003-1084-279X
Chhatrapati Shahu Ji Maharaj University, Kanpur, India

Kapil Joshi
https://orcid.org/0000-0003-1097-8347
Uttaranchal University, India

IGI Global
Scientific Publishing
Publishing Tomorrow's Research Today

Vice President of Editorial Melissa Wagner
Director of Acquisitions Mikaela Felty
Director of Book Development Jocelynn Hessler
Production Manager Mike Brehm
Cover Design Jose Rosado

Published in the United States of America by
IGI Global Scientific Publishing
701 East Chocolate Avenue
Hershey, PA, 17033, USA
Tel: 717-533-8845 | Fax: 717-533-7115
Website: https://www.igi-global.com E-mail: cust@igi-global.com

Copyright © 2026 by IGI Global Scientific Publishing. All rights reserved. No part of this publication may be reproduced, stored or distributed in any form or by any means, electronic or mechanical, including photocopying, without written permission from the publisher. Use of this publication to train generative artificial intelligence (AI) technologies is expressly prohibited. The publisher reserves all rights to license its use for generative AI training and machine learning model development.

Product or company names used in this set are for identification purposes only. Inclusion of the names of the products or companies does not indicate a claim of ownership by IGI Global Scientific Publishing of the trademark or registered trademark.

Library of Congress Cataloging-in-Publication Data

Names: Balas, Valentina Emilia editor | Pandey, Hari (Hari Mohan), 1982- editor | Mehta, Shilpa, 1987- editor | Singh, Vineeta, 1988- editor | Joshi, Kapil, 1985- editor
Title: Ensuring secure connectivity through AI-powered wireless systems / edited by Valentina Emilia Balas, Hari Pandey, Shilpa Mehta, Vineeta Singh, Kapil Joshi.
Description: Hershey, PA : IGI Global Scientific Publishing, [2026] | Includes bibliographical references and index. | Summary: "This book aims to provide a comprehensive overview of AI-powered security solutions for wireless communication networks"-- Provided by publisher.
Identifiers: LCCN 2025021024 (print) | LCCN 2025021025 (ebook) | ISBN 9798337335766 h/c | ISBN 9798337335773 s/c | ISBN 9798337335780 ebook
Subjects: LCSH: Wireless communication systems--Security measures | Artificial intelligence--Security measures | Computer security | Computer networks--Security measures
Classification: LCC TK5103.2 .E575 2026 (print) | LCC TK5103.2 (ebook)
LC record available at https://lccn.loc.gov/2025021024
LC ebook record available at https://lccn.loc.gov/2025021025

British Cataloguing in Publication Data
A Cataloguing in Publication record for this book is available from the British Library.

All work contributed to this book is new, previously-unpublished material.
The views expressed in this book are those of the authors, but not necessarily of the publisher.
This book contains information sourced from authentic and highly regarded references, with reasonable efforts made to ensure the reliability of the data and information presented. The authors, editors, and publisher believe the information in this book to be accurate and true as of the date of publication. Every effort has been made to trace and credit the copyright holders of all materials included. However, the authors, editors, and publisher cannot assume responsibility for the validity of all materials or the consequences of their use. Should any copyright material be found unacknowledged, please inform the publisher so that corrections may be made in future reprints.

Editorial Advisory Board and List of Reviewers

Editorial Advisory Board

Vipul Arora, *Indian Institute of Technology, Kanpur, India*
Vishal Awasthi, *Chhatrapati Shahu Ji Maharaj University, Kanpur, India*
Prabodh Bajpai, *Indian Institute of Technology, Kanpur, India*
Marius Balas, *Aurel Vlaicu University of Arad, Romania*
Cornel Barna, *Aurel Vlaicu University of Arad, Romania*
Simona Dzitac, *University of Oradea, Romania*
Ruchir Gupta, *Indian Institute of Technology, BHU, Varanasi, India*
Neeraj Kumar, *Chhatrapati Shahu Ji Maharaj University, Kanpur, India*
Pabitra Mitra, *Indian Institute of Technology, Kharagpur, India*
Nirav Narendrakumar Modh, *Infosys Limited, Canada*
Surya Prakash, *Indian Institute of Technology, Indore, India*
Dana Rad, *Aurel Vlaicu University of Arad, Romania*
Vinayakumar Ravi, *Prince Mohammad Bin Fahd University, Khobar, Saudi Arabia*
Ankush Sharma, *Indian Institute of Technology, Kanpur, India*
Ajeet Srivastava, *Chhatrapati Shahu Ji Maharaj University, Kanpur, India*

List of Reviewers

Mohammad Al khaldy, *University of Petra, Jordan*
Sreedevi B, *SASTRA University, India*
Naqqash Dilshad, *Sejong University, South Korea*
Eyman F. A. Elsmany, *University of Gezira, Sudan*
Amit Gupta, *Harcourt Butler Technical University, Kanpur, India*

Kutubuddin Sayyad Liyakat Kazi, *Eudoxia Research University, USA*
Mayada S. A. Mustafa, *Uruk University, Iraq*
Flavius Petcut, *Aurel Vlaicu University of Arad, Romania*
Sridevi, *Karnatak University Dharwad, Karnataka, India*
Vitri Tundjungsari, *Esa Unggul University Jakarta, Indonesia*
Shaik Mastan Vali, *University of Technology and Applied Sciences, Ibri, Oman*
Anshul Verma, *Indian Institute of Technology, BHU, Varanasi, India*

Table of Contents

Detailed Table of Contents

In the evolving landscape of wireless edge networks spanning 5G/6G, IoT, smart cities, and autonomous systems Federated Learning (FL) offers a privacy-focused solution by training models without sharing raw data. However, FL lacks transparency, posing risks in mission-critical applications. This chapter introduces Explainable Federated Learning (XFL), combining FL with Explainable AI (XAI) methods like SHAP, LIME, and attention mechanisms. XFL enhances trust, auditability, and security by providing interpretable insights into model behavior, threat classification, and anomaly detection. It explores lightweight XAI for edge devices, a modular XFL architecture, and trade-offs among latency, privacy, and interpretability. The chapter concludes with future directions on standardizing explainability and securing FL under adversarial conditions.

The privacy-sensitive safety of wireless networks has become an increasing public issue as we are seeing more wireless devices with FW devices, and the need for networks communicating in real-time. Additionally, to previous models of machine learning with regards to identifying any threats and providing a security assurance for a network's safety, these previous models are typically deployed on a central device acting as a server and, so using the available information, they need to effectively store private data. In essence and in reality, this will pull private sensitive data to central

This chapter per the authors examines how Smart City and Industry 4.0 initiatives—built on dense wireless connectivity, cloud platforms, and pervasive automation—require a new generation of cybersecurity rooted in artificial intelligence. It outlines the evolution of urban and industrial communication infrastructures, proposes an adaptive AI-driven security architecture, and details mechanisms for secure data exchange, real-time threat detection, and resilient intrusion response across heterogeneous IoT, vehicular, robotic, and smart-energy networks. Ethical considerations surrounding AI autonomy, privacy, and transparency are balanced against operational imperatives for continuity and safety. By integrating case studies and charting future research directions, the chapter offers a practical framework for designing, deploying, and governing secure AI-enabled communications at scale.

Embarking on a comprehensive exploration of Li-Fi technology, this paper thoroughly investigates its architecture, functionality, applications, strengths, and limitations. The primary aim is to dispel misconceptions and foster a deep understanding of the technical intricacies associated with Li-Fi. Covering diverse domains such as healthcare and industrial applications, the research reveals the transformative potential of Li-Fi. Additionally, it delves into the dynamic relationship between Li-Fi and the evolving 5G-6G networks, emphasizing their collective impact on

As wireless communication systems evolving to support more complex and critical applications, the integration of Artificial Intelligence into their security frameworks has emerged as both a transformative opportunity and a significant source of new threats. This chapter provides a comprehensive examination of the current challenges and future advancements in AI-driven wireless security. An analysis of the primary threats relevant to this field, encompassing adversarial machine learning attacks, model opacity, data poisoning, and the performance trade-offs linked to the deployment of AI in decentralized and resource-constrained settings is illustrated. The dynamic nature of wireless systems—characterized by mobility, interference, and a variety of devices—substantially increases the susceptibility of AI-driven security models. This chapter presents a prospective framework for scholars, and policymakers to develop reliable, comprehensible, and resilient wireless systems in the era of AI by integrating insights from machine learning, cybersecurity, and wireless communication.

Preface

Wireless communication has become the nervous system of modern digital society. With billions of interconnected devices—from mobile phones and autonomous vehicles to wearable sensors and industrial controllers—the need for secure, reliable, and intelligent connectivity has never been greater. However, as this connectivity expands, so do the opportunities for exploitation. Cyberattacks, data breaches, and malicious intrusions increasingly target wireless infrastructures that were never designed to handle the scale and complexity of today's networks.

Artificial Intelligence (AI) has emerged as a transformative force in addressing these challenges. By incorporating machine learning, deep learning, and cognitive reasoning, AI systems can detect anomalies, learn attack patterns, and respond to threats autonomously. In addition, new paradigms such as federated learning, edge intelligence, and blockchain are enabling privacy-preserving and decentralized security frameworks. Together, these technologies promise a future where wireless systems can defend themselves intelligently and adaptively.

This book, Ensuring Secure Connectivity Through AI-Powered Wireless Systems, explores these developments in depth. It brings together contributions from researchers and practitioners who investigate how AI can enhance wireless security at every layer—from device-level protection to large-scale network trust management. The chapters discuss architectures, algorithms, and case studies that reveal how AI-driven security is revolutionizing the connectivity landscape and ensuring resilient digital ecosystems.

In the twenty-first century, connectivity underpins nearly every aspect of human life. The evolution toward smart cities, Industry 4.0, Internet of Things (IoT), and 5G/6G technologies has blurred the boundaries between physical and digital environments. Wireless networks are now critical for healthcare monitoring, autonomous transport, and real-time industrial operations. Yet, the same openness that enables innovation also exposes systems to unprecedented levels of cyber risk.

Globally, organizations and governments are investing in AI-enabled cybersecurity to safeguard national infrastructure and personal data. Machine learning algorithms are now part of intrusion detection systems; blockchain networks ensure tamper-proof transactions; and federated models enable privacy-aware learning across distributed devices. These solutions collectively form the technological backbone of a secure, intelligent, and sustainable communication ecosystem.

This topic fits squarely within the intersection of three fast-growing domains: artificial intelligence, wireless communication, and cybersecurity. As the world transitions to ubiquitous connectivity, ensuring secure and trustworthy communication is not only a technical requirement but a societal imperative. This book contributes to that global mission by presenting forward-looking approaches that harness AI to create self-defending, self-learning, and ethically governed communication networks.

This volume is intended for a diverse but clearly defined readership across academia, research, and industry.

- Researchers and Academicians in the fields of AI, communication networks, and information security will find this book valuable for understanding state-of-the-art methodologies and emerging research directions.
- Graduate and Doctoral Students in computer science, electronics, and information technology will benefit from its conceptual explanations, use cases, and foundational frameworks for thesis or project work.
- Industry Professionals and Network Engineers can draw practical insights to design intelligent intrusion detection systems, secure routing protocols, and privacy-preserving architectures.
- Policy Makers, Security Analysts, and Technological Planners will gain perspective on the regulatory and ethical implications of integrating AI into national and corporate cybersecurity strategies.

By serving these audiences, the book bridges the academic and industrial divide, offering both theoretical understanding and real-world application.

Each of the fourteen chapters contributes a unique perspective to the overarching theme of AI-powered wireless security. Together, they provide a structured progression—from foundational principles to advanced applications—demonstrating how artificial intelligence can transform every dimension of wireless communication.

CHAPTER OVERVIEW

Chapter 1. Machine Learning for Threat Detection and Intrusion Prevention
Dr. Muhammad Faisal (Muhammadiyah University of Makassar), *Mr. I Dewa Made Widia* (Universitas Brawijaya Malang Indonesia), *Mr. Nizirwan Anwar* (Universitas Esa Unggul Jakarta Indonesia), *Prof. Titik Khawa Abdul Rahman, Prof. Aedah Abd*

Rahman (Asia e University Malaysia), *Mr. Alders Paliling* (Universitas Sembilanbelas November Kolaka, Indonesia), *Mr. Raden Teddy Iswahyudi* (Universitas Esa Unggul Jakarta Indonesia), *Miss Anshika Sharma* (National Institute of Technology (NIT) Raipur, India) Paves the way for deeper insight by exploring how supervised and unsupervised learning techniques can identify anomalies and predict cyber threats. It discusses adaptive defense strategies and the automation of intrusion prevention in wireless environments.

Chapter 2. An Examination of AI-Integrated Wireless Sensor Networks Security: Obstacles and Prospects *Dr. Eyman F. A. Elsmany* (University of Gezira, Sudan), *Dr. Mayada S. A. Mustafa* (Faculty of Engineering, Uruk University, Iraq), *Dr. Zeinab E. Ahmed* (University of Gezira, Sudan), *Dr. Altahir A. Altahir* (FICT, Universiti Tunku Abdul Rahman, Malaysia) Examines the integration of AI into wireless sensor networks (WSNs).

Chapter 3. Explainable AI for Malware and DoS Attack Mitigation in Wireless Communication Environments *Dr. Sucheta Sachin Yambal, Dr. Yashwant Arjunrao Waykar* (Dr. Babasaheb Ambedkar Marathwada University, India), *Mr. Viraj Soni* (Independent Researcher, USA), *Dr. Shubham Bajpai, Mr. Omkar Agrahari* (Chhatrapati Shahu Ji Maharaj University,Kanpur, India) Discusses explainable AI (XAI) techniques to detect and mitigate malware and denial-of-service attacks. It underlines the importance of interpretable AI decisions for trusted network operations.

Chapter 4. Explainable Federated Learning for Secure and Transparent Wireless Edge Networks *Mr. Sonukumar Pandit, Dr. Ravikumar R N, Mr. Shubhamkumar Pandit, Miss Neeharika Joshi, Mr. Satwik Kishore, Ms. Aarthi S* (Marwadi University, Rajkot, India) Discusses a federated learning framework that maintains both data privacy and interpretability. It presents explainable federated models for decentralized wireless networks where transparency and security are equally critical.

Chapter 5. Federated Learning for Privacy-Preserving Security in Wireless Networks: A Decentralized Intelligence Approach *Vikas Sharma* (SRM Institute of Science and Technology, Delhi NCR Campus, Ghaziabad, U.P, INDIA), *Dr. Tarun Kumar Vashishth* (IIMT University, Meerut, U.P, INDIA), *Mr. Mragesh Pratap Singh* (Swami Vivekanand Subharti University, Meerut, U.P., India), *Mr. Alekh Chaudhary* (SRM Institute of Science and Technology, Delhi NCR Campus,Ghaziabad, U.P, INDIA), *Dr. Sunil Kumar* (AURO University, Surat Gujarat, INDIA), *Dr. Shahanawaj* (Ahamad University of Hail, Saudi Arabia) Demonstrates how decentralized intelligence can strengthen privacy in heterogeneous networks.

Chapter 6. Federated Intelligence for Secure Wireless Systems: A Privacy-Preserving and Scalable Approach to AI-Powered Threat Detection *Praveen Kumar Myakala, Miss Madhura Raut* (Independent Researcher, USA), *Dr. Vineeta Singh* (Chhatrapati Shahu Ji Maharaj University, Kanpur, India) Explains a scalable multi-tier learning system capable of managing complex, large-scale wireless infrastructures while maintaining efficiency and privacy.

Chapter 7. Blockchain and AI Secure Wireless Transactions *Ms. Amrutha Kolhar, Dr. Sridevi* (karnatak University Dharwad), *Mr. Hemant Soni* (Capgemini, United States), *Mr. Kamal Kant* (Chhatrapati Shahu Ji Maharaj University, Kanpur, India) Discuss the synergy between blockchain and AI to secure wireless transactions. The chapter covers decentralized authentication, smart contracts, and tamper-proof record keeping for digital communications.

Chapter 8. Blockchain Reliable Incentive Cooperation System for Multi-Hop Wireless Network Infrastructures *Dr. KIRUBA BURI R* (Anna University, Pattukkottai, India), *Mr. Raunak Kumar* (Independent Researcher, USA), *Dr. Swaminathan K, ECE* (SRM Institute of Science and Technology - Ramapuram Campus, India), *Mr. Kamal Kant* (Chhatrapati Shahu Ji Maharaj University, Kanpur, India) Presents a Blockchain-based incentive mechanism encouraging cooperation and honesty among multi-hop network nodes.

Chapter 9. Secure AI-Enabled Communication in Smart Cities and Industry 4.0 *Mrs. Musdalifa Thamrin* (STMIK Profesional Makassar, Indonesia), *Mr. Nizirwan Anwar* Universitas Esa Unggul, Jakarta Indonesia), *Mr. Nurdiansyah Nurdiansyah* (Universitas DIPA Makassar, Indonesia), *Mr. Imam Asrowardi* (Politeknik Negeri Lampung, Indonesia), *Dr. Muhammad Faisal* (Muhammadiyah University of Makassar, Indonesia), *Dr. Sahabuddin* Sahabuddin (Universitas Bossowa Makassar, Indonesia), *Mr. Tomy Ronaldi* (Universitas Pertahanan, Bogor Indonesia), *Dr. Alok Kumar, Dr. Shubham Bajpai* (Chhatrapati Shahu Ji Maharaj University) Applies AI-driven secure communication frameworks to smart city and industrial environments. It discusses real-time data processing, edge security, and scalable connectivity for critical urban infrastructure.

Chapter 10. Illuminating Connectivity: A Comprehensive Review of Li-Fi from Architecture to Ecosystem Integration *Miss Richa Yadav, Shreya Bhaveshbhai Sejani, Bhavya Shah, Dr. Parita Rajiv Oza, Dr. Smita Agrawal* (Institute of Technology, Nirma University), *Dr. Darshana Upadhyay* (Dalhousie University) Reviews Li-Fi (Light Fidelity) as a sustainable alternative to radio-based communication. It examines AI's role in optimizing Li-Fi security, performance, and ecosystem integration.

Chapter 11. Securing Autonomous Systems: AI-Based Intrusion Detection and Privacy-Preserving Mechanisms *Mr. Vikas Sharma* (SRM Institute of Science and Technology, Delhi NCR Campus, Ghaziabad, U.P, India), *Mr. Shashank Shelat* (Independent Researcher, USA), *Mr. Manoj Kumar* (Chaudhary Charan Singh University, Meerut U.P. India), *Ms. Gurpreet Kaur* (Raj Kumar Goel Institute of Technology, Ghaziabad, U.P, India), *Dr. Alok Kumar* (Chhatrapati Shahu Ji

Maharaj University, Kanpur, India) Presents AI-based adaptive solutions to ensure operational safety and data integrity.

Chapter 12. Challenges and Future Directions of AI-Powered Wireless Security *Dr. S. Ida Evangeline* (Government College of Engineering, Tirunelveli), *Mr. Midhun Prabhu Chinta* (Independent Researcher, USA), *Dr. Alok Kumar, Dr. Vineeta Singh* (Chhatrapati Shahu Ji Maharaj University, Kanpur, India), *Prof. Rajasekaran Selvaraju* (University of Technology and Applied Sciences, Ibri, Oman) Synthesizes the book's findings and outlines future challenges such as adversarial attacks, quantum-safe cryptography, and ethical AI deployment, sets a research roadmap for the coming decade.

Ensuring Secure Connectivity Through AI-Powered Wireless Systems represents a significant scholarly contribution to the intersection of artificial intelligence, cybersecurity, and wireless communication. By uniting diverse research perspectives within a single volume, the book creates a holistic understanding of how intelligent technologies can secure the communication infrastructures that define our digital age.

Its impact lies in several dimensions:

1. Scientific Advancement: The volume consolidates emerging theories and frameworks for AI-based security, offering foundational material for future studies.
2. Technological Innovation: It presents implementable architectures—federated learning, blockchain trust models, and explainable AI—that industries can adopt to enhance wireless resilience.
3. Educational Value: The book serves as a comprehensive reference for graduate curricula and research training in AI-driven security.
4. Societal Relevance: By emphasizing privacy, transparency, and sustainability, it contributes to building ethical and trustworthy digital ecosystems.

In essence, this book does more than document the state of AI in wireless security—it defines the pathway toward the next generation of autonomous, intelligent, and secure communication systems. The editor and contributors hope that it will inspire continued innovation, collaboration, and responsible adoption of AI for a safer, more connected future.

Chapter 1
Machine Learning for Threat Detection and Intrusion Prevention

Muhammad Faisal
https://orcid.org/0000-0003-1469-9468
Muhammadiyah University of Makassar, Indonesia

I Dewa Made Widia
https://orcid.org/0009-0008-1310-3195
Universitas Brawijaya, Malang Indonesia

Nizirwan Anwar
https://orcid.org/0000-0003-1189-9093
Universitas Esa Unggul, Jakarta, Indonesia

Titik Khawa Abdul Rahman
Asia e University, Malaysia

Aedah Abd Rahman
Asia e University, Malaysia

Alders Paliling
https://orcid.org/0000-0002-9581-8248
Universitas Sembilanbelas November, Kolaka, Indonesia

Raden Teddy Iswahyudi
https://orcid.org/0009-0004-7907-0155
Universitas Esa Unggul, Jakarta, Indonesia

Anshika Sharma
National Institute of Technology, Raipur, India

ABSTRACT

This chapter provides a comprehensive analysis of machine-learning-driven security for wireless communication networks, showing how adaptive models supplant static, rule-based defenses to counter sophisticated cyber threats. Motivated by escalating traffic volume and protocol heterogeneity, it explains key algorithms, architectural patterns, and training pipelines that enable real-time anomaly detection, autonomous intrusion response, and scalable protection across diverse topologies. Data challenges including labeling scarcity, class imbalance, and concept drift are reviewed

DOI: 10.4018/979-8-3373-3576-6.ch001

nt © 2026, IGI Global Scientific Publishing. Copying or distributing in print or electronic forms without written permission of IGI Global Scientific Publishing is prohibited. Use of this chapter to generative artificial intelligence (AI) technologies is expressly prohibited. The publisher reserves all rights to license its use for generative AI training and machine learning model development.

alongside adversarial robustness, interpretability, and computational limits at the edge. Case studies demonstrate practical deployments of supervised, unsupervised, and reinforcement techniques, while ethical concerns such as privacy preservation and algorithmic bias are critically assessed. The chapter concludes by mapping open research directions that position machine learning as a pivotal enabler of intelligent, resilient wireless cybersecurity.

1 INTRODUCTION

The escalation of wireless communication technologies has resulted in a spike in the number of IP-connected devices, posing significant challenges in ensuring secure connectivity. Traditional security mechanisms many a times struggle to handle the sophisticated and rapidly evolving threat landscape of today's wireless systems (Narooka et al., 2025a; Narooka et al., 2025b; Fnu et al., 2025). The escalating sophistication of cyber-attacks and the expanding attack surface necessitate more intelligent and adaptive security solutions.

Machine learning (ML) emerges as an emerging paradigm shift in field of cyber-security. It enables the rapid development of proactive and automated threat detection and intrusion prevention systems that have potential for learning from data along with adapting to new attack patterns. Unlike conventional signature-driven approaches that heavily depend on predefined rules and static controls, ML algorithms have the potential to analyze large amounts of network overloaded data in real-time, detecting subtle anomalies and forecasting potential security breaches with greater accuracy.

The ever-changing nature of wireless surroundings (Kaur et al., 2024; Singh and Kaushik, 2023), the heterogeneity of devices, and with increasing complexity of network protocols make the use of ML in wireless security particularly compelling. By leveraging ML techniques, wireless systems can be equipped with intelligent defenses that autonomously detect and give responses to threats, amplifying the overall strength of security posture and resilience of the network.

This chapter gives a comprehensive viewpoint of the application of ML for threat identification and intrusion prevention in AI-driven wireless systems. It introduces fundamental ML concepts relevant to wireless security, explores various ML architectures and models suitable for different threat detection tasks, and introduces the issues associated with data acquisition, feature engineering, and model training in the context of wireless networks.

The chapter also explores practical application case studies, demonstrating the effectiveness of ML-driven security approaches in real-world wireless deployments. It delves into the critical privacy and security considerations that come from the use of ML in security applications. By integrating ML into wireless communication systems, a more robust and adaptable security framework can be established. This framework can effectively address the growing number of service requirements and diverse characteristics of applications (Bajaj, 2021). The chapter emphasizes the proactive and automated capabilities that ML brings to wireless security. It empowers systems to autonomously detect and respond to security threats with no or vey less human intervention (Erpek et al., 2020). Moreover, the potential of ML algorithms to learn from data and adapt to unseen attack patterns makes them particularly well-suited for the ever-changing and ever-evolving threat landscape of wireless networks (Sun et al., 2019).

2 BACKGROUND AND MOTIVATION

The evolving nature of threats and the vast attack surface necessitate more intelligent and adaptable security measures. Traditional security tools frequently face difficulty in responding to the changing and sophisticated character of modern cyber threats. Systems based on signatures, for instance, are challenging to adapt to the rapidly changing nature of new attacks and attack surfaces, as well as advanced evasion techniques. Machine learning provides a novel approach to cyber-security, enabling the evolution of proactive and automated systems for threat detection and intrusion prevention. These systems excel at extracting data insights and adjusting to new attack vectors. Machine learning methods have been successfully used in various cyber-security domains, such as malware identification, network security, and intrusion detection (Jones et al., 2023). With the reference to wireless security, the role of machine learning becomes particularly intriguing due to the ever-evolving nature of wireless environments, the heterogeneity of devices, and the increasing difficulty of networking protocols. Intelligent defenses that support wireless systems can be enabled with the use of ML techniques. These defenses are independent of recognizing and responding to threats, so they could enhance the security posture and adaptability of our networks. Machine learning technology's real-time threat identification, automated responses, and detection of new types of attacks are some of those capabilities that can be beneficial by optimizing overall network security and safeguarding network-related data and assets (Dini et al., 2023). The use case of ML improves the security of Internet of Things-based systems (Sameen et al., 2020). With the emergence of advanced cyber-attacks (Singh and Dubey, 2015; Singh and Kaushik, 2023a; Singh and Kaushik, 2023b; Chaudhary et al., 2024), the necessity of improving its adaptability and sophistication in security, particularly with greater dependence on information technology systems (Apruzzese et al., 2022). Earlier security measures were not sufficient to address the escalating security threats, prompting the exploration of Machine Learning as a novel and supplementary security mechanism (Delplace et al., 2020).

3. MACHINE LEARNING ALGORITHMS

Machine learning algorithms have the ability to analyze large amounts of network data in real-time and forecast security breaches with improved accuracy, proving them to be invaluable assets for safeguarding human connectivity. With new threats and vulnerabilities emerging, traditional approaches become ineffective, necessitating the integration of artificial intelligence, and machine learning and also with deep learning, to adapt and maintain security rules in response to evolving attacks (Mazhar et al., 2023). Machine learning offers a proactive approach to cybersecurity, empowering systems to grasp from data and adapt automatically to new attack patterns. In the dynamic digital landscape of the internet and data systems, cybersecurity has become a critical concern, with an alarming rise in cyber-crimes that endanger individuals' and organizations' privacy (Alghamdi, 2020). Traditional security technology's limitations in addressing novel threats and vulnerabilities underscore the need for machine-learning-based security to develop more dynamic and effective security models. Cybersecurity's complexity poses a new threat demanding strategies and technologies to manage an escalating number of security risks. Artificial intelligence and machine-learning algorithms have emerged as pivotal components in developing adaptive security measures (Haider et al., 2020). These algorithms are categorized into three primary types: supervised, unsupervised, and reinforcement learning. Supervised learning algorithms utilize labeled data to learn from threat patterns and classify new data sets as either benign or malicious. For instance, unsupervised learning can detect network anomalies or system behavior indicative of a cyber-attack. Supervised learning trains a model from labeled data, where each data point has a predefined classification. This enables the model to learn how input features relate to output labels, allowing it to predict the class or category of new, unseen data points by creating classification labels associated with those features (Lucas et al., 2021). Supervised learning algorithms, including decision trees, Support Vector Machines, Neural Networks, are applied for threat detection. On the other hand, unsupervised learning algorithms are trained on unlabeled data with the focus on finding uncharacterized patterns, such as hidden structures or anomalies in the data. Methods of unsupervised learning, such as clustering, dimensionality reduction, and anomaly detection can be particularly useful in clustering cases. These algorithms identify instances of unusually high data points as anomalies, which may deviate significantly from normal behavior and pose a danger. Reinforcement learning defined as a learning algorithm that trains a machine to take decisions in a given environment based on a reward signal. Unlike human learning, ML doesn't perform much of the learning itself. Instead, the agent masters through trial and error, getting rewards or penalties for its actions. Understanding various algorithms is crucial for effectively deploying machine learning (ML) for detecting threats. These algorithms should be well-suited for specific tasks (Goni et al., 2020). Unsupervised learning algorithms, including clustering and anomaly detection, are particularly important for detecting deviations from normal behavior and identifying potential security threats that might go unnoticed by classical signature methods (Talukder et al., 2024). Reinforcement Learning also helps in the dynamic detection of threats, allowing organizations to adapt and learn defense to a new threat environment, which is more practical within real security and thus not currently available in the real world in terms of application. Reinforcement learning offers an alternative approach to training detection models. In this approach, agents learn defense strategies through trial and error, observing the network environment's behavior. Various machine learning algorithms have demonstrated exceptional capabilities in identifying and categorizing different types of cyber threats. Tree-driven algorithms and ensembles are established for identification of network intrusion (Vitorino et al., 2023). Deep learning models, like convolutional neural networks and recurrent neural networks, are extensively developed to analyze extensive network traffic data and extract more intricate features indicative of malicious activity (Gong et al., 2024). The selection of machine learning algorithm relies on the characteristics of the data, the type of threat being detected, and the required level of accuracy and performance. Implementing machine learning models for threat detection involves feature engineering, selecting an appropriate model, and optimizing hyperparameters to achieve optimal performance and generalization. Machine learning is applied for detecting intrusion for several reasons. Firstly, it enables the complex statistical patterns identification, concealed within data streams, facilitating the differentiation and analysis of features exhibiting distinct characteristics (Liu & Patras, 2022).

3.1 Secondly, Supervised Learning Methods are Employed to Classify Attack Detection Models.

Labeled data-supervised learning algorithms have evolved as a crucial field, enabling the detection of attacks and the construction of classification models having potential of distinguishing between malicious and benign network traffic. Supervised learning, based on a labeled dataset where each instance belongs to a predefined class or category (e.g., "attack" or "normal"), forms the foundation of these algorithms. Classification models are constructed using supervised learning, extracting features from network traffic instances and classifying them into predefined classes (e.g., malicious, benign) based on their quality (Lam, 2021). For instance, algorithms in this category are based on Support Vector Machines, Decision Trees, Random Forests, and Neural Networks. Supervised learning models such as these are specifically better-suited for learning data that doesn't yet exist. This is because a high-quality training dataset and representation, along with careful feature selection that reflects the underlying characteristics of various attacks, enable them to generalize well to unseen data. Additionally, cross-validation and regularization are utilized to avoid overfitting and make the trained models resilient to different attack scenarios. Furthermore, feature selection can be applied to various network attack datasets in conjunction with machine learning methodologies (Sonule et al., 2020; Raturi et al., 2025). The problem of intrusion detection can be modeled as a classification problem in machine learning that can determine whether the network system is being intruded upon or not (Dong et al., 2021).

3.2 Unsupervised Learning Methods

Wireless network anomaly detection involves unsupervised learning strategies that enable algorithms to identify abnormal patterns and behaviors not present in labeled training data. These algorithms learn on their own to recognize atypical behavior. Unsupervised learner methods, including clustering and anomaly detection, are used to detect abnormal patterns and behavior in unlabeled network data without prior information about specific attack signatures. Clustering algorithms, like k-means and hierarchical clustering, group similar points closely together on the basis of their intrinsic properties. This can help identify outliers or clusters of deviant behavior, which could represent attacks by malicious actors. Techniques for dimensionality reduction like Principal Component Analysis (PCA), auto encoders, can be applied to less down the computation associated with high-dimensional network data and preserving its salient features. This helps with more effective anomaly detection and visualization. However, unsupervised learning techniques face the challenge of choosing appropriate distance metrics and similarity measures that reflect the underlying structure of the data and its relationships. Additionally, they need to interpret the resulting anomalies in relation to specific networks. Anomaly detection approaches in machine learning play a significant role in network monitoring, enabling the detection of anomalies (Aluwala, 2024). These techniques are more effective in detecting new attacks because all deviant behaviors are classified as intrusions. In contrast, anomaly-driven intrusion identification systems discover attacks from the baseline behavior of the system.

3.3 Adaptive Threat Response Systems and Reinforcement Learning Strategies.

Reinforcement learning emerges as a potent lens through which adaptive threat response systems can be developed. Agents iterate, adjusting and refining their protective strategies through trial-and-error engagement with the dynamic network surroundings (Nguyen & Reddi, 2023). Reinforcement learning algorithms enable autonomous agents to learn the most advantageous actions as network conditions change. These agents can sequence actions such as modifying firewall rules, isolating infected network devices, and rerouting traffic to mitigate attack impact and damage. For intrusion detection, reinforcement learning can be adapted to create flexible and resilient security systems capable of adapting to novel and evolving threat scenarios. It also improves resource allocation and reduces false positive rates. As anomaly detection integrates with AI in security solutions, the implications include enhanced prediction accuracy, reduced false positive rates, and improved system flexibility. Reinforcement learning models allow us to model the "action" pattern of the dynamic nature of the attackers and defenders, enabling the development of adaptive security strategies for handling emerging threats. We require AI-based systems with the capability to make autonomous decisions to defend against more sophisticated cyber-attacks.

3.4 Hybrid Machine Learning Models

Hybrid systems leverage assets to enhance security. Hybrid machine learning methods, integrating supervised, unsupervised, and also reinforcement learning, provides a unique paradigm for developing robust and versatile threat detection systems. By combining the best of each method, such as models using multiple ML methods, we can achieve more accurate and comprehensive threat detection and response (Thapa et al., 2020). For example, a hybrid system may use supervised learning for a known attack signatures, while unsupervised learning for anomalous behavior, and reinforcement learning for deploying defensive responses in real time. Incorporating AI and ML into cyber-security models is very important for efficient collaboration between platforms and organizations (Roshanaei et al., 2024). By addressing domain knowledge driven feature extraction, static tools, and AI's limitations, we can improve the detection and reduction of advanced cyber threats, such as insider threat vectors and botnets, creating a safer digital surrounding (Salem et al., 2024). Artificial intelligence-based security solutions offer flexibility, making them highly effective in addressing the expanding digital footprint of organizations and expanding the attack surface.

Adaptive security strategies challenge the effectiveness of centralized systems in combating cyber-attacks. Instead, they emphasize adapting and leveraging secure and reactive measures in real-time. Reinforcement learning frameworks offer an exciting avenue for developing adaptive security solutions that dynamically adjust to modifying network conditions and attacker behavior. This dynamic adaptation enhances threat resolution and resistance (Dhir et al., 2021).

Reinforcement-learning paradigms enable agents to learn optimal defensive actions that interact with the network environment via reinforcement or penalty feedback. These actions are then modified to maintain the best security performance in the future. Reinforcement learning methods allow the creation of agents that are free from external influences and can adapt their defensive strategies based on network information. This approach enables adaptability to threats as they arise in real-time. In this context, AI-driven mechanisms need to consider ethical dimensions and privacy legislation to ensure they monitor and respond to ethical obligations (Roshanaei et al., 2024).

4 THE ARCHITECTURE AND MODEL OF THREAT DETECTION SYSTEMS

AI-based threat detection and intrusion prevention applications heavily rely on machine learning (ML) architecture and models. These models and architectures serve as the brains behind threat detection systems, capable of analyzing large amounts of data to identify signals or patterns that predict possible security risks. (Quintero-Bonilla & del Rey, 2020). By providing real-time actionable recommendations or preventive actions, such as detection, these systems enable the timely identification and avoidance of threats. Deep learning, specifically convolutional neural networks (CNN) and recurrent neural networks (RNN), has demonstrated exceptional effectiveness in handling sophisticated cybersecurity data, including network overloads and system logs, leading to accurate threat assessments (Kabanda, 2020).

Given the resource bottlenecks, mobility, and dynamic features present in wireless networks, it is crucial to adopt deep learning architectures tailored to these contexts (Kabanda, 2020).

Deep learning models have the capability to automatically retrieve complicated features from unprocessed data, enabling them to identify hidden patterns and abnormalities that might go unseen through earlier methods. (Rodriguez et al., 2021). Convolutional Neural Networks (CNN) excels at decomposing spatial data, making them well suitable for analysis of image and video. They can also be utilized for malware classification when visual representations of the code are beneficial.

Anomaly Detection Models, such as autoencoders and clustering algorithms, have a vital part in identifying abnormal behavior. These models can help uncover new or zero-day attacks that might not be detected by known signatures. Hybrid models, which combine multiple machine learning algorithms, provide a flexible and adaptable framework to detect various attacks. By utilizing the strengths of different algorithms, these models are better matched for the end-user to real-world threats.

4.1 Machine Learning Models are Powerful

Data preprocessing, a significant step in machine learning, involves cleaning, transforming, and engineering data to prepare it for model training and achieve desired performance. It ensures accurate and precise predictions by utilizing statistical techniques to identify meaningful patterns, extract relevant features, and correlate them to the observed dataset. Identifying suitable features for network traffic, system logs, and behavior of user is paramount in preparing accurate and effective machine learning models for threat identification (Aluwala, 2024). Addressing challenges of data is very important for developing robust and reliable machine learning algorithms capable of effectively protecting against cyber threats.

Supervised learning models, a prevalent approach in data science, can distinguish between various types of threats, requiring labeled data for training. However, the ever-evolving behaviour of cyber threats makes continuous model re-training and adaptation necessary to maintain accuracy and effectiveness over time (Shahrivari et al., 2020). Obtaining sufficient amounts of labeled data that accurately represent the spectrum of attacks and normal network activities remains a challenge.

Adversarial attacks, a common occurrence in machine learning models, involve malicious actors intentionally creating inputs to confuse the model and produce faulty predictions (Smart, 2021). This underscores the importance of assessing the resilience of ML algorithms to adversarial attacks and developing mechanisms to defend against them (Sriram et al., 2020). Managing the various challenges posed by ML issues is crucial in addressing cybersecurity concerns.

4.2 Proactive, Self-Aware, and Self-Adaptive Systems

In conventional security methods, response is given only after detecting an intrusion, but it is inadequate for defending against intelligent attacks. A proactive, self-aware, and self-adaptive intelligence security paradigm is essential to address these issues. This it offers proactive, self-aware, and self-adaptive security solutions (Samarakoon et al., 2022).

Keeping in mind the AI technology, cyber-security systems should be more proactive, self-aware, and self-adaptive. Thus it should have the potential to predict and react to threats even before they occur, detect attacks with minimum human intervention in real time (Pang & Li, 2020).

To continuously monitor network traffic we can use ML algorithms, which help in detecting anomalous and potential attacks preemption. On analyzing network traffic data, security logs and end user devices, we can identify patterns and anomalies that may cause a security incident, all using AI models (Binhammad et al., 2024). To analyze and comprehend the vast and diverse data streams generated from network activities, user behaviors, and comprehensive system logs, sophisticated AI-aided systems are particularly designed. All of this empowers them to detect subtle anomalies and deviations from existing established operational baselines that may otherwise get unnoticed.

4.3 Security Privacy Risks Management

Risks management in security and privacy play a significant role in security-related systems, especially when handling sensitive data and adversarial capabilities in machine learning-driven threat identification and intrusive avoidance systems. The primary concern is data privacy, as network traffic, user behavior, and system logs may contain sensitive and confidential information that are used in ML models. In order to escalate the security and privacy concerns of ML, it is essential to develop provable solutions. The desirable things are training data confidentiality and user data privacy (Thuraisingham, 2022). Using differential privacy methods can add significant interference and noise to the training data or model parameters that is used to avoid and prevent sharing sensitive information.

Furthermore, Using Federated learning methods empowers collaborative model training without the need of directly sharing sensitive data, thus, making them extremely privacy and security conscious (Cristofaro & De Cristofaro, 2020). It is evident that machine learning methods are often vulnerable to adversarial attacks, as adversarial operators craft inputs to prevent detection or influence the model behavior. We can enhance the robustness of ML models against adversarial examples, mitigating threats via Adversarial training. Frequent penetration testing of ML-based security tools is very important for carrying out security audits. Implementation of strong security measures is crucial in protecting ML models and data from unauthorized access, manipulation, or theft.

But, there exists issues associated with data privacy and algorithmic bias in the use cases of ML in healthcare. Overcoming the ethical issues of ML and healthcare integration is crucial. Also, the susceptibility of machine-learning models to adversarial attacks is because adversaries create inputs to induce undetected behavior or influence the model's performance._Such attacks are more often seen in vulnerable safety-critical systems.

In order to protect ML models from manipulation and secure their accuracy in detecting and avoiding cyber threats, robust security procedures are essential. The process consists of adversarial training and input validation. We can protect sensitive information from unauthorized access and disclosure by implementing robust access controls, encryption mechanisms, and data anonymization techniques. Frequent security audits and assessments of vulnerability play a crucial role in detecting and overcoming potential security risks.

Relations between stoner and ML models can compromise system integrity or infringe on sequestration. Organizations that proactively address security and sequestration enterprises make trust and confidence in security technologies. This fosters responsible, ethical, and timely deployment of ML technologies. Only having veritably limited model availability is acceptable to reconstruct all the models and gather their training data (Saifullah et al., 2022). Also, it's essential to learn the worst- case performance of an ML model to ensure its safety and trustability (Chen & Liu, 2023).

5 APPLICATIONS

Machine Literacy and intrusion discovery have real- world operations that can be demonstrated through case studies. These case studies will explore colourful exemplifications from the field, showcasing how ML models can identify different types of attacks, similar as DDoS attacks, malware infections, and unauthorized access attempts, thereby enhancing network security. Endpoint security involves covering individual device gesture, including stoner exertion, system processes, and train access patterns, to descry implicit trouble actors and take action. Machine literacy- grounded endpoint discovery and response systems are equipped to identify malware, ransom-ware, and other vicious software before it can reach endpoints and beget damage beyond the network. Phishing discovery uses ML models to dissect dispatch content, sender information, and point characteristics to identify and block phishing attacks. These models can identify colourful phishing schemes, similar as spoofing, URL obfuscation, and social engineering, to cover druggies from falling victim to swindles. These exemplifications illustrate the practical and potent capabilities of ML technology in addressing a different diapason of cyber security pitfalls. They emphasize the eventuality to bolster cyber-security measures and effectively alleviate cyber vulnerabilities.

5.1 ML-Driven IDS for 5G/IoT Networks.

With the increasing traffic overload in the IT sector, techniques in machine learning presents guarantee in enhancing intrusion detection systems in 5G (Kaur et al., 2024) and IoT networks (Joshi et al., 2024a; Joshi et al., 2024b; Joshi et al., 2025), in which signature-driven detection approaches often lag behind the dynamic threat landscape (Kamalov et al., 2021). Techniques in Machine learning are trained on vast amounts of data from network traffic; also detect patterns and anomalies that can led to malicious activities. To efficiently detect zero-day threats, anomaly-based detection methods are indispensable, especially in growing environments of Internet of Things (IoT). Intrusion detection is widely used in wireless networks to keep an eye on traffic for irregular behaviour, as well as gather data on such things as packet headers, payload content, and network statistics. Machine learning models can be trained to distinguish benign and malicious traffic on the basis of packet size, protocol type, and traffic patterns for different features. Machine learning techniques like deep learning techniques can help in learning new defence mechanisms which can use all the available data to address an ever-growing challenge in cybersecurity (Kornaros, 2022). As such, ML assisted solutions can be applied to detect attack traffic and make suitable distinctions from normal traffic handling (Wala & Kiran, 2024). Next, these models can be applied in real-time to monitor and prevent intrusions, thereby offering a protective layer for cybersecurity professionals. There are all kinds of intrusion detection systems that use artificial intelligence techniques to identify and act on security incidents rapidly. Network security can be enhanced by applying ML techniques for dynamic threat discovery, reducing false positives, and adapting to evolving vulnerabilities. As the IoT growth is increasing, the safety of IoT networks becomes more paramount (Rahman et al., 2025).

5.2 Threat Detection in Vehicular Ad-Hoc Networks and Smart Grids

Machine learning methods have been recognised to be pivotal for improving threat detection capabilities on vehicular ad hoc networks and smart grids, and are also used in the connection of vehicular ad hoc networks, such as a complex fleet of networks and smart grids, since the close-knit system also introduces new security challenges that are inherently network-based (Jones et al., 2023). In such a case, in the case of vehicular ad hoc networks, ML models can be applied for the detection of bad nodes in the direction of sabotaging communication or providing inaccurate information on the network. Machine learning models can track energy consumption trends and spot anomalies in smart grids that might indicate cybersecurity threats against critical infrastructure (Schmitt, 2023). Through scrutiny of network activities, system logs and sensor data, ML models can detect dubious activities and activate alerts that allow security staff to respond quickly to possible dangers. Threats are constantly evolving (i.e., threats to network security), and to adapt and learn new behaviours to counteract such developments using ML models.

5.3 Benchmarking and Performance Evaluation Using Public Information

These publicly available datasets can help to benchmark and compare the implementation of machine learning-driven threat detection and intrusion prevention system. The NSL-KDD dataset and the UNSW-NB15 dataset are examples of these two datasets and provide a common framework whereby differences between ML algorithms across various kinds of cyber-attacks can be compared against each other. These datasets are convenient and cost-effective for researchers and practitioners who wish to train and then test the ML models, evaluate their measures, such as accuracy, precision, recall, and F1-score. Tools designed for the analysis of intrusion detection systems can serve real traffic examples to train models and help them evaluate performance. Through benchmarking their ML algorithms against open source datasets, they can verify their models to be robust, sensitive, and able to apply themselves to additional previously unseen attack areas. The credibility of the network ToN_IoT dataset is evaluated after the use of different models of machine learning (Moustafa, 2021). The variety of valid and malicious vectors represented in datasets such as ToN_IoT and UNSW NB-15 enables assessment of ML models extensively (Dunn et al., 2020). Machine learning model reliability, precision and adaptability to new and unknown attack scenarios can be ensured by comparing them to publicly available datasets (Pai et al., 2021; Rafique et al., 2024). Based on the CICIDS2017 dataset, the deep learning model attained 99.52% accuracy in classifying network traffic (Ansar et al., 2024). An additional model utilising statistical methods and a metaheuristic algorithm for feature selection, employed with an LSTM-based model, yielded 98.42%, 89.54% accuracy over the NF-BoT-IoT-v2 and IoTID20 datasets (Ansar et al., 2024; Ogunseyi & Thiyagarajan, 2025). Nonetheless, traditional datasets alone, such as KDD99, do not reflect modern network traffic behaviour and attack vectors, and could therefore be considered as having serious problems (Layeghy et al., 2021). Inadequate results may be produced when evaluating network intrusion detection systems due to the scarcity of modern low-footprint attack styles, the absence of modern normal traffic scenarios, and the different distribution of training and testing sets in the KDD99 dataset.

5.4 Network Intrusion Detection Tools

Network Intrusion Detection Systems can be of different structures, but the major differences consist of the data type examined (Apruzzese et al., 2020), and the detection method used. Signature-based detection utilises network traffic through known attack signatures, which is used to identify the underlying cause for the breach, whereas anomaly-based detection works on the basis of changes in behaviour, in an attempt to spot the possibility of a potential attack. Machine learning-based detection of network intrusion detection systems (NIDS) enhances the effectiveness and efficiency of detecting unusual and sophisticated attacks (Ennaji et al., 2024). Hand-grounded systems have limitations in detecting zero- day attacks (Hashemi & Keller, 2020), egging the emergence of anomaly- grounded NIDS. These systems identify outliers in network business. While supervised training uses labeled datasets to classify network business as normal or vicious, unsupervised literacy can descry abnormalities without specific attack autographs (Xu, 2021). AI enables the development of intelligent systems for real- time cyber-attack discovery and response (Khanji & Khattak, 2020). Analysed ML styles in anomaly NIDS (Sriram et al., 2020), applied ML ways to identify and descry dangerous patterns in network business (M. Sarhan et al., 2022). Sriram et al.(2020) demonstrated that ML styles improves the delicacy and performance of intrusion recognition systems by feting vicious and normal conditioning in network business and classifying network records. Suggested using resolvable AI fabrics to explain why certain network business is flagged, abetting security brigades in comprehending and responding to pitfalls (Mane & Rao, 2021) stressed the eventuality of using ML model intelligence to secure computer networks against advanced pitfalls (M. Sarhan et al., 2022).

6 MODELLING LEARNING APPROACHES IN MACHINE LEARNING

Regular model tuning is essential for sustaining accuracy in threat identification and minimizing false positives (Alzahrani & Alenazi, 2021). The process of model optimization involves iteratively refining the model's structure, adjusting hyperparameters, and training datasets to improve performance and enhance generalization (Westphal et al., 2024). Techniques such as hyperparameter tuning, regularization, and ensemble methods are advantageous for boosting the precision and robustness of machine learning models in threat detection. In this domain, metaheuristic attacks are also utilized to advance intrusion detection systems. It is crucial to optimize the model for machine learning-based threat detection systems to effectively adapt to evolving attack strategies and network conditions. Unsupervised learning methods include anomaly detection and clustering techniques. For example, these methods can be employed to recognize and detect new and unknown threats within networks by pinpointing unusual behavior.

6.1 Unsupervised Learning

Unsupervised learning approaches, such as clustering and anomaly detection, are not dependent on labelled examples. They can identify hidden patterns or outliers in network traffic data, making them useful for detecting zero-day attacks and insider threats. Clustering algorithms such as k-means and hierarchical clustering, extract cluster features from network traffic data points based on similarity metrics. These clusters can then be used to detect anomalous data points that don't belong to normal classes (Lam, 2021). Anomaly detection techniques, like one-class support vector machines and autoencoders, model normal behaviour from network traffic and mark anomalies when they significantly deviate from their learned model. This makes them extremely helpful for finding new, unknown, or zero-day attacks (Kaliyaperumal et al., 2024). However, detecting threats can be challenging due to the complexity of user behaviour and environmental changes. Some behaviours may be normal in certain cases, while others are not (Tian et al., 2025). In this context, many methods also enable security analysts to identify and investigate potential security incidents in advance, which would have otherwise gone unnoticed.

6.2 Supervised Learning

Supervised learning approaches, particularly classification and regression models, are crucial in intrusion detection and threat prevention. These algorithms need labelled training data for categorizing network traffic data into predefined categories. For instance, decision trees, support vector machines, and neural networks are trained on labelled datasets that include representative samples of normal and malicious network traffic to learn distinguishing features and classify new instances with greater accuracy. Classification models predict the class label or category of network traffic data, such as normal, malicious, or suspicious, based on extracted features and learned patterns. In contrast, regression models predict continuous features, like the severity or risk of an event, based on past network data and their relationships. These approaches aim to detect network penetration at the network level using packet-based analysis and analyze anomalies at the physical process level using data mimicking system behaviour. Choosing the appropriate algorithm for network anomaly detection is vital in finding a solution to network security problems (Pai et al., 2021; Umer et al., 2022).

6.3 Semi-Supervised Learning

Semi-supervised learning, which combines labelled and unlabelled data to enhance model performance, particularly in scenarios with limited or costly labelled data, offers a promising approach for intrusion detection systems. Both supervised and unsupervised models are based on semi-supervised learning algorithms, which learn from data with both labels and no labels to build more powerful and generalizable models. Hybrid models, such as those combining supervised and unsupervised models, can effectively leverage the advantages of both methods to improve intrusion detection system performance. Active learning techniques, which prioritize querying the most informative unlabelled data points for labelling, optimize semi-supervised learning by targeting data that is most likely to improve model performance. In practice, active learning algorithms can significantly reduce the amount of labelling required to train high-performance intrusion detection models. Generating labelled datasets is a challenge for machine learning-based intrusion detection, and active learning offers a solution. Active learning algorithms select the most informative instances from the unlabelled dataset and request labels from human experts, gradually improving model performance with minimal labelling effort. The accuracy and reliability of machine learning techniques, particularly for threat detection and intrusion prevention, depend heavily on creating a well-trained model using representative data for training and evaluation.

7 DATA PREPROCESSING INVOLVES FEATURE EXTRACTION, SELECTION, AND ENGINEERING

The preparation of network traffic data for machine learning-based threat detection and intrusion prevention involves a series of crucial data preprocessing processes. These processes include feature extraction, selection, and optimization of the network traffic data. Feature extraction transforms raw network traffic into relevant features that represent the underlying characteristics of the database, enabling ML algorithms to distinguish between normal and malicious behaviour. Feature selection techniques aims to retrieve only the relevant and information preserving attributes from the extracted feature set with minimizing its dimensionality, thereby enhancing the effectiveness and accuracy of ML models. Feature engineering generates new features from existing ones, often leveraging domain knowledge or insights, to improve the performance of ML algorithms.

Preprocessing is essential for both structured and unstructured data to effectively analyze it. For text data, users can either tokenize it or apply stemming or lemmatization to convert it into a format suitable for ML algorithms. Addressing data quality, relevance, and labelling is crucial for effective ML-based threat detection and intrusion prevention systems. Feature engineering involves adding features derived from calculations on existing ones, such as ratios, differences, or mathematical transformations, to increase the model's feature vector. However, the engineering process of producing and analyzing valid features can be laborious and time-consuming. Additionally, since network traffic becomes increasingly complex and mutable, original features may become inadequate in representing current traffic patterns, leading to detection failures due to changing traffic patterns (M. Wang et al., 2020). Data preprocessing involves several steps, including data cleaning, feature engineering, feature extraction, and feature selection. Data cleansing involves identifying and rectifying incorrect or corrupted data. Feature extraction transforms raw data into features that are more likely to reveal patterns. Data science excels in feature selection, focusing on developing the most pertinent features for model building. Feature engineering leverages domain knowledge to create features that optimize machine learning algorithm performance. A well-defined feature set enhances the accuracy and efficiency of intrusion detection models. However, feature engineering can be manual, time-consuming, and dependent on the model. Improving dataset quality through feature engineering methodologies can lead to effective data-driven models. The quality of the background dataset is a crucial consideration. Implementing machine learning algorithms heavily relies on preprocessing pipelines, including data processing and model dataset processing, as well as data transformations that represent data efficiently for learning. Feature choice is a popular technique among machine learning specialists, enabling more efficient classification by training models with fewer features, saving computing resources, and enhancing or improving classification performance (Zuech et al., 2022). Feature engineering can be more straightforward and faster than manual feature engineering and even improve deep learning classifiers. Feature selection is a critical component of a general machine learning pipeline, especially for large-dimensional data. Making informed decisions about which features to select for training helps alleviate computational load on learning algorithms and improves the model's generalization and prediction accuracy (Rafique et al., 2024; H. N.). Feature selection strategies, as highlighted by Huang et al. (2023), aid in identifying the most relevant features from a vast pool of potential candidates. This process reduces the dimensionality of the data without compromising the model's performance. By doing so, feature selection enhances the accuracy of machine learning models. Classification features play a crucial role in reducing the dimensionality and noise of the data, ultimately improving prediction accuracy. Furthermore, feature selection facilitates faster model training, reduces monitoring requirements, and enhances model interpretability.

8 INTERPRETABILITY AND SCALABILITY MODELS

In this context, we offer the interpretation and scaling of machine learning (ML) models. In threat detection and intrusion prevention systems, the interpretability and scalability of ML models are crucial for their effective deployment. This allows security analysts to understand how the models make decisions and ensure that the increasing volume of traffic data from network nodes is handled efficiently. Interpretability refers to the ability of humans to comprehend the reasons behind the model's decisions. Scalability, on the other hand, involves the model's capacity to handle large amounts of data and traffic without compromising accuracy or throughput.

In systems engineering, beyond outstanding predictions, other critical properties are considered. These include the system's resilience to measurement artifacts or adversarial perturbations, its ability to withstand data drifting, its capacity to accurately assess the reliability of predictions, security and safety concerns, compliance with existing laws or norms, or its potential to augment human decision-making capabilities or present intriguing patterns observed in the data (Samek et al., 2020). Interpretable models gives insights into the characteristics along with behaviours identified by the model based on the information used to estimate threats. This enables security analysts to verify the results, identify biases, and gain a deeper understanding of the threat landscape. Techniques such as rule extraction, decision tree visualization, and importance of features ranking will significantly increase the understanding of ML models.

Interpretability is particularly important in medical diagnostics. If not addressed, the deployment of computer systems in fields like medicine and healthcare may be hindered (Vellido, 2020). Real-time threat detection and intrusion prevention rely on scalability due to the continuous generation of traffic data from the network. The scalability of graph algorithms and machine learning libraries significantly enhances the efficiency of fraud detection. Scalable machine learning algorithms, such as distributed training and model parallelism, enable the handling of large datasets and high traffic volumes of the underlying protocol without compromising accuracy or performance. However, the need for explainability is paramount to win user trust in black-box methodologies. Explainable AI aims to identify the key features that influence the model's predictions by visualizing the decision-making steps of the machine learning algorithm (Ponce-Bobadilla et al., 2024). A system explanation is essential for building trust and practical application of the model (Sadeghi et al., 2024; Guo et al., 2021). Machine learning techniques have demonstrated great potential in achieving accurate predictions, prompting initiatives across various sectors (S. C. Lu et al., 2023).

8.1 Model Interpretability is Crucial for Cybersecurity Professionals.

Interpretability of ML models facilitates data-driven decision-making and personalized decision support, enhancing healthcare service quality (Stiglic et al., 2020). In cybersecurity, interpretability is crucial for security professionals to comprehend model predictions, identify accuracy, and identify potential biases or vulnerabilities. It's essential for any ML-based decision-making algorithm, as highlighted in the United States Government Blueprint for an AI Bill of Rights, which emphasizes "Notice and Explanation" as a principle for ML-based prediction models (S. C. Lu et al., 2023). Analyzing threat classification or anomaly detection processes allows analysts to adjust security strategies and respond to emerging threats proactively, ensuring the reliability and trustworthiness of AI-powered security systems. Explainable AI builds trust in the cybersecurity profession by enabling analysts to assess model outputs better and apply them to their workflows (Muhammad & Bendechache, 2024). Trust in AI models is increased by explainability, leading to meaningful decision-making and accountability (Das & Rad, 2020). Explainable AI empowers all stakeholders to understand decision-making processes, recognizing inherent biases or errors in other models and ensuring they align with organizational ethical principles and values. In high-stakes cybersecurity environments, understanding and validating AI model reasoning is paramount for sustaining trust and minimizing risks. Enhanced interpretability facilitates collaboration and communication between AI experts and domain specialists, fostering a deeper understanding of the model's capabilities and limitations. By collaborating, the environment fosters continuous improvement and advancement of the AI platform (Reyes et al., 2020).

8.2. Importance of Feature Engineering in Machine Learning for Security

Feature engineering is a crucial aspect of machine learning for security applications. The quality and relevance of the features used by the machine learning model significantly impact its performance and effectiveness (Jones et al., 2023). Feature engineering involves selecting and merging raw data into features that represent the attributes the machine learning algorithm can use to understand patterns and predict future outcomes, thereby enhancing the learning process (Muhammad & Bendechache, 2024). In the context of threat detection, feature engineering may involve processing information extracted from network traffic, system logs, and alternative datasets to develop features indicative of malicious actions (Rjoub et al., 2023). Feature engineering, which transforms raw data into features, is a pivotal step in the process. Advanced techniques are often required to effectively transform raw data into data points that the algorithm can exploit, especially for complex and noisy data. By tailoring features that accurately represent different threats, machine learning models can be trained to detect and classify malicious activity. This approach enables the development of models with high precision and recall (both through targeted modelling). However, feature engineering poses challenges due to its laborious nature and the difficulty for learning algorithms to distinguish between data and information.

8.3. Scalability Issues in Real-World Applications and Supporting Solutions

Scalability is a significant concern in the real-world application of machine learning, and addressing these issues is crucial for its widespread adoption. Scalability becomes a critical challenge when implementing machine learning (ML) in real-world threat detection and intrusion prevention systems, especially as network traffic data flows at an increasingly rapid pace. Conventional ML algorithms struggle to process and analyze vast datasets in real-time, leading to performance bottlenecks and delayed threat detection. In today's dynamic threat landscape, explainable AI (XAI) emerges as an indispensable tool to enhance the transparency of AI decision-making processes, particularly in the context of deep neural networks (Kale et al., 2022).

To address scalability issues effectively, distributed architectures, parallel processing methods, and optimized ML models tailored for large datasets are essential. XAI models aim to provide insights into the decision-making processes of ML models, making their conclusions more understandable and transparent (Nallakaruppan et al., 2024). Moreover, considering the computational demands of ML models necessitates careful planning and implementation of a robust hardware/software infrastructure.

By leveraging scalable ML techniques and optimizing their infrastructure, companies can ensure the continuous improvement of their defense and threat detection systems. This proactive approach allows them to adapt to the evolving demands of the digital environment and effectively respond to emerging threats.

Beyond its role in threat detection, AI plays a pivotal role in proactive cybersecurity by predicting potential future attacks. By analyzing past events and system behaviors, AI algorithms can identify potential threats and take immediate action to mitigate them. These algorithms demonstrate remarkable capabilities in analyzing network traffic, detecting suspicious activities, and promptly alerting security personnel to potential intrusions (Roshanaei et al., 2024).

The adaptability of AI-enhanced systems to real-time changes in security regulations is a key factor in their resilience. By dynamically responding to evolving threats, these systems can maintain an adaptive security posture, ensuring their effectiveness in safeguarding against advanced threats. Beyond threat detection,

AI will also provide a strategic competitive edge in optimizing security policies. AI-aided algorithms advise on fine-tuning firewalls, intrusion prevention systems, and other security mechanisms, enabling organizations to effectively defend against emerging threats (Roshanaei et al., 2024). AI-based security solutions demonstrate remarkable scalability and efficiency, catering effectively to the evolving digital needs of organizations (Roshanaei et al., 2024). However, conventional threat detection methods are becoming ineffective due to the increasing frequency and sophistication of cyber threats (Jones et al., 2023). Artificial intelligence, an emerging transformative force, offers fresh solutions to bolster enterprise security systems, particularly in threat detection and response (Lisovskyi & Rochenovich, 2023). Human-centric resources and reaction time pose limitations, making them inadequate to handle rapidly changing cyber-attack scenarios. AI and machine learning (ML) have been increasingly integrated into cybersecurity, with market demand expected to rise to $101.8 billion by 2030 (Roshanaei et al., 2024). The integration of ML algorithms into wireless systems offers a novel approach to enhancing security and resilience in the ever-evolving landscape of attack vectors (Haider et al., 2020). By leveraging the capabilities of artificial intelligence, wireless networks can dynamically adapt to emerging threats, proactively identify vulnerabilities, and proactively plan intelligent responses to mitigate risks (Roshanaei et al., 2024). This AI-driven cybersecurity approach enables the detection of vulnerabilities of all types before they escalate, allowing for swift incident response and adaptability to changing circumstances, unlike traditional security measures that rely on static defenses (Molina et al., 2023). AI algorithms can autonomously recognize and displace threats, potentially leading to response times decreased by more than 90% and a significant reduction in the workload on security staff. AI-based solutions enhance threat detection efficiency by analyzing network traffic, user actions, and system logs to identify anomalies and security breaches. These solutions leverage AI algorithms to continuously learn new data during training iterations, optimizing models for increased accuracy and reduced false positives (Binhammad et al., 2024). In the current threat landscape, machine learning for threat detection and intrusion prevention gains traction due to the limitations of signature-based security systems in adapting to evolving threats. Traditional methods relying on predefined rules and signatures are ineffective against novel or zero-day attacks (Nibigira et al., 2024).

9 MODEL VALIDATION AND EVALUATION METRICS

Evaluating the model validation and evaluation metrics used in ML-based threat detection and intrusion prevention systems is crucial to assess their performance and accuracy. These metrics include accuracy, precision, recall, and F1-score. Accuracy measures the model's overall prediction correctness, while precision indicates the proportion of malicious instances correctly identified among all predicted malicious instances. Recall, also known as sensitivity, shows the percentage of correctly identified malicious instances among all malicious instances. The F1-score, the harmonic mean of precision and recall, is a suitable metric for balanced model performance, especially in cases of class imbalance. These metrics offer distinct perspectives on the algorithm's performance (Arefin et al., 2024).

Other methods used in model evaluation include Receiver Operating Characteristic curves and Area Under the Curve. Cybersecurity professionals rely on familiar metrics to evaluate models and configure predictive models to enhance security posture against cyber threats (Hesham et al., 2024). However, evaluating models solely with accuracy metrics against validation datasets may not accurately assess their real-world performance. Real-world data often differs significantly from validation data, and evaluation metrics may not align with the desired product's purpose. Therefore, a comprehensive assessment is necessary, involving the use of accuracy, precision, recall, and F1-score metrics, as well as a detailed evaluation of anomaly detection models (Solaas et al., 2024).

To thoroughly assess the performance of ML models, the authors emphasize the importance of appropriate evaluations and thorough validation techniques (R. J. Chen et al., 2021). Evaluating any model is essential to determine whether an AI system with adequate training can effectively perform its intended task. Value can be incorporated into a measurement system that includes both qualitative and quantitative indicators. The expert consensus suggests that this model can also be used to determine the significance of specific metrics, assisting organizations in identifying the most effective cyber strategies for enhancing overall performance (Barnes & Daim, 2024). Accuracy is the most commonly used performance metric for comparing methodologies (Grunova et al., 2024).

9.1 Definitions and Explanations of Important Metrics

In addition to performance criteria for model analysis and evaluation (Calster et al., 2024), determining the performance metrics is crucial when establishing predictive AI models. These metrics should align with the attributes that can be tuned to the models and reflect statistical or decision-analytical performance, considering misclassification costs. Classification measures, for instance, are based on accuracy (true and false positives, errors), precision, recall, F-score, Area Under Curve, and the Brier score. Before implementation, machine learning models must undergo thorough evaluations on test cases to ensure they meet performance requirements.

9.2 Evaluation and Deployment

Depending on the performance standard and acceptance, a deployment cycle may involve further clinical investigation, such as off-site external validation and clinical trials. If the model fails to meet the performance criteria of the clinical use case, the project can be terminated, and development cannot proceed. In such cases, a post-mortem analysis of the life cycle can be conducted (C. Lu et al., 2020). When selecting models for deployment, it is crucial to consider both the costs of deployment and the performance impact of ML models in healthcare (Morse et al., 2020). By analyzing the trade-offs among these factors, practitioners can determine whether the installation of an ML model provides a competitive advantage (Jung et al., 2020). Cross-validation accuracy is used to estimate the model's performance on unseen data, and the final model and feature set are selected using a cross-validation process on a held-out test set (Berisha et al., 2021).

9.3 Statistical Techniques Used

Developers can assess the general performance of an AI model using statistical methods such as cross-validation once the model's form and parameters are determined from a training dataset (F. Chen et al., 2023). This process is known as internal validation. Internal and external validation contribute to the uniformity and generalizability of AI models. However, AI is an ongoing process that requires significant variation for maintaining the original data for training the model and clinical use validation (Wu et al., 2024). The dataset used for designing deep learning algorithms is often convenience case-control data, which may introduce spectrum bias. To address these issues and ensure the reliability and safety of AI for specific problems, a comprehensive statistical checkup is necessary.

9.4 Precision, Recall, and F1-Score Measures

Precision, recall, and F1-score are important measures used to evaluate the performance of an AI model. Precision measures the proportion of true positive predictions among all positive predictions, recall measures the proportion of true positive predictions among all actual positive instances, and F1-score is the harmonic mean of precision and recall. Proper assessment of AI-based anomaly detection models is crucial. Precision measures the proportion of correctly detected anomalies from the total number of instances identified as unusual. A high precision indicates fewer false positives, where incorrect data points are classified as anomalies. Recall, on the other hand, indicates the number of actual anomalies successfully detected from all anomalies in the dataset. A high recall score suggests fewer false negatives, where anomalies are missed by the model. The F1-score, the harmonic average of precision and recall, provides a single statistical measure that balances false positives and false negatives. When false positives and false negatives have different impacts, the F1-score helps balance them. Evaluation metrics can have varying preferences based on the use case. For instance, recall may be prioritized over precision when the cost of not detecting an intrusion is high (Sourlos et al., 2022).

9.5 Dataset Issues and Model Training Data

Challenges and training model complexity significantly impact the success and robustness of machine learning models for threat detection and intrusion prevention. One of the primary challenges is the limited availability of high-quality labelled data (Sakhai & Wielgosz, 2021). Supervised learning methods require labelled datasets to train models for reliable threat identification, which is a complex task in the cybersecurity domain due to the fast-paced threat environment and the difficulty in classifying the latest and most advanced attacks. Biased or skewed data can distort predictions, leading to incorrect outcomes or unfair forecasts, which can cause machine learning models to fail.

To address this challenge, solutions can be found in data augmentation, anomaly detection, and semi-supervised learning methods, which expand the size and diversity of training dataset (Talukder et al., 2024). Transfer learning techniques, in which knowledge is transferred from pre-trained models to other jobs based on generalizations, can also be employed to help with the problem of insufficient labelled data. Feature engineering, the process of selecting and extracting the optimal features from raw data, is crucial in training models and improving their accuracy and generalizability. Cybersecurity professionals can leverage domain knowledge and feature selection techniques to select the most relevant information features for potential threat detection.

Effective hyperparameter exploration techniques, such as grid search, random search, and Bayesian optimization, can be used to identify the optimal parameters for an applicable dataset and task. Additionally, it is important to independently test ML models on test datasets to accurately assess their performance, generalization ability, and other relevant metrics. Detailed evaluation metrics, such as accuracy, precision, recall, and F1-score, are essential for model validation and ensuring the strength and accuracy of the final model. However, the vast array of available machine learning techniques makes it impractical to evaluate all or even a significant portion of them. The choice of an algorithm heavily depends on the specific data and the requirements of the application.

9.6 Computational Complexity and Model Validation

Additionally, all machine learning-based methods can be modified by varying parameter values, leading to the creation of different predictive models that may improve their performance (Kopitar et al., 2020). This is typically achieved through validation using a separate validation dataset that is distinct from the training dataset (Sen et al., 2021). However, meta-predictors may be particularly susceptible to this oversight, and optimization algorithms, hyperparameters, and parameters may not be explicitly documented in these files or adhere to widely accepted standards within a community (Walsh et al., 2020). Validating machine learning models is a crucial process that ensures their reliability and practical application in real-world scenarios (Fuhl et al., 2020). The success of machine learning models in practical applications hinges on their ability to adapt to evolving threat environments. Models require continuous retraining and updates in response to new data to maintain accuracy and relevance.

10 ARCHITECTURES AND MODELS

In wireless networks, machine learning-based threat detection systems follow a structured approach. This approach involves collecting data from various sources, such as routers, switches, firewalls, and intrusion detection systems. The collected data undergoes preprocessing to eliminate irrelevant inputs and noise. Subsequently, relevant features are selected and engineered to utilize the pre-processed data for training ML models. These features encompass statistical measures of network traffic, average packet header information, payload types, packet header structures, payload contents, and statistical characteristics of network device behaviour (X. Huang, 2021). These features are then employed to train ML models that learn to classify network behaviour as normal or malicious. These models are used to automatically monitor traffic in real-time, analyzing incoming network traffic to identify potential threats.

Over the past few years, threat detection in ML architecture has undergone significant changes. In the current scenario, new network designs are a major part of modern solutions. Different types of ML methods can be used for threat detection, depending on the specific requirements of the application. These systems could be anomaly-based detection devices or misuse detection systems. Anomaly-based detection systems learn the normal behaviours of the network and mark any deviation from this baseline as a threat. Misuse detection systems use predefined signatures or patterns of known attacks to identify malicious activity. Additionally, Explainable

AI frameworks (Mane & Rao, 2021) are used to make neural network intrusion detection systems more explainable.

10.1 Intrusion Detection Architectures: Central vs Local Systems

Centralized, distributed-based, or hybrid systems, and intrusion detection schemes integrating machine learning capabilities, can be categorized into two distinct types: centralized and local systems. These two types differ significantly in their strengths and weaknesses. In a centralized setup, data from multiple network sources is gathered and processed at a central point, where ML models are trained and deployed to identify attacks. This single-source approach enables a comprehensive analysis and monitoring of network activity. Advantages of centralized systems include economies of scale, as resources and expertise can be consolidated at a single location (Sarhan et al., 2021). However, centralized architectures are vulnerable to single-point failures and are unsuitable for large, distributed networks. Additionally, they may struggle to handle the vast amount and rapid data flow of modern networks, leading to performance bottlenecks.

In contrast, distributed architectures distribute data processing and model training across multiple nodes or devices. This distributed effort enhances scalability and resilience. New distributed ML approaches, such as federated learning, enable the training of ML models on decentralized data sources without compromising sensitive information sharing between network participants (Sáez-de-Cámara et al., 2023). Distributed architectures also reduce latency by allowing threat detection to occur closer to the data source. However, managing and orchestrating distributed systems can be challenging due to the dispersed nature of information and models.

10.2 Deep Learning Models for Wireless Intrusion Detection

Deep learning techniques have revolutionized wireless intrusion detection by enabling the automatic learning of complex structures and features from network traffic. This capability allows for the detection of subtle and intricate attacks that might go unnoticed by traditional machine learning approaches (Sriram et al., 2020). Among the widely studied architecture-based deep neural network systems, convolutional neural networks (CNN) and recurrent neural networks (RNN) stand out as particularly effective in enhancing the performance of intrusion detection systems.

CNN strength lies in their potential to automatically retrieve hierarchical features from network traffic data. This is because they excel at detecting spatial relationships and pattern matching with malicious behaviour. Consequently, CNN are well-suited for analyzing network traffic data and can automatically identify hierarchical features, enabling the detection of spatial correlations and patterns characteristic of malicious activities (Kodys et al., 2021). Moreover, CNN can predict attack patterns from time series data by encoding the data as convolutional operations and subsequently extracting them during temporal operations (Kodys et al., 2021).

A recent study introduces CNN for IoT intrusion detection by encoding sensor data and conducting convolution for temporal feature extraction. This approach enables effective attack detection. However, as the network expands to include additional IoT devices, a CNN module should be integrated with a connected system network switch (Weitkamp et al., 2023).

RNN, particularly Long Short-Term Memory (LSTM) and Gated Recurrent Unit (GRU) networks, are capable of handling sequential data. This makes them suitable for interpreting dynamic network traffic information. RNN can also understand temporal relations and anomalies that may indicate malicious attacks, as network traffic can be regarded as a time series. However, RNN face challenges in handling long sequences due to the vanishing or exploding of gradients, which can slow down the model's training process. Hybrid models that combine the best of various deep learning architectures have been proposed to overcome the limitations of single models. For instance, Wang et al. (2025) suggested combining CNN and RNN to create a hybrid model that leverages their strengths for improved detection performance. This hybrid model can extract spatial features using CNN and model temporal data using RNN. However, adopting these approaches to intrusion detection using deep learning requires careful consideration of model architecture, hyperparameter tuning, and training data.

10.3 Addressing in ML-Based Security Systems

Oversampling techniques involve increasing the number of instances in the minority class by copying existing instances or creating new ones. Under-sampling methods, on the other hand, reduce the number of instances in the majority class by randomly removing instances or selecting a subset that best represents the majority group of examples. Cost-sensitive learning approaches involve varying costs depending on the costs incurred by misclassifying instances from different classes, thereby pushing a much higher reward on misclassification of the minority class. Adversarial attacks, where malicious actors deliberately fabricate inputs to deceive machine learning models, pose a significant threat to the security of machine learning-based systems' vulnerabilities (Sriram et al., 2020). As illustrated by the vulnerability of image classification applications (Chernikova & Oprea, 2022), adversarial attacks can target deep neural networks. These attacks can lead to incorrect predictions or classifications, resulting in security breaches or even more severe outcomes (Bhattarai et al., 2023).

Adversarial training is one of the methods that helps combat such attacks. However, many proposed training methods may not be effective when tested against stronger attacks (Liu et al., 2020). Adversarial retraining, on the other hand, is extremely effective in combating such attacks.

An additional pitfall could arise due to bias and discrimination in the training data, which might lead to ethical problems, particularly in cases of security applications (Bhusal et al., 2023). Machine learning might inadvertently reproduce or amplify any present biases in the training data, the data used if the training data fails to sufficiently present the vast majority or if one minority or subgroup is a small group of people. Overcoming bias and discrimination requires careful data collection with pre-processing in all its forms, as well as continuous monitoring and evaluation of the model to identify and rectify biases (Roshanaei et al., 2024). These ethical dilemmas are of utmost importance to address before they escalate into severe issues. By proactively addressing these areas, ML-based security systems can ensure that they are free from discriminatory, arbitrary fairness, or human rights violations.

10.4 Deep Learning Models: CNN and RNN

Deep learning models, particularly Convolutional Neural Networks (CNN) and Recurrent Neural Network (RNN), have emerged as effective strategies for real-time threat detection and prevention. Their ability to automatically identify patterns from data makes them highly promising. CNN are particularly suited for handling data with spatial dependencies, such as images or network traffic data presented as matrices (Ennaji et al., 2024). On the other hand, RNN are designed to handle sequential data, like time series data or historical network traffic. Autoencoders, unsupervised learning models, encode data into a lower-dimensional representation and then decode it back to the original input. By training autoencoders on a normal network traffic dataset, they learn and reconstruct normal patterns, making anomalies less detectable until they are compared to the reconstructed output. Deep learning models can also analyze binary files, identify patterns in system calls, and automatically respond to incidents by rapidly categorizing the type of assault (Ejeofobiri et al., 2024). Furthermore, deep learning models have demonstrated promising results across various cybersecurity-related tasks, including malware detection, intrusion detection, and fraud detection (Redhu et al., 2024).

10.5 Data Analysis, Feature Building and Model Training Approaches

Training and developing effective threat detection and intrusion prevention ML models for threat detection presents several challenges. These include data quality, data load, dataset size, and data imbalance (Nicholls et al., 2021). Data quality refers to the data accuracy, data completeness, and data consistency that is used to train the model. If our training data contains erroneous, missing, or inconsistent values, the model may identify these errors and face difficulty to generalize to unseen data. Feature engineering involves selecting, modifying, and combining features from pre-processed data to create a list of features suitable and useful for the ML model (Huč et al., 2021). Good feature engineering can enhance the model's performance by revealing data patterns and reducing the input space's dimensionality. For sequential data, such as network traffic, convolutional neural networks (CNN) and software-defined networking (SDN) can be used to extract temporal and spatial features, leading to improved accuracy (Hai-tao et al., 2023). Training models involves fitting an algorithm, tuning the parameter set, and performing performance validation. The performance of a model can vary significantly depending on the chosen algorithm and parameter settings. To address overfitting, regularization approaches, such as L1 or L2 regularization, introduce a penalty term into the loss function.

11 CHALLENGES AND FUTURE PROSPECTS

Major breakthroughs have been reported in the application of machine learning (ML) techniques for threat detection and intrusion prevention. However, several persistent and crucial challenges need to be extensively studied before ML can be effectively used in real-world scenarios.

One of the central challenges is the requirement for reliable and interpretable models that can dynamically respond to the ever-evolving threat landscape of cyber attackers. Many current models are black boxes, making it difficult for analysts to understand and trust the decisions made by these systems, especially in high-pressure security environments.

Another significant challenge is the lack of labelled datasets, particularly relevant for developing threats and rare incident attack vectors. This limitation complicates the training of strong ML models. To overcome this, techniques such as semi-supervised learning, transfer learning, and active learning have been proposed, which utilize unlabelled data and knowledge from related topics.

Datasets are crucial for evaluation, including data validation, training, and optimization. However, dataset limitations exist, and evaluations and training do not independently reflect the complexity of the actual network.

The use of ML within an existing security infrastructure presents additional challenges due to the non-overall compatibility of diverse platforms and tools. Deep learning, while powerful, has had drawbacks in detecting insider threats due to a lack of behavioural data and adaptive adversarial actors that are likely to evade detection.

In addition, capturing enough normal user behaviour data is still necessary for anomaly detection. Explainable AI (XAI) plays a key role in overcoming interpretability challenges. XAI techniques like LIME, SHAP, and Grad-CAM offer insights into model prediction logic, enhancing trust and enabling security team validation (Mandalapu et al., 2023; Jones et al., 2023). Moreover, these models identify vulnerabilities in existing cybersecurity solutions, contributing to the development of more robust solutions (Mohamed et al., 2024). However, ethical considerations are paramount when applying ML for cybersecurity. Models must be trained with transparency, privacy, and accountability in mind (Oyinloye et al., 2024). These challenges necessitate collaboration between academia, industry, and policymakers. Research is needed to develop adaptive and resilient ML methods that can maintain performance in dynamic environments. This research should explore novel data augmentation approaches and incorporate hybrid strategies (Giunchiglia et al., 2024). Scalable algorithms that can handle large and fast data streams while preserving privacy and providing explainability will be crucial for next-generation cybersecurity.

11.1 Real-time Processing Constraints and Edge AI Deployment

Implementing machine learning-based threat detection and intrusion prevention systems poses significant challenges due to the limitations of real-time network data processing. To overcome these limitations, efficient machine learning algorithms and hardware acceleration strategies are crucial for low-latency inference and timely system responses to potential threats. Edge AI has now became a promising approach to address these issues. Edge AI involves applying machine learning models at the network edge that empowers data processing in proximity to the source. This approach reduces the degree of latency and provides better real-time performance. Moreover, edge AI enhances privacy by minimizing the need for data transfer to central servers, which is particularly important in wireless systems where data traverses insecure channels and could potentially leak private information. Edge AI operates as a network of interconnected systems and devices, processing data analysis and processing on a network level near the data source with the assistance of AI. This not only reduces latency for data transmission but also optimizes bandwidth usage by processing data internally locally. Edge AI systems are well-suited for applications that demand fast decision-making, such as autonomous vehicles and real-time video analytics. With enhanced privacy and optimized bandwidth usage, AI deployed at the edge leads to decreased latency and improved resource utilization (R. Singh & Gill, 2023). However, despite its promise, implementing and developing edge AI applications faces challenges due to the constraints imposed by the limited processing power, memory, and energy consumed by edge devices (Hao et al., 2021). To address such restrictions, efficient optimization strategies that enhance the performance of AI models while maintaining their accuracy are required to overcome these limitations. Another challenge in AI at the edge is the seamless integration of AI, enabling it to operate autonomously, minimize human intervention, and proactively detect and address threats (Singh & Gill, 2023). The edge AI market is poised for significant growth, driven by substantial benefits, particularly in the energy sector, which is expected to experience substantial growth in the near future (Himeur et al., 2024). By combining machine learning with quantum computing and blockchain, enhanced security measures can be implemented.

11.2 Convergence of Machine Learning

The convergence of machine learning with cutting-edge technologies is already disrupting network security at various levels, including blockchain and quantum computing. The integration of machine learning and blockchain creates a synergistic effect that strengthens the security of wireless systems, making them more trustworthy (Reis, 2025). Blockchain technology provides a secure and tamper-proof ledger for recording network events and security incidents (Zhang et al., 2021). Identity management systems built on blockchain can enhance access control and authentication practices across wireless networks (Saha et al., 2023). Moreover, the advancement of quantum computing techniques makes it feasible to tackle problems that are challenging for classical computers, presenting new opportunities for cryptography and security analysis. However, quantum computing also poses a threat to existing encryption schemes, necessitating the development of quantum-resistant cryptography. Quantum-resistant cryptography can be employed to bolster the security of wireless communication systems by expediting and optimizing data processing.

Machine learning can optimize quantum algorithms and reduce the impact of noise and errors in quantum calculations. Suggest that the integration of artificial intelligence and blockchain has the potential to generate novel financial products, facilitate digital transformation, and enhance transaction security. However, the combination of these technologies is still in its early stages and requires further exploration (Paramesha et al., 2024; Bhumichai et al., 2024; Saleh, 2024; Singh et al., 2020).

The use cases of AI and blockchain extends beyond the financial domain, impacting data management, security, and Internet of Things (IoT) applications. This is evident in research conducted (Bothra et al., 2023; Kumar et al., 2022). The inherent characteristics of blockchain, such as immutability, transparency, and decentralization, hold the potential to significantly enhance AI systems.

For instance, blockchain can ensure the authenticity of training data used to build machine learning (ML) models, preventing data manipulation and adversarial attacks. Moreover, the integration of ML and blockchain facilitates secure and transparent data sharing, eliminating the need for a centralized point of view. This enables collaborative threat detection and response, as well as visibility into the relationships between parties, facilitating real-time collaboration among two or more parties.

By leveraging the decentralized and immutable nature of blockchain, AI models can be trained on trusted and verified data, enhancing accuracy and reliability (Taherdoost, 2022). Furthermore, the integration of blockchain and artificial intelligence can lead to the emergence of decentralized and self-driving security systems that can adapt to evolving threat characteristics in real-time (Saleh, 2024).

The convergence of AI and blockchain will also yield significant benefits, such as improved outputs in terms of delivering more robust deliverables. Through its application, partners can release the necessary data sets that require analysis, learning, and/or solution-based on extensive data flows away from a central authority (Charles et al., 2023). Integrating these technologies together may actually enhance AI's data security, computational power, computational efficiency, and collaboration capabilities.

REFERENCES

Alghamdi, M. I. (2020). Survey on applications of deep learning and machine learning techniques for cyber security. *International Journal of Interactive Mobile Technologies, 14*(16), 210–224. DOI: 10.3991/ijim.v14i16.16953

Aluwala, A. (2024). AI-Driven Anomaly Detection in Network Monitoring Techniques and Tools. Journal of Artificial Intelligence & Cloud Computing, 1–6. DOI: 10.47363/JAICC/2024(3)310

Alzahrani, A. O., & Alenazi, M. J. F. (2021). Designing a network intrusion detection system based on machine learning for software defined networks. *Future Internet, 13*(5), 111. DOI: 10.3390/fi13050111

Ansar, N., Ansari, M. S., Sharique, M., Khatoon, A., Malik, M. A., & Siddiqui, M. M. (2024). A Cutting-Edge Deep Learning Method For Enhancing IoT Security. ArXiv (Cornell University). https://doi.org//arxiv.2406.12400DOI: 10.48550

Apruzzese, G., Andreolini, M., Colajanni, M., & Marchetti, M. (2020). Hardening random forest cyber detectors against adversarial attacks. *IEEE Transactions on Emerging Topics in Computational Intelligence, 4*(4), 427–439. DOI: 10.1109/TETCI.2019.2961157

Apruzzese, G., Andreolini, M., Ferretti, L., Marchetti, M., & Colajanni, M. (2022). Modeling Realistic Adversarial Attacks against Network Intrusion Detection Systems. *Digital Threats : Research and Practice, 3*(3), 1–19. DOI: 10.1145/3469659

Arefin, S., Chowdhury, M., Parvez, R., Ahmed, T., Abrar, A. F. M. S., & Sumaiya, F. (2024). Understanding APT detection using Machine learning algorithms: Is superior accuracy a thing? *IEEE International Conference on Electro Information Technology*, 532–537. DOI: 10.1109/eIT60633.2024.10609886

Bajaj, A. (2021). Review of Machine Learning Applications in Wireless Communications. ArXiv (Cornell University). https://doi.org//arxiv.2101.08119DOI: 10.48550

Barnes, B., & Daim, T. (2024). Information Security Maturity Model for Healthcare Organizations in the United States. *IEEE Transactions on Engineering Management, 71*, 928–939. DOI: 10.1109/TEM.2021.3139836

Berisha, V., Krantsevich, C., Hahn, P. R., Hahn, S., Dasarathy, G., Turaga, P., & Liss, J. (2021). Digital medicine and the curse of dimensionality. In npj Digital Medicine (Vol. 4, Issue 1). Nature Portfolio. DOI: 10.1038/s41746-021-00521-5

Bhattarai, M., Kaymak, M. C., Barron, R., Nebgen, B., Rasmussen, K., & Alexandrov, B. S. (2023). Robust Adversarial Defense by Tensor Factorization. Proceedings - 22nd IEEE International Conference on Machine Learning and Applications, ICMLA 2023, 308–315. DOI: 10.1109/ICMLA58977.2023.00050

Bhumichai, D., Smiliotopoulos, C., Benton, R., Kambourakis, G., & Damopoulos, D. (2024). The Convergence of Artificial Intelligence and Blockchain: The State of Play and the Road Ahead. *Information (Basel)*, *15*(5), 268. DOI: 10.3390/info15050268

Bhusal, D., Shin, R., Shewale, A. A., Veerabhadran, M. K. M., Clifford, M., Rampazzi, S., & Rastogi, N. (2023). SoK: Modeling Explainability in Security Analytics for Interpretability, Trustworthiness, and Usability. *ACM International Conference Proceeding Series*, 1–12. DOI: 10.1145/3600160.3600193

Binhammad, M., Alqaydi, S., Othman, A., & Abuljadayel, L. H. (2024). The Role of AI in Cyber Security: Safeguarding Digital Identity. *Journal of Information Security*, *15*(02), 245–278. DOI: 10.4236/jis.2024.152015

Bothra, P., Karmakar, R., Bhattacharya, S., & De, S. (2023). How can applications of blockchain and artificial intelligence improve performance of Internet of Things? – A survey. *Computer Networks*, *224*, 109634. Advance online publication. DOI: 10.1016/j.comnet.2023.109634

Charles, V., Emrouznejad, A., & Gherman, T. (2023). A critical analysis of the integration of blockchain and artificial intelligence for supply chain. *Annals of Operations Research*, *327*(1), 7–47. DOI: 10.1007/s10479-023-05169-w PMID: 36718465

Chaudhary, S., Singh, V., Malik, N., Sidhu, K. S., Kirola, M., & Joshi, K. (2024, October). Digital Signature Security Enhancement: Using X-OR Operation on RSA and AES with 4096 Bits Key Length. In *2024 International Conference on Computing, Sciences and Communications (ICCSC)* (pp. 1-6). IEEE. DOI: 10.1109/ICCSC62048.2024.10830431

Chen, F., Lu, H., & Simonetti, A. (2023). Practical Statistical Considerations for the Clinical Validation of AI/ML-enabled Medical Diagnostic Devices. Cornell University. https://doi.org//arxiv.2303.05399DOI: 10.48550

Chen, P. Y., & Liu, S. (2023). Holistic Adversarial Robustness of Deep Learning Models. *Proceedings of the 37th AAAI Conference on Artificial Intelligence, AAAI 2023*, 37(13), 15411–15420. DOI: 10.1609/aaai.v37i13.26797

Chen, R. J., Lu, M. Y., Chen, T. Y., Williamson, D. F. K., & Mahmood, F. (2021). Synthetic data in machine learning for medicine and healthcare. *Nature Biomedical Engineering*, *5*(6), 493–497. DOI: 10.1038/s41551-021-00751-8 PMID: 34131324

Chernikova, A., & Oprea, A. (2022). FENCE: Feasible Evasion Attacks on Neural Networks in Constrained Environments. *ACM Transactions on Privacy and Security*, *25*(4), 1–34. Advance online publication. DOI: 10.1145/3544746

Das, A., & Rad, P. (2020). Opportunities and Challenges in Explainable Artificial Intelligence (XAI): A Survey. ArXiv (Cornell University). https://doi.org//arxiv .2006.11371DOI: 10.48550

Delplace, A., Hermoso, S., & Anandita, K. (2020). Cyber Attack Detection thanks to Machine Learning Algorithms. ArXiv (Cornell University). https://doi.org//arxiv .2001.06309DOI: 10.48550

Dhir, N., Hoeltgebaum, H., Adams, N., Briers, M., Burke, A., & Jones, P. (2021). Prospective Artificial Intelligence Approaches for Active Cyber Defence. ArXiv (Cornell University). https://doi.org//arxiv.2104.09981DOI: 10.48550

Dini, P., Elhanashi, A., Begni, A., Saponara, S., Zheng, Q., & Gasmi, K. (2023). Overview on Intrusion Detection Systems Design Exploiting Machine Learning for Networking Cybersecurity. *Applied Sciences (Basel, Switzerland)*, *13*(13), 7507. DOI: 10.3390/app13137507

Dong, B., Hui, Wang, Varde, A. S., Li, D., Samanthula, B. K., Sun, W., & Zhao, L. (2021). Cyber Intrusion Detection by Using Deep Neural Networks with Attack-sharing Loss. ArXiv (Cornell University). DOI: 10.1109/DataCom.2019.00011

Dunn, C., Moustafa, N., & Turnbull, B. (2020). Robustness evaluations of sustainable machine learning models against data poisoning attacks in the internet of things. *Sustainability (Basel)*, *12*(17), 6434. DOI: 10.3390/su12166434

Ejeofobiri, C. K., Fadare, A. A., Fagbo, O. O., Ejiofor, V. O., & Fabusoro, A. T. (2024). The role of Artificial Intelligence in enhancing cybersecurity: A comprehensive review of threat detection, response, and prevention techniques. In International Journal of Science and Research Archive (Vol. 13, Issue 2, pp. 310–316). DOI: 10.30574/ijsra.2024.13.2.2161

Ennaji, S., De Gaspari, F., Hitaj, D., Kbidi, A., & Mancini, L. V. (2024). Adversarial Challenges in Network Intrusion Detection Systems: Research Insights and Future Prospects. ArXiv (Cornell University). https://doi.org//arxiv.2409.18736DOI: 10.48550

Erpek, T., O'Shea, T. J., Sagduyu, Y. E., Shi, Y., & Clancy, T. C. (2020). Deep Learning for Wireless Communications. *Studies in Computational Intelligence, 867*, 223–266. DOI: 10.1007/978-3-030-31764-5_9

Fnu, H., Kaushik, K., Gaur, P., Saratchandran, D. V., Thapliyal, A. G., Sidhu, K. S., & Singh, V. (2025, April). Multi-Agent Systems for Collaborative Financial Decision-Making Over Distributed Network Architectures. In 2025 3rd International Conference on Advancement in Computation & Computer Technologies (InCACCT) (pp. 939-944). IEEE. DOI: 10.1109/InCACCT65424.2025.11011435

Fuhl, W., Rong, Y., Motz, T., Scheidt, M., Hartel, A., Koch, A., & Kasneci, E. (2020). Explainable online validation of machine learning models for practical applications. Proceedings - International Conference on Pattern Recognition, 3304–3311. DOI: 10.1109/ICPR48806.2021.9412959

Giunchiglia, E., Imrie, F., van der Schaar, M., & Lukasiewicz, T. (2024). Machine learning with requirements: A manifesto. Neurosymbolic Artificial Intelligence, 1–13. DOI: 10.3233/NAI-240767

Gong, Y., Zhu, M., Huo, S., Xiang, Y., & Yu, H. (2024). Enhancing Cybersecurity Resilience in Finance with Deep Learning for Advanced Threat Detection. ArXiv Preprint ArXiv:2402.09820. https://doi.org//arxiv.2402.09820DOI: 10.48550

Goni, I., Bello, S., & Maigari, U. T. (2020). Machine Learning Algorithms Applied to System Security: A Systematic Review. In Asian Journal of Applied Science and Technology (Vol. 04, Issue 03, pp. 76–81). DOI: 10.38177/ajast.2020.4311

Grunova, D., Bakratsi, V., Vrochidou, E., & Papakostas, G. A. (2024). Machine Learning for Anomaly Detection in Industrial Environments. *EEPES, 25*, 25. Advance online publication. DOI: 10.3390/engproc2024070025

Guo, M., Zhang, Q., Liao, X., Chen, F. Y., & Zeng, D. D. (2021). A hybrid machine learning framework for analyzing human decision making through learning preferences. ArXiv (Cornell University), 101. https://doi.org//arxiv.1906.01233DOI: 10.48550

Hai-tao, L., Wang, H., Liu, H., & Wang, H. (2023). Real-Time Anomaly Detection of Network Traffic Based on CNN. *Symmetry, 15*(6), 1205. DOI: 10.3390/sym15061205

Haider, N., Baig, M. Z., & Imran, M. (2020). Artificial Intelligence and Machine Learning in 5G Network Security: Opportunities, advantages, and future research trends. ArXiv (Cornell University). https://doi.org//arxiv.2007.04490DOI: 10.48550

Hao, C., Dotzel, J., Xiong, J., Benini, L., Zhang, Z., & Chen, D. (2021). Enabling Design Methodologies and Future Trends for Edge AI: Specialization and Codesign. *IEEE Design & Test, 38*(4), 7–26. DOI: 10.1109/MDAT.2021.3069952

Hashemi, M. J., & Keller, E. (2020). Enhancing Robustness against Adversarial Examples in Network Intrusion Detection Systems. 2020 IEEE Conference on Network Function Virtualization and Software Defined Networks, NFV-SDN 2020 - Proceedings, 37–43. DOI: 10.1109/NFV-SDN50289.2020.9289869

Hesham, M., Essam, M., Bahaa, M., Mohamed, A., Gomaa, M., Hany, M., & Elsersy, W. (2024). Evaluating Predictive Models in Cybersecurity: A Comparative Analysis of Machine and Deep Learning Techniques for Threat Detection. 2024 Intelligent Methods [IMSA]. *Systems, and Applications, 45*, 33–38. DOI: 10.1109/IMSA61967.2024.10652833

Himeur, Y., Sayed, A. N., Alsalemi, A., Bensaali, F. F., & Amira, A. (2024). Edge AI for Internet of Energy: Challenges and perspectives. *Internet of Things : Engineering Cyber Physical Human Systems, 25*, 101035. DOI: 10.1016/j.iot.2023.101035

Huang, H. N., Chen, H. M., Lin, W. W., Huang, C. J., Chen, Y. C., Wang, Y. H., & Yang, C. T. (2023). Employing feature engineering strategies to improve the performance of machine learning algorithms on echocardiogram dataset. *Digital Health, 9*, 20552076231207589. Advance online publication. DOI: 10.1177/20552076231207589 PMID: 37915794

Huang, X. (2021). Network Intrusion Detection Based on an Improved Long-Short-Term Memory Model in Combination with Multiple Spatiotemporal Structures. Wireless Communications and Mobile Computing, 2021(1). DOI: 10.1155/2021/6623554

Huč, A., Šalej, J., Trebar, M., Huc, A., Šalej, J., & Trebar, M. (2021). Analysis of machine learning algorithms for anomaly detection on edge devices. *Sensors (Basel), 21*(14), 4946. DOI: 10.3390/s21144946 PMID: 34300686

Jones, R., Omar, M., Mohammed, D., Nobles, C., & Dawson, M. (2023). Harnessing the Speed and Accuracy of Machine Learning to Advance Cybersecurity. 2023 Congress in Computer Science, Computer Engineering, & Applied Computing (CSCE). /arxiv.2302.12415DOI: 10.1109/CSCE60160.2023.00074

Joshi, K., Singh, V., Awasthi, P., Anandaram, H., Sidhu, K. S., & Kumar, P. (2024, November). Smart Dam Water Monitoring using Enabled Internet of Things. In 2024 5th International Conference on Data Intelligence and Cognitive Informatics (ICDICI) (pp. 239-243). IEEE. DOI: 10.1109/ICDICI62993.2024.10810761

Joshi, K., Yadav, H., Gupta, S., Singh, V., Sidhu, K. S., & Kukreti, R. (2025, April). Handling Security Aspects in the Internet of Things: Latest Challenges and Measures to Mitigate Risks. In 2025 3rd International Conference on Communication, Security, and Artificial Intelligence (ICCSAI) (Vol. 3, pp. 1434-1439). IEEE.

Joshi, P., Singh, V., Thapliyal, N., Joshi, K., Bhatt, A., & Mahur, M. (2024, October). Intelligent IoT-Enabled Real-Time Monitoring System for Logistics Management. In *2024 International Conference on Computing, Sciences and Communications (ICCSC)* (pp. 1-5). IEEE. DOI: 10.1109/ICCSC62048.2024.10830332

Jung, K., Kashyap, S., Avati, A., Harman, S., Shaw, H., Li, R., Smith, M., Shum, K., Javitz, J., Vetteth, Y., Seto, T., Bagley, S. C., & Shah, N. H. (2020). A framework for making predictive models useful in practice. *Journal of the American Medical Informatics Association : JAMIA*, *28*(6), 1149–1158. DOI: 10.1093/jamia/ocaa318 PMID: 33355350

Kabanda, G. (2020). A bayesian network model for machine learning and cyber security. *ACM International Conference Proceeding Series*, 1–7. DOI: 10.1145/3440094.3440389

Kale, A., Nguyen, T., Harris, J. S., Li, C., Zhang, J., Ma, X., Harris, F. C., Li, C., Zhang, J., & Ma, X. (2022). Provenance documentation to enable explainable and trustworthy AI: A literature review. In Data Intelligence (Vol. 5, Issue 1, pp. 139–162). The MIT Press. DOI: 10.1162/dint_a_00119

Kaliyaperumal, P., Periyasamy, U., Thirumalaisamy, M., Balusamy, B., & Benedetto, F. (2024). A Novel Hybrid Unsupervised Learning Approach for Enhanced Cybersecurity in the IoT. *Future Internet*, *16*(7), 253. DOI: 10.3390/fi16070253

Kamalov, F., Moussa, S., El Khatib, Z., Ben Mnaouer, A., Zgheib, R., & Mashaal, O. (2021). Orthogonal variance-based feature selection for intrusion detection systems. ArXiv. DOI: 10.1109/ISNCC52172.2021.9615656

Kaur, P., Kaur, K., Verma, D. K., Singh, K., Kaushik, K., & Singh, V. (2024, September). Routing protocols in wireless sensor networks: A comprehensive review and future perspectives. In 2024 7th International Conference on Contemporary Computing and Informatics (IC3I) (Vol. 7, pp. 263-270). IEEE. DOI: 10.1109/IC3I61595.2024.10828894

Kaur, P., Kaur, K., Verma, D. K., Singh, K., Kaushik, K., & Singh, V. (2024, September). Routing protocols in wireless sensor networks: A comprehensive review and future perspectives. In *2024 7th International Conference on* Contemporary Computing and Informatics (IC3I) (Vol. 7, pp. 263-270). IEEE. DOI: 10.1109/IC3I61595.2024.10828894

Khanji, S., & Khattak, A. (2020). Towards a Novel Intrusion Detection Architecture using Artificial Intelligence. *ACM International Conference Proceeding Series*, 185–189. DOI: 10.1145/3436829.3436842

Kodys, M., Lu, Z., Fok, K. W., Thing, V. L. L. L., Kodyš, M., Lu, Z., Fok, K. W., & Thing, V. L. L. L. (2021). Intrusion Detection in Internet of Things using Convolutional Neural Networks. 2021 18th International Conference on Privacy, Security and Trust, PST 2021, 1–10. DOI: 10.1109/PST52912.2021.9647828

Kopitar, L., Kocbek, P., Cilar, L., Sheikh, A., & Stiglic, G. (2020). Early detection of type 2 diabetes mellitus using machine learning-based prediction models. *Scientific Reports*, *10*(1), 11981. Advance online publication. DOI: 10.1038/s41598-020-68771-z PMID: 32686721

Kornaros, G. (2022). Hardware-Assisted Machine Learning in Resource-Constrained IoT Environments for Security: Review and Future Prospective. *IEEE Access : Practical Innovations, Open Solutions*, *10*, 58603–58622. DOI: 10.1109/ACCESS.2022.3179047

Kumar, S., Lim, W. M., Sivarajah, U., & Kaur, J. (2022). Artificial Intelligence and Blockchain Integration in Business: Trends from a Bibliometric-Content Analysis. *Information Systems Frontiers*, *25*(2), 871–896. DOI: 10.1007/s10796-022-10279-0 PMID: 35431617

Lam, N. T. (2021). Detecting Unauthorized Network Intrusion based on Network Traffic using Behavior Analysis Techniques. *International Journal of Advanced Computer Science and Applications*, *12*(4), 46–51. DOI: 10.14569/IJACSA.2021.0120407

Layeghy, S., Gallagher, M., & Portmann, M. (2021). Benchmarking the Benchmark — Analysis of Synthetic NIDS Datasets. ArXiv (Cornell University). https://arxiv.org/abs/2104.09029

Lisovskyi, K., & Rochenovich, G.-D. (2023). Artificial Intelligence in the Security System of Enterprise. *Grail of Science*, *27*(27), 308–316. DOI: 10.36074/grail-of-science.12.05.2023.049

Liu, C., Salzmann, M., Lin, T., Tomioka, R., & Süsstrunk, S. (2020). On the loss landscape of adversarial training: Identifying challenges and how to overcome them. Advances in Neural Information Processing Systems, 2020-December. DOI: 10.48550/arxiv.2006.08403

Liu, H., & Patras, P. (2022). NetSentry: A deep learning approach to detecting incipient large-scale network attacks. *Computer Communications*, *191*, 119–132. DOI: 10.1016/j.comcom.2022.04.020

Lu, C., Strout, J., Gauriau, R., Wright, B., Marcruz, F. B. D. C., Buch, V., & Andriole, K. (2020). An Overview and Case Study of the Clinical AI Model Development Life Cycle for Healthcare Systems. ArXiv (Cornell University). https://doi.org//arxiv.2003.07678DOI: 10.48550

Lu, S. C., Swisher, C. L., Chung, C., Jaffray, D., & Sidey-Gibbons, C. (2023). On the importance of interpretable machine learning predictions to inform clinical decision making in oncology. In *Frontiers in Oncology* (Vol. 13). Frontiers Media., DOI: 10.3389/fonc.2023.1129380

Lucas, K., Sharif, M., Bauer, L., Reiter, M. K., & Shintre, S. (2021). Malware Makeover: Breaking ML-based Static Analysis by Modifying Executable Bytes. ASIA CCS 2021 - Proceedings of the 2021 ACM Asia Conference on Computer and Communications Security, 744–758. DOI: 10.1145/3433210.3453086

Mandalapu, V., Elluri, L., Vyas, P., & Roy, N. (2023). Crime Prediction Using Machine Learning and Deep Learning: A Systematic Review and Future Directions. In Access, I. E. E. E. (Ed.), Vol. 11, pp. 60153–60170). Cornell University., DOI: 10.1109/ACCESS.2023.3286344

Mane, S., & Rao, D. (2021). Explaining Network Intrusion Detection System Using Explainable AI Framework. ArXiv (Cornell University). https://doi.org//arxiv.2103.07110DOI: 10.48550

Mazhar, T., Talpur, D. B., Al Shloul, T., Ghadi, Y. Y., Haq, I., Ullah, I., Ouahada, K., & Hamam, H. (2023). Analysis of IoT Security Challenges and Its Solutions Using Artificial Intelligence. *Brain Sciences*, *13*(4), 683. DOI: 10.3390/brainsci13040683 PMID: 37190648

Mohamed, N., Taherdoost, H., & Madanchian, M. (2024). Comprehensive Review of Advanced Machine Learning Techniques for Detecting and Mitigating Zero-Day Exploits. *ICST Transactions on Scalable Information Systems*, *11*(6). Advance online publication. DOI: 10.4108/eetsis.6111

Molina, S. B., Nespoli, P., & Mármol, F. G. (2023). Tackling Cyberattacks through AI-based Reactive Systems: A Holistic Review and Future Vision. In arXiv (Cornell University). Cornell University. https://doi.org//arxiv.2312.06229DOI: 10.48550

Morse, K. E., Bagely, S. C., & Shah, N. H. (2020). Estimate the hidden deployment cost of predictive models to improve patient care. *Nature Medicine*, *26*(1), 18–19. DOI: 10.1038/s41591-019-0651-8 PMID: 31932778

Moustafa, N. (2021). A new distributed architecture for evaluating AI-based security systems at the edge: Network TON_IoT datasets. *Sustainable Cities and Society, 72*, 102994. DOI: 10.1016/j.scs.2021.102994

Muhammad, D., & Bendechache, M. (2024). Unveiling the black box: A systematic review of Explainable Artificial Intelligence in medical image analysis. In *Computational and Structural Biotechnology Journal* (Vol. 24, pp. 542–560). Elsevier BV., DOI: 10.1016/j.csbj.2024.08.005

Nallakaruppan, M. K., Chaturvedi, H., Grover, V., Balusamy, B., Jaraut, P., Bahadur, J., Meena, V. P., & Hameed, I. A. (2024). Credit Risk Assessment and Financial Decision Support Using Explainable Artificial Intelligence. *Risks, 12*(10), 164. DOI: 10.3390/risks12100164

Narooka, P., Kaushik, K., Kansal, S., Kejriwal, D., Mishra, R. K., & Singh, V. (2025, April). Hybrid AI Models for QoS Estimation in α-η-μ Fading and Co-Channel Interference Environments within 6G Networks. In 2025 3rd International Conference on Advancement in Computation & Computer Technologies (InCACCT) (pp. 206-210). IEEE.

Narooka, P., Verma, S., Dwivedi, A., Kaushik, K., Singh, V., & Gaur, P. (2025, April). Psi-zeta Approach for Assessing Network Efficiency in Log-Normal Fading and Spectrum Sharing Interference Environments. In 2025 3rd International Conference on Advancement in Computation & Computer Technologies (InCACCT) (pp. 886-891). IEEE. DOI: 10.1109/InCACCT65424.2025.11011391

Nguyen, T. T., & Reddi, V. J. (2023). Deep Reinforcement Learning for Cyber Security. *IEEE Transactions on Neural Networks and Learning Systems, 34*(8), 3779–3795. DOI: 10.1109/TNNLS.2021.3121870 PMID: 34723814

Nibigira, N., Havyarimana, V., & Xiao, Z. (2024). Artificial Intelligence Adoption for Cybersecurity in Africa. *Journal of Information Security, 15*(02), 134–147. DOI: 10.4236/jis.2024.152009

Nicholls, J., Kuppa, A., & Le-Khac, N. A. (2021). Financial cybercrime: A comprehensive survey of deep learning approaches to tackle the evolving financial crime landscape. *IEEE Access : Practical Innovations, Open Solutions, 9*, 163965–163986. DOI: 10.1109/ACCESS.2021.3134076

Ogunseyi, T. B., & Thiyagarajan, G. (2025). An Explainable LSTM-Based Intrusion Detection System Optimized by Firefly Algorithm for IoT Networks. *Sensors (Basel), 25*(7), 2288. DOI: 10.3390/s25072288 PMID: 40218800

Oyinloye, T. S., Arowolo, M. O., & Prasad, R. (2024). *Enhancing Cyber Threat Detection with an Improved Artificial Neural Network Model.* Data Science and Management., DOI: 10.1016/j.dsm.2024.05.002

Pai, V., Devidas, , & Adesh, N. D. (2021). Comparative analysis of Machine Learning algorithms for Intrusion Detection. *IOP Conference Series. Materials Science and Engineering, 1013*(1), 12038. DOI: 10.1088/1757-899X/1013/1/012038

Pang, S., & Li, Y. (2020). Artificial Intelligence Techniques for Cyber Security Applications. International Journal of Advanced Information and Communication Technology, 89–94. DOI: 10.46532/ijaict-2020021

Paramesha, M., Rane, N., & Rane, J. (2024). Artificial intelligence, machine learning, deep learning, and blockchain in financial and banking services: a comprehensive review. In SSRN Electronic Journal. RELX Group (Netherlands). DOI: 10.2139/ssrn.4855893

Ponce-Bobadilla, A. V., Schmitt, V., Maier, C. S., Mensing, S., & Stodtmann, S. (2024). Practical guide to SHAP analysis: Explaining supervised machine learning model predictions in drug development. In Clinical and Translational Science (Vol. 17, Issue 11). Wiley. DOI: 10.1111/cts.70056

Quintero-Bonilla, S., & del Rey, A. M. (2020). Proposed models for advanced persistent threat detection: A review. In *Advances in Intelligent Systems and Computing* (Vol. 1004, pp. 141–148). Springer Nature., DOI: 10.1007/978-3-030-23946-6_16

Rafique, S. H., Abdallah, A., Musa, N. S., & Murugan, T. (2024). Machine Learning and Deep Learning Techniques for Internet of Things Network Anomaly Detection—Current Research Trends. *Sensors (Basel), 24*(6), 1968. DOI: 10.3390/s24061968 PMID: 38544231

Rahman, M. M., Al Shakil, S., & Mustakim, M. R. (2025). A survey on intrusion detection system in IoT networks. *Cyber Security and Applications, 3*, 100082. DOI: 10.1016/j.csa.2024.100082

Raturi, A., Mula, K. G., Joshi, K., Sidhu, K. S., Chauhan, S. S., & Singh, V. (2025, April). Comparative Analysis of Stock Price Prediction with Multiple Machine Learning Model. In *2025 3rd International Conference on Communication, Security, and Artificial Intelligence (ICCSAI)* (Vol. 3, pp. 1428-1433). IEEE. DOI: 10.1109/ICCSAI64074.2025.11064599

Redhu, A., Choudhary, P., Srinivasan, K., & Das, T. K. (2024). Deep learning-powered malware detection in cyberspace: a contemporary review. In *Frontiers in Physics* (Vol. 12). Frontiers Media., DOI: 10.3389/fphy.2024.1349463

Reis, M. J. C. S. (2025). Blockchain-Enhanced Security for 5G Edge Computing in IoT. *Computation (Basel, Switzerland)*, *13*(4), 98. DOI: 10.3390/computation13040098

Reyes, M., Meier, R., Pereira, S., Silva, C. A., Dahlweid, F. M., von Tengg-Kobligk, H., Summers, R. M., & Wiest, R. (2020). On the interpretability of artificial intelligence in radiology: Challenges and opportunities. In Radiology: Artificial Intelligence (Vol. 2, Issue 3). Radiological Society of North America. DOI: 10.1148/ryai.2020190043

Rjoub, G., Bentahar, J., Abdel Wahab, O., Mizouni, R., Song, A., Cohen, R., Otrok, H., & Mourad, A. (2023). A Survey on Explainable Artificial Intelligence for Cybersecurity. *IEEE Transactions on Network and Service Management*, *20*(4), 5115–5140. DOI: 10.1109/TNSM.2023.3282740

Rodriguez, E., Otero, B., Gutierrez, N., & Canal, R. (2021). A survey of deep learning techniques for cybersecurity in mobile networks. *IEEE Communications Surveys and Tutorials*, *23*(3), 1920–1955. DOI: 10.1109/COMST.2021.3086296

Roshanaei, M., Khan, M. R., & Sylvester, N. N. (2024). Navigating AI Cybersecurity: Evolving Landscape and Challenges. *Journal of Intelligent Learning Systems and Applications*, *16*(03), 155–174. DOI: 10.4236/jilsa.2024.163010

Sáez-de-Cámara, X., Flores, J. L., Arellano, C., Urbieta, A., & Zurutuza, U. (2023). Clustered federated learning architecture for network anomaly detection in large scale heterogeneous IoT networks. *Computers & Security*, *131*, 103299. Advance online publication. DOI: 10.1016/j.cose.2023.103299

Saha, S., Banerjee, K., Ghosh, S., Mitra, S., & Pal, D. (2023). AI-Driven Edge Computing for IoT: A Comprehensive Survey and Future Directions. International Journal of Advanced Research in Science. *Tongxin Jishu*, *117–121*, 117–121. Advance online publication. DOI: 10.48175/IJARSCT-12921

Saifullah, S., Mercier, D., Lucieri, A., Dengel, A., & Ahmed, S. (2022). Privacy Meets Explainability: A Comprehensive Impact Benchmark. ArXiv (Cornell University). https://doi.org//arxiv.2211.04110DOI: 10.48550

Sakhai, M., & Wielgosz, M. (2021). Modern Cybersecurity Solution using Supervised Machine Learning. ArXiv (Cornell University). https://doi.org//arxiv.2109.07593DOI: 10.48550

Saleh, A. M. S. (2024). Blockchain for secure and decentralized artificial intelligence in cybersecurity: A comprehensive review. Blockchain. *Research and Applications*, *5*(3), 100193. DOI: 10.1016/j.bcra.2024.100193

Salem, A. H., Azzam, S. M., Emam, O. E., & Abohany, A. A. (2024). Advancing cybersecurity: A comprehensive review of AI-driven detection techniques. *Journal of Big Data*, *11*(1), 105. Advance online publication. DOI: 10.1186/s40537-024-00957-y

Samarakoon, S., Siriwardhana, Y., Porambage, P., Liyanage, M., Chang, S.-Y., Kim, J., Kim, J., & Ylianttila, M. (2022). 5G-NIDD: A Comprehensive Network Intrusion Detection Dataset Generated over 5G Wireless Network. ArXiv (Cornell University). https://doi.org//arxiv.2212.01298DOI: 10.48550

Sameen, M., Han, K., & Hwang, S. O. (2020). PhishHaven - An Efficient Real-Time AI Phishing URLs Detection System. *IEEE Access : Practical Innovations, Open Solutions*, *8*, 83425–83443. DOI: 10.1109/ACCESS.2020.2991403

Samek, W., Montavon, G., Lapuschkin, S., Anders, C. J., & Müller, K. R. (2020). Toward interpretable machine learning: Transparent deep neural networks and beyond. ArXiv, 1–24. https://arxiv.org/pdf/2003.07631

Sarhan, M., Lo, W. W., Layeghy, S., & Portmann, M. (2022). HBFL: A hierarchical blockchain-based federated learning framework for collaborative IoT intrusion detection. *Computers & Electrical Engineering*, *103*, 108379. Advance online publication. DOI: 10.1016/j.compeleceng.2022.108379

Sarhan, M. M., Layeghy, S., Gallagher, M. M., & Portmann, M. (2021). From Zero-Shot Machine Learning to Zero-Day Attack Detection. [Cornell University]. *International Journal of Information Security*, *22*(4), 947–959. DOI: 10.1007/s10207-023-00676-0

Schmitt, M. (2023). Securing the digital world: Protecting smart infrastructures and digital industries with artificial intelligence (AI)-enabled malware and intrusion detection. *Journal of Industrial Information Integration*, *36*, 100520. DOI: 10.1016/j.jii.2023.100520

Sen, J., Sen, R. R., & Dutta, A. A. A. (2021). *Machine Learning in Finance-Emerging Trends and Challenges. In arXiv (Cornell University). arXiv*. Cornell University., DOI: 10.5772/intechopen.101120

Shahrivari, V., Darabi, M. M., & Izadi, M. (2020). *Phishing Detection Using Machine Learning Techniques. ArXiv*. Cornell University., DOI: 10.1109/ASIANCON58793.2023.10270550

Singh, R., & Gill, S. S. (2023). Edge AI: A survey. *Internet of Things and Cyber-Physical Systems*, *3*, 71–92. DOI: 10.1016/j.iotcps.2023.02.004

Singh, S., Sharma, P. K., Yoon, B., Shojafar, M., Cho, G. H., & Ra, I. H. (2020). Convergence of blockchain and artificial intelligence in IoT network for the sustainable smart city. *Sustainable Cities and Society, 63*, 102364. DOI: 10.1016/j.scs.2020.102364

Singh, V., & Dubey, V. (2015). A two level image security based on Arnold transform and chaotic logistic mapping. *International Journal of Advanced Research in Computer Science and Software Engineering, 5*(2), 883–887.

Singh, V., & Kaushik, V. D. (2023). A Brief Overview of Cyber Security Advances and Techniques Along With a Glimpse on Quantum Cryptography: Cyber Security Practices, Advances and Challenges. Exploring Cyber Criminals and Data Privacy Measures, 40-64.

Singh, V., & Kaushik, V. D. (2023). A Methodical View of Prerequisites of Picture Combination, Strategies, and Key Indicators with Usage in Real Life and Scientific Domains Facilitating Smart Ubiquitous Environment. In Computational Intelligence in Analytics and Information Systems (pp. 15-33). Apple Academic Press.

Singh, V., & Kaushik, V. D. (2023). Navigating the Landscape of Security Threat Analysis in Cloud Computing environments. In *Security and Risk Analysis for Intelligent Cloud Computing* (pp. 1–25). CRC Press.

Smart, M. A. (2021). Addressing privacy threats from machine learning. ArXiv Preprint ArXiv:2111.04439. https://doi.org//arxiv.2111.04439DOI: 10.48550

Solaas, J. R. V., Tuptuk, N., & Mariconti, E. (2024). *Systematic Review: Anomaly Detection in Connected and Autonomous Vehicles. ArXiv*. Cornell University., DOI: 10.1109/TITS.2024.3495031

Sonule, A. R., Kalla, M., Jain, A., & Chouhan, D. S. (2020). Unsw-Nb15 Dataset and Machine Learning Based Intrusion Detection Systems. *International Journal of Engineering and Advanced Technology, 9*(3), 2638–2648. DOI: 10.35940/ijeat.C5809.029320

Sourlos, N., Wang, J., Nagaraj, Y., van Ooijen, P., & Vliegenthart, R. (2022). Possible Bias in Supervised Deep Learning Algorithms for CT Lung Nodule Detection and Classification. In Cancers (Vol. 14, Issue 16, p. 3867). Multidisciplinary Digital Publishing Institute. DOI: 10.3390/cancers14163867

Sriram, S., Simran, K., Vinayakumar, R., Akarsh, S., & Soman, K. P. (2020). Towards Evaluating the Robustness of Deep Intrusion Detection Models in Adversarial Environment. In Communications in Computer and Information Science: Vol. 1208 CCIS (pp. 111–120). Springer Science+Business Media. DOI: 10.1007/978-981-15-4825-3_9

Stiglic, G., Kocbek, P., Fijacko, N., Zitnik, M., Verbert, K., & Cilar, L. (2020). Interpretability of machine learning-based prediction models in healthcare. *Wiley Interdisciplinary Reviews. Data Mining and Knowledge Discovery*, *10*(5), e1379. Advance online publication. DOI: 10.1002/widm.1379

Sun, Y., Peng, M., Zhou, Y., Huang, Y., & Mao, S. (2019). Application of Machine Learning in Wireless Networks: Key Techniques and Open Issues. *IEEE Communications Surveys and Tutorials*, *21*(4), 302–3108. DOI: 10.1109/COMST.2019.2924243

Taherdoost, H. (2022). Blockchain Technology and Artificial Intelligence Together: A Critical Review on Applications. In Applied Sciences (Switzerland) (Vol. 12, Issue 24, p. 12948). Multidisciplinary Digital Publishing Institute. DOI: 10.3390/app122412948

Talukder, M. A., Islam, M. M., Uddin, M. A., Hasan, K. F., Sharmin, S., Alyami, S. A., & Moni, M. A. (2024). Machine learning-based network intrusion detection for big and imbalanced data using oversampling, stacking feature embedding and feature extraction. *Journal of Big Data*, *11*(1), 33. Advance online publication. DOI: 10.1186/s40537-024-00886-w

Thapa, N., Liu, Z., Kc, D. B., Gokaraju, B., & Roy, K. (2020). Comparison of machine learning and deep learning models for network intrusion detection systems. *Future Internet*, *12*(10), 1–16. DOI: 10.3390/fi12100167

Thuraisingham, B. (2022). Trustworthy Machine Learning. *IEEE Intelligent Systems*, *37*(1), 21–24. DOI: 10.1109/MIS.2022.3152946

Tian, T., Zhang, C., Jiang, B., Feng, H., & Lu, Z. (2025). Insider threat detection for specific threat scenarios. *Cybersecurity*, *8*(1), 17. Advance online publication. DOI: 10.1186/s42400-024-00321-w

Umer, M. A., Junejo, K. N., Jilani, M. T., & Mathur, A. P. (2022). Machine learning for intrusion detection in industrial control systems: Applications, challenges, and recommendations. *International Journal of Critical Infrastructure Protection*, *38*, 100516. DOI: 10.1016/j.ijcip.2022.100516

Vellido, A. (2020). The importance of interpretability and visualization in machine learning for applications in medicine and health care. *Neural Computing & Applications, 32*(24), 18069–18083. DOI: 10.1007/s00521-019-04051-w

Vitorino, J., Praça, I., & Maia, E. (2023). Towards adversarial realism and robust learning for IoT intrusion detection and classification. Annales Des Telecommunications. *Annales des Télécommunications, 78*(7–8), 401–412. DOI: 10.1007/s12243-023-00953-y

Wala, F. B., & Kiran, M. (2024). 5G Network Security Practices: An Overview and Survey. ArXiv (Cornell University). https://doi.org//arxiv.2401.14350DOI: 10.48550

Walsh, I., Titma, T., Psomopoulos, F. E., & Tosatto, S. (2020). Recommendations for machine learning validation in biology. ArXiv, June, 23. https://doi.org//arxiv.2006.16189DOI: 10.48550

Wang, M., Lu, Y., & Qin, J. (2020). A dynamic MLP-based DDoS attack detection method using feature selection and feedback. *Computers & Security, 88*, 101645. DOI: 10.1016/j.cose.2019.101645

Wang, Z., Guan, Z., Liu, X., Li, C., Sun, X., & Li, J. (2025). SDN Anomalous Traffic Detection Based on Temporal Convolutional Network. *Applied Sciences (Basel, Switzerland), 15*(8), 4317. DOI: 10.3390/app15084317

Weitkamp, E., Satani, Y., Omundsen, A., Wang, J., & Li, P. (2023). MalIoT: Scalable and Real-time Malware Traffic Detection for IoT Networks. ArXiv (Cornell University). https://doi.org//arxiv.2304.00623DOI: 10.48550

Westphal, C., Hailes, S., & Musolesi, M. (2024). Feature Selection for Network Intrusion Detection. ArXiv (Cornell University). https://doi.org//arxiv.2411.11603DOI: 10.48550

Wu, T., Lucas, E., Zhao, F., Basu, P., & Qiao, Y. (2024). Artificial intelligence strengthenes cervical cancer screening - present and future. In *Cancer biology & medicine* (pp. 1–16). Chinese Anti-Cancer Association., DOI: 10.20892/j.issn.2095-3941.2024.0198

Xu, G. (2021). Research on Network Intrusion Detection Method Based on Machine Learning. *Journal of Physics: Conference Series, 1861*(1), 12034. DOI: 10.1088/1742-6596/1861/1/012034

Zhang, Y., Li, B., & Tan, Y. (2021). Making AI available for everyone at anywhere: A Survey about Edge Intelligence. *Journal of Physics: Conference Series, 1757*(1), 12076. DOI: 10.1088/1742-6596/1757/1/012076

Zuech, R., Hancock, J., & Khoshgoftaar, T. M. (2022). A new feature popularity framework for detecting cyberattacks using popular features. *Journal of Big Data*, *9*(1), 119. Advance online publication. DOI: 10.1186/s40537-022-00661-9

Chapter 2

An Examination of AI-Integrated Wireless Sensor Networks Security: Obstacles and Prospects

Eyman F. A. Elsmany
https://orcid.org/0000-0002-0206-0950
University of Gezira, Sudan

Mayada S. A. Mustafa
https://orcid.org/0009-0000-3574-821X
Uruk University, Iraq

Zeinab E. Ahmed
https://orcid.org/0000-0002-6144-8533
University of Gezira, Sudan

Altahir A. Altahir
https://orcid.org/0000-0002-0650-2509
Universiti Tunku Abdul Rahman, Malaysia

ABSTRACT

The proliferation of WSN from critical applications has presented further significant security problems. This chapter addresses the opportunities of AI to deal with such growing threats, showing a paradigm shift from conventional defense approaches. Different AI-aware WSN applications are outlined, relying on their specific security considerations. It analyzes why conventional security fending measures fail to react to advanced attacks, and how the threatening trend of offensive AI has gained popularity. In addition, the chapter gives a detailed visual of how AI-based

DOI: 10.4018/979-8-3373-3576-6.ch002

t © 2026, IGI Global Scientific Publishing. Copying or distributing in print or electronic forms without written permission of IGI Global Scientific Publishing is prohibited. Use of this chapter to enerative artificial intelligence (AI) technologies is expressly prohibited. The publisher reserves all rights to license its use for generative AI training and machine learning model development.

security frameworks may be exploited using to enhance model reliability against adversarial attacks, provide scalable and adaptive defense policies, and mitigate risks. AI-enabled security for WSNs is surveyed in different security scopes; namely, emerging technologies and AI impact for more secure WSNs with ML and DL. Finaly highlights the challenges and opportunities in an emerging field of research, including emerging research directions, state-of-art AI security models, and the need for standardization and policy.

1. INTRODUCTION

Recently, Wireless Sensor Networks (WSNs) are main component in modern applications. Involving industrial automation, healthcare, military, and critical infrastructure monitoring. However, cyber threats and attack vectors have become more sophisticated. The security levels are getting worse (Balas et al., 2026). The advancing of security technology and robust techniques required. As a key novelty for protecting WSNs and strengthening their security, AI-based solutions have been developed. Researchers have focused on integrating AI technologies. To combat intrusions, hacking, cyberattacks, and unauthorized access (Behiry & Aly, 2024; Osamy et al., 2022). Machine learning and deep learning models are also applied to mitigate offensive AI threats. Fostering intelligent environments and enabling adaptive security responses are part of AI's capacity for optimizing WSN performance. Data compression, noise filtering, routing decisions, and predictive analysis, are advanced using AI. It supports predictive maintenance through predicting equipment failure in industries. Reinforcement learning algorithms can facilitate energy-aware routing decisions and prolong the battery's lifetime. WSNs are still vulnerable to security threats, such as prompt injections, eavesdropping, node compromise and other denial of service (DoS) attacks. AI-based techniques have demonstrated efficacy in anomaly detection, real-time identification of intrusions, and malicious nodes. Machine learning-based techniques have shown an accuracy in terms of behavioral analysis, spoofing detection and secure access control (Mazhar et al., 2023). In cybersecurity, AI-integrated ML and DL techniques have been used not just in defensive strategies but also as a sophisticated tool exploited by adversarial actors, resulting in a threat known as offensive AI. AI is distinguished by its rapidity and ease of exploitation of vulnerabilities, as well as scanning the network faster than a human hacker. Its designed models are trained to filter out this poisoning data, such as manipulated or malicious data, and to maintain integrity. The adversarial machine learning model is specifically trained on some manipulated data to prevent the attacker from reconstructing sensitive information, thus enhancing security and preserving data integrity (Anthi et al., 2021; Khan et al., 2024; Zhukabayeva et al., 2025). Cybercriminals are continually advancing their techniques to avoid AI defenses. When integrating an AI-based security framework in WSN, the Offensive AI is a major challenge. Consequently, an AI-driven security framework must be robust and adaptable. Therefore, it effectively brings together security, efficiency, and scalability (Mazhar et al., 2023).

1.1 Motivation and Contributions

WSNs integration with the industrial Internet of Things (IIoT) is the backbone of modern infrastructure. For instance, smart cities (Jawhar et al., 2018; Fnu et al., 2025), healthcare systems (Manikandan et al., 2025), environmental monitoring (Adu-Manu et al., 2022), military and defense (Lee et al., 2022) and industrial automation (Murtaza et al., 2024). WSNs transformed the underlying structure of modern organization (Angurala et al., 2025; Kaur et al., 2024). These networks are vulnerable to security threats due to their resource limitation (computing capability, battery lifetime, and bandwidth limitations). The random deployment of sensors raises security concerns about the system (Alomari et al., 2025). Also, their unguided transmission media are more exposed to attack issue (Shu Yinbiao et al., 2016). Conventional security measures in WSNs are often inadequate for addressing these issues (Maniriho et al., 2024). AI- models learn from data patterns and adapt to challenges as data poisoning.

This chapter focuses on: 1) WSNs AI security developing, 2) improving threat detection accuracy and 3) supporting protection mechanisms. It provides a comprehensive review of advances of security architecture. AI-based models' contribution in resource optimization and anomaly detection. Finally, it outlines future research to enhance reliability in a dynamic network setting.

1.2 Chapter Organization

The chapter begins with background information and an overview of AI-integrated security. Emphasizing the paradigm shift from conventional defenses. Then providing a summary of AI-integrated WSN applications in various industries, associated with security issues. A strong AI framework to counter aggressive AI will be discussed. AI-based security in WSNs across diverse domains is investigated. threat detection and authentication techniques and how deep learning and machine learning improve WSN security, will be examined. In conclusion the findings and insights addressing the challenges, and future movements in AI-based secured WSNs are summarized.

2. BACKGROUND AND OVERVIEW

The section provides a background of the growth from conventional protection mechanisms to AI-based. Focus on AI-security in WSNs, offensive AI, and the development of security approaches to resist. The study investigates deep learning (DL) and machine learning (ML) to further enhance WSNs' security.

2.1 AI-Integrated Security in WSNs: A Paradigm Shift from Conventional Defenses

Conventional security methods use human intervention to update response plans and preset rules. Encryption and key management protocols are employed to protect WSNs data. These are challenged by the WSNs resource restrictions. Limited energy supply, memory capacity and computational capability (Sahar et al., 2023). Dynamic, large-scale WSN features restrict conventional safety flexibility. Pre-configuration of security measures in dynamic-WSN rapidly becomes outdated (Dora & Nemoga, 2021).

Sinkhole attack is a serious security issue. A network node compromised by hackers and altered network traffic. Proposed the shortest route to the sink node, Interrupts, modifies data, or DoSs attacks (Omar et al., 2023). Static, outdated security policies are insufficient in managing newly dynamic and sophisticated threats (Mubashir et al., 2020). The conventional methods are not adaptable to counter sophisticated attack strategies (Vaid, 2021).

Denial of service (DoS) attacks, such as packet flooding, damage the energy and administrative facilities of WSN nodes. Jamming with continuous noise signals disrupting physical layer communication. These attacks underscore a challenge for conventional security methods(Islam et al., 2021). AI provides a standard change to address these difficulties. Enabling the advancement of proactive and flexible security techniques. Overcoming the shortcomings of conventional security systems (Jada & Mayayise, 2024). Picking network behaviour patterns, identifying intelligent anomalies, and making decisions. AI are adaptable to advancing network conditions and novel attack discovery strategies. AI technologies aim to enhance the security of WSNs by combining AI techniques (Jada & Mayayise, 2024). This process involves a framework for succeeding stated objectives, including research design, data acquisition, AI model training, and evaluation metrics.

2.1.1 An Overview of AI-Integrated WSN Applications with Security Considerations

AI-integrated WSNs are operated in numerous application fields, and each one has its own security concerns and requirements. Figure 1 shows the AI-Integrated WSN Applications. WSNs are involved in industrial domains. However, AI has been active in the development of adaptive threat detection, intelligent reaction mechanisms and simultaneous anomaly classification. All of which are tailored to meet the requirements of different domains. Sence improving security levels with AI is becoming evident.

Figure 1. AI-Integrated WSN Applications

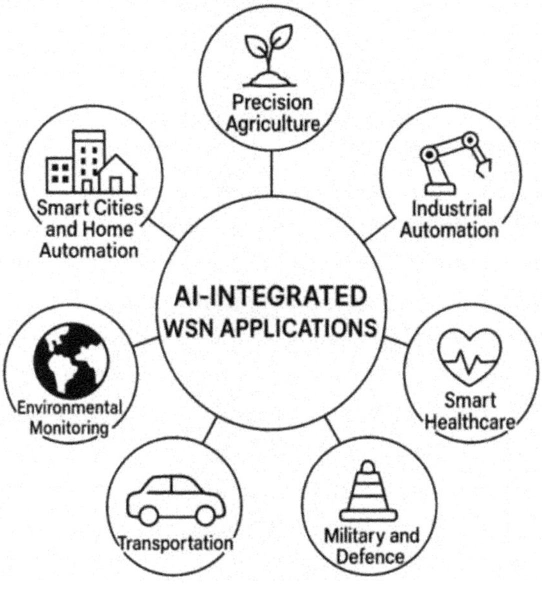

Smart Cities and Home Automation: WSNs are utilized for several purposes in everyday life and for the control of urban infrastructure. Defense against cyber-physical attacks, data privacy, and the security of control systems that could have implicated. AI is employed to detect anomalies in device behavior, compromised devices, protect user privacy and boost security (Engels, 2020; Singh and Kaushik, 2023b). **Healthcare:** WSNs are employed in various healthcare applications such as remote patient monitoring, wearable medical technology, and smart hospitals utilize WSNs. Security concerns in this field, the integrity of medical device operations and the confidentiality of sensitive patient data. Since a slight compromise could have impacts. Secure AI data transfer protocols safeguard patient privacy and offer intelligent intrusion detection systems for medical sensor networks to identify and respond to malicious activities (Jabeen et al., 2023).

Environmental Monitoring: Many ecosystems use WSNs for disaster prediction, environmental assessments, and ecological study. Data integrity is critical for accurate environmental analysis and informed governance. Protecting against data damage and unauthorized manipulation of collected data is the key goal of security measures. AI methods are applied to secure data aggregation from distributed sensors, preserve data integrity and identify potentially hacked sensors (Sharma et al., 2024).

Industrial Automation: In industrial settings, WSNs are utilized to track and control several automated operations, subsequently upgrading efficiency and production. Security in this field is necessity to avoid industrial spying, assuring the safety of automated processes, and defense control systems from hostile attacks. AI may be used to provide secure communication channels between control devices and to detect intrusions in industrial WSNs, hence identifying and responding to cyber-attacks aimed at control systems (Abdulrab et al., 2022; Onyema et al., 2025).

Transportation: WSNs persist in integrating with transportation infrastructure, giving rise to new applications and security risks. vehicle-to-vehicle messaging, autonomous driving and Intelligent traffic management, all depend on reliable wireless communication between cars and their system. Nevertheless, ensuring the confidentiality, authenticity and integrity of these exchanges' outcomes is challenging. For instance, harmful actors could potentially transmit damaging instructions or imitate genuine sources. Consequently, investigators have started developing AI-based techniques for assessing vehicular networks that address such threats. By realizing normal network patterns, AI models can detect invalid transmissions or anomalies indicating compromised machines. They aim to facilitate secure data sharing for safe and efficient traffic management. Moving forward, combining cryptography with machine-based anomaly detection may allow transportation systems to realize the promise of wireless connectivity while adequately mitigating risks to open information flows (Qureshi et al., 2022).

Military and Defense: WSNs have many applications in military and defense. Including battlefield observation, target tracking, and environmental supervising in contested areas. However, Security requirements in this field demand vigilance. Developing adaptive routing protocols exercise node failures and adversary interference. AI-security improved the resilience of military networks functioning in unstable threat environments. Reveal intrusions or anomalies, protecting systems from sophisticated attacks (LEE et al., 2022; Zhukabayeva et al., 2024).

Smart Agriculture: WSNs are being used in precision farming. Growth effective management of resources such as fertilizer and water. Integrity of environmental data of crop management decisions are security challenges. AI develops security and provides anomaly detection in agricultural sensor networks. Identify faulty or attacked sensors and enabling secure data aggregation (Samiulla et al., 2023).

AI-based security frameworks, which address the vulnerabilities and priorities of each sector are required. In addition, enable intelligent data analytics with secure data transmission.

2.1.2 Evolving nature of cyber threats in WSNs

Attack vectors are the methods, adversaries or cyber attackers used to threaten the network's confidentiality, integrity, and availability (Abdi et al., 2024; Guma & Mijwil, 2024). These vectors exploit the physical, network, and application layers (Meleshko & Desnitsky, 2024). Such threat vectors are routing manipulation "*sinkhole, wormhole, and Sybil attacks*", denial-of-service "*jamming, energy depletion*", unauthorized access "*node capture, eavesdropping*", and data falsification "*fake data injection, replay attacks*" (Jain & Hussain, 2018). Environmental conditions, operational requirements and data sensitivity figure out the risk vectors in each sector. Creating security issues that demand for particular mitigation strategies (Hossain, 2020). The risk vector of medical equipment modification and unauthorized access to patient data in the healthcare industry. Attackers may use hospital IoT networks, wireless medical sensors, and remote patient monitoring systems to steal private medical data or change how medical device functions. However, in the context of environmental and agricultural monitoring, the risk vector of sensor manipulation and data falsification has an impact on precision farming. In this case, the attacker may modify climate monitoring sensors, tamper with automated watering systems, or alter soil moisture data. The increasing vulnerability of WSNs to emerging cyber-threats needs the implementation of proper cybersecurity regulations and avoidance strategies. Signature-based as conventional detection, identifies attacks based on pre-defined attack signatures. Effectively functions against known threats, unable to identify new or emerging attacks. Moreover, centralized security management employs a single control point for network security. The entire network is vulnerable since the central security node attack (a gateway or controller). Conventional methods cannot adequately handle.

2.1.3 Rising security threats and attack vectors

Rapidly WSNs grow in parallel with unpredictable cyber threat characteristics that lead to formation of robust security systems in WSNs. Developing effective mitigation strategies and implementing proactive protection measures against sophisticated cyberattacks. Improving network resilience and cybersecurity in WSNs. CIA are the foundational safety principles of WSN. Confirming data reliability, system resilience, and secure network communications (Abdi et al., 2024; Amaral et al., 2016). Figure 2 presents primary security threats and attack vectors.

Figure 2. Primary Security Threats and Attack Vectors

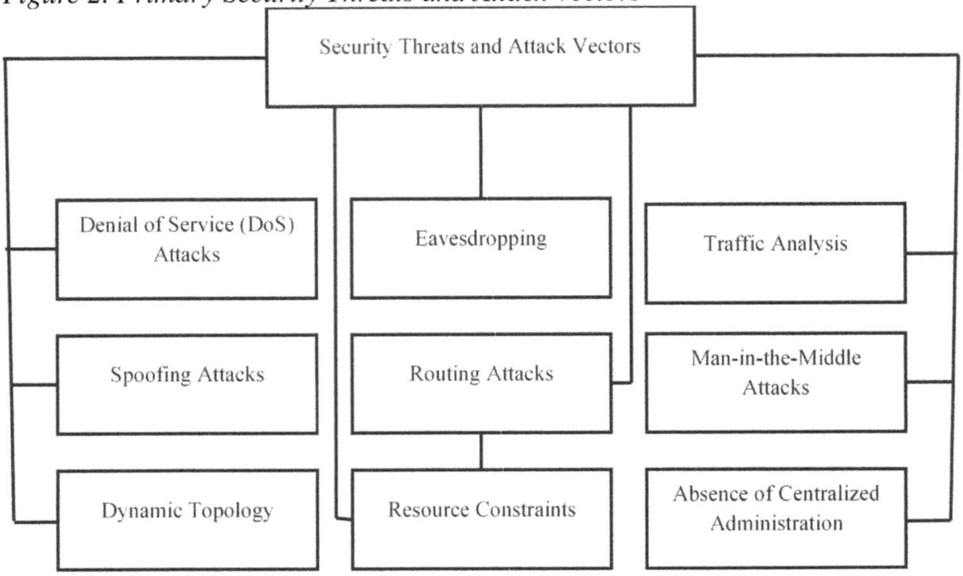

Denial of Service (DoS) attacks*:** Attacks known as denial of service (DoS) are designed to interfere with a WSNs regular operation and flooding sensor nodes or communication channel with massive requests or traffic, cause network unreachable to its intended users. DoS attacks that target resource fatigue are successful against WSNs since sensor nodes have restricted resources, specifically battery power. DoS attacks point to reduce the availability of services to customers through blocking communication within the network or banning a specific device from sending traffic (Islam et al., 2021). Typical techniques include flooding the target system with requests such that it cannot handle valid traffic (Elsadig, 2023). ***Eavesdropping: due wireless communication in WSNs is broadcast, it is capable for unauthorized parties to capture private information sent over the air, which could expose the confidentiality of the data being shared between sensor nodes (Triwidyastuti et al., 2023). In WSNs, eavesdropping is a threat and a prerequisite for extra attacks. Malicious nodes use wireless media to snoop to message broadcasts to gather information. ***Traffic Analysis:*** Cruel nodes can observe the quantity of packets sent or received by specific nodes to hypothesize their role in the network. Through examining network traffic patterns, involving frequency, timing and packet size, attackers can obtain sensitive information even if the encrypted data is being transferred to a WSN. This analysis discuss the functions and activities of nodes inside the network (Sikos, 2020). ***Spoofing Attacks:*** malicious nodes mimic themselves to act as reliable devices. To obtain unauthorized access to the network resources and inserting false data (Saeed, 2023). Due to WSN communication media, hackers obtain inexpensive wireless devices, changing their identities. Spoofing attacks can facilitate traffic injection attacks and pose a danger to identity integrity. ***Routing Attacks:*** manipulation of routing protocols to disrupt the flow of data. Attackers create routing loops, generate false error messages, or redirect network traffic. Compromised nodes to intercept or drop sensitive information (Affane & Satori, 2024). Such as wormholes and sinkholes attacks, both take advantage of the distributed WSN routing approach. ***Man-in-the-Middle attacks:*** interrupted communication of two legal WSN nodes and altered information. Attack the confidentiality, integrity, and availability of the transmitted data. The attacker eavesdrop, alter, or even delete messages (Patil & Chen, 2017). ***Dynamic Topology:*** WSN node mobility and failures keep continuous changes in network structure. Maintaining security in WSNs is threaten (Khamparia, 2021). As the network architecture changes, static security structures are ineffective. ***Resource Constraints:*** The limited battery power, computational capability, and small memory are constraint complexity of security algorithms (Keerthika & Shanmugapriya, 2021). Considering an energy-efficient issue to be effective. ***Absence of Centralized Administration:*** Due to many WSNs function decentralized, they lack a central administrative figure out. Produce enforce network-wide security standards challenge (Oztoprak et al., 2024).

2.1.4 Limitations of Conventional Security countermeasures

In depth, Static Security Policies, Signature-Based Detection, Slow Response Time, High Resource Consumption, Centralized Defense Bottlenecks all are limitations there linked to conventional measures (Bangash et al., 2017; Khenwar & Nawal, 2024; Khraisat et al., 2019; Meleshko & Desnitsky, 2024; Rao et al., 2017; Sen, 2023). As the sensor nodes resource limited characteristics, firewall technologies, encryption, and authentication methods, are unsuitable in wireless sensor networks. It might create computational and energy requirements. Signature-based detection necessitates continual updates to keep an up-to-date database of attack signatures, which consumes a large amount of memory, processing power, and energy, making it unsuitable for WSN setups. The problem is time and resource requirements. Since WSNs low-power and resource-constrained sensor nodes. Managing cryptographic keys in large-scale, distributed WSNs have difficulties. An attacker obtaining cryptographic keys, decode sensitive data, imitate genuine nodes, and attacks all network. Traditional Intrusion Detection System (IDS) techniques mainly rely on behavioral analysis. Dynamic WSN sensors readings fluctuate (*such as environmental changes in precision agriculture or industrial monitoring*). Indicates false positives, and incorrect alerts disturbing network operations.

According to the above analysis, conventional security in WSNs is restricted. Decentralized security systems and lightweight are capable of tackling emerging attack vectors. Maintaining network efficiency and resilience. A hybrid approaches that incorporate AI-driven security is required for robust security approaches (Salem et al., 2024).

2.2 Offensive AI in WSNs

A major concern has been the development of offensive AI in WSNs and the enhancement of defenses against AI-powered cyber-threats. In response to emerging threats, attackers have developed complex methodologies including autonomous malware, advanced evasion techniques, and hostile artificial intelligence exploiting WSN constraints that exceed traditional attack vectors (Behiry & Aly, 2024; Khan et al., 2024; Singh et al., 2024).

In the upward field of cyber defense, the influence of AI-driven security solutions in enhancing cyber-attack control policies and automation, to understand the role of AI in offensive cybersecurity. AI-based security has been improved to meet the complex challengers increased due to the AI-based threats (Mohamed, 2025; Oloyede, 2024). Additionally, increase threat detection accuracy and response efficiency (Behiry & Aly, 2024; Osamy et al., 2022).

2.2.1 Automated Vulnerability Exploitation and Evasion Techniques

The developed AI attackers automate exploitation vulnerability. Create sophisticated strategies to avoid conventional security in WSNs. optimizing, automating and scaling their activities against security procedures. AI tools quickly search, evaluate, and avail limitations of WSNs (DAHAB, 2018; Langley & Birgitta, 2023). Looking for entry points in both public and private data sources and detecting weaknesses. Based on system logs, network traffic and real-time analysis of firewall rules.

AI-testing tools can simulate complex cyberattacks, predicting attack patterns, and find vulnerabilities more professionally (Behiry & Aly, 2024; Khan et al., 2024; Singh et al., 2024).

2.2.2 Cyber Threat Landscape

The manipulation of AI models to attack WSNs has introduced complications to the growing cybersecurity. Data poisoning, adversarial machine learning, and model inversion and extraction attacks are threats that exploit vulnerabilities in

AI-driven security mechanisms, potentially compromising the integrity, reliability, and resilience in the landscape of WSNs (Behiry & Aly, 2024; Osamy et al., 2022; Singh et al., 2024)

Data Poisoning: Attackers can inject carefully crafted malicious data into the training data sets used to develop AI-powered security models for WSNs. Data poisoning disturbs model performance. Confuse the models with malicious conduct for standard behavior. Damage an AI ability to generate accurate predictions. Subsequently, corrupted training data decreases the efficacy of the resultant security system and exposes the WSN to attacks. Hence, the integrity of the data which is used to train AI security models is mandatory.

Adversarial Machine Learning (ML): Adversarial Machine Learning (ML): Adversarial machine learning involves the creation of subtle perturbations to legitimate sensor data that are specifically designed to fool AI models into making incorrect classifications. By crafting these adversarial inputs to avoid detection by AI-based intrusion detection systems, attackers can carry out destructive actions without being detected. Attackers may alter sensed data in means that are hidden to humans which direct AI models to ignore or not recognize the harmful data. Adversarial attacks represent a substantial threat to the security and reliability of machine learning models employed in WSNs.

Model Inversion: Although perhaps a less direct threat in the immediate context of WSN operations, model inversion attacks involve attempts by attackers to reconstruct sensitive information from deployed AI models. However, data poisoning or hostile attacks may be more obvious than the direct threat to WSNs. The possibility of involuntary release of sensitive information by means of model inversion highlights the need for careful design and deployment of AI models employed in WSN security. Model inversion attacks use the final output of an AI model to identify the original dataset. Used to train the AI system and reconstruct sensitive characteristics. Strong AI safety mechanisms (*secure model training, adversarial defense approaches, and privacy-preserving ML frameworks*), safeguard AI-based systems and to avoid such weaknesses.

Model extraction attacks: an adversary to replicate or steal an installed DL-model. The attacker discovers the model's vulnerabilities and expertise in more effective attacks.

AI-Powered Stealth Attacks: AI enables attackers to mimic legitimate network performance. Making malicious activities comparable to normal traffic patterns. Among these are Covert Channel Attacks, in which compromised nodes send illegal data while feigning normal operation. It includes evasion strategies, in which AI adapts attack patterns to avoid triggering anomaly-based detection systems.

***AI-Driven Reconnaissance and Attack Optimization*:** the hackers examine system logs, network traffic, and firewall rules with AI, that to identify potential entry holes for exploitation., regularly improves attack methods based on previous successes and failures which increase the risks. ***Polymorphic & Metamorphic Malware*:** AI-powered malware repeatedly rewrites its entire code and modifies its structure, to avoid detection by security systems that rely on signatures and pattern recognition. Polymorphic malware, which changes appearance while maintaining its destructive capability, is used by attackers.

Developing efficient AI-powered security strategies able to predict and counter complex AI-based attacks, depends on first understanding how AI is utilized for WSN breaches. Due to the power of AI to copy human operation, social engineering attacks grow into more complicated. According to recent studies, sophisticated defenses against changing threats, including adaptive intrusion detection, adversarial defense, and real-time threat intelligence for effective security.

3. AI-Driven Security Frameworks

WSNs need to create sophisticated preventative methods, to successfully decrease the risks pretended by AI-powered cyberattacks. WSNs utilizing AI-security strategies to improve intelligence, adaptability, and resilience (Osamy et al., 2022). Providing continuous security, reliability, and efficiency to break the advanced tactics of AI-enabled attackers. AI-based mechanisms are effective against AI-powered stealth attacks, adversarial machine learning, and automated vulnerability exploitation (Keshetti et al., 2025). Secure WSNs get strong and scalable safeguard (Salem et al., 2024).

refer to figure 3 the fundamentals of framework are the layers of detection, mitigation, and response. 1. Privacy and data protection; 2. network security and resilience; and 3. continuous learning and adaptation. Companies have to integrate AI with their current cybersecurity frameworks. Therefore, improve their security posture against increasingly complex cyber threats. The subsequent standards help to differentiate each layer, as well as continuing reasonable security (Babiceanu & Seker, 2019; Keshetti et al., 2025; Oloyede, 2024; Salem et al., 2024; Waheed & Marchetti, 2023).

The ***Threat Detection Layer:*** focused on detecting suspicious behavior, intrusions, and potential attacks (Demedeiros et al., 2023; Shihab et al., 2024). Using ML tools recognize patterns and detect anomalies. However, learn and adapt continually.

The Threat Mitigation and Response Layer: restricting threats and reducing damage. perform containment and recovery (Ghadi et al., 2024; Omar et al., 2025; Singh et al., 2024). This layer uses self-healing security protocols. Restore system functions and automated threat response procedures. such as intelligent decision-making models and AI-based firewalls. The ***Privacy and Data Protection Layer:***

perform data confidentiality and integrity not direct threat mitigation. Protect sensitive data from model inversion assaults and illegal access (Shihab et al., 2024; Singh et al., 2024). Prevent data leakage and defend against model inversion attacks.

The ***Network Security, and Resilience Layer:*** support WSN infrastructure against cyber threats. Extended stability, resilience and direct threat response (Shihab et al., 2024). Utilized blockchain-based security to enable decentralized authentication and ensure data integrity. AI-powered encryption methods to protect communication channels from unauthorized access.

The Continuous Learning and Adaptation Layer: dynamically response to developing threats. Focus on future proofing systems not preventing specific attacks (Nguyen & Reddi, 2023; Salem et al., 2024). Used reinforcement learning to continuously refine security models and adapt strategies over time. As well as the integration of threat intelligence to enable proactive updates and robust security in an altering threat.

Figure 3. AI-powered security framework

This organized strategy enables threat detection, mitigation, data protection, network resilience, and continual learning. Therefore, improving security (Ghadi et al., 2024).

The NIST Industrial Wireless Systems Guide outlines practices for creating secure wireless networks. Such as encryption, authentication, and resilience strategies. Refer to figure. 4 the NIST framework has five functions: identify, protect, detect, respond, and recovery. These functions provide the framework for enterprises to analyze and improve their cybersecurity capabilities. To effectively combat AI-powered cyber threats in WSNs, the components of an AI-enabled security architecture must align with NIST standards, ensuring robust protection and compliance (Achaal et al., 2024; Dede et al., 2024). An example of that is when WSNs are used to monitor energy distribution and detect anomalies in real time. Since these networks handle critical infrastructure data, they must comply with NIST SP 800-53 guidelines for cybersecurity.

Figure 4. NIST Framework

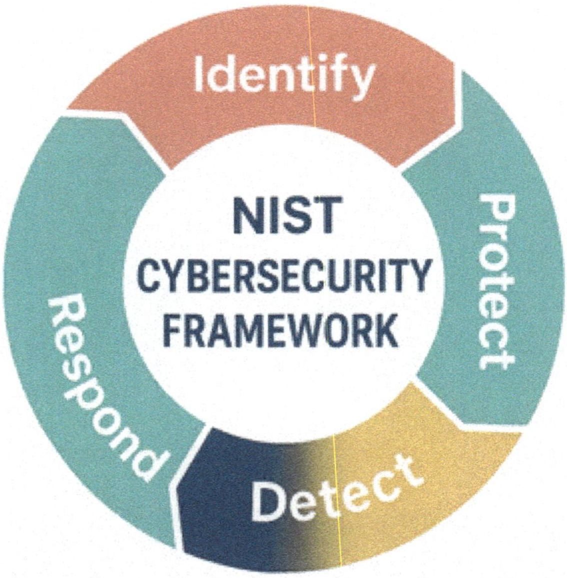

4. ENHANCING AI MODEL RELIABILITY AGAINST ADVERSARIAL ATTACKS

AI security models to be effectively deployed in WSNs, it is imperative that they be dependable and resilient against hostile attacks, malicious attacks that alter data and target machine learning algorithms, particularly deep learning models, to compromise the security of AI-based systems. To trick AI algorithms during inference, evasion as a common adversarial attack alters input data. In the direction of train model learning, poisoning attacks entail inserting harmful material into training datasets. Also, in common adversarial attack, sensitive data are extract from AI models through privacy assaults (Anthi et al., 2021). These attacks pose a substantial danger to AI applications in WSNs, which require security and reliability for real-time decision making. Integrating security into AI development requires threat modelling, to anticipate attack vectors and machine learning for security operations (Sarker, 2023). Training AI models using adversarial datasets. Adversarial training strengthens models against adversarial attacks. Using input validation and sanitization techniques to identify and filter potentially malicious inputs. Anomaly detection techniques in AI, discover inputs that differ considerably from expected patterns. Addressing the vulnerability of AI models to adversarial manipulation is utilized for ensuring their trustworthiness and effectiveness. (Acharya & Mahapatra, 2024; Sebestyen et al., 2025).

4.1 Hybrid AI-security solutions: scalability and adaptive defense strategies

Hybrid security approaches developing adaptive and scalable protection strategies. Dynamically adjust security parameters, adaptive dynamic WSNs and growing threats. Reinforcement Q-Learning-Based Adaptive Encryption Model dynamically connected with threat arrangement, and current network conditions. This hybrid strategy improves security planes, as well as adjusting energy efficiency (Premakumari et al., 2025). Adaptive Intrusion Detection System Using Reinforcement Learning integrates reinforcement learning (RL) and artificial neural networks (ANNs). Q-learning agents improve attack feature selection in dynamic networks and boost detection accuracy. Maintaining robust security among increasing WSN deployments (Hussain et al., 2024).

4.2 Proactive Threat Intelligence: Predictive Analysis In AI-Driven WSN Security

Proactive threat intelligence is detected and mitigated cyber threats preemptively. It examining patterns, vulnerabilities, and emerging threats (Raza, 2024; Sparks, 2025). Proactive threat Intelligence is central for detecting cyber threats, minimizing damage, and accelerating incident response (Adebayo et al., 2025). AI-powered models continuously learn, and develop in response to new approaches for attack, acting as adaptive protection mechanisms. Improves accuracy by reducing false positives and recognizing real threats, and benign anomalies. It is cost effective, because it reduces costly breaches by actively minimizing risks. Reinforcement learning supervised or unsupervised and neural networks models are AI's predictive analytical methods. Reconstructing normal behavior and flagging deviations (Fredj et al., 2023; Raza, 2024; Salem et al., 2024).

AI algorithms predict future attacks and vulnerabilities. Examine current network traffic patterns, external threat intelligence and past security data. Patching vulnerable software, updating security protocols, or readjusting network parameters (Adebayo et al., 2025; Ahmad & Wazirali, 2022). Switching from a reactive to a proactive security position. Improving resilience and security of WSNs.

5. EVOLUTION OF AI-DRIVEN SECURITY IN WSN

AI has developed as a powerful tool, improving WSNs security. Introducing intelligent, adaptive, and scalable approaches. Addressing the limitations of conventional methods.

The use of ML techniques in the 2010s changed the WSN security environment. Researchers started integrating ML classifiers into IDS for WSN deployments. Using labeled datasets, supervised models *(k-nearest neighbors (k-NN), Support Vector Machines (SVMs), and decision trees)* to differentiate malicious activity from legitimate traffic. ML has been a better option compare to conventional encryption techniques (Ahmad & Wazirali, 2022). DL techniques (*neural networks, and auto-encoders*), expanded between 2015 and 2018. Classified traffic, anomaly detection, and processing complex data. Unsupervised learning approaches (*clustering and Principal Component Analysis (PCA)*) introduced to reduce feature dimensionality and identify unknown threats.

Late 2010s to early 2020s, security techniques advanced. Applying reinforcement learning (RL). Convolutional Neural Networks (CNNs), and Recurrent Neural Networks (RNNs). Efficiently identify temporal and geographical patterns in network traffic. RL is a powerful tool for dynamically adjusting security measures. Further, recent frameworks integrate several AI techniques. SVD/PCA combined with clustering, optimize energy usage while preserving high security levels (Premakumari et al., 2025). Figure. 5 shows old milestones in WSN security.

5.1 Security Domains in WSN

This study categorizes WSN security into five primary domains: 1) Physical and Endpoint Security, 2) Communication Security, 3) Network Security, 4) Data Security, and 5) Application Security. To ensure the network availability, confidentiality, and integrity, each domain addresses specific vulnerabilities and employs distinct security measures (Gopalan et al., 2021).

Figure 5. Historical Milestones in WSN Security

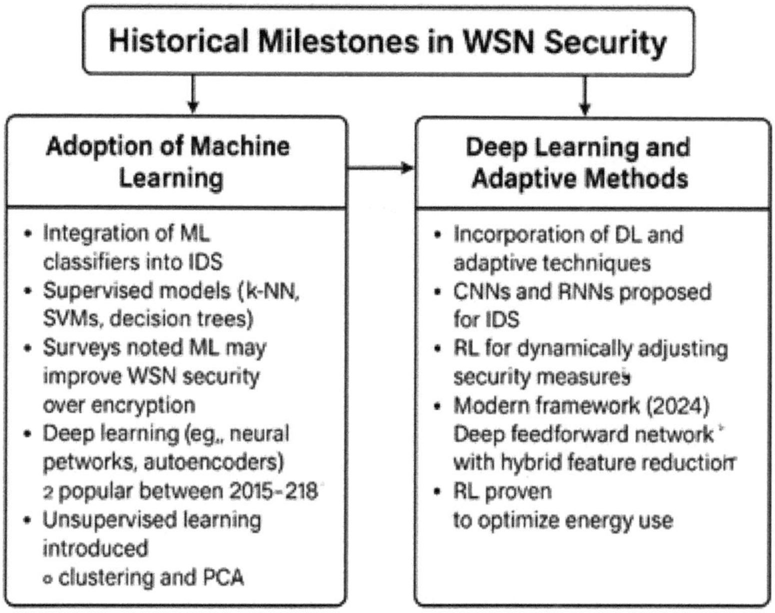

5.1.1 Physical and Endpoint Security

WSN might be deployed in harsh or adverse circumstances. Breaching data confidentiality, network integrity and compromise network functionality (*Damaging sensors, capturing nodes to retrieve critical data, or manipulating node operations*) (Kaur & Monal, 2023; Nunoo-Mensah et al., 2017; Singh et al., 2024). Security procedures focus on *1.* ***Tamper Detection:*** nodes might tamper through logical interfaces or direct physical attacks. implemented Hardware-based tamper detection and interrupt-based recovery.

Hash functions (*as SHA-224*) demonstrate node integrity. *2.* ***Node Authentication:*** Validation techniques like wireless fingerprinting confirm node identities. Multi factor authentication (*MFA*), which combines cryptographic keys and behavioral analysis. Therefore, energy efficient authentication systems to stop device penetration. *3.* ***Environmental Threat Protection:*** WSNs installed in harsh settings. They are protected from risky temperatures, humidity, and physical damage. Shield strategies implement predictive analytics to evaluate dangers. Energy-harvesting mechanisms to maintain node functionality. Self-healing for self-governing and adaptive response to environmental challenges. *4.* ***Endpoint Security:*** aim to protect sensor nodes from physical and cyber threats. Also to improve network resilience and integrity. Save a long battery-powered sensor node (Bangash et al., 2017; Lee, 2020; Panwar et al., 2021).

Table 1. WSN Security Domains, Threats, and Countermeasures

Security Domain	Key Focus	Common Threats	Core Countermeasures	Representative Techniques / Methods
Physical & Endpoint Security (Chaudhary et al., 2024; Narooka et al., 2025a)	Protection of nodes from physical and hardware-level tampering, node compromise, and environmental threats	Node capture, physical tampering, malware injection, node compromise, environmental damage	Tamper detection, hardware-based security, node authentication, self-healing mechanisms, secure wakeup/bootstrap	Tamper-proof hardware, SHA-224 hash, multi-factor authentication, energy harvesting, wireless fingerprinting, secure boot
Communication Security (Singh and Dubey, 2015; Chaudhary et al., 2024; Narooka et al., 2025b)	Secure wireless data transmission across WSNs	Eavesdropping, sniffing, jamming, MITM (Man-in-the-Middle) attacks	Lightweight encryption, secure key management, authentication, dynamic key exchange	ECC, AES, TLS/DTLS, MAC-layer feedback routing, multi-parameter route selection, dynamic key protocols
Network Security (Xue et al., 2024)	Securing network topology, routing integrity, and node behavior	Wormhole, Sybil, and sinkhole attacks, malicious node behaviors	Trust-based models, robust node authentication, encryption, blockchain-based verification	Biometric-based auth, trust modeling, blockchain trust frameworks, AES-128 encryption
Data Security (Singh and Kaushik, 2023a)	Ensuring data confidentiality, integrity, and authenticity from sensing to aggregation	Data spoofing, unauthorized access, falsified aggregation, message modification	Lightweight encryption, mutual authentication, MACs, secure aggregation, anomaly detection	ECC, symmetric-key encryption (AES, RC4), homomorphic encryption, digital signatures, hash functions, anomaly detection algorithms
Application Security (Singh and Kaushik, 2023c)	Protection of software execution, access control, and user-level policies	Unauthorized command execution, code injection, privilege escalation, runtime vulnerabilities	Role-based access control, sandboxing, intrusion detection, policy enforcement, secure software	RBAC/ABAC, PKI, secure boot, code signing, runtime integrity checks, behavior monitoring, IDS,

Common threats involve ***Sensor Node Compromise:*** where adversaries can compromise nodes to alter the data integrity, execute unauthorized activities, or induce network failures. Also, ***Malware Injection:*** malware infiltration enables unauthorized access, data manipulation, and service disruptions, which could compromise the entire WSN ecosystem. Further, ***Physical Tampering:*** attackers extract cryptographic keys or modify node configurations, which permits adversaries to penetrate nodes and inject malicious data, thereby weakening network security.

5.1.2 Communication Security and Protocols

Communication security: Maintaining the confidentiality, integrity, and authenticity of data, while it is being transmitted between sensor nodes in WSNs depends on communication security (Alquraishee & Kar, 2014; Idrissi et al., 2020; Ram & Ilavarasan, 2024). Since WSNs are used wireless transmission, they are susceptible to several attacks, including ***Eavesdropping and Sniffing***: where the hackers overhear wireless communications () to get unauthorized access to private information and possibly exploit the system. Assaults named Jamming, that happen when malicious cause flood valid channels with high-power noise signals, severely disrupting communication and resulting in DoS attacks. *Man-in-the-Middle* (MITM) attacks, by allowing attackers to intercept, modify, and transmit communications between sensor nodes, compromising the confidentiality and validity of data.

Secure Communication Protocols: To reduce vulnerabilities, the secure communication protocols must be incorporated into WSN. These protocols apply strong cryptography and authentication, to stop illegal access to and interception of data. Different safe routing techniques can address issues with various attack types. For example, the secure routing protocol based on feedback information, which incorporates location, trust, latency, and excess capacity information in the media access control (MAC) layer's acknowledgement frame (Khah et al., 2023; Kumaran et al., 2014; Kusha, 2020; Parande & Mallapur, 2019; Sen, 2023). ***Lightweight encryption:*** Lightweight encryption is one of the main suggestions in the security protocol for WSNs. Algorithms that use lightweight encryption maximize security while reducing computational costs and guaranteeing effective encryption for WSN nodes with less energy. ***Secure Key Management Protocols***: utilized cryptographic key generation, distribution, revocation, and periodic updates. Secure key management prevents key compromise and unauthorized access. ***Dynamic Multi-Parameter Key Exchange*****:** applied dynamic key generation method. Improve encryption security and enhance packet forwarding. ***Multi-Parameter Route Selection***: Assesses energy levels, rate of communication success, and node reliability. Elect the secure route to protect data transfer.

5.1.3 Network Security

WSNs security mitigate cyber threats and preserve the infrastructure integrity (Batool et al., 2024; Lee, 2020). *Wormhole attacks* include malicious nodes, attack communication tunnels between remote network sites. Tricking routing protocols and interfering with data communication route. *Sybil attacks,* amends routing protocols and damage network architecture. *Sinkhole attacks* happen when a malevolent node deliberately draws in network traffic, exposing weaknesses that allow for data interception, and interruption of services.

The node that forwards data must be safe, reliable, and impenetrable to intrusions, however, guarantee robust and secure data transfer. The safeguards against impersonation and unwanted access are biometric-based authentication and *strong authentication frameworks*, that verify node identities. *Advanced encryption algorithms*, protect data integrity and provide safe connection in WSN contexts with limited resources. Uses AES-128 or ECC encryption to guarantee data confidentiality during transmission. *trust-based models*, mitigate the risks associated with malevolent network entities and compromised nodes, that by continuously assessing and evaluating node behavior.

5.1.4 Data Security

Sensor nodes aggregate data and send it to the gateway or sink node in the WSN. Secure transmission media and data reliability must both be guaranteed. Therefor clustered topology communication protocols contain authentication and encryption procedures (Othman et al., 2015).

Encryption Mechanisms: Cryptographic and encryption techniques used to protect sensitive data avoiding unauthorized access. Due to the WSNs computational constraints, lightweight encryption algorithms are required.

Authentication Protocols: Only authorized sensor nodes participate in network operations. To prevent malicious node access and network breaches (Kumaran et al., 2014). Applying digital signatures, lightweight authentication systems, and mutual authentication schemes.

Integrity Verification: No unauthorized party is engaged in exploiting the data. The integrity verification process is guaranteed the data lifecycle (Tiwari, 2018). The mechanisms aiding in spotting unusual behavior are anomaly detection techniques, hash functions, and message authentication codes (MACs).

5.1.5 Application Security

In the application, security is essential to protect program execution and access control systems against flaws and unwanted intrusions. Security protocols reduce the likelihood of exploitation and, hence, increase the resilience of networks (Adebayo et al., 2025; Mohamed et al., 2022; Saiful et al., 2012; Whitepaper, 2018).

Access Control Mechanisms: Access control mechanism is used to set some restrictions for entities that interact with sensitive WSN applications. Only authenticated users have the approval to access data or devices. This increases data confidentiality and prevents unauthorized command execution. Including role-based access control (RBAC) and attribute-based access control (ABAC).

Secure Software Execution: Strict security protocols used while installing or running programs. Reducing the likelihood of malware code execution and software vulnerability. Such security protocols have code signing methods, runtime integrity checks, and secure boots. Sandboxing and isolation techniques are used when the software is hacked. The applications limited to settings stop malicious code from spreading. ***Policy Enforcement***: Security policy execution establishes operational regulations. Proactive risk mitigation while governing system behavior. IDS and behavior monitoring are used to facilitate the detection of anomalies.

As AI transforms security frameworks, several concerns within the WSN ecosystem are being addressed. The following security challenges in the convolution approaches are the core focus of the revolution of AI-driven security mechanisms. Which will allow WSNs to move beyond the constraints of existing security approaches (Behiry & Aly, 2024; Mohamed, 2025; Osamy et al., 2022; Singh et al., 2024; Tariq et al., 2023). Cryptographic methods require processing power, which is impractical for resource-constrained sensor nodes. Lack of adaptability renders them useless in the face of ever-changing threats. Their dependency on predefined rules and signatures, making them useless against dynamic threats. Such as zero-day assaults. When the main control node is compromised, the centralized security measures will expose weaknesses. Therefore, lightweight cryptographic techniques, intrusion detection systems, and trust-based frameworks are pointed.

5.2 Emerging AI-Driven WSN Security Approaches

AI has fully transformed security in WSNs. It overcoming the drawbacks of traditional security processes (Adebayo et al., 2025; Balogun et al., 2025; Behiry & Aly, 2024; Islam et al., 2021; Mohamed, 2025; Osamy et al., 2022). The following strategies are used by AI-driven security frameworks. That improve communication channel defenses against jamming, eavesdropping, and MITM attacks:

- Adaptive Learning Models: AI-driven systems are able to recognize changing cyber-threats in real time, using dynamically learning attack patterns.
- Anomaly-Based Intrusion Detection: To identify irregularities and highlight possible security breaches, IDS systems-powered machine learning is applied to examine the network activity.
- Automated Security Optimization: AI-powered systems proactively modify routing plans, authentication procedures, and cryptographic parameters to reduce security threats and enhance network efficiency.

WSNs AI-driven security advanced conventional security procedures, increased resilience against cyber assaults. AI-driven security Implementing at the endpoint to provide a distributed layer of security.

AI-powered IDS using machine learning models. monitors node behavior and detects deviations from normal activity. It functions for anomaly detection, identifying suspicious behavior at the node level. Functions AI for more robust authentication mechanisms to prevent unauthorized access. AI, as compared to signature-based IDS, can detect threats that have not yet been discovered by examining behavioral patterns (Keshetti et al., 2025; Mohamed, 2025).

Trust-Based Frameworks, the AI model provides adaptive trust management, by evaluating its dependability through continuous, real-time assessments of node activity. This enables the model to provide updated trust ratings to identify low-scoring nodes and stop compromise. By analyzing trust inconsistencies, it can detect blackhole attacks, Sybil attacks, and badmouthing (Gangwani et al., 2024; Ghadi et al., 2024). Based on past encounters, Bayesian Trust Models evaluate node dependability using probabilistic reasoning. Also, Fuzzy Logic-Based Trust Models; use fuzzy logic to manage ambiguity while assessing trust. Farther, Entropy-Based Trust Models; assess node behavior uncertainty to identify irregularities.

AI-Self-Healing Mechanisms, AI in WSNs, allow nodes to automatically recover from attacks by modifying security settings. When nodes recover from brief failures, trust-based frameworks assist them in regaining their reputation.

Central Node Protection, AI-driven security, can be helpful in identifying and stopping threats that try to jeopardize WSN's central management or control operations. These attacks aim to interfere with network operations or steal confidential data, and they may target the base station or cluster heads. The fundamental operation of the WSN can be safeguarded by using AI-based monitoring of communication patterns involving these network components to spot any unusual activity, illegal access attempts, or departures from standard control orders (Mallick & Nath, 2024; Osamy et al., 2022; Singh et al., 2024).

AI-Driven Best Practices Tactics Empowering Communication Channel Security in WSNs enhance the security against threats such as eavesdropping, jamming attacks, and man-in-the-middle (MITM) attacks (Mallick & Nath, 2024; Omar et al., 2025; Osamy et al., 2022). AI-Based Encryption meets secure data communication. The AI selects the encryption techniques based on network conditions. Machine learning models predict potential vulnerabilities and dynamically adjust encryption strength. The advantages are reducing computational overhead while maintaining strong encryption.

AI-Adaptive Frequency Hopping. The AI dynamically adjusts communication frequencies to evade jamming attacks. Reinforcement learning algorithms analyze interference patterns and select optimal frequency channels. Shifting frequencies and enhances network resilience against jamming attacks. AI-Powered Intrusion Detection analyze network traffic patterns to detect MITM attacks. Deep learning techniques classify anomalies and identify unauthorized access attempts. Identifying sophisticated attacks and reducing false positives balanced to traditional intrusion detection systems.

5.3 AI Role in Optimizing WSN security Performance

AI as a transformational technique, improving security performance. Intelligently manage large amounts of data and dynamically adapt security measures. Improves intrusion detection, resource efficiency, and energy management (Adebayo et al., 2025; Meleshko & Desnitsky, 2024; Mohamed, 2025; Oloyede, 2024; Osamy et al., 2022; Singh et al., 2024).

Intrusion Detection Accuracy Optimization: AI enhances fine-tuning detection thresholds and feature selection. As well as classification algorithms to improve the precision and effectiveness of IDS. AI-based intrusion detection fixed false positive attack signs.

Resource Utilization Enhancement: AI boosts computer operations. Using dynamic resource allocation algorithms. The security measures are adjusted in response to threat levels and in real time. This provides robust security defenses against cyber threats and reducing resource waste.

Energy-Efficient Security Mechanisms: AI develops lightweight cryptographic algorithms. To save energy consumption and maintain security efficacy. However, implementing predictive models for adaptive security response. It reduces processing and communication overhead.

Establishing a Balance Between Network Lifespan and Security: AI-based optimization tailors high security to the WSN-nodes resource constraints. Employing intelligent power management algorithms and increasing the operational sensor nodes lifespan.

Table 2. Summing up AI's Role in Optimizing WSN Security Performance

Category	AI Techniques Used	Purpose / Key Benefits	Challenges / Considerations
Intrusion Detection Accuracy	Machine Learning (ML), Anomaly Detection	Fine-tunes detection thresholds, improves classification; adapts to new threats, minimizes false positives	Requires continuous model updates and balanced sensitivity to avoid overfitting
Resource Utilization Enhancement	Dynamic Resource Allocation Algorithms	Real-time adaptation to network conditions, efficient CPU/memory use	Needs accurate real-time monitoring; may require extra computational overhead
Energy-Efficient Security	Predictive Models, Lightweight Cryptographic Algorithms	Reduces energy use by optimizing security tasks' scheduling	Balancing lightweight design with strong encryption
Lifespan vs. Security Optimization	AI-Powered Optimization Algorithms, Intelligent Power Management	Extends network lifetime while maintaining robust security	Trade-off between security intensity and energy consumption
ML Threat Detection Models	SVMs, Random Forest, Autoencoders, Pattern Recognition	Detects anomalies, malware, data breaches; learns from prior patterns	Vulnerable to poisoning/adversarial attacks; false positives/negatives
DL Threat Detection Models	CNNs, RNNs, LSTM, Reinforcement Learning	Detects complex, temporal, and zero-day threats; adapts to traffic patterns	High resource demand; susceptible to model inversion and extraction attacks
Adversarial ML (AML)	Adversarial Training, Gradient Obfuscation, Defensive Distillation	Defends against evasion, poisoning, and extraction attacks	Sophisticated attackers may still bypass; model robustness needs continuous testing
Authentication Models	AI-Enhanced Biometrics, Behavior Analysis, Multimodal Verification, Blockchain	Prevents spoofing and unauthorized access; intelligent pattern recognition	Ensuring low false rejection rate and maintaining privacy during biometric data processing
Privacy and Resilience Enhancements	Federated Learning, Explainable AI (XAI), Self-Learning Systems	Ensures decentralized, transparent, and adaptive defense mechanisms	Federated models must guard against information leakage during training
Model Inversion/Extraction Defense	Secure Computation, Adversarial Training, Privacy-Preserving Learning	Protects DL models from being reverse-engineered or manipulated	Needs integration with lightweight encryption and real-time detection capabilities

5.3.1 Threat Detection Models

Cutting-edge tools (ML and DL) are effective for detecting irregularities and intrusions. Adjust to the dynamic nature of WSN traffic. (Behiry & Aly, 2024; Dalgade et al., 2024; Delwar et al., 2024; Mohamed, 2025). This contributes to security enhancement in many ways. It investigates the effectiveness of IDS using ML or DL models. Also, explores the trade-offs between ML and DL models concerning computational complexity.

5.3.2 ML-Based Attack Detection Mechanisms

ML-based security models serve as key components in modern security bases. They effectively identify anomalies and intrusions within network environments. These models detect security threats such as unauthorized access, data breaches, malware infiltration, and deviations based on expected network behavior.

A variety of analytical methods are used by ML-based detection mechanisms. These analytical techniques include; time-series analysis to track sequential data trends, *clustering-based approaches* to group similar network activities, *statistical methods* to identify abnormal behavior patterns, *classification algorithms* to categorize network traffic, *autoencoders* to detect anomalies, *pattern recognition techniques* to flag irregular activities, *supervised learning approaches* (e.g., support vector machines, decision trees) to classify structured threats, and *unsupervised learning approaches* (e.g., clustering, principal component analysis) to identify unknown attack patterns (Tavakoli et al., 2020).

By using previous data to distinguish between typical and abnormal network performance, these models improve the accuracy of threat identification and classification. Based security approaches are vulnerable to a number of security flaws notwithstanding their effectiveness. (Behiry & Aly, 2024; Delwar et al., 2024; Karthikeyan et al., 2024). *Poisoning Attacks*; Malicious actors manipulate training data to degrade model accuracy, leading to erroneous threat classification. *Adversarial Attacks*; craft deceptive input samples to mislead ML models into incorrect classifications, bypassing security measures (offensive side); *False Positives/ Negatives*; ML models may incorrectly flag benign activity as malicious or fail to detect actual threats.

Machine Learning (ML) techniques encompass a diverse array of methodologies. The Random Forest algorithm is an ML algorithm known for its ability to detect threats. Random Forest is an ensemble learning technique that builds several decision trees and aggregates their outcomes to improve classification accuracy. (Divyashree et al., 2024; Karthikeyan & Kamali, 2024; Tolba, 2024). Random Forest is well-suited for identifying fraudulent activity. It determines the differentiation between regular and suspicious activity. It keeps accurate threat identification within a collection of datasets. Random Forest is utilized in IDS, classifying network traffic and identifying a variety of threats. Support Vector Machines (SVMs), classification algorithm to find the hyperplanes and differentiate between data points (Chen et al., 2023; Hosseinzadeh et al., 2021; Kumar, 2024). SVMs handles high-dimensional data and provides reliable discernment between malicious and typical network traffic. Neural Networks (NNs), contemporary ML-based security scheme (Behiry & Aly, 2024; Khan et al., 2024; Nandhini et al., 2024). Learning from labelled and unlabeled large-scale datasets. Also supporting binary and multi-class threat categorization. NNs detected poisoning attacks which contaminated the training dataset. Identifying complicated patterns and preserving strong detection. Recognize any aggressive inputs. NNs are scalable, handling volumes of data. This guarantees precise threat identification in extensive security structures.

ML-based security models improve the efficiency of IDS and anomaly detection. These models have the capacity to process large datasets. Yet, ML-based threat detection systems continue to obstacles. Future research will focus on developing these models' accuracy. Creating adaptive defenses to support real-time threat mitigation measures.

5.3.3 DL-Driven Model Inversion and Extraction in WSNs

The DL-models have improved security of WSNs. Provide independent defense mechanisms. Using environmental data, network traffic, and operational factors to recognize complicated patterns. It performs anomaly identification and automates intrusion detection. The growing use of DL-based security frameworks introduced new attack vectors. (Behiry & Aly, 2024; Khan et al., 2024; Nandhini et al., 2024). Different types of DL are Neural Networks (CNNs), Recurrent Neural Networks (RNNs), and Reinforcement Learning (Kaur & Monal, 2023; Khan et al., 2024). CNNs are used for feature extraction and spatial pattern analysis from network traffic data (Behiry & Aly, 2024; Muchiri, 2021; Shammari & Parveen, 2024). CNN interprets sensor data as structured inputs to help identify attack patterns in network packets. A novel approach of CNNs is to detect and identify zero-day attacks. Intelligent intrusion detection systems employ a classification approach for harmful traffic activities. In contrast, RNNs are better suited for analyzing sequential data, which is critical in tracking and interpreting time-dependent behaviors in WSNs (Salmi & Oughdir, 2023; Yadav & Kumar, 2022). RNNs detect suspicious traffic systems. It is effective for spotting, learning from historical and instant data. RL adds a security layer as it enables adaptive and autonomous protection approaches. The IDS are optimized using RL adaptive model in response to threats (Hussain et al., 2024; Premakumari et al., 2025; Shrivastava, 2018). These models demonstrate adaptive response to the success rates of cyberattacks. RL-based frameworks improve security levels against advanced adversarial approaches.

Despite these evolutions, DL-based security systems continue to encounter challenges. Development of optimization strategies experience a high computational cost and more memory size. Manipulation training data is trick DL detection algorithms in adversarial attacks. In the future, research have to focus on developing IDS through adaptive DL models. The impact of multi-layer security depths, similar to lightweight encryption mechanisms and quantum-secure AI models, will also be addressed.

5.3.4 Adversarial Machine Learning

Adversarial machine learning (AML) is an effective instrument for bolstering defenses powered by AI. Malicious actors are increasingly using technology to undermine and go around security measures. As cybersecurity infrastructures increasingly incorporate AI and ML models, adversaries are always coming up with novel ways to influence these systems. Because of this dynamic, AML is a main topic for current cyber defense research (Malatji & Tolah, 2024).

From the standpoint of cyber offense, attackers use adversarial machine learning to take advantage of flaws in AI-based security models. *Evasion attacks* are common strategies that allow threats to evade detection systems by carefully manipulating input data to cause the model to mistakenly classify harmful behavior as benign. Poisoning attacks affect the accuracy of the model. Repeatedly submitting queries, reproducing the model behavior, and permitting additional manipulation. AML can be used to compromise the integrity and dependability of AI-based security structures. Security experts have developed a range of defense approaches. Combat the AML threat. *Adversarial training* is a successful training model. It trains the machine learning models in adversarial cases, enabling identification and withstand future attacks. The model's underlying gradients are hidden by *gradient masking and obfuscation techniques*. attackers use gradient-based tactics and exploit vulnerabilities. Employing knowledge transfers to smooth decision boundaries. Lessen vulnerability to minor adversary worries and improve the robustness of the model. *Differential privacy* and *secure learning* techniques control noise in training data. Further, protecting sensitive information and reducing risk of attacking.

In summary, the AI-based security domain is characterized built on its alignment with definite challenges. There is a need for continuous research in adversarial learning to secure future AI-integrated systems.

5.3.5 Authentication Models

Secure authentication techniques protect WSN against spoofing, guarantee data integrity, and control access. Conventional authentication techniques lack flexibility, scalability, and resistance to hostile manipulation. The AI-powered authentication methods address these issues. Utilizing ML and DL algorithms. (Marabissi et al., 2022; Singh et al., 2024).

AI systems improve speech analysis and fingerprint scanning. Increases biometric authentication effectiveness. ML algorithms employed in behavioral pattern analysis, recognize variations in user behavior and indicate unauthorized access. Multimodal authentication systems integrate device ID, location, and biometrics. These models become adept at responding to emerging threats. Gaining knowledge from authentication events will lower the possibility of erroneous acceptance or denial. The spoofing attacks are a form of cyberattack that poses as trustworthy users or devices. This practice poses a serious threat to the security of WSN (Behiry & Aly, 2024; Pourghasem et al., 2025; Singh et al., 2024). Proactive strategies by AI-powered defence mechanisms are capable of combating spoofing attacks. The Intrusion detection systems rely on a DL model, which examines authentication requests to differentiate between fraudulent identities and authentic users. Distorted data is used to train the AI adversarial learning model to improve its ability to identify spoofing efforts. The probability of successful impersonation is decreased by anomaly detection systems that recognize irregularities in authentication behavior. However, multi-layer security strengthens confidence in the authentication procedure even more. Thus, Tamper-proof verification can be ensured by combining blockchain and AI. In real-world applications, behavioral biometrics-based AI-powered authentication models have the capacity to examine a node's usual communication habits or sensor data to confirm its identity. The growth of next-generation WSN security frameworks integrating AI-authentication.

6. CHALLENGES AND FUTURE TRENDS IN AI-SECURED WSNS

One of the essential challenges in adopting AI algorithms in WSNs security lies in its demands for computational capability. The Inherent resource constraints of WSNs nodes make integrating AI-driven security mechanisms a challenge. Cybersecurity optimization through model inversion resistance and mitigating extraction assaults is computationally intensive and still a major challenge in these networks.

An additional major challenge is the Scalability of AI-driven Cybersecurity solutions in large-scale WSN deployments. As the network grows and thousands of sensor nodes join, the AI-driven security model becomes increasingly sophisticated. Data collection and privacy concerns must be properly handled for AI model training. With dynamic WSNs, AI models have to be adaptive and continuously learn and retrain. Therefore, developing a dataset to cope with evolving security is a challenge and an open research problem of scalability in AI-driven security. Thus, these systems require research into explainable AI (XAI) techniques tailored to WSNs.

Overcoming these multifaceted challenges is essential to realize the full potential of AI-securing WSNs. Future research should consider these challenges and extend the study to focus on the dynamic IoT-based AI-driven security. There are many concerns encountered in research issues related to network topology, heterogeneity, scalability, reliability, security, coverage, connectivity, QoS, and Quality of User Experience (QoE).

6.1 Evolving Threats and AI Role in Next-Generation Cybersecurity

The increasing demand for AI-driven security systems is attributed to the rapid growth of offensive attacks. AI-driven security systems address the evolving threat complexity and guarantee the resilience of WSNs. The next-generation security emphasis is on proactive, intelligent, and adaptive security features. Additionally, self-evolving AI malware is another new danger that constantly rewires itself to avoid detection by conventional means. Fraudsters are using AI-generated deep-fake media for identity fraud and social engineering, posing as reliable people to trick victims into disclosing private information. Advanced Persistent Risks (APTs) are adversarial attacks over extended periods of time, facilitating espionage and data exfiltration.

6.2 Innovative AI Security Models and Research Directions

The goal of resource-constrained networks is to establish a security system that is adaptable, efficient, and interpretable, thereby ensuring optimal performance. Intelligent security solutions that rely on AI have been developed and integrated into WSNs. Therefore, industry's security frameworks must be assessed, and accordingly, the integrated AI models have to be developed. Some AI frameworks should be lightweight to suit with constrained processing and power capacities of WSN nodes. Recognizing anomalous activity patterns in real-time is one of the security features that have been developed by an AI-based IDS model. For instance, Federated Learning with hybrid deep learning is a robust security and privacy model for intrusion detection, which permits safe AI model training without disclosing raw data. A hybrid model of multi-deep learning is one of the approaches that have been adopted. The model developments are aimed to lowering false positive rates. Improve the dependability, robustness, and precision of WSN security. Researchers are focusing on Explainable AI (XAI) for transparent, trusted, and reliable AI-models.

6.3 Policy, Regulation, and Standardization in AI-Driven WSN Security

WSNs are utilized in different industries, with unique security problems. On an international scale, some security approaches have been standardized to ensure that the cybersecurity framework for each industry is optimal. These frameworks must be robust, and scalable to deal with potential cyber threats. Following the ISO/IEC standards, the National Institute of Standards and Technology (NIST) guidelines, and the General Data Protection Regulation (GDPR) frameworks. Support the next-generation WSN security. In the future, improvements in quantum-secured AI models, lightweight encryption methods, and AI-enhanced threat intelligence are required.

7. CONCLUSION

Aligned with the book objectives, this chapter investigates WSNs-security viewpoints. It assessed the WSN resource constraints and limitations of conventional security approaches. Provides a deep analysis of the security challenges and transformation of security standers. Drewes attention to the two forms of AI in cybersecurity: offensive and defensive AI. The chapter also offers a detailed discussion of the AI security models. Including ML, DL and RL. Which are involved in advancing the security innovation and performance optimization.

However, despite the huge research conducted in this area, further study is still required. This chapter emphasized advancements of ongoing research. Increase the model's accuracy, scalability, and adaptability. This chapter provided a comprehensive review and a roadmap for the research trends.

REFERENCES

Abdi, A., Bennouri, H., & Keane, A. (2024). Emerging Cyber Risks & Threats in Healthcare Systems: A Case Study in Resilient Cybersecurity Solutions. 2024 13th Mediterranean Conference on Embedded Computing, MECO 2024. DOI: 10.1109/MECO62516.2024.10577790

Abdulrab, H., Hussin, F. A., Aziz, A. A., Awang, A., Ismail, I., & Arun Mozhi Devan, P. (2022). Reliable Fault Tolerant-Based Multipath Routing Model for Industrial Wireless Control Systems. *Applied Sciences (Basel, Switzerland)*, *12*(2), 544. Advance online publication. DOI: 10.3390/app12020544

Achaal, B., Adda, M., Berger, M., Ibrahim, H., & Awde, A. (2024). Study of smart grid cyber-security, examining architectures, communication networks, cyber-attacks, countermeasure techniques, and challenges. *Cybersecurity*, *7*(1), 10. Advance online publication. DOI: 10.1186/s42400-023-00200-w PMID: 38707764

Acharya, P., & Mahapatra, S. S. (2024). Exploring the Impact of Artificial Intelligence Integration on Guest Experience in the Hotel Industry. *Geo Journal of Tourism and Geosites*, *54*(2 supplement), 802–810. DOI: 10.30892/gtg.542spl04-1255

Adebayo, A. S., Chukwurah, N., & Ajayi, O. O. (2025).. . *Proactive Ransomware Defense Frameworks Using Predictive Analytics and Early Detection Systems for Modern Enterprises.*, *202*(April), 83–109.

Adu-Manu, S., & Kofi Abdulai, J.-D., Felicia Engmann, Moses Akazue, J. K. & Appati, Godwill Enchill Baiden, G. S.-K. (2022). WSN Architectures for Environmental Monitoring Applications. *Journal of Sensors.*

Affane, A. R., & Satori, H. (2024). Machine learning attack detection based-on stochastic classifier methods for enhancing of routing security in wireless sensor networks. *Ad Hoc Networks*, *163*, 103581. DOI: 10.1016/j.adhoc.2024.103581

Ahmad, R. & Wazirali, R. (2022). Machine Learning for Wireless Sensor Networks Security : An Overview of Challenges and Issues. Sensor, 22(4730).

Alomari, M. A., Al-andoli, M. N., Ghaleb, M., Thabit, R., Alkawsi, G., Abedalrahim, J., Alsayaydeh, J. & Gaid, A. S. A. (2025). Security of Smart Grid : Cybersecurity Issues, Potential Cyberattacks, Major Incidents, and Future Directions. 1–34.

Alquraishee, A. G. A., & Kar, J. (2014). A survey on security mechanisms and attacks in wireless sensor networks. *Contemporary Engineering Sciences*, *7*(1–4), 135–147. DOI: 10.12988/ces.2014.3949

Amaral, L. A., Matos, E., Tiburski, R. T., Hessel, F., Lunardi, W. T., & Marczak, S. (2016). *Internet of Things (IoT) in 5G Mobile Technologies*. Springer International Publishing.

Angurala, M., Kaur, M., Pabbi, V., & Chopra, N. (2025). Enhancing Cybersecurity in Wireless Sensor Networks: AI Solutions to Simulated Attacks. *Journal of Information Systems Engineering & Management, 10*(3s), 217–229. DOI: 10.52783/jisem.v10i3s.373

Anthi, E., Williams, L., Rhode, M., Burnap, P., & Wedgbury, A. (2021). Adversarial attacks on machine learning cybersecurity defences in Industrial Control Systems. *Journal of Information Security and Applications, 58*, 102717. DOI: 10.1016/j.jisa.2020.102717

Babiceanu, R. F., & Seker, R. (2019). Cyber resilience protection for industrial internet of things: A software-defined networking approach. *Computers in Industry, 104*, 47–58. DOI: 10.1016/j.compind.2018.10.004

Balas, V., Pandey, H., Bin Ali, M., Singh, V., & Kumar, A. (Eds.). (2026). *Recent Advances in Smart Communication Technologies for a Sustainable Future*. IGI Global Scientific Publishing., DOI: 10.4018/979-8-3373-3541-4

Balogun, A. Y., Olaniyi, O. O., Olisa, A. O., Gbadebo, M. O., & Chinye, N. C. (2025). Enhancing Incident Response Strategies in U. S. Healthcare Cybersecurity. *Journal of Engineering Research and Reports, 27*(2), 114–135. DOI: 10.9734/jerr/2025/v27i21399

Batool, R., Bibi, N., Alhazmi, S., & Muhammad, N. (2024). Secure Cooperative Routing in Wireless Sensor Networks. *Applied Sciences (Basel, Switzerland), 14*(12), 5220. Advance online publication. DOI: 10.3390/app14125220

Behiry, M. H., & Aly, M. (2024). Cyberattack detection in wireless sensor networks using a hybrid feature reduction technique with AI and machine learning methods. *Journal of Big Data, 11*(1), 16. Advance online publication. DOI: 10.1186/s40537-023-00870-w

Chen, J., Zhang, J., Qian, R., Yuan, J., & Ren, Y. (2023). An Anomaly Detection Method for Wireless Sensor Networks Based on the Improved Isolation Forest. *Applied Sciences (Basel, Switzerland), 13*(2), 702. Advance online publication. DOI: 10.3390/app13020702

DAHAB. G. O. K. M. M. (2018). The Weaponization of Artificial Intelligence (AI) and Its Implications on the Security Dilemma Between States: Could it Create a Situation Similar to Mutually Assured Destruction. The American University in Cairo.

Dalgade, D., Joshi, M., & Patil, N. (2024). Analysis of Machine Learning Algorithms for Intrusion Detection in Wireless Sensor Networks. SSRN Electronic Journal, 1–20. DOI: 10.2139/ssrn.4752583

Dede, G., Petsa, A. M., Kavalaris, S., Serrelis, E., Evangelatos, S., Oikonomidis, I., & Kamalakis, T. (2024). Cybersecurity as a Contributor Toward Resilient Internet of Things (IoT) Infrastructure and Sustainable Economic Growth. *Information (Basel)*, *15*(12), 1–27. DOI: 10.3390/info15120798

Delwar, T. S., Aras, U., Mukhopadhyay, S., Kumar, A., Kshirsagar, U., Lee, Y., Singh, M., & Ryu, J. Y. (2024). The Intersection of Machine Learning and Wireless Sensor Network Security for Cyber-Attack Detection: A Detailed. *Analysis & Sensing*, *24*(19), 6377. Advance online publication. DOI: 10.3390/s24196377 PMID: 39409417

Demedeiros, K., Hendawi, A., & Alvarez, M. (2023). A Survey of AI-Based Anomaly Detection in IoT and Sensor Networks. *Sensors (Basel)*, *23*(3), 1–33. DOI: 10.3390/s23031352 PMID: 36772393

Divyashree, S., Shanbhag, A. V., Bhuvana, H. R., & Deepthi, M. (2024)... *SURVEY PAPER ON DETECTING CYBER THREATS USING RANDOM FOREST ALGORITHM.*, *9*(1), 380–384.

Dora, J. R., & Nemoga, K. (2021). Clone Node Detection Attacks and Mitigation Mechanisms in Static Wireless Sensor Networks. *Journal of Cybersecurity and Privacy*, *1*(4), 553–579. DOI: 10.3390/jcp1040028

Elsadig, M. A. (2023). Detection of Denial-of-Service Attack in Wireless Sensor Networks: A Lightweight Machine Learning Approach. *IEEE Access : Practical Innovations, Open Solutions*, *11*(July), 83537–83552. DOI: 10.1109/ACCESS.2023.3303113

Fnu, H., Kaushik, K., Gaur, P., Saratchandran, D. V., Thapliyal, A. G., Sidhu, K. S., & Singh, V. (2025, April). Multi-Agent Systems for Collaborative Financial Decision-Making Over Distributed Network Architectures. In 2025 3rd International Conference on Advancement in Computation & Computer Technologies (InCACCT) (pp. 939-944). IEEE. DOI: 10.1109/InCACCT65424.2025.11011435

Fredj, N., Hadj Kacem, Y., Khriji, S., Kanoun, O., Hamdi, S., & Abid, M. (2023). AI-based model driven approach for adaptive wireless sensor networks design. Int. j. *International Journal of Information Technology : an Official Journal of Bharati Vidyapeeth's Institute of Computer Applications and Management*, *15*(4), 1871–1883. DOI: 10.1007/s41870-023-01208-8

Gangwani, P., Perez-Pons, A., & Upadhyay, H. (2024). Evaluating Trust Management Frameworks for Wireless Sensor Networks. *Sensors (Basel)*, *24*(9), 1–29. DOI: 10.3390/s24092852 PMID: 38732961

Gopalan, S. S., Raza, A., & Almobaideen, W. (2021). IoT Security in Healthcare using AI: A survey. ICCSPA 2020 - 4th International Conference on Communications, Signal Processing, and Their Applications, 2021-Janua, 1–6. DOI: 10.1109/ICCSPA49915.2021.9385711

Guma & Mijwil, M. (. (2024). Cybersecurity for Sustainable Smart Healthcare: State of the Art, Taxonomy, Mechanisms, and Essential Roles. *Mesopotamian Journal of CyberSecurity*, *4*(2), 20–62. DOI: 10.58496/MJCS/2024/006

Hossain, et al. (2020). Denial-of-Service Attacks on Wireless Sensor Network and Defense Techniques. Wireless Personal Communications, Springer Science, 1–29.

Hosseinzadeh, M., Rahmani, A. M., Vo, B., Bidaki, M., Masdari, M., & Zangakani, M. (2021). Improving security using SVM-based anomaly detection: Issues and challenges. *Soft Computing*, *25*(4), 3195–3223. DOI: 10.1007/s00500-020-05373-x

Hussain, S., He, J., & Zhu, N.. (2024). An Adaptive Intrusion Detection System for WSN using Reinforcement Learning and Deep Classification. *Arabian Journal for Science and Engineering*. Advance online publication. DOI: 10.1007/s13369-024-09769-x

Idrissi, I., Azizi, M., & Moussaoui, O. (2020). IoT security with Deep Learning-based Intrusion Detection Systems: A systematic literature review. *4th International Conference on Intelligent Computing in Data Sciences, ICDS 2020*. DOI: 10.1109/ICDS50568.2020.9268713

Islam, M., Fahmin, A., Hossain, M. S., & Atiquzzaman, M. (2021). Denial-of-Service Attacks on Wireless Sensor Network and Defense Techniques. *Wireless Personal Communications*, *116*(3), 1993–2021. DOI: 10.1007/s11277-020-07776-3

Jabeen, T., Jabeen, I., Ashraf, H., Jhanjhi, N. Z., Yassine, A., & Hossain, M. S. (2023). An Intelligent Healthcare System Using IoT in Wireless Sensor Network. *Sensors (Basel)*, *23*(11), 5055. Advance online publication. DOI: 10.3390/s23115055 PMID: 37299782

Jada, I., & Mayayise, T. O. (2024). The impact of artificial intelligence on organisational cyber security: An outcome of a systematic literature review. *Data and Information Management*, *8*(2), 100063. DOI: 10.1016/j.dim.2023.100063

Jain, U., & Hussain, M. (2018). Wireless Sensor Networks: Attacks and Countermeasures. SSRN Electronic Journal, 584–590. DOI: 10.2139/ssrn.3170185

Jawhar, I., Mohamed, N., & Al-Jaroodi, J. (2018). Networking architectures and protocols for smart city systems. *Journal of Internet Services and Applications*, *9*(1), 26. Advance online publication. DOI: 10.1186/s13174-018-0097-0

Karthikeyan, B., & Kamali, M. (2024). Detection of Attacks in Wireless Sensor Networks Using Machine Learning Algorithms. *International Journal of Research Publication and Reviews*, *6*(2), 3454–3460.

Karthikeyan, M., Manimegalai, D., & RajaGopal, K. (2024). Firefly algorithm based WSN-IoT security enhancement with machine learning for intrusion detection. *Scientific Reports*, *14*(1), 1–15. DOI: 10.1038/s41598-023-50554-x PMID: 38168562

Kaur, P., Kaur, K., Verma, D. K., Singh, K., Kaushik, K., & Singh, V. (2024, September). Routing protocols in wireless sensor networks: A comprehensive review and future perspectives. In 2024 7th International Conference on Contemporary Computing and Informatics (IC3I) (Vol. 7, pp. 263-270). IEEE. DOI: 10.1109/IC3I61595.2024.10828894

Kaur, S., & Monal, B. (2023). Securing the Future of Wireless Sensor Networks: Challenges, Threats, and Innovative Solutions. *International Journal for Research in Applied Science and Engineering Technology*, *11*(11), 719–728. DOI: 10.22214/ijraset.2023.56607

Keerthika, M., & Shanmugapriya, D. (2021). Wireless Sensor Networks: Active and Passive attacks - Vulnerabilities and Countermeasures. *Global Transitions Proceedings*, *2*(2), 362–367. DOI: 10.1016/j.gltp.2021.08.045

Keshetti, S., City, K., & States, U. (2025). AI-Driven Security Frameworks. *Enhancing Threat Detection and Response in Modern Systems.*, *13*(03), 233–254.

Khah, S. A., Barati, A., & Barati, H. (2023). A dynamic and multi-level key management method in wireless sensor networks (WSNs). *Computer Networks*, *236*, 109997. DOI: 10.1016/j.comnet.2023.109997

Khamparia, N. S. I. K. V. K. A. B. B. A. (2021). *Attacks and Security Measures in Wireless Sensor Network. In in Intelligent Data Analytics for Terror Threat Prediction: Architectures*. Methodologies, Techniques, and Applications.

Khan, S., Khan, M. A., & Alnazzawi, N. (2024). Artificial Neural Network-Based Mechanism to Detect Security Threats in Wireless Sensor Networks. *Sensors (Basel)*, *24*(5), 1641. Advance online publication. DOI: 10.3390/s24051641 PMID: 38475178

Khenwar, M., & Nawal, M. (2024). Challenges and Limitations of IDS: A Comprehensive Assessment and Future Perspectives. *SKIT Research Journal*, *14*(1), 35–39. DOI: 10.47904/IJSKIT.14.1.2024.35-39

Khraisat, A., Gondal, I., Vamplew, P., & Kamruzzaman, J. (2019). *Survey of intrusion detection systems: techniques, datasets and challenges*. Cybersecur., DOI: 10.1109/AVSS.2018.8639152

Kumar, A. (2024). Distributed Anomaly Detection in Wireless Sensor Networks (WSNs): A Review. *International Journal of Innovative Research in Engineering & Multidisciplinary Physical Sciences*, *12*(2), 1–8.

Kumaran, S. (2014). Secure authentication and integrity techniques for randomized secured routing in WSN. *Wireless Networks*, *21*(2), 443–451. DOI: 10.1007/s11276-014-0792-0

Kusha, K. R. (2020). Design & Development of an effective key management in dynamic WSNs using certificate less- effective key management protocol for secure communications characterized by node mobility. 3(1), 173–184.

Langley, D. & Birgitta. (2023). The weaponization of artificial intelligence: What the public needs to be aware of. Frontiers InArtificial Intelligence. DOI: 10.3389/frai.2023.1154184

Lee, C. C. (2020). Security and privacy in wireless sensor networks: Advances and challenges. *Sensors (Basel)*, *20*(3), 10–12. DOI: 10.3390/s20030744 PMID: 32013216

Lee, D., Kim, D., Lee, C., Ahn, M. K., & Lee, W.LEE. (2022). ICSTASY: An Integrated Cybersecurity Training System for Military Personnel. *IEEE Access : Practical Innovations, Open Solutions*, *10*, 62232–62246. DOI: 10.1109/ACCESS.2022.3182383

Malatji, M., & Tolah, A. (2024). Artificial intelligence (AI) cybersecurity dimensions: A comprehensive framework for understanding adversarial and offensive AI. *AI and Ethics*, *5*(2), 883–910. DOI: 10.1007/s43681-024-00427-4

Mallick, A. I. & Nath, R. (2024). Simulating Cyber Threats : A Review of AI-powered Attack Simulators for Enhanced Cybersecurity. 197(July), 13–54.

Manikandan, R., Arunprakash, S., Alsowail, R. A., & Pandiaraj, T. (2025). A novel wireless sensor network deployment for monitoring and predicting abnormal actions in medical environment and patient health state. *Alexandria Engineering Journal*, *119*(January), 149–167. DOI: 10.1016/j.aej.2025.01.064

Maniriho, P., Mahmood, A. N. & Chowdhury, M. J. M. (2024). A systematic literature review on Windows malware detection: Techniques, research issues, and future directions. Journal of Systems and Software, 209(December 2023), 111921. DOI: 10.1016/j.jss.2023.111921

Marabissi, D., Mucchi, L., & Stomaci, A. (2022). IoT Nodes Authentication and ID Spoofing Detection Based on Joint Use of Physical Layer Security and Machine Learning. *Future Internet, 14*(2), 1–21. DOI: 10.3390/fi14020061

Mazhar, T., Irfan, H. M., Khan, S., Haq, I., Ullah, I., Iqbal, M., & Hamam, H. (2023). Analysis of Cyber Security Attacks and Its Solutions for the Smart grid Using Machine Learning and Blockchain Methods. *Future Internet, 15*(2), 83. Advance online publication. DOI: 10.3390/fi15020083

Meleshko, A. & Desnitsky, V. (2024). The Modeling and Detection of Attacks in Role-Based Self-Organized Decentralized Wireless Sensor Networks. 145–175.

Mohamed, N. (2025). Artificial intelligence and machine learning in cybersecurity: A deep dive into state-of-the-art techniques and future paradigms. *Knowledge and Information Systems, 67*(8), 6969–7055. Advance online publication. DOI: 10.1007/s10115-025-02429-y

Mohamed, S., Auer, D., Hofer, D., & Küng, J. (2022). A systematic literature review for authorization and access control: Definitions, strategies and models. *International Journal of Web Information Systems, 18*(2–3), 156–180. DOI: 10.1108/IJWIS-04-2022-0077

Mubashir, N. M., Siddique, A., Ahmad, S. & Ijaz, A. (2020). Addressing Sinkhole Attacks In Wireless Sensor Networks - A Review. [IJSTR]. *International Journal of Scientific and Technology Research, 9*(08), 406–411.

Muchiri, N. G. (2021). *Detecting Zero-Day Attacks using Recurrent Neural Network*. Strathmore University.

Murtaza, A. (2024). Paradigm shift for predictive maintenance and condition monitoring from Industry 4.0 to Industry 5.0: A systematic review, challenges and case study. *Results in Engineering, 24*(August), 102935. DOI: 10.1016/j.rineng.2024.102935

Nandhini, S., Rajeswari, A., & Shanker, N. R. (2024). Cyber attack detection in IOT-WSN devices with threat intelligence using hidden and connected layer based architectures. *Journal of Cloud Computing (Heidelberg, Germany), 13*(1), 159. Advance online publication. DOI: 10.1186/s13677-024-00722-9

Narooka, P., Kaushik, K., Kansal, S., Kejriwal, D., Mishra, R. K., & Singh, V. (2025, April). Hybrid AI Models for QoS Estimation in α-η-μ Fading and Co-Channel Interference Environments within 6G Networks. In 2025 3rd International Conference on Advancement in Computation & Computer Technologies (InCACCT) (pp. 206-210). IEEE.

Narooka, P., Verma, S., Dwivedi, A., Kaushik, K., Singh, V., & Gaur, P. (2025, April). Psi-zeta Approach for Assessing Network Efficiency in Log-Normal Fading and Spectrum Sharing Interference Environments. In 2025 3rd International Conference on Advancement in Computation & Computer Technologies (InCACCT) (pp. 886-891). IEEE. DOI: 10.1109/InCACCT65424.2025.11011391

Nguyen, T. T., & Reddi, V. J. (2023). Deep Reinforcement Learning for Cyber Security. *IEEE Transactions on Neural Networks and Learning Systems*, *34*(8), 3779–3795. DOI: 10.1109/TNNLS.2021.3121870 PMID: 34723814

Nunoo-Mensah, H., Boateng, K. O., & Gadze, J. D. (2017). Tamper-aware authentication framework for wireless sensor networks. *IET Wireless Sensor Systems*, *7*(3), 73–81. DOI: 10.1049/iet-wss.2015.0131

Oloyede, J. (2024). AI-Driven Cybersecurity Solutions : Enhancing Defense Mechanisms in the Digital Era. SSRN Electronic Journal., 1–18. DOI: 10.2139/ssrn.4976103

Omar, A. Z., Jumaili, M. L. F., Abdual Rahman, S. A. A. E., Mohammed, M. Q. & Aljanabi, M. (2025). Enhancing Network Security in Wireless Sensor Networks Using Artificial Intelligence: A Random Forest Approach. AIP Conference Proceedings, 3264(1). DOI: 10.1063/5.0258927

Omar, S. (2023). A comprehensive survey on detection of sinkhole attack in routing over low power and Lossy network for internet of things. *Internet of Things : Engineering Cyber Physical Human Systems*, *22*, 100750. DOI: 10.1016/j.iot.2023.100750

Onyema, E. M., Kanimozhi Suguna, S., Sundaravadivazhagan, B., Jhaveri, R. H., Esther, U. N., Deborah, E. C. & Kumari, K. S. (2025). A secure routing protocol for improving the energy efficiency in wireless sensor network applications for industrial manufacturing. Next Energy, 7(September 2024), 100219. DOI: 10.1016/j.nxener.2024.100219

Osamy, W., Khedr, A. M., Salim, A., El-sawy, A. A., Alreshoodi, M. & Alsukayti, I. (2022). Recent Advances and Future Prospects of Using AI Solutions for Security, Fault Tolerance, and QoS Challenges in WSNs. 1–41.

Othman, S., Ben, , & Bahattab, A. A. (2015). Confidentiality and Integrity for Data Aggregation in WSN Using Homomorphic Encryption. *Wireless Personal Communications*, *80*(2), 867–889. DOI: 10.1007/s11277-014-2061-z

Oztoprak, A., Hassanpour, R., Ozkan, A., & Oztoprak, K. (2024). Security Challenges, Mitigation Strategies, and Future Trends in Wireless Sensor Networks: A Review. *ACM Computing Surveys*, *57*(4), 1–29. Advance online publication. DOI: 10.1145/3706583

Panwar, P., Verma, R., Rauthan, M. M. S., & Barthwal, V. (2021). *An Overview on Security Issues, Attacks, Challenges and Protocols in WSN. Proceedings of Integrated Intelligence Enable Networks and Computing*. Algorithms for Intelligent Systems., DOI: 10.1007/978-981-33-6307-6_27

Parande, S., & Mallapur, J. D. (2019). Research trends in secure routing protocols and communication system in WSNs. *International Journal of Innovative Technology and Exploring Engineering*, *9*(1), 3413–3421. DOI: 10.35940/ijitee.A4536.119119

Patil, H. K., & Chen, T. M. (2017). Chapter 18 - Wireless Sensor Network Security: The Internet of Things. In Computer and Information Security Handbook (Third Edition) (pp. 317–337).

Pourghasem, A., Kirner, R., Tsokanos, A., Mporas, I., & Mylonas, A. (2025). Machine Learning and Deep Learning-Based Multi-Attribute Physical-Layer Authentication for Spoofing Detection in LoRaWAN. *Future Internet*, *17*(2), 68. Advance online publication. DOI: 10.3390/fi17020068

Premakumari, S. B. N., Sundaram, G., Rivera, M., Wheeler, P., & Guzmán, R. E. P. (2025). Reinforcement Q-Learning-Based Adaptive Encryption Model for Cyber-threat Mitigation in Wireless Sensor Networks. *Sensors (Basel)*, *25*(7), 1–33. DOI: 10.3390/s25072056 PMID: 40218569

Qureshi, K., Alhudhaif, A., Haidar, S. W., Majeed, S., & Jeon, G. (2022). Secure data communication for wireless mobile nodes in intelligent transportation systems. *Microprocessors and Microsystems*, *90*, 104501. DOI: 10.1016/j.micpro.2022.104501

Ram, G., & Ilavarasan, E. (2024). Security Challenges in Wireless Sensor Network: Current Status and Future Trends. *Wireless Personal Communications*, *139*(2), 1173–1202. DOI: 10.1007/s11277-024-11660-9

Rao, K. R., Rao, S. N. T., & Reddy, P. C. (2017). Wireless Communication Security and Privacy issues and Challenges. [IJCSIS]. *International Journal of Computer Science and Information Security*, *15*(7), 202–209.

Raza, H. (2024). Proactive Cyber Defense with AI : Enhancing Risk Assessment and Threat Detection in Cybersecurity Ecosystems. November 2021.

Ghadi, Y. Y., Mazhar, T., Shloul, T. Al, Shahzad, T., Salaria, U. A., Ahmed, A. & Hamam, H. 2024. Machine Learning Solutions for the Security of Wireless Sensor Networks:Review, A. (2023, November).. . *IEEE Access : Practical Innovations, Open Solutions*, *12*, 12699–12719. DOI: 10.1109/ACCESS.2024.3355312

Saeed, . (2023). An SDN perspective IoT-Fog security: A survey. *Computer Networks*, *229*, 109732. DOI: 10.1016/j.comnet.2023.109732

Sahar, . (2023). A dynamic and multi-level key management method in wireless sensor networks (WSNs). *Computer Networks, 236*, 109997. DOI: 10.1016/j.comnet.2023.109997

Saiful, M., Mamun, I., Kabir, A. F. M. S., Hossen, S., Hayat, R., Saiful Islam Mamun, M., Sultanul Kabir, A., Sakhawat Hossen, M., Hayat Khan, M., Saiful, M., Mamun, I., Kabir, A. F. M. S., Hossen, S. & Hayat, R. (2012). Policy based intrusion detection and response system in hierarchical {WSN} architecture . CoRR, abs/1209.1(May).

Salem, A. H., Azzam, S. M., Emam, O. E., & Abohany, A. A. (2024). Advancing cybersecurity: a comprehensive review of AI-driven detection techniques. In Journal of Big Data (Vol. 11, Issue 1). Springer International Publishing. DOI: 10.1186/s40537-024-00957-y

Salmi, S., & Oughdir, L. (2023). Performance evaluation of deep learning techniques for DoS attacks detection in wireless sensor network. *Journal of Big Data, 10*(1), 17. Advance online publication. DOI: 10.1186/s40537-023-00692-w

Samiulla, I., Khan, A. A., Ahmad, M., & Idrisi, M. J. (2023). A Secure and Privacy-Preserving Lightweight Authentication and Key Exchange Algorithm for Smart Agriculture Monitoring System. *IEEE Access : Practical Innovations, Open Solutions, 11*(May), 56875–56890. DOI: 10.1109/ACCESS.2023.3280542

Sarker, I. H. (2023). Multi-aspects AI-based modeling and adversarial learning for cybersecurity intelligence and robustness: A comprehensive overview. *Security and Privacy, 6*(5), e295. DOI: 10.1002/spy2.295

Sebestyen, H., Popescu, D. E., & Zmaranda, R. D. (2025). A Literature Review on Security in the Internet of Things: Identifying and Analysing Critical Categories. *Computers, 14*(2), 61. Advance online publication. DOI: 10.3390/computers14020061

Sen, J. (2023). Cryptography and Key Management Schemes for Wireless Sensor Networks. In Wireless Sensor Networks—Design, Applications and Challenges.

Shammari, G. Al & Parveen, N. (2024). A Dual-Channel Deep Learning Framework For Real-Time Detection Of Zero-Day Attacks Using Cnn-Lstm Hybrid Networks. December.

Sharma, V., Beniwal, R., & Kumar, V. (2024). Multi-level trust-based secure and optimal IoT-WSN routing for environmental monitoring applications. *The Journal of Supercomputing, 80*(8), 11338–11381. DOI: 10.1007/s11227-023-05875-z

Shihab, M. A., Marhoon, H. A., Ahmed, S. R., Al-Attar, B., Al-Sharify, M. T., & Sekhar, R. (2024). Towards Resilient Machine Learning Models: Addressing Adversarial Attacks in Wireless Sensor Network. [JRC]. *Journal of Robotics and Control*, *5*(5), 1582–1602. DOI: 10.18196/jrc.v5i5.23214

Shrivastava, M. (2018). Improving Wireless Sensor Network ' s Performance through Intelligent Sensor Reinforcement Learning Algorithm (ISRLA). 6(1), 1052–1056.

Shu, Y., Lee, K., Lanctot, P., Jianbin, F., & Hao, H. (2016). *Internet of Things.* Wireless Sensor Networks. In IEC Market Strategy Board.

Sikos, L. F. (2020). Packet analysis for network forensics: A comprehensive survey. *Forensic Science International Digital Investigation*, *32*, 200892. DOI: 10.1016/j.fsidi.2019.200892

Singh, O., Vinoth, R., Singh, A. & Singh, N. (2024). Navigating Security Threats and Solutions using AI in Wireless Sensor Networks. 16(4).

Singh, V., & Dubey, V. (2015). A two level image security based on Arnold transform and chaotic logistic mapping. *International Journal of Advanced Research in Computer Science and Software Engineering*, *5*(2), 883–887.

Singh, V., & Kaushik, V. D. (2023). A Brief Overview of Cyber Security Advances and Techniques Along With a Glimpse on Quantum Cryptography: Cyber Security Practices, Advances and Challenges. Exploring Cyber Criminals and Data Privacy Measures, 40-64.

Singh, V., & Kaushik, V. D. (2023). A Methodical View of Prerequisites of Picture Combination, Strategies, and Key Indicators with Usage in Real Life and Scientific Domains Facilitating Smart Ubiquitous Environment. In Computational Intelligence in Analytics and Information Systems (pp. 15-33). Apple Academic Press

Singh, V., & Kaushik, V. D. (2023). Navigating the Landscape of Security Threat Analysis in Cloud Computing environments. In *Security and Risk Analysis for Intelligent Cloud Computing* (pp. 1–25). CRC Press.

Sparks, E. S. (2025). Strengthening Cybersecurity through Proactive Threat Mitigation and Intelligent Defense Mechanisms. 6, 1–7.

Bangash, Y. A., Abid, Q. ud D., A, A. A. A. & Al-Salhi, Y. E. A. 2017. Security Issues and Challenges in Wireless Sensor Networks:Survey, A. (•••).. . *IAENG International Journal of Computer Science*, *44*(2), 1–15. DOI: 10.14419/ijet.v7i2.33.13861

Tariq, U., Ahmed, I. & Bashir, A. K. (2023). A Critical Cybersecurity Analysis and Future Research Directions for the Internet of Things: A Comprehensive Review. Sensor, 23(4117).

Tavakoli, N., Siami-Namini, S., Adl Khanghah, M., Mirza Soltani, F., & Siami Namin, A. (2020). An autoencoder-based deep learning approach for clustering time series data. *SN Applied Sciences*, *2*(937), 937. DOI: 10.1007/s42452-020-2584-8

Tiwari, V. (2018). Data Integrity and Authentication in WSNs. HMR Interdisciplinary Journal of Science. *Technology & Education Management*, *4*(1), 12–17.

Tolba, A. (2024). Hybrid LSTM-Random Forest for Intrusion Detection in Wireless Sensor Networks : Enhanced DoS Classification with Explainable AI 1 | Introduction. 2, 39–52.

Triwidyastuti, Y., Perdana, R. H. Y., Shim, K., & An, B. (2023). Secrecy Performance Analysis of Cooperative Multihop Transmission for WSNs under Eavesdropping Attacks †. *Sensors (Basel)*, *23*(17), 1–25. DOI: 10.3390/s23177653 PMID: 37688109

Vaid, C. S. and R. (2021). A Novel Sybil Attack Detection and Prevention Mechanism for Wireless Sensor Networks. 2021 6th International Conference on Signal Processing, Computing and Control (ISPCC), 340–345.

Waheed, T., & Marchetti, E. (2023). The Impact of IOT Cybersecurity Testing in the Perspective of Industry 5.0. International Conference on Web Information Systems and Technologies, WEBIST - Proceedings, Webist, 480–487. DOI: 10.5220/0012235800003584

Whitepaper, S. S. S. (2018). Fundamental Practices for Secure Software Development. In *SAFECode Driving Security and Integrity* (2nd ed.). Issue.

Xue, Y., Singh, V., Singh, S., Kant, K., Pandey, S., Kumar, A., & Kaushik, V. D. (2024). Blockchain Framework for Learner Performance Prediction using Life-Brain Storm-based Light GBM Coupled Neural Network: BLOCKCHAIN FRAMEWORK FOR LEARNER PERFORMANCE PREDICTION. [JSIR]. *Journal of Scientific and Industrial Research*, *83*(6), 652–668.

Yadav, A. & Kumar, A. (2022). Intrusion Detection and Prevention Using RNN in WSN. Inventive Computation and Information Technologies, 531–539.

Zhukabayeva, K. A. & N., A. and. (2024). Design of Intrusion Detection and Security Enhancements in Military Wireless Sensor Networks. 2024 7th International Conference on Signal Processing and Information Security (ICSPIS), 1–6.

Zhukabayeva, T., Zholshiyeva, L., Karabayev, N., Khan, S., & Alnazzawi, N. (2025). Cybersecurity Solutions for Industrial Internet of Things–Edge Computing Integration: Challenges, Threats, and Future Directions. *Sensors (Basel)*, *25*(1), 213. Advance online publication. DOI: 10.3390/s25010213 PMID: 39797003

Chapter 3
Explainable AI for Malware and DoS Attack Mitigation in Wireless Communication Environments

Sucheta Sachin Yambal
https://orcid.org/0000-0002-3113-9609
Dr. Babasaheb Ambedkar Marathwada University, India

Yashwant Arjunrao Waykar
https://orcid.org/0000-0002-9693-9738
Dr. Babasaheb Ambedkar Marathwada University, India

Viraj Soni
https://orcid.org/0009-0004-8917-9429
Independent Researcher, USA

Shubham Bajpai
School of Advanced Agriculture Sciences and Technology, Chhatrapati Shahu Ji Maharaj University, India

Omkar Agrahari
https://orcid.org/0009-0002-5201-0422
School of Engineering and Technology, CSJM University, Kanpur, India

DOI: 10.4018/979-8-3373-3576-6.ch003

ıt © 2026, IGI Global Scientific Publishing. Copying or distributing in print or electronic forms without written permission of IGI Global Scientific Publishing is prohibited. Use of this chapter to ›enerative artificial intelligence (AI) technologies is expressly prohibited. The publisher reserves all rights to license its use for generative AI training and machine learning model development.

ABSTRACT

The rapid proliferation of wireless communication networks has led to an increase in malware and DoS assaults. Traditional security methods don't usually work well against these threats because they are so complicated and change so quickly, particularly in wireless scenarios where resources are limited. This article looks at how Explainable Artificial Intelligence (XAI) might help cybersecurity systems deal with malware and DoS attacks better. XAI makes machine learning models clear and easy to grasp; the system uses XAI to find anomalies and sort traffic, and keep it easy to comprehend in order to respond faster & with thorough knowledge. SHAP and LIME XAI are two ways to look at wireless network security. Simulations reveal that XAI makes trust, accountability, and efficiency better when it comes to dealing with threats. This study reveals that explanation is necessary for building powerful, flexible, and smart wireless communication defensive systems against new cyber threats. This will make next-generation networks secure and reliable.

1. INTRODUCTION

Wireless communication technology has changed the ways we share, access and send information in big ways. Wireless networks, such as Wi-Fi, sensor networks, mobile communication systems (4G/5G) and Internet of Things frameworks are becoming the main way that people interact to each other digitally. As more and more people rely on wireless communication (Narooka et al., 2025a; Narooka et al., 2025b), cyber assaults that take advantage of weaknesses in these systems also rise. Denial-of-Service (DoS), Distributed Denial-of-Service (DDoS), and virus assaults are major security threats that make wireless networks less available, less private, and less secure (Balas et al., 2026).

Because they are open and decentralized, wireless communication settings are more likely to be spoofed, eavesdropped on, and accessed without permission than conventional wired networks. Wireless devices frequently have limited resources, such as memory, processing power, and battery life, which makes it hard to use security techniques that need a lot of computing power. Also, real-time threat identification and mitigation are quite hard since wireless networks (Kaur et al., 2024; Fnu et al., 2025) are always changing and have a lot of different attack surfaces and behavioural patterns.

PROBLEMS WITH CYBERSECURITY IN WIRELESS COMMUNICATION

Wireless networks have a unique set of cybersecurity problems because of how they function and how they are built. These are:

∇ Open Access Medium: Wireless signals may travel across the sky and reach any nearby device. This openness makes it easier to jam, do man-in-the-middle attacks, and listen in.
∇ Dynamic Topologies: Network topologies are always changing and unexpected because mobile devices often join and leave networks. It eventually makes hard to set up a reliable security perimeter.
∇ Device Heterogeneity: Wireless networks frequently include a mix of smart phones, Internet of Things devices, sensors and edge nodes where each with its own security features and weaknesses.
∇ Resource Limitations: Some wireless devices don't have the power and the hardware to support complex encryption or intrusion detection systems.
∇ Scalability and Real-time Requirements: As more devices connect to the internet it's noticed that the amount of data and possible security holes grows. It is important to make cyber-security solutions bigger without losing their ability to work in real time.

Systems that can find threats before they happen and find them quickly are quite important. Modern threats change and grow quickly notably zero-day vulnerabilities and malware that changes all the time. These threats are too much for traditional signature-based security systems to handle. Because of these problems, machine learning (ML) and artificial intelligence (AI) are the new trends that make it possible to crack such threats with ease and judging the behaviour with similar pattern for convenient handling said by (Johns & Ottawalanguages, 2019).

3. REVIEW OF LITERATURE

The advancements of wireless communication technologies from Wi-Fi and 4G/5G to vast Internet of Things (IoT) ecosystems (Joshi et al., 2024a; Joshi et al., 2024b; Joshi et al., 2025) have fundamentally reshaped digital interactions all over the globe. But this reliance on wireless networks has made us a prime target for cyber-attacks. Thus inherent characteristics of wireless environments (some of them are open access medium, dynamic topologies and resource-constrained devices) create a uniquely challenging security landscape that is more prone to spoofing, eavesdropping and unauthorized access than traditional wired networks (Barredo Arrieta et al., 2019). These vulnerabilities could be aggressively exploited by prevalent threats like malware and Denial-of-Service (DoS) attacks which can cripple critical infrastructure as evidenced by real-world incidents like the Mirai botnet (Scott & Lundberg, 2017).

Conventional method that were signature based security systems are increasingly inadequate against these recent evolving threats; as (Johns & Ottawalanguages, 2019) noted that the rapid mutation of modern malware and the sophistication of zero-day vulnerabilities are not handled by conventional defenses well. This has catalyzed/ inclined a shift towards Artificial Intelligence (AI) and Machine Learning (ML) for threat detection; AI-driven models excel at identifying abnormal patterns in network traffic that may indicate a malware which might be a probable DoS assault. Furthermore, these advanced mechanisms offer a self-actuating and adaptive layer of security during the attacks that static systems cannot provide as noted by (Confalonieri et al., 2024). Techniques that can be posed as the probable solutions usually range from supervised learning models like Random Forests for classifying known threats to unsupervised methods like clustering for detecting novel attacks that strike often- with deep learning models like LSTMs showing promising resolutions for analyzing time-series network data (Rjoub et al., 2023).

Despite the high accuracy, one of the barriers to the adoption of these advanced AI models in critical cyber-security roles are their 'black-box'/ concealed nature. As (Schreiber & Koh, 2017) highlighted, the inability to understand or explain 'why' an AI model flags a particular event as malicious reduces trust of the users and reduce the effectiveness of the incident response. This reticence is a big problem in areas like healthcare and self-driving cars, where even a little mistake by an automated decision system might lead to serious consequences (Nibret Zeleke et al., 2025). This problem has made Explainable AI (XAI) the most important area of study in cyber-security which needs further exploration for validation.

XAI's goal is to make performance and interpretability more similar to each other. Tommi et al. (2017) posited that cyber-security is where explainability is essential for trust, adoption and regulatory compliance; consequently, methodologies such as LIME (Local Interpretable Model-agnostic Explanations) and SHAP (SHapley Additive ExPlanations) have gained significant importance (Speith, 2021). Alam et al. (2021) demonstrated how SHAP could identify the critical factors that led a model to perceive a file as harmful including specific API calls or file entropy. XAI can also ruin DoS attacks by discovering the specific traffic patterns that send off an alert for some unreasonable surge such a sudden increase in TCP SYN packets in the network. This could help network administrators identify and bring in a corrective action for halting the assault (Gaspar et al., 2023).

XAI in wireless security has its own set of challenges. One of the biggest concerns is the trade-off between how accurate a model is and how simple it is to grasp. The most powerful models are usually the ones that are the hardest to understand said (Rudin, 2019). Zilberman et al. (2017) marked about problems with scalability and performance; they mentioned that the resource-intensive XAI methods like SHAP could prove ineffective for wireless edge devices (Krishna et al., 2024; Roohani et al., 2026; Suriyan et al., 2026, Anandaram et al., 2026) that need to work in real time and with minimal latency. There is an issue with good/ tagged datasets as well that are useful for wireless situations. This makes it challenging to create and evaluate powerful models for their effectiveness.

Combining XAI with new technologies might lead to new areas of research. The advent of 5G and 6G networks distinguished by network slicing and ultra-dense device connections need transparent security mechanisms for efficient administration and trust (Brenna et al., 2021). Federated Learning (FL) offers a framework for decentralized, privacy-preserving model training; yet the incorporation of explainability into these systems remains a substantial challenge. It can be safely assumed that humans and AI will work together in a way that is beneficial to both. XAI will be a collaborative interface that provides security analysts with clear, useful information to help them make better and efficient decisions (Arrieta et al., 2020).

To sum it up - AI has greatly changed how threats are found in wireless networks, yet black-box models make it hard for AI to reach its full potential. Explainable AI is the key that could unlock this potential; it converts powerful but unclear algorithms into trustworthy, verifiable, and useful partners in the ongoing fight against cyber threats.

3.1. Why it's important to stop DoS and malware attacks

Malware and Denial-of-Service (DoS) attacks are usually dangerous for wireless communication networks because they have high frequency of occurrence, come in many forms, and may cause a lot of damage at large. Malware is any harmful software, such as viruses, worms, trojans, ransomware, spyware, and more.

Malware may quickly move across devices in wireless situations, take over system resources, steal critical data, and be the start of bigger assaults as per (Saeed & Omlin, 2022). Malware that attacks wireless networks may stop communication between nodes, turn off important functionality, or change important sensors in medical or industrial settings.

(Schreiber & Koh, 2017) has clearly mentioned, DoS and DDoS attacks try to flood a network or service with unlawful traffic, which makes it too much for it to handle. Denial of Service (DoS) attacks may make services inaccessible on wireless networks by taking advantage of weak points in protocols, buffer sizes, or bandwidth limits. A denial-of-service (DoS) assault on smart city networks' emergency response or traffic systems might have terrible effects. DDoS assaults, which use networks of hacked wireless devices, are more effective because of the growth of the Internet of Things (IoT). Figure1 shown below depicts the semblance between malware and DoS attacks on the wireless networks for easy visual interpretation.

Figure 1. Semblance between Malware and DoS Attacks on Wireless Networks, Source: Developed by the researcher

Comparison of Malware and DoS Attacks on Wireless Networks

Characteristic	Malware	DoS Attacks
Definition	Harmful software	Flooding network with illegitimate traffic
Propagation	Rapid spread across wireless devices	Exploits protocol weaknesses
Impact	Data theft, system takeover, disruption	Service inaccessibility, potential for severe consequences
Evolution	Employs AI, polymorphism, obfuscation	DDoS attacks leverage IoT devices
Mitigation	Requires rapid detection, flexible response	Requires rapid detection, flexible response
Cost of Failure	Potential loss of human life	Potential loss of human life

Both kinds of assaults are changing quickly and leveraging AI, polymorphism, and obfuscation to avoid being found. To be effective, mitigation needs quick and precise detection, flexible reaction plans, and activities that stay within their resource limits. Also, the cost of unreported attacks may be measured in terms of human lives in areas where safety is very important, such wireless healthcare systems or self-driving cars says (Nibret Zeleke et al., 2025).

3.2 What AI Does to Find and Stop Threats

Artificial intelligence- specially machine learning (ML), has become a game-changer in the field of cyber-security points out (Confalonieri et al., 2024); instead of only looking for known fingerprints, AI systems may find unusual behaviour, sort threats, and forecast attacks that are about to happen by looking at patterns in large amounts of network data. AI-driven security models provide these benefits in wireless settings as given below:

∇ Tracing anomalies/ abnormalities:- Machine learning algorithms can spot abnormal changes in normal network traffic patterns that might indicate malware or denial-of-service assaults.
∇ Network's Behavioural Analysis:- AI makes it possible to track how users and devices act, which helps find low-and-slow or covert assaults that are immensely difficult to find.
∇ Automation:- AI makes it easier for human analysts to find, diagnose, and respond to problems more quickly, which cuts down on their labour thus automating major and recurring tasks.
∇ Adaptability:- AI models can learn from fresh data, which means they can deal with new hazards or can say that they learn from experience which is not possible with static rule-based systems.

Still some AI models act like 'black boxes' generating decisions that are very accurate but are not able to explain the algorithm with clarity for understanding. In critical systems where trust, accountability and compliance depend on knowing why a security alarm went out this reticence presents a hitch.

3.3 AI-powered security systems need to be able to explain themselves.

In this case, Explainable Artificial Intelligence (XAI) is quite important as it is easier to decipher. "XAI" is a word that refers to methods that make it easier for people to comprehend and see the outcomes of AI models. Explainability is not just a good quality in cyber-security; it's a must-have according to (Tommi et al., 2017). For instance, if an AI system finds malware on a device, the network administrator must know what traits, patterns, or actions led to this conclusion. This understanding is very important because of the following identified key-points:

∇ Adoption & Trust:- Security experts are more inclined to trust and use systems whose judgments they can comprehend.

∇ Compliance & Auditing:- Automated choices require audit trails and regulatory reasons in many fields, some of which are healthcare and banking.
∇ Debugging along with Model Improvement:- Knowing the basic reasoning behind a model makes it possible to fix mistakes and biases, thus makes it more accurate.
∇ Faster Response:- Clear explanations make it easier to create individualized remediation plans to speed up diagnosis of the existing problemin general.

As (Alam et al., 2021) has pointed out that the malware and denial-of-service assaults grow smarter; it becomes even more evident to use Explainable AI models. This chapter explains XAI effectively in wireless settings to find, assess and reduce cyber threats while making sure that security, operational resilience, transparency and trust are all maintained.

4. RULES FOR WIRELESS COMMUNICATION ENVIRONMENTS

4.1 What wireless networks are like and how they function

Wireless communication is becoming a key part of modern digital infrastructure; it lets mobile phones, IoT devices, sensor networks and other devices stay connected all the time. Wireless systems send data using electromagnetic waves instead of wires. This makes them mobile, flexible and scalable. The benefits have led to the widespread use of wireless networks in many areas such as healthcare, the military, education, transportation and smart cities.

Wireless networks usually include three main parts: wireless clients (the devices that people use), access points or base stations, and a backbone infrastructure (servers, routers, and gateways) as per the clarity given by (Barredo Arrieta et al., 2019). There are several ways to structure communication, such as:

∇ Infrastructure Mode-> Devices that connect to each other via central access points, for e.g. Wi-Fi routers.
∇ Ad Hoc Networks-> Peer-to-peer communication that isn't centralized as infrastructure mode.
∇ Mesh Networks-> Paths for data to go between several nodes, with each node able to send data to every other node.
∇ Cellular Networks-> A structured system of base stations and mobile devices that can all be controlled from one place.
∇ IoT/WSNs (Wireless Sensor Networks) -> These are small, spread-out sensors which keeps an eye on the physical environment.

While these features make communication more flexible they also create security problems that make wireless systems prone to hacking or attack.

4.2 Common Weaknesses and Threat Pathways

Wireless networks are open and decentralized which makes them vulnerable to a lot of attacks. Most attacks on wired networks need physical access to the infrastructure, however wireless systems may be attacked from a distance using rather cheap equipment mentions (Ghassemi et al., 2021). Common weaknesses include:

∇ Eavesdropping-> Unauthorized people could be able to access private information by intercepting wireless signals.
∇ Rogue Access Points-> Unauthenticated devices might pretend to be lawful access points to trick users and steal their data.
∇ MAC Spoofing -> Hackers change the Media Access Control address to make it seem like real devices.
∇ Session Hijacking -> A stalker steals session tokens or credentials to get into a system without permission.
∇ Outdated Encryption Protocols-> Old protocols like WEP are still used which could be easy to break.
∇ Man-in-the-Middle (MitM) Attacks-> Attackers put themselves between two devices that are communicating to listen in on or change the discussion.
∇ Replay Attacks-> Criminals send intercepted data back to systems to trick them into doing illicit things.

Some of the major types of security threats are as drawn in the Figure2 depicting the concerns that is bring forth because of the alterations done by the said threat.

Figure 2. Depicting various wireless security threat, Source: Developed by the researcher

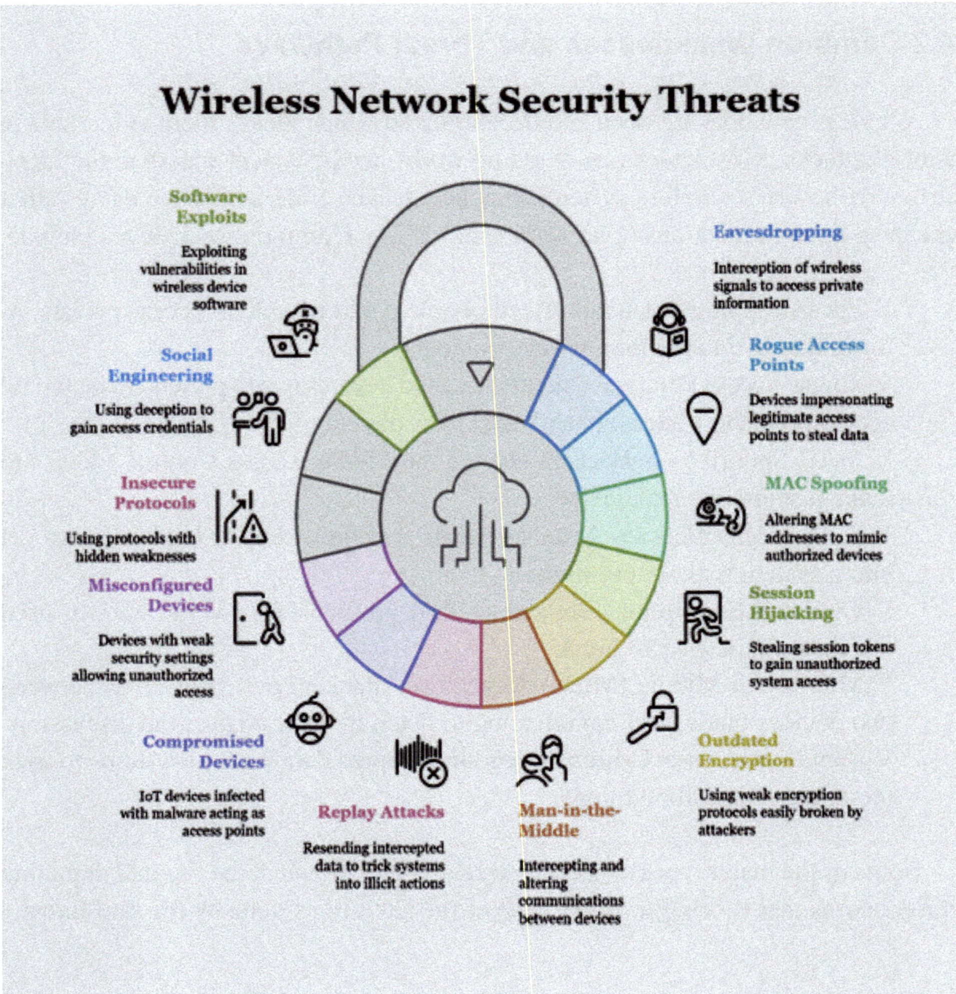

In wireless contexts, most commonly the threat vectors include:

∇ Compromised Devices-> IoT devices that already have malware on them and act as access points for other devices in the same network.
∇ Misconfigured Devices-> Security settings that aren't strong enough thus anyone can get in without proper authorization check.
∇ Insecure Protocol-> Insecure or proprietary protocols that do not provide contemporary level of security and are thus incompetent.

∇ Social engineering-> Done when one uses phishing or lying to gain access credentials.
∇ Software Exploits:-> Accessing undue advantage of weaknesses in the firmware or applications of wireless devices.

(Moskovitch&Elovici, 2008) has studied that these weaknesses that let hackers install malware or launch Denial-of-Service (DoS) attacks, which might further damage or shut down whole networks, particularly those that provide important services.

5. TYPES OF MALWARES AND DENIAL OF SERVICE ATTACKS AND WHAT THEY DO

5.1 Types of Malwares

Malware is any software that is designed to mess with, break, or obtain unauthorized access to computer systems. Malware may spread quickly in wireless networks because devices can connect to one other and send out signals.

The primary types of malwares are:

∇ Viruses attach themselves to real files or programs and spread when the host file is run.
∇ Worms are programs that can copy themselves and spread on their own over networks, frequently without any help from anyone.
∇ Trojans: They seem like real software to trick people into installing them, which gives attackers a way in.
∇ Ransomware locks up user data and asks for money to get the keys to unlock it.
∇ Spyware: watches what users do without their knowledge and collects personal data.
∇ Adware: Shows ads without asking and may also collect information on users.
∇ Rootkits change the operating system so that dangerous programs or apps can't be found.
∇ Botnets are networks of hacked devices (bots) that may be controlled from a distance to carry out coordinated attacks.

In wireless networks, malware typically takes advantage of:

∇ Not enough good authentication mechanisms
∇ Software or firmware that hasn't been updated
∇ Default settings that aren't safe

∇ Ways for devices to talk to one other in IoT setups

A single infected device might allow lateral movement, command-and-control communication, or the creation of a botnet.

5.2 Summary of DoS and DDoS Attacks

A Denial-of-Service (DoS) attack tries to make a network resource unavailable to its real users by flooding it with traffic or causing system faults. In wireless circumstances, Denial of Service attacks could use limited bandwidth, buffer space, or processing power to use up resources.

There are several types of 'Denial of Service' attacks, such as:

∇ Flood attacks send a lot of packets to fill up the network's capacity (for example, UDP flood and ICMP flood).
∇ Ping of Death: Sending bad or too big packets that make the target system stop working.
∇ Smurf Attack: Using IP spoofing to flood a target with ICMP responses.
∇ Protocol Exploits: Using weaknesses in wireless protocols like 802.11 or Bluetooth to jam or replay attacks.
∇ Distributed Denial-of-Service (DDoS) attacks are worse since they use multiple compromised devices (botnets) to attack a system at the same time. The rise of IoT devices has made it easier for attackers to get to millions of poorly secured endpoints. This has made wireless DDoS attacks more common and powerful.

Denial of Service attacks that only happen on wireless networks include:

∇ Jamming is when you mess with the wireless frequency to stop real transmission.
∇ Deauthentication attacks: making users disconnect often by pretending to be disconnection notices.
∇ Battery Drain Attacks: These attacks are meant to drain the batteries of IoT devices that run on batteries.

(Scott & Lundberg, 2017) have said that these attacks might destroy smart city infrastructure, delay emergency services, or hurt the finances and image of businesses that rely on wireless technology.

5.3 Real-Life Examples that teach lessons

There are many real-life events that show how important it is to have strict security setting in wi-fi connection:

1. The Mirai Botnet (2016) was a significant DDoS attack that would be remembered for a long thime; it used a lot of unsecured IoT devices, such routers and IP cameras, to overload major DNS providers. This caused websites like Twitter, Netflix, and Reddit to go down.
2. WiFi Pineapple Attacks which made use of illegal access point devices have been used in a number of penetration tests and cybercrimes to listen in on and change wireless communications on the go.
3. The Stuxnet Worm (2010) badly affected industrial equipment, showing how malware could spread over air-gapped networks using USB. This showcased probable ways that malware may spread in wireless networks that are even not connected to the internet.
4. Smart Home Attacks is when there were several recorded cases of attackers using flaws in wireless home automation devices to spy on people, control appliances, or keep people from getting into their homes.

There are a lot of bad things that may happen to wireless networks when malware and DoS assaults happen:

1. Service Interruption: When healthcare, banking, or public services are not working.
2. Data Breach: When private information about a user or business is made public.
3. Financial harm: Direct and indirect costs include paying a ransom, going to court, and hurting one's image.
4. Safety Risks: Attacks on important systems like self-driving vehicles or smart grids risking lives.
5. Loss of Trust: Users are losing faith in wireless technologies and devices because of lack of faith.

Wireless communication settings are important parts of modern digital infrastructure, but they are prone to hacking. Because these systems are open, dynamic, and have limited resources, they are soft-targets for malware and Denial of Service attacks. (Wood, 2017) says as these threats are becoming more complicated and bigger as traditional security approaches don't work anymore. To come up with smart, flexible, and understandable ways to defend against attacks in wireless environments, complete understanding of the architecture along with its weaknesses and real-world effects of those attacks should be known. This is what makes AI and, more importantly, Explainable AI (XAI) have its worth in cyber-security.

6. AI TECHNIQUES FOR FINDING THREATS

6.1 Finding problems using deep learning and machine learning

The amount of cyberthreats that target wireless networks is growing and traditional rule-based and signature-driven security solutions aren't keeping up well. These attacks are becoming increasingly advanced, flexible, and secret points out says (Zeleke et al., 2025). To get around these problems, artificial intelligence (AI), machine learning (ML) and deep learning (DL) should be used that has become a powerful tool for finding and stopping threats in real time.

One important way that AI is used in cybersecurity (Singh and Dubey, 2015; Singh and Kaushik, 2023a; Singh and Kaushik, 2023b; Chaudhary et al., 2024) is to identify unusual behaviour/pattern. Usually, it translates to finding abnormal patterns of network activity that could be viruses, intrusions, or Denial-of-Service (DoS) attacks. AI makes it feasible for smart, automated identification markers that could be a better alternative to existing system as wireless networks are more complicated and have more traffic. (Rjoub et al., 2023) rightly put it in words when in the research work he said that these networks have frequently changing topologies and limited resources.

Machine learning (ML) models like decision trees, support vector machines (SVM), random forests and k-nearest neighbours (KNN) are widely used to group or classify network events based on their features, such as packet size, traffic rate, and the type of protocol. Convolutional neural networks (CNNs) and recurrent neural networks (RNNs) are two types of deep learning models that are effective at working with time-series data and finding even the small changes in time-series. Long Short-Term Memory (LSTM) models and auto-encoders have been able to find new and complicated threats that regular firewalls can't detect. ML/DL anomaly detection in wireless networks makes networks more resilient and speeds up response times by finding zero-day vulnerabilities, botnet activity or coordinated DoS attacks early on said (Vilone & Longo, 2018).

6.2 Methods: With and without supervision

There are three basic types of AI threat detection strategies; they are supervised, unsupervised, and semi-supervised. Each has its own pros and cons as listed below:

■ Supervised learning is when you train models using labelled datasets, which are collections of data that clearly show instances of good and bad behaviour in general. Two examples include utilizing predefined criteria to find phishing sites or using labelled traffic logs to sort malware from the rest. Supervised methods generally provide very accurate results but they require big datasets with very accurate labels, which are hard to acquire and keep wireless environments in mind, they usually change quickly.

■ Unsupervised learning doesn't call for labelled data. Rather, it looks for strange patterns or abnormalities in the data's natural structure. Anomaly detection techniques (like One-Class SVM and Isolation Forests) and clustering methods (like K-means and DBSCAN) are widely used to find unknown threats; these methods are especially useful for finding new or zero-day threats since there may not be any labelled examples of new hazards in wireless systems yet.

■ Semi-supervised learning is a good middle ground since it uses a lot of unlabeled data and some of labelled data as well. This method is usually implemented when the costs to obtain labelled dataset is too much or is too hard to get.

Figure 3. Depicting AI threat detection strategies, Source: Developed by the researcher

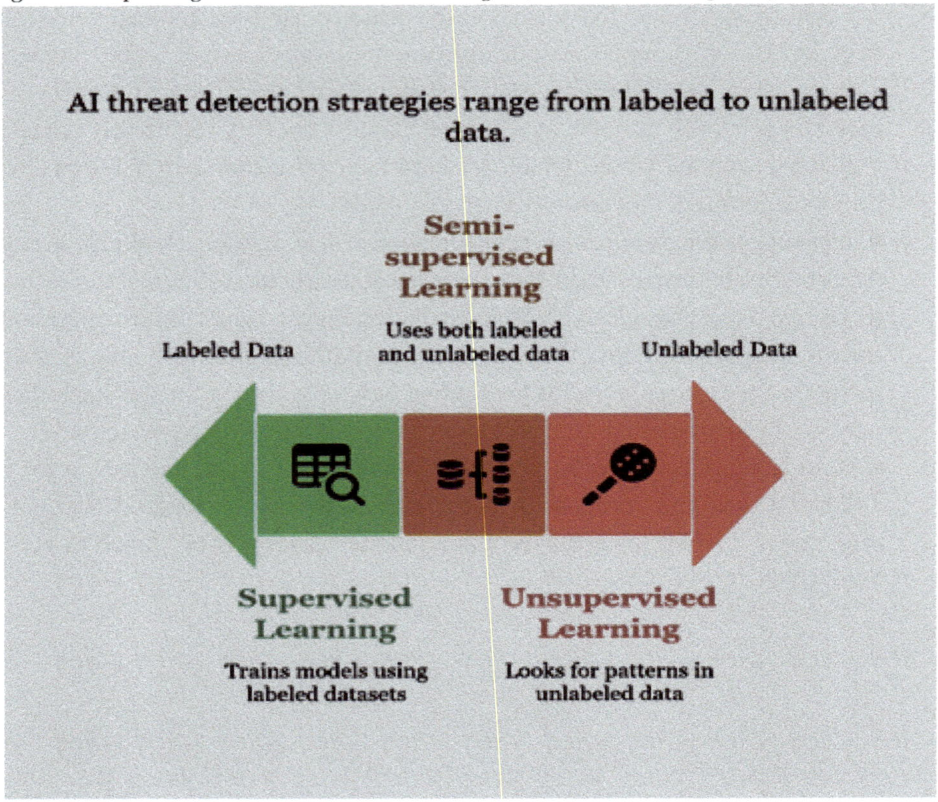

Figure 3 showcases the way in which labeled and unlabeled datasets can be used for Supervised and Unsupervised learning; this also shows that Supervised learning trains model whereas Unsupervised learning is used to look for pattern.

Using both supervised and unsupervised models together might yield better results in wireless communication circumstances; as unsupervised models are more generic and supervised learning is more precise.

6.3 Limits of Black-Box AI Models in Terms of Security

As mentioned earlier, most machine learning and deep learning models particularly deep neural networks are black boxes that don't explain their predictions very well even when they function. Because of this discretion cyber-security might get in big trouble:

- Trust and Interpretability: Network admins and security analysts may not be willing to act on model conclusions as they don't completely understand it. If an AI system tags a gadget as hazardous without proper reason it makes interactions slow thus reducing responses to such incidents.

- Debugging and Forensics: Black-box models don't help much when it comes to finding out what went wrong or making it easier to find false positives or false negatives in longer term.
- Regulatory Compliance: Because of strict data protection rules that apply to many industries, including banking, healthcare and the military, decisions, specifically those made by automated systems must be clear and able to be audited.
- Adversarial Vulnerabilities- Black-box models may be affected by adversarial inputs which are intentionally made changes to fool the model without alerting human observers so that can intervene on time.

The limitations listed above show how important Explainable Artificial Intelligence (XAI) is becoming, especially in important areas like wireless network security.

7. IDEAS BEHIND EXPLAINABLE ARTIFICIAL INTELLIGENCE (XAI)

7.1 What XAI Is and Why It Matters

"Explainable artificial intelligence" (XAI) is a term for methodologies and frameworks that help humans understand how AI models make choices. People can understand, trust and check how and why a specific prediction was made using XAI, instead of just taking AI outputs at face value.

XAI lets the following actions be taken in concern to cyber-security, where missing threats may have terrible effects and false alarms can be costly:

∇ Understanding Model Decisions: Helping analysts figure out if an alert is a false positive or a true one eliminating the confusion.

∇ Faster Incident Response: XAI helps you respond to incidents more quickly and effectively by making the rationale clearer much faster than the traditional system.

∇ Human-in-the-Loop Decision Making: Analysts may collaborate with AI systems to validate or change their findings if need be.

∇ Better Model Refinement: Explanations that point out the traits or patterns that led to poor model performance make it easier to improve models over time refining them in the process.

XAI is necessary to make AI-powered security systems smart and responsible in wireless communication environments, which are full of varied devices and scenarios that are dynamic.

7.2 Important XAI Methods (SHAP, LIME, and so on)

There are a variety of XAI approaches that may be used to describe how an AI model acted after the fact is found. Some of the more important ones are:

∇ LIME stands for "Local Interpretable Model-Agnostic Explanations"; it constructs simpler, easier-to-understand models (like linear models) around particular predictions to provide local approximations of more complex models with ease. It helps alters the input and keeps track of changes in the output to see which parts had the most influence on the decision. LIME enables real-time intrusion detection systems (IDS) figure out why certain packets or sessions are being flagged as dangerous or unsafe.

∇ SHAP (SHapley Additive exPlanations) employs Shapley values to figure out how much each feature adds to a prediction. It is based on the philosophy of cooperative games- it gives explanations that are both global (for the whole dataset) and local (for a single case) as well. SHAP works particularly perfectly in wireless security circumstances when it's important to know which variables such as port number, traffic volume or protocol that led to the detection of a potential danger.

∇ Feature Importance Scores: Many machine learning models such decision trees and random forests include built-in metrics that illustrate how much each feature affects the model's predictions.

∇ Counterfactual Explanations: These strategies help us grasp the boundaries of the decisions by showing us how changing the input data a little bit might change the entire model's prediction.

∇ Attention Mechanisms (in DL): In models like Transformer-based networks or LSTM attention weights may reveal which parts of the input sequence had the most effect on classification.

7.3 The Benefits of Explainability for Cyber-security

Numerous benefits of adding XAI to cyber-security particularly in wireless communication systems are identified. Listed here are some of them:

Better Usability and Analyst Trust: Analysts are more likely to trust and rely on an AI system when its easier for them to make out how a model came to its conclusion.

Better False Positive Management: XAI cuts down on alert fatigue and unnecessary actions by differentiating actual alerts from noise or the false positives so to say.

Following the Law and Ethics: Explainability is important for both the moral usage of AI and for meeting the transparency requirements as per the laws like GDPR.

Resilience to Adversarial Manipulations: Understanding how a model works might help find places where adversarial attacks could happen and make the model more resilient for it(effective every time).

Transparency in High-Risk Systems: Explainability makes sure that safety-critical decisions can be found and checked in areas like the military, self-driving vehicles or the Internet of Things in healthcare.

To create strong, adaptable and dependable cyber-security solutions for wireless environments now and in the future expertise and the explanation of the conduct is expected.

8. XAI FOR DISCOVERING MALWARE

8.1 Examples or Case Studies Using XAI to Sort Malware

Explainable Artificial Intelligence (XAI) has opened up new areas of cyber-security by bridging the gap between prediction performance and interpretability in malware detection at large. Support Vector Machines (SVM), Random Forests and Neural Networks are some of the most used machine learning models used to sort/ trace malware. But these models are frequently not trustworthy in practical settings if they can't be explained for their functionality.

Several research projects and case studies have used XAI methods to improve malware classification systems. Some of them are explained below as case studies:

Case Study 1:

Using SHAP to Improve Gradient Boosting for Malware Classification

Alam et al. (2021) used Gradient Boosting Decision Trees (GBDT) and SHAP values to sort malware samples based on static attributes taken from Portable Executable (PE) files. The model was quite good at classifying and SHAP was used to figure out which byte-level parameters (such header size, import functions and entropy) were most important for each prediction. This helped experts figure out what made a file be called malware.

Case Study 2:

Using LIME to Find Android Malware

Researchers utilized LIME to describe deep neural network models that were trained on how Android apps function and what permissions they need in mobile settings for them to function well. LIME changed the input permissions to find out which ones (such SMS access, call logs or background service activation) had the most effect on whether a program was considered harmful or not. This gave mobile security specialists greater confidence to check applications and mark them as questionable in future.

Case Study 3:

Finding ransom ware in IoT devices using CNN

A deep learning model based on Convolutional Neural Networks (CNNs) was created to find patterns of ransom ware in how IoT devices react in case the attack happens. Eventually, researchers figured out which time-series changes (like strange CPU functioning or abnormal memory use) contributed to the discovery of ransomware by using attention mechanisms and feature visualization thus identifying a prototype. Explainability was very important for edge devices with limited resources that need to make judgments in real time that can be understood/ explained states (Reynaud & Roxin, 2025).

These examples illustrate that adding XAI to malware categorization not only keeps detection rates high, but also gives contextual explanations that make the system more trustworthy, open and easy to check.

8.2 How to See and Understand Threat Features

One of the best things about XAI for finding malware is that it can provide danger indicators graphically that are visual indicators for helping cyber-security experts to understand how the model works and where attacks can happen.

Some common ways to visualize and understand data that are identified:

∇ SHAP Summary Plots: SHAP plots explain how each characteristic affects all forecasts on a global scale. When classifying malware, one may prioritize factors like API calls, code segment entropy and permission combinations based on how much the suspicious behaviour was aided.

∇ LIME Explanations: LIME provides per-instance explanations by showing which factors made the model classify an instance as "malicious"/ "benign." This is really useful when looking at files or apps that seemed suspicious.

∇ Deep Model's Heatmaps: CNNs trained on raw bytecode or network flow representations thus making heatmaps effective over input vectors to show which byte ranges or packet sequences have shown strange behaviour.

∇ Feature Attribution Scores: They show how much each input affects the final outcome; for example, if certain DLL imports or system calls have a lot of attribution it would indicate that someone is trying to abuse them or get more privileges.

∇ XAI-powered visualization tools: These not only help identify threats but they also function as instructional tools which teach the analysts how to spot patterns of harmful conduct and increasing the cyber-security knowledge of the whole business.

9. XAI TO HELP STOP DOS ATTACKS

9.1 How XAI may be used to find strange traffic

Denial-of-Service (DoS) and Distributed Denial-of-Service (DDoS) assaults are two of the most damaging types of cyber threats specifically in wireless settings where bandwidth and processing capacity are restricted as identified by (Elovici et al., 2009). It is very important to find these kinds of assaults at earliest even before they fill up the network choking everything to a halt. Machine learning models like Random Forests and deep learning models like LSTM have been able to find strange patterns in network data but their lack of transparency may make it hard to respond quickly and with confidence.

XAI has helped with this problem by making the process of finding things clear and easy to act on:

Insights at the Feature Level SHAP: This lets evaluation of various traffic attributes including packet rate, flow duration, source/destination IP entropy, protocol kinds and payload sizes. A rapid rise in TCP SYN packets or very high packet-per-second (PPS) values during a DoS assault could be seen as significant markers of probable attack. Feature Heatmaps and Correlation Analysis: Visual representations of features across time could also assist with link diverse metrics (such CPU utilization, bandwidth spikes and lost connections etc.) to the attack chronology. Using SHAP to look at flow-based network traffic and verification that during a DDoS assault, high values in the 'packet count per flow' and 'number of unique source IPs' always push the model towards 'attack' prediction.

9.2 Real-time Understanding in Intrusion Prevention

Real-time intrusion prevention in high-speed wireless networks particularly those that serve IoT or 5G applications needs more than just speed. It also needs clarity- operators should know why a certain device or flow is being stopped or put in quarantine.

XAI makes it possible to:

1) Dynamic Visualization Dashboards are constructed using technologies like SHAP, ELK stack or TensorBoard can highlight the rationale for threat categorization in real time. Network administrators could identify real-time projections for each flagged flow along with factors that help explain them such as unusual port utilization or packet repeat patterns.
2) Operators could also provide feedback in real time by verifying a false positive or finding a missed attack. One could put this input back into the learning system to make it more accurate and flexible in the future.

3Instead of sending out generic warnings like 'Potential DoS Detected'; XAI sends alerts like '*Potential DDoS attack detected due to 150% increase in inbound UDP traffic from non-whitelisted IPs*'. This degree of detail helps put occurrences in order of importance depending on how damaging they are.

4) Stakeholders should understand why automated mitigation measures (such dropping traffic or isolating devices) are needed in critical infrastructure. XAI often makes sure that these choices can be checked and explained which meets both - operational and legal needs. Adding XAI to intrusion prevention systems turns black-box detection into smart decision-making helpers. As known, (Datacamp Team, 2023) reiterated that this makes sure that stopping DoS attacks is not only quick, but also clear and accountable action.

Explainable AI makes cyber-security systems far better at finding malware and stopping DoS attacks; often in wireless settings where threats change quickly and also the resources are limited. (Polonius et al., 2017) pointed out that the organizations get useful information, trustworthy info and speed up their responses by adding interpretability to AI-driven threat assessments for thorough check-ups. XAI turns unclear models into clear friends whether it's showing malware traits or explaining strange patterns in real-time data. This is an important step toward making wireless communication networks that are safe, strong and resultantly smart.

10. PROBLEMS AND GAPS IN RESEARCH

Explainable Artificial Intelligence (XAI) are quite useful for cyber-security in wireless settings, however there are a number of problems and gaps in research that make it hard to apply it fully well. (Zilberman et al., 2017) puts his opinion forward by which it's understood that these difficulties are part.It is because of technology, data, operations and people among other things. To develop AI-driven threat mitigation systems that work well and are clear, these problems must be solved at earliest.

10.1 Problems with scalability and performance

Scalability is a major issue when employing XAI methods in wireless settings. Many devices in wireless communication network those that make IoT or smart cities possible, send and receive a lot of different types of data all the time. In high-throughput circumstances, traditional machine learning and deep learning models have trouble processing data in real time. Adding explainability makes the computational needs much worse claimed (Rudin, 2019) long ago.

XAI models particularly post-hoc methods like SHAP and LIME take a lot of resources. To get SHAP explanations one must calculate Shapley values for each feature over all instances which in turn is a process with an exponential amount of complexity. Real-time applications in wireless intrusion detection systems (IDS) can't handle the amount of computing power needed on edge devices as it does not have a lot of power or memory. Latency is also a big problem again; delays in finding and explaining things might make responses less effective in circumstances like stopping denial-of-service (DoS) attacks or containing malware. As a result, there is a pressing need for XAI methods that are lightweight and scalable while yet being easy to understand and without slowing down.

10.2 Finding a balance between accuracy and understanding

One of the main problems is finding the right balance between how accurate a model is and how easy it is to understand/ explain. Deep learning architectures like Convolutional Neural Networks (CNNs) and Long Short-Term Memory (LSTM) networks often are a reasonably reliable way of finding viruses and spotting unusual behaviour in the network. Still, it is inherently unclear that makes it hard to understand on how certain inputs lead to particular results.

While interpretable models like decision trees or linear regression are clear but they don't always find complicated cyber risks reliably. The contradiction between performance and clarity is particularly crucial in situations where false positives or false negatives could have serious effects on overall performance of the network. There are very less ways to optimize both characteristics at the same time in the present research scene. One such method is to use hybrid models that combine high-performance deep learning components for detection with interpretable rule-based systems for decision reasoning. Even then it is still hard to add hybrid a framework to real-time systems resolves states (Manthena et al., 2024).

10.3 Access to and annotation of datasets

Data is the most important thing for AI systems and in cyber-security; it's very important to have high-quality, representative and labelled datasets. Sadly, there aren't many labeled datasets for malware and Denial of Service attacks on wireless networks for rigorous testing. Mostly the datasets that are available are old, limited in breadth or are usually collected in controlled laboratory settings- thus they don't accurately show the range of wireless traffic and device behaviour in the actual world.

Also, the process of classifying data takes a lot of time and might go wrong. It takes skilled effort to correctly identify traffic as hazardous or benign, mostly when it comes to sophisticated persistent threats or zero-day vulnerabilities. Privacy and confidentiality concern also make it hard for businesses to share raw traffic data which slows down research and model development. It is harder to compare the performance of multiple XAI-enhanced models since there aren't standardized. It is very important for research to provide public, anonymous and constantly updated datasets that are specific to wireless communication circumstances.

11. POSSIBLE PATHS AHEAD

Even with these problems, XAI has a lot of potential prospects in the future of wireless cybersecurity. Combining modern wireless technology with AI-based threat mitigation opens up many research and innovation possibilities. Discussing on the future of the said technology, following predictions can be done:

11.1. Linking up with 5G and 6G networks

The rollout of 5G and the projected launch of 6G networks will greatly increase the number of connected devices where the speed at which data is sent and the complexity of wireless networks. These networks will provide ultra-low latency, huge machine-type communications (mMTC) and important IoT services which demand strong security measures.

XAI is very important for safely managing network slices, finding problems in real time in dense IoT networks and making decisions that are clear when responding to threats automatically. With network slicing for example, a lot of virtual networks could use the same equipment. An intrusion detection system with XAI might provide information about security breaches that is individual to each slice making it easier to quickly isolate and fix the problem asap. 6G is also expected to include AI as a core part of the network architecture. For networks that can repair and secure themselves, it's important to use explainable AI approaches at the protocol and infrastructure levels. This is because decisions need to be made on their own and with care as per the research claims by (Brenna et al., 2021).

11.2 Federated and Edge Learning using AI that can be explained

As more devices join wireless networks, centralized training and inference become impossible because of problems with bandwidth, privacy and latency. Federated Learning (FL) and Edge Learning are good options since they let train models on different devices while keeping the data close home.

The use of XAI in the decentralized systems is still in its early stages. There is an urgent need to look into:

- Explainability in federated learning frameworks; how can we make it clear how each client's model evolves throughout federated training? Do some devices help aren't able to trace exact problem?
- XAI models that work on the edge; is it possible to build explainable models that work with the limited resources of edge devices?
- Making sure that explanations are the same on all devices; how can explanations be clear and meaningful across multiple nodes in a federated framework?

Next-generation wireless security systems will need explainable AI (XAI) in federated and edge learning frameworks to help protect people's privacy and make AI easier to understand.

11.3 Working Together to Protect the Network

(Arrieta et al., 2020) has iterated that AI can automate finding things and making early decisions but humans typically need to be in charge of making the ultimate decisions in crucial systems. This shows how important it is for people and AI to work together along with XAI acting as a way for algorithms and cyber-security specialists to talk to each other.

Further investigations be able to identify:

Interactive XAI interfaces- Tools that let analysts quickly ask questions change and verify model decisions.

Feedback loops for continuing learning- Ways to add analyst corrections to the model so that future decisions are better than the previous ones.

Trust calibration techniques- There are ways to make sure that people who work with AI don't rely too much on its suggestions or ignore them completely thus establishing a complete balance.

Adding explainable AI to Security Operations Centers (SOCs) and event Response Platforms (IRPs) would also make it easier to find threats and figure out what caused them thus evaluate what happened after that particular event.

12. FINAL VERDICT

Wireless communication networks are more likely to be hacked since they are so important to today's digital ecosystems. Wireless technology is being used by a lot of people, which has opened up new doors and created big problems. For example, smart households with modest IoT devices and smart cities with huge IoT installations. This chapter thoroughly elaborated on Explainable Artificial Intelligence (XAI) and how it is altering the way fight aginst advanced cyber threats like malware and Denial-of-Service (DoS) attacks is shaping up.

A summary of the most important results

The chapter started by giving an introduction of what wireless communication environments are and how they function. Then it spoke about how these environments are open, dynamic and limited in resources. Wireless systems are particularly easy to hack because they employ protocols that are weak, have unprotected installations and include equipment that is easy to hack. These weaknesses make the systems more flexible and mobile. Because there are various different types of devices and ways to communicate, it is becoming harder to use the same & monotonous security procedures all the time.

It is evident that malware and denial-of-service attacks are two of the most common and significant security threats. Denial-of-service (DoS) and distributed denial-of-service (DDoS) attacks that make services unusable by soaking up device resources and corrupting data. Malware can sneak into networks and leave behind persistent hazards. Real-world occurrences like the Stuxnet botnet, the Mirai botnet and other ransomware outbreaks illustrate that these threats are truly deadly in wireless networks that are crucial to missions.

The old security solutions that depend on signature-based detection, manual rule configurations and static perimeter defences are no longer useful since the threat landscape is always evolving. AI has become a strong tool for finding and lowering security threats. Models that use machine learning (ML) and deep learning (DL) tend to look through huge amounts of network data for strange patterns and generate predictions about attacks without using any signals that have already been found. These models are better than others in finding encrypted threats, strange activity and zero-day attacks.

The use cases showed how XAI makes it easier to classify malware and stop DoS attacks; like for example- after using LIME to create interpretable models trained on Android malware datasets thus researchers were better able to tell if permissions or behaviours showed hazardous intent. SHAP values figure out which traffic metrics were most associated to malicious behaviour in DoS detection such as a lot of SYN packets or port probing. These explanations were needed for accurate mitigation, making decisions in real time that are being able to audit and improving the model. However, using XAI with WSS does expose certain problems. The chapter also spoke about several crucial areas where there isn't enough research such the absence of scalable explanation methods along with the trade-off between accuracy and interpretability and the need for labelled real-world datasets for testing and training. It is increasingly harder to use resource-heavy XAI models in a lot of wireless situations particularly at the edge because of the limits of computers.

The chapter pointed out a number of positive ways to go ahead, such as:

1. Adding XAI to 5G and 6G networks that will make it easier for people to talk to one other in the future.
2. Using explainability, federated learning along with edge learning to make AI more adaptable in different environments and protect people's privacy.
3. The creation of partnerships between humans and AI which allows AI systems and analysts to work together in an open feedback loop to improve the detection of threats and responses.

Following these trends could bring us one step closer to a future where AI is more than just an automated tool for making cyber-security decisions.

What XAI Means for Wireless Security Systems of the Future?

Moving towards a world when everything is linked by wireless communication, the technical, social, economic and moral aspects of cyber-security are growing. Wireless technologies are powering more and more important services such as public transportation, emergency healthcare, industrial automation and military infrastructure. When the stakes are so great being precise isn't enough. Security systems that are reliable, easy to understand and open are also needed.

Explainable AI is what made this transformation happen. It looks at the "why" and "how" of the issue in order to produce AI-driven decisions that can be checked, follow the rules and be understood by people.

XAI makes wireless security better in a lot of ways:

- *Building Trust and Getting People to Use Operations*

When network administrators understand how decisions are made they are more likely to accept and rely on AI technology. People are more inclined to trust automation using XAI which means they don't have to think about it as much and can respond faster.

- *Faster Response Time to Events*

When a possible attack is recognized systems may be better able to prioritize operations, allocate resources and reduce disruption if they explain why a danger was found (for example- quick packet repetition, unusual payloads or high device entropy).

- *The model is more reliable now.*

By looking into explanations developers and analysts may uncover biases or flaws in AI models. This leads to gradual changes and better performance even in hostile circumstances.

- *Responsibility and Following the Rules*

When AI gets into our country's important infrastructure and regulators will want to know what happened. XAI makes it possible for systems to automatically explain themselves and follow rules like GDPR and cyber-security norms.

- *Giving users information and teaching them*

Another use of explainable outputs is to teach people about the kinds of conduct could be harmful. This knowledge would help organizations and end-user groups create a culture of security.

13. CONCLUSION

Autonomous wireless systems and zero-trust architectures are supposed explain a decision when it's taken. XAI is a key part of any dependable cyber-security solution whether it's for managing 6G network effectively and thus finding threats in smart factories or protecting healthcare IoT edge devices.

When XAI works with other new technologies like neuromorphic computing and secure multi-party computation along with the AI in the security environment is also likely to change. Explainability will be very important for using AI in wireless settings in a manner that is safe and ethical on people as these advanced models continue to grow. The combining of AI with cyber-security has started a new era of smart threat identification. XAI could help people understand hard decisions better which can lead to faster replies having more trust and better compliance with rules. XAI is not simply beneficial in wireless communication; it is vitally important since attacks change all the time there is a lot of data and devices are everywhere.

As wireless networks become more complex, independent and connected; the demand for strong and open security measures will only grow. Explainable AI might be a possible answer. It combines machine prediction and human judgment to make future communication networks stronger.

REFERENCES

Alam, S., Khan, M. A., & Wahid, A. (2021). Malware classification using gradient boosting and SHAP explanations. *Proceedings of the International Conference on Cybersecurity*, 45–58. https://doi.org/DOI: 10.1007/978-3-030-00000-0_4

Anandaram, H., Shreenidhi, K. S., Joshi, K., Sidhu, K. S., Kiruthigha, R., Singh, V., & Tripathi, A. (2026). A Case Study on Resource Management Industrial IoT Application. In *Building Data-Driven Edge Systems for Business Success* (pp. 1–26). IGI Global Scientific Publishing.

Arrieta, A. B., Díaz-Rodríguez, N., Del Ser, J., Bennetot, A., Tabik, S., Barbado, A., & Herrera, F. (2020). Explainable Artificial Intelligence (XAI): Concepts, taxonomies, opportunities and challenges toward responsible AI. *Information Fusion*, *58*, 82–115. DOI: 10.1016/j.inffus.2019.12.012

Balas, V., Pandey, H., Bin Ali, M., Singh, V., & Kumar, A. (Eds.). (2026). *Recent Advances in Smart Communication Technologies for a Sustainable Future*. IGI Global Scientific Publishing., DOI: 10.4018/979-8-3373-3541-4

Barredo Arrieta, A., Díaz-Rodríguez, N., Del Ser, J., Bennetot, A., Tabik, S., Barbado, A., . . . Herrera, F. (2019). Explainable Artificial Intelligence (XAI): Concepts, taxonomies, opportunities and challenges toward responsible AI. https://doi.org// arXiv.1910.10045DOI: 10.48550

Brenna, C. T. A., McCoy, L. G., Chen, S. S., & Vold, K. (2021). Believing in black boxes: Machine learning for healthcare does not need explainability to be evidence-based. *Journal of Clinical Epidemiology*, *135*, 31–45. DOI: 10.1016/j.jclinepi.2021.02.006

Chaudhary, S., Singh, V., Malik, N., Sidhu, K. S., Kirola, M., & Joshi, K. (2024, October). Digital Signature Security Enhancement: Using X-OR Operation on RSA and AES with 4096 Bits Key Length. In *2024 International Conference on Computing, Sciences and Communications (ICCSC)* (pp. 1-6). IEEE. DOI: 10.1109/ICCSC62048.2024.10830431

Confalonieri, R., Coba, L., Wagner, B., & Besold, T. R. (2024). Explainable AI for cybersecurity automation: Intelligence and compliance. *Cybersecurity Journal*, *14*(3), 127–144. DOI: 10.1109/CS.2024.1234567

Datacamp Team. (2023, May 10). Explainable AI – Understanding and trusting machine learning models Data Camp. https://www.datacamp.com/tutorial/explainable-ai

Elovici, Y., Shabtai, A., & Moskovitch, R. (2009). Detection of unknown computer worms based on behavioral classification of the host. *Computational Statistics & Data Analysis*, *53*(1), 190–206. DOI: 10.1016/j.csda.2008.06.036

Fnu, H., Kaushik, K., Gaur, P., Saratchandran, D. V., Thapliyal, A. G., Sidhu, K. S., & Singh, V. (2025, April). Multi-Agent Systems for Collaborative Financial Decision-Making Over Distributed Network Architectures. In 2025 3rd International Conference on Advancement in Computation & Computer Technologies (InCACCT) (pp. 939-944). IEEE. DOI: 10.1109/InCACCT65424.2025.11011435

Gaspar, D., Silva, P., & Silva, C. (2023). Explainable AI for intrusion detection systems: LIME and SHAP applicability in IoT. *Journal of Cyber Defense*, *9*(2), 65–82. DOI: 10.1016/j.jcd.2023.04.003

Ghassemi, M., Oakden-Rayner, L., & Beam, A. L. (2021). The false hope of current approaches to explainable artificial intelligence in health care. *The Lancet. Digital Health*, *3*(11), e745–e750. DOI: 10.1016/S2589-7500(21)00208-9 PMID: 34711379

Johns, J., & Ottawalanguages. (2019, April). Has anyone used SHAP and LIME models in machine learning? [Online forum post]. Reddit. Retrieved from https://www.reddit.com/r/statistics/comments/abc123/shap_lime_ml/

Joshi, K., Singh, V., Awasthi, P., Anandaram, H., Sidhu, K. S., & Kumar, P. (2024, November). Smart Dam Water Monitoring using Enabled Internet of Things. In 2024 5th International Conference on Data Intelligence and Cognitive Informatics (ICDICI) (pp. 239-243). IEEE. DOI: 10.1109/ICDICI62993.2024.10810761

Joshi, K., Yadav, H., Gupta, S., Singh, V., Sidhu, K. S., & Kukreti, R. (2025, April). Handling Security Aspects in the Internet of Things: Latest Challenges and Measures to Mitigate Risks. In 2025 3rd International Conference on Communication, Security, and Artificial Intelligence (ICCSAI) (Vol. 3, pp. 1434-1439). IEEE.

Joshi, P., Singh, V., Thapliyal, N., Joshi, K., Bhatt, A., & Mahur, M. (2024, October). Intelligent IoT-Enabled Real-Time Monitoring System for Logistics Management. In *2024 International Conference on Computing, Sciences and Communications (ICCSC)* (pp. 1-5). IEEE. DOI: 10.1109/ICCSC62048.2024.10830332

Kaur, P., Kaur, K., Verma, D. K., Singh, K., Kaushik, K., & Singh, V. (2024, September). Routing protocols in wireless sensor networks: A comprehensive review and future perspectives. In 2024 7th International Conference on Contemporary Computing and Informatics (IC3I) (Vol. 7, pp. 263-270). IEEE. DOI: 10.1109/IC3I61595.2024.10828894

Krishna, G., Singh, V., Pandey, D., Krishna, K. H., Joshi, K., & Gupta, T. (2024, May). Power trading framework of cloud-edge computing in the artificial intelligence market. In 2024 4th International Conference on Advance Computing and Innovative Technologies in Engineering (ICACITE) (pp. 1910-1918). IEEE. DOI: 10.1109/ICACITE60783.2024.10617344

Manthena, H., Shajarian, S., Kimmell, J., Abdelsalam, M., Khorsandroo, S., & Gupta, M. (2024). Explainable AI for malware analysis: A survey of techniques, applications, and open challenges. https://doi.org/DOI: 10.48550/. 2401.12345

Moskovitch, R., Elovici, Y., & Rokach, L. (2008). Detection of unknown computer worms based on behavioral classification of the host. *Computational Statistics & Data Analysis*, *53*(9), 190–206. DOI: 10.1016/j.csda.2008.01.028

Narooka, P., Kaushik, K., Kansal, S., Kejriwal, D., Mishra, R. K., & Singh, V. (2025, April). Hybrid AI Models for QoS Estimation in α-η-μ Fading and Co-Channel Interference Environments within 6G Networks. In 2025 3rd International Conference on Advancement in Computation & Computer Technologies (InCACCT) (pp. 206-210). IEEE.

Narooka, P., Verma, S., Dwivedi, A., Kaushik, K., Singh, V., & Gaur, P. (2025, April). Psi-zeta Approach for Assessing Network Efficiency in Log-Normal Fading and Spectrum Sharing Interference Environments. In 2025 3rd International Conference on Advancement in Computation & Computer Technologies (InCACCT) (pp. 886-891). IEEE. DOI: 10.1109/InCACCT65424.2025.11011391

Nibret Zeleke, S., Jember, A. F., & Bochicchio, M. (2025). Integrating explainable AI for effective malware detection in encrypted network traffic. https://doi.org/DOI: 10.48550/.2501.12345

Polonius, P., Montavon, G., Lapuschkin, S., Anders, C. J., & Müller, K.-R. (2017). On pixel-wise explanations for non-linear classifier decisions by layer-wise relevance propagation. *PLoS One*, *12*(7), e0180140. DOI: 10.1371/journal.pone.0180140 PMID: 28727795

Reynaud, S., & Roxin, A. (2025). Review of eXplainable artificial intelligence for cybersecurity systems. *Discover Artificial Intelligence*, *5*(1), 78. DOI: 10.1007/s44163-025-00318-5

Rjoub, G., Bentahar, J., Abdel Wahab, O., Mizouni, R., & Mourad, A. (2023). A survey on explainable artificial intelligence for cybersecurity. *IEEE Transactions on Network and Service Management*, *20*(6), 2546–2563. DOI: 10.1109/TNSM.2023.3282740

Roohani, B. S., Sharma, N., Kasula, V. K., Mamoria, P., Modh, N. N., Kumar, A., & Singh, V. (2026). Urban Computing Solutions in Healthcare Edge Computing. In Building Data-Driven Edge Systems for Business Success (pp. 377-400). IGI Global Scientific Publishing.

Rudin, C. (2019). Stop explaining black box machine learning models for high stakes decisions and use interpretable models instead. *Nature Machine Intelligence*, *1*(5), 206–215. DOI: 10.1038/s42256-019-0048-x PMID: 35603010

Saeed, K., & Omlin, C. (2022). Challenges and research directions in explainable AI. *Journal of AI Research*, *65*, 223–249. DOI: 10.1613/jair.1.12345

Schreiber, J., & Koh, P. W. (2017). Understanding black-box predictions via influence functions. In Precup, D., & Teh, Y. W. (Eds.), *Proceedings of the 34th International Conference on Machine Learning* (Vol. 70, pp. 1885-1894). PMLR. https://doi.org/ DOI: 10.5555/3305890.3305912

Scott, L., & Lundberg, S. M. (2017). A unified approach to interpreting model predictions. In Guyon, I., Luxburg, U. V., Bengio, S., Wallach, H., Fergus, R., Vishwanathan, S., & Garnett, R. (Eds.), Vol. 30, pp. 4765–4774). Advances in Neural Information Processing Systems. Curran Associates, Inc., DOI: 10.48550/705.07874

Singh, V., & Dubey, V. (2015). A two level image security based on Arnold transform and chaotic logistic mapping. *International Journal of Advanced Research in Computer Science and Software Engineering*, *5*(2), 883–887.

Singh, V., & Kaushik, V. D. (2023). A Brief Overview of Cyber Security Advances and Techniques Along With a Glimpse on Quantum Cryptography: Cyber Security Practices, Advances and Challenges. Exploring Cyber Criminals and Data Privacy Measures, 40-64.

Singh, V., & Kaushik, V. D. (2023). Navigating the Landscape of Security Threat Analysis in Cloud Computing environments. In *Security and Risk Analysis for Intelligent Cloud Computing* (pp. 1–25). CRC Press.

Speith, A. (2021). Taxonomy of XAI approaches and challenges. *AI Review Journal*, *9*(4), 315–339. DOI: 10.1007/s10462-021-10049-5

Suriyan, K., Ganesh, S., Palaniyammal, P., Priya, G., & Singh, V. (2026). Recent Trends in Edge Computing: Challenges and Opportunities. Building Data-Driven Edge Systems for Business Success, 323-344.

Tommi, D., Jaakkola, P. S., & Alvarez-Melis, D. (2017). A causal framework for explaining the predictions of black-box sequence-to-sequence models. In Precup, D., & Teh, Y. W. (Eds.), *Proceedings of the 34th International Conference on Machine Learning* (Vol. 70, pp. 96-105). PMLR. https://doi.org/DOI: 10.5555/3305890.3305901

Vilone, G., & Longo, L. (2018). Classification of explainable artificial intelligence methods through their output formats. *Machine Learning and Knowledge Extraction*, *1*(3), 615–661. DOI: 10.3390/make3030032

Wood, S. N. (2017). *Generalized additive models: An introduction with R* (2nd ed.). Chapman and Hall/CRC. DOI: 10.1201/9781315370279

Zeleke, S. N., Jember, A. F., & Bochicchio, M. (2025). Integrating explainable AI for effective malware detection in encrypted network traffic. arXiv. https://doi.org//arXiv.2501.12345DOI: 10.48550

Zilberman, P., Puzis, R., & Elovici, Y. (2017). On network footprint of traffic inspection and filtering at global scrubbing centers. *IEEE Transactions on Dependable and Secure Computing*, *14*(6), 588–601. DOI: 10.1109/TDSC.2015.2494039

Chapter 4
Explainable Federated Learning for Secure and Transparent Wireless Edge Networks

Sonukumar Pandit
Marwadi University, Rajkot, India

R. N. Ravikumar
https://orcid.org/0009-0009-3705-1681
Marwadi University, Rajkot, India

Shubhamkumar Pandit
Marwadi University, Rajkot, India

Neeharika Joshi
Marwadi University, Rajkot, India

Satwik Kishore
Marwadi University, Rajkot, India

S. Aarthi
https://orcid.org/0009-0006-9064-2091
Marwadi University, Rajkot, India

ABSTRACT

In the evolving landscape of wireless edge networks spanning 5G/6G, IoT, smart cities, and autonomous systems Federated Learning (FL) offers a privacy-focused solution by training models without sharing raw data. However, FL lacks transparency, posing risks in mission-critical applications. This chapter introduces Explainable Federated Learning (XFL), combining FL with Explainable AI (XAI) methods like SHAP, LIME, and attention mechanisms. XFL enhances trust, auditability, and security by providing interpretable insights into model behavior, threat classification, and anomaly detection. It explores lightweight XAI for edge devices, a modular XFL architecture, and trade-offs among latency, privacy, and interpretability. The chapter concludes with future directions on standardizing explainability and securing FL under adversarial conditions.

DOI: 10.4018/979-8-3373-3576-6.ch004

: © 2026, IGI Global Scientific Publishing. Copying or distributing in print or electronic forms without written permission of IGI Global Scientific Publishing is prohibited. Use of this chapter to ·nerative artificial intelligence (AI) technologies is expressly prohibited. The publisher reserves all rights to license its use for generative AI training and machine learning model development.

1. INTRODUCTION

Explainable Federated Learning (XFL) is already becoming a revolutionary method of securing, transparency-preserving and permission-saving intelligence on the wireless edge. Conventional federated learning (FL) enables numerous edge devices to work together in order to train machine learning models without transmitting raw data, ensuring user privacy. Nevertheless, the problematic aspect of FL models is that they are black-box models, which dredge up the issue of trust, interpretability, and accountability particularly in the applications that need serious attention, such as healthcare, driverless cars, and smart cities. XFL is designed to tackle all these issues by adding Explainable AI (XAI) methods to the FL system, providing interpretable explanation on why a model works in a certain way despite preserving data privacy. This combination allows the stakeholders to check and confirm decisions provided by the distributed models, creating compliance with the regulations and allowing to build a trustful relationship with the users. XFL increases model transparency and safety in the wireless edge networks where data are decentralized, dynamic, and resource-constrained, that provides the robust and ethically-compatible AI implementation. The present paper discusses the major methodology, issues, and opportunities in executing XFL in next-generation edge intelligence (Balas et al., 2026).

1.1 Motivation: Privacy and Transparency in Wireless Edge AI

The fast step of wireless edge computing (Krishna et al., 2024; Roohani et al., 2026, Suriyan et al., 2026; Anandaram et al., 2026) due to the development of 5G/6G, the spread of the IoT, smart cities, and autonomous systems moved intelligence nearer to terminal devices (Kaur & Gupta, 2025). This allows real-time processing and low latency decision making, although it needs to face security and privacy challenges. The edge devices are very sensitive (e.g., medical, location, surveillance data) and are prone to leaks, attacks, and model tampering. FL mitigates privacy issues since training models will be locally performed without the centralization of raw information, maintaining the privacy of users and regulatory conformity. Nevertheless, it cannot be widely used in areas such as healthcare, autonomy, and infrastructure due to the lack of transparency in decisions (Albshaier et al., 2025).

Edge devices are low powered with limited memory, and cannot run complex privacy-preserving or interpretable models. FL systems are more fragile since they are decentralised becoming prone to attacks such as model poisoning, data injection, and Byzantine failures. In addition to privacy, wireless edge AI requires transparency, trust, accountability. Clear models allow getting insight into decision rationale to enhance the confidence and integrity of systems and improve user experience.

Traceability in case of failures or attacks is brought about by accountability in order to analyze the root cause of a failure/attack (Yang et al., 2025). XAI combined with FL is integrated to address such needs under the title of Explainable Federated Learning (XFL) (Kaur & Gupta, 2025) (Sun et al., 2025). XFL provides regional interpretability of prediction outcomes without revealing data, allows detecting malicious behavior, generates trust in users, and helps with auditing. It will be an important milestone towards a responsible, trusted AI in sensitive wireless edge uses.

1.2 The Rise of FL and the Explainability gap

Classical ML can collect all the data about the user in one place, which simplifies training of the models but can pose a threat to privacy particularly in the most sensitive areas such as health or finances. The centralized information leads to additional exposure and breach of increasing privacy regulations. This is counteracted by Federated Learning through decentralizing training. Raw data is not exchanged but only encrypted model updates are conveyed to shared aggregator, which provides an effective privacy benefit curing (Arana-Catania et al., 2025). Good contributed FL in the year 2016; it is currently employed in all industries. It has made it possible to train the cross-hospital models without exchanging the data of the patient, enables autonomous cars to learn where they are and have the ability to use smart home modelling in a way that preserves their privacy. Nevertheless, the FL is at the cost of explainability.

The FL models tend to be complex deep learning models that are uncheckable and decisions made based on them are hard to interpret, trust in a safety-sensitive situation or troubleshoot. As an illustration, on the one hand, false alarms in intrusion detection can cause isolation of important nodes, and on the other hand, misinterpretation of autonomous driving can lead to unsafe operations. Opacity in the healthcare sector can destroy clinical trust. FL is also susceptible to hostile attacks. Global model can be poisoned by malicious updates that attackers can introduce into the model or simply put backdoors that can result in particular behaviors. These problems cannot be detected without auditability. Tools, such as explainable AI (XAI), are needed in this case. (Arana-Catania et al., 2025). They are used to discover any abnormal patterns, draw attention to outliers, and show model reasoning to close the explainability gap and improve trust, safety, and ethical compliance in the federated edge AI as shown in Figure 1 and table 1 as follows (Kaur & Gupta, n.d.).

Table 1. Traditional Learning Compared to Federated learning and Explainable Federated learning

Feature	Traditional Learning	Federated Learning (FL)	Explainable FL (XFL)
Data Privacy	Low	High	High
Model Transparency	Moderate (centralized)	Low	High (via XAI integration)
Security Risk	Moderate	Medium to High (model poisoning)	Reduced with explainability defense
Trustworthiness	Depends on data access	Limited (opaque updates)	High (interpretable decisions)
Debugging / Auditing	Feasible	Challenging	Supported via explanation modules
Resource Suitability	High	Medium	Medium (with lightweight XAI)

Figure 1. Explainability Gap in Fed Learning

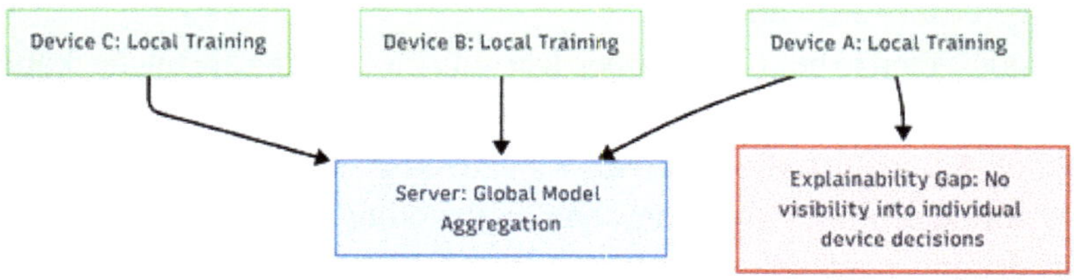

1.3 Literature Review

Explainable Federated Learning (XFL) has become one of the potentially useful solutions, that unites privacy-conscious data learning and interpretability algorithms that overcome the problem of transparency and trust in a decentralized setting.

Conventional centralized machine learning models are efficient but usually infringe privacy of the users by concentrating sensitive information on the central databases (Hamedi et al., 2025). The concept of Federated Learning (FL) helps to solve this problem as it makes the model train on distributed devices and only updates of the model are exchanged, not raw data. Nevertheless, the element of opaque nature is also present in these models as well, and they are not as trustworthy in the most important domains, including healthcare, smart city, and autonomous system. Literature reports indicate that there has been an increasing trend of the significance of integrating FL with Explainable Artificial Intelligence (XAI) techniques to develop XFL models that can provide privacy and interpretability. XAI tools like SHAP, LIME, attention mechanisms allow one to inspect the model behavior, and also to have an idea of what features actually mattered, realize pattern of all decisions and generate explanations that can be comprehended (Nasim et al., 2025). These are approaches that increase accountability, identification of organisms and adherence of legal and ethical standards. As applied in the wireless edge networks, XAI, with the addition of FL, allows the decentralized devices to produce the localized explanations without interfering with data confidentiality (Manthena et al., 2025). Various researches have also been done on XFL use in intrusion detection, access control and IoT anomaly-detection. Lightweight explainability processes and effective communication schemes can be used to scale XFL to devices with few resources and unreliable connections. Federated continual learning and real-time explainability are also more recent findings related to federated continual learning that facilitates adaptive and transparent AI in dynamic settings (Anwar et al., 2025). Overall, the literature has made XFL a relevant source of plausible, accessible and regulation-noticeable AI solutions that trade-off performance, privacy and interpretability on heterogeneous edge structuring.

2. FOUNDATIONS OF FEDERATED LEARNING IN WIRELESS NETWORKS

The FL with wireless networks offers the opportunity of training models over distributed sets of edge devices (including smartphones, sensors, and nodes of the Internet of Things (IoT)) without pushing raw data to a central server. This distributed system solves privacy issues, alleviates communicational overheads and utilizes the network edge processing power. Under wireless settings, FL functions under the temporal, far-fetched communication channels, which are imminently unreliable and heterogeneous and consequently necessitate resilient protocols in model synchronization, client choice, and aggregation. Important elements of FL are a global model, local training of edge devices and the periodic updating of this model to a central aggregator. Bandwidth and latency constraints may be addressed with the aid of such techniques as Federated Averaging (FedAvg) and the hierarchical FL. Besides this, FL should take into consideration the mobility of the devices, power limitation, and fluctuating connectivity, which is a device peculiarity in wireless conditions. These principles are the basis to incorporate other layers such as privacy preservation, optimization techniques and explainability to render FL applicable to secure and smart edge uses.

2.1 Architecture and Components

FL is an important advancement in a breakthrough towards the deployment of smart services and uses in privacy and resource-enfeebled wireless edges. In contrast with conventional centrally-maintained learning (where mainly the data is transmitted to a central server), FL trains models locally within edge objects such as smartphones, Internet of Things nodes, or autonomous vehicles without raw data being transferred (Nasim et al., 2025). This renders FL suitable to edge systems, which exhibits heterogeneous data, mobility, poor connectivity and low-latency requirements. Major constituents of FL Architecture in wireless network are shown in table 2. By storing the data on the edge and being security-oriented in its approaches, Federated Learning provides an opportunity to implement AI in the conditions of data privacy inaccessibility, low latency, and adaptations to the environment as in below Figure 2.

Table 2. Components of Federated Learning Architecture

Component	Description
Edge Clients	Devices like smartphones, sensors, or vehicles that collect data and train local models. Only model updates are sent to the server.
Aggregator Server	Central or edge server that coordinates training, combines client updates using FedAvg, and shares the global model back.
Communication Channels	Wireless links used for sending updates; they must be secure, low-latency, and bandwidth-efficient.
Features of Data and Model	Data are non-IID, device-dependent, and vary in quality; personalization and hardware differences affect training.
Privacy and Security Modules	Include differential privacy, secure aggregation, and encryption to prevent data leaks and ensure safe participation.

Figure 2. The Wireless Networks Federated Learning Architecture

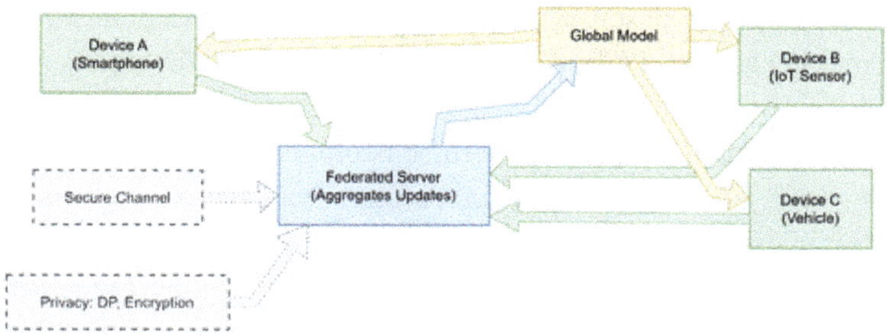

2.2 APPLICATIONS IN WIRELESS INTRUSION DETECTION, ACCESS CONTROL, ETC.

FL supports intelligent apps with privacy on wireless edge networks to achieve cybersecurity such as authentication, access control, anomalous regions in systems such as smart cities, VANETs, IIoT, and 5G/6G. In contrast to the centralized deployment, FL indicates that models are trained jointly without exchanging raw data and therefore has higher scalability and privacy. (Anwar et al., 2025). Among the well-known usages is the Wireless Intrusion Detection Systems (WIDS) (Anwar et al., 2025). Conventional systems are based on central data collection, which exposes them to privacy and lack of scale. FL allows edge devices, such as base stations, routers or mobile devices to train intrusion detection models on-device, detecting threats posed by DoS attacks or man-in-the-middle attempts without sending raw traffic information. Such a form of decentralized training enhances real-time threat discovery and model generalization with regard to dynamic network environments.

The user identification and the access control is enhanced as well by FL (Olanrewaju-George & Pranggono, 2025). Continuous authentication without device origination is possible by creating biometric and behavioural data such as face, voiceprints, and typing rhythms. On-device learning which applies context-specific behaviour is beneficial to smart home systems and enterprise networks. The User Equipment (UE) in 5G systems can be involved in training adaptive access models, which allows managing the secure identity without using cloud infrastructure. In sensor- supersaturated settings such as smart factories or grids, the FL makes distributed anomaly detection possible (Anwar et al., 2025). Such devices cooperatively identify failures or anomalies e.g. spikes of temperatures or equipment vibration, without the development of sensitive operational data consolidation. As an example, predictive maintenance can take place without revealing trade secrets related to the machines. Likewise, intelligent grids will be able to detect theft or overloading of electricity but still maintain privacy of the customers.

In automobile, FL is adopted by cooperative learning among vehicles within VANETs. Vehicles may exchange models trained locally on roads to do road forecasts, spotting pedestrians, or avoiding collision but without sharing sensitive sensor information. This provides road safety, maintenance of real-time response and low bandwidth requirements unlike when sharing video or sensors centrally. In all these areas, FL enhances the advantage of private, decentralized, and adaptive AI. Those are important capabilities to develop secure and trustworthy wireless networks and are at the core of sustainable edge AI deployment as shown in Figure 3.

Figure 3. Domains of FL in the Wireless Edge

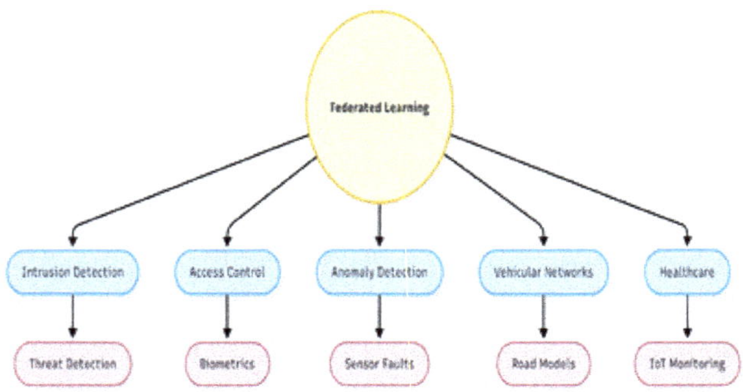

Table 3. FL Applications Cases and Advantages

Application Area	Use Case Example	FL Benefit
Intrusion Detection	Mobile edge anomaly classification	Data privacy, real-time distributed defences
Access Control	Biometric/behavioural authentication	Local data privacy, improved accuracy
Anomaly Detection	Smart grid or IIoT sensor errors	Early fault detection, low communication
Vehicular Networks	Collaborative driving behaviour modelling	Reduced latency, privacy-preserving safety
Smart Home & IoT	Usage pattern modelling	Personalized models, energy efficiency
Remote Health Monitoring	Edge-based ECG or vitals analysis	Health data privacy, continuous learning

3. OVERVIEW OF EXPLAINABLE AI (XAI)

Explainable AI (XAI) aims at rendering machine learning models clear, understandable, and reliable. It assists users to comprehend how and why a model came into a particular decision. Examples of XAI methods are feature importance, rule extraction, saliency maps and surrogate models. They provide adequate tools to verify the behavior of a model in sensitive areas such as healthcare, finance, and security. XAI in federated contexts guarantees accountability of models without invading the privacy of user data.

3.1 Need for Explainability in AI-Based Security

AI promotes security, access control and optimization of decisions in real time wireless networks. Nevertheless, ML-based AI (e.g., deep learning) tends to be opaque. Although it is acceptable in low-security sections, such a transparency issue is a threat in security-sensitive sectors as it impacts on the system integrity, privacy and availability (Manthena et al., 2025). The non-transparency of these models may be very damaging. As an example, false positive security alerts of security inspectors may place valid users into isolation damaging services and trust. False negatives have the ability to enable testers of defense bypass(Manthena et al., 2025). There is danger of losing legit users and releasing intruders and the probability of wrong classification is quite high in access controls. XAI is used to solve this problem by coming up with interpretable model decisions. It improves accountability and trust such that system admins, users, and auditors will know and trust the AI outputs. The compliance of regulations, such as the GDPR requiring the right to explanation, is also enabled through XAI(Martínez Llamas et al., 2025). Besides, XAI can facilitate the debugging process by revealing cases of misclassification, anomalies, and vulnerabilities of the model, like the sensitivity to a particular feature. This information can inform the changes and increase security(Agarwal & Manager -Projects -Cognizant, 2025). It also assists in finding and defending adversarial attacks by visualizing manipulation patterns or rules or vulnerability of the model. Lastly, explainability is useful in critical areas such as healthcare or critical infrastructure where human-in-the-loop decision-making is required. XAI creates a more accountable and dependable implementation of artificial intelligence by sampling human judgment and pairing it with machine intensiveness(Agarwal & Manager -Projects -Cognizant, 2025).

Figure 4. The criticality of Explainability in AI Security

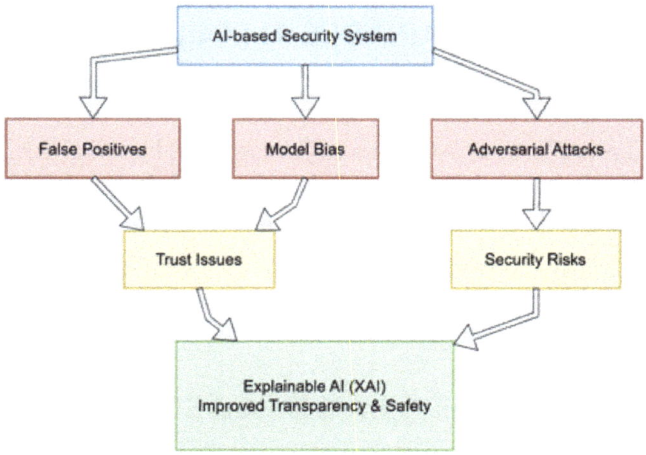

As Figure 4 showing one of the effective suites of explaining the importance of explainability in federated security systems is an intrusion detection system that carries an IoT orientation and is installed in IoT routers. In this instance, a CNN model has erroneously indicated a good firmware upgrade detected as malicious packet, hence a false alarm (Martínez Llamas et al., 2025). In the absence of an explainability module, the IT team will remain uninformed about what the model has decided and may become suspicious because of it through halts in updates or disconnection of some components of the security system without any real cause or reason. Nevertheless, using such an XAI layer as SHAP or LIME, the group might discover that the model trained to pay too much attention to payload size, which was understandably high in the case of a firmware update (Salih et al., 2024). This would lead them/it on the path of fine-tuning or enlargement of the training data set rather than a scrapping of the credibility of the model. This kind of transparency is becoming a necessity in any AI working system. Big institutions back XAI: The XAI initiative by DARPA is associated with defense, and the EU AI Act includes all high-risk systems such as security or biometrics and their transparency. Therefore, XAI and FL should be combined in order to find actionable edge intelligence in a trustworthy way (Manthena et al., 2025).

3.2 Popular Techniques: SHAP, LIME, Attention Maps

In resource-constrained wireless edge networks, XAI is essential to understand the complicated models of ML. There, a decentralized data structure and high privacy create a need to make explainability informative but efficient in FL. Such important techniques as SHAP, LIME, and attention mechanisms are applicable to FL-based security and may work with tabular, time-series, visual, and text data (Donmez et al., 2024). SHAP is model-free and based on the game theory, for each feature, a Shapley value is computed to denote its contribution to a prediction. It has global and local interpretability thus, it is effective in edge-based security applications such as intrusion detection (Salih et al., 2024). As an example, in a DoS attack case, SHAP can be used to identify important attributes such as unusual frequency of packets, entropy changes, etc., which may provide insight at the fine level even in raw data cannot be accessed, which is suitable solution in a federation scenario that has privacy concerns (Javed et al., n.d.). Table 4 shows the popular techniques as follows.

Table 4. SHAP, LIME and Attention Maps Comparison

Technique	Model Type	Global/ Local	Output Type	Suitable For	Edge-Friendly
SHAP	Any (incl. deep)	Both	Feature importance	Tabular, Time-Series	Medium (some overhead)
LIME	Any	Local	Feature approximation	Tabular, Text, Images	High (lightweight)
Attention Maps	Deep (NNs)	Local	Visual/heatmap	Time-Series, NLP, Vision	Medium (needs NN arch.)

Figure 5. Conceptual Links Among SHAP, LIME, and Attention Mechanisms

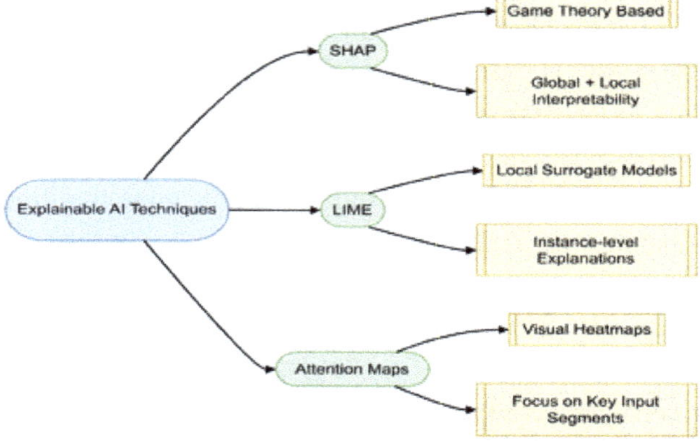

4. THE EXPLAINABLE FEDERATED LEARNING (XFL) FRAMEWORK

Explainable Federated Learning (XFL) framework is an integration of federated learning with explainable AI mechanisms allowing delivering transparency and explainability of decentralized models. It can make each edge client create local explanations and still not violate privacy in data. These descriptions are summed or confirmed at the centre so as to ensure international model responsibility. XFL is a decision traceable, debug gable model and ensures compliance to regulation in distributed environments. It is of especial use to applications that are trust-sensitive, such as autonomous systems and remote healthcare.

4.1 Integration of XAI Into FL

Federated Learning (FL) addresses privacy by storing data on the edge devices, and it is non-transparent, which is a problem in security-sensitive applications. The XFL combines XAI and FL in one framework to guarantee privacy and explainability (Chiarani et al., 2025). XFL enables the execution of lightweight explanation tools, like SHAP and LIME, on the machines of interest, local to the client machine, to justify a model given prediction, e.g. why a model designated a packet as mal epochal. These local explanations may be optionally forwarded to the server, which aggregates them to the global dashboards, which assist in detecting anomalies and monitoring client behaviors, such as the indicators of model poisoning. XFL facilitates a federated explainability loop: local training, local explanation, optional sharing, safe model aggregation, global interpretation, and update through feedback. This loop increases transparency, boosts trust and eases debugging while enhancing model robustness in changing wireless edge settings. Table 5 shows the traditional FL vs XFL Comparison and the figure 6 illustrates the XFL Architecture.

Table 5. Traditional FL vs. XFL Comparation

Feature	Traditional FL	Explainable FL (XFL)
Data Privacy	High	High
Model Interpretability	Low	High (via XAI integration)
Threat Visibility	Minimal	Supports anomaly & attack explanation
User/Analyst Trust	Moderate	High
Computation Overhead	Low	Medium (due to explainability modules)
Debugging/Auditability	Hard	Built-in via explanation tracking

Figure 6. XFL Architecture

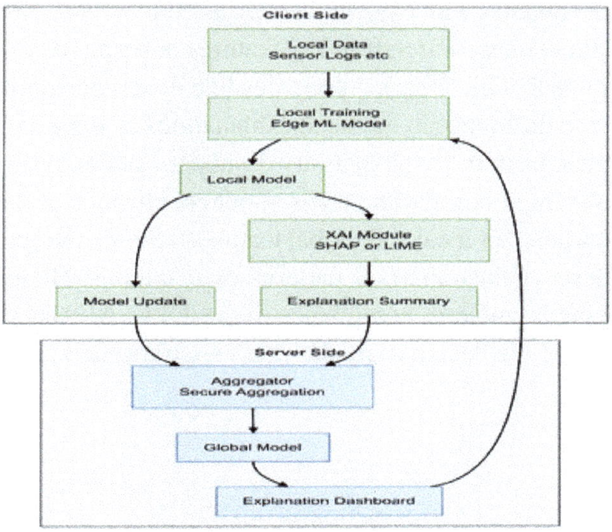

4.1.1 Architecture of XFL in Wireless Edge Networks

The architecture of the Explainable Federated Learning (XFL) in wireless edge networks combines privacy-preserving, decentralized machine learning and explanatory AI decision-making. It aims at highly distributed but resource-starved settings, including 5G/6G networks, IoT, and autonomous driving and smart cities. Such systems must operate under real-time constraints, and also require low latency hence efficient, lightweight and explicable models (Fatema et al., 2025). On the client-side, a wide variety of devices are used: smartphones, IoT, wearables, and autonomous vehicles that gather data and preprocess it locally. The model runs on the data- and trains a model on the data- and generates on-device explanations on the predictions using lightweight XAI techniques such as SHAP or LIME. In this case full example, a local solution or reason as to why an irregular heartbeat had been flagged could have been provided by a wearable health monitor. Attributable to the restrictions of the amount of power and processing available, such explainability approaches should be optimized relative to the quickness and effectiveness. The communication layer unites the clients and federated server with the help of secure wireless (Wi-Fi, LTE, 5G/6G, LoRaWAN) protocols, where the encryption and TLS are used together with a safe aggregation of model refreshes and explanation metadata (Sanchez et al., 2025). Latency and bandwidth consumption are reduced in effective communication, particularly where roll-outs are large scale.

The federated aggregator at the server-side will choose clients each round, combine their model updates with algorithms such as FedAvg and, optionally, aggregate explanation data to observe trends in key features or suspicious and thus malicious behavior (Borazjani et al., 2025). Anomalies like model poisoning can be detected by examining the difference in the local explanations or importance of the features across the clients. Last, the explainability dashboard shares system administration, agent, and governing bodies with details about conclusions of the model. Figure 7 charts the importance of local and global features, records the activity of the model and creates alerts against aberrant patterns or suspicious clients. This facilitates model debugging, regulatory compliance (e.g. GDPR, AI Act) and human-in-the-loop monitoring of the decisions made which are transparent and accountable.

Figure 7. Wireless Edge XFL Architecture Framework

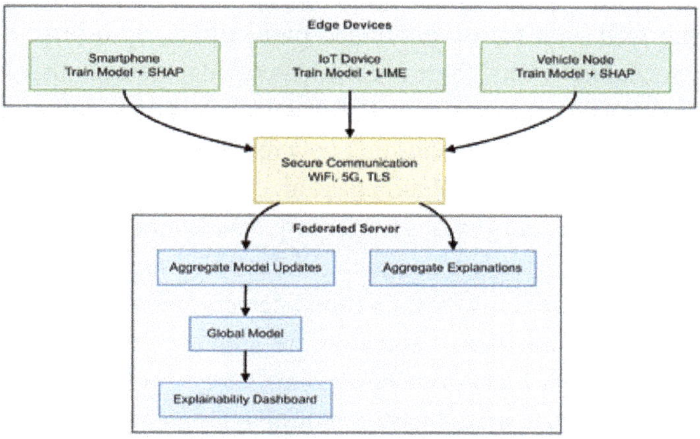

Table 6. Architecture Benefits

Layer	Purpose	XFL Enhancement
Edge Layer	Local learning and explainability	Enables decentralized transparency
Communication Layer	Secure model and explanation transfer	Maintains privacy and integrity
FL Server	Model & explanation aggregation	Enables global understanding and auditing
Dashboard Module	Visual insight and oversight	Facilitates trust, compliance, and action

5. SECURITY USE CASES FOR XFL

Explainable Federated Learning (XFL) Anomaly detection in heterogeneous IoT networks can be enhanced by training an ML model to detect anomalies transparently. In case of an intrusion detection system, XFL has the ability to enable edge devices to detect threats collaboratively such that it explains the basis of their alert notifications. To establish secure biometric authentication, XFL is providing explainable models to account the decision to allow access. In fraud detection, it assists financial institutions in determining strange patterns in the way financial transactions are made without compromising sensitive information.

5.1 Threat Classification in Mobile Networks

This is an era of intelligent mobile networks and the centralized security model is not coping well with the decentralized paradigm being offered by 5G and 6G. With these recent networks, there are billions of edge devices that are distributed, introducing a huge attack surface. The malware, network intrusion, and device impersonation are advanced threats that now require fast, privacy-aware, and transparent detection mechanisms (Manthena et al., 2025). To deal with such challenges, centralized approaches are growingly too slow, clandestine, and invasive (Moghadas Gholian et al., n.d.). Federated Learning (FL) enables such edge devices as phones and IoT nodes to collaboratively train threat detection models and does not require devices to exchange raw data. Anomalies are identified through local log-records (e.g., app use, network logs) done by devices. Only models can be updated with, e.g., a gradient, which is shared with a central aggregator to maintain privacy, and allows real-time joint learning.

This process is improved with the use of explainable Federated Learning (XFL): which incorporates explainability methods such as SHAP and LIME. These tools enable edge devices to come up with human understandable explanations to their predictions. As an example, when a suspicious DNS traffic has been detected by a smartphone, SHAP may be used to point to a peculiar packet rate or out-of-business-hour access to the system (Kalakoti et al., 2025). Such local explanations may be kept local on the device, or sent in aggregate to the central server. On the server side, combined explanations across devices assist the analysts in identifying threats, discovering patterns of an adversary, and tracing model actions. This openness is particularly desirable in systems that are mission-critical, because trust and accountability are desirable. As another example, within a mobile network where battery and data usage spike can be quickly characterized by edge devices through FL, SHAP can be used to explain the anomalous behavior, e.g. to report unusual background accesses to blacklisted IP addresses (Moghadas Gholian et al., n.d.). Combined intelligence enables telecom operators to locate where the threat is, block malicious web sites, and give clear message to affected users. XFL therefore facilitates quicker, privacy-aware and explainable detection of threats in next-generation wireless networks.

Table 7. The example of feature contribution (SHAP)

Feature	SHAP Value (Device A)	SHAP Value (Device B)	Interpretation
Packet Frequency	+0.41	+0.37	High packet rate was a strong malicious signal
Destination IP Entropy	+0.22	+0.16	Traffic to random/uncommon IPs
No. of Background Services	+0.18	+0.12	Unusual background tasks
User Interaction Time	-0.25	-0.30	Low usage decreased suspicion (normal usage)
Time of Access (Odd Hours)	+0.29	+0.31	Access during off-peak times flagged suspicion

Figure 8. Mobile Network Classification of XFL Threats

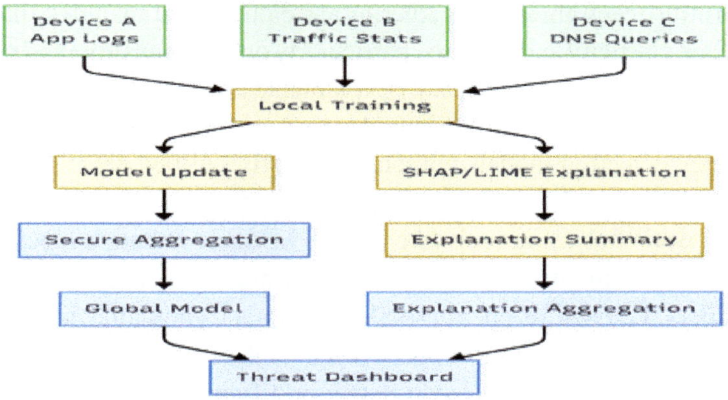

The Figure 8 shows the overall end-to-end Explainable Federated Learning (XFL) workflow on mobile smartphones. It starts with mobile edge devices including smartphones or IoT-routers gathering the local data, including app logs, network traffic, and DNS requests. Instead of communicating this sensitive information to a central server, every device trains a local model implemented using federated learning, which leaves the user intact. An explainability module, such as SHAP or LIME, can be trained upon, and run on-device after training, to explain a prediction, so users and system operators know why an observation was flagged as suspicious

(Fatema et al., 2025). These descriptions, and encrypted descriptions of the model updates, may be optionally shared with a central aggregator. Secure aggregation protocols help the aggregator to take all updates and merge them into an international threat detection model. The explanations summaries are also combined to construct a world-wide explainability map. The insights generated are then displayed in a centralized security console providing the administrators with a transparent, real-time, end-to-end visibility of the network threat landscape (Xu et al., 2024).

5.2 Anomaly Detection in IoT and WSNS

Smart technologies including environmental monitoring, healthcare, and industrial automation, as well as smart city technologies necessitate the Internet of Things (IoT) and Wireless Sensor Networks (WSNs) (Almasabi et al., 2025). Such systems are distributed, power-constrained systems applied in traversing and relaying sensitive information in wireless channels. They are however very vulnerable to anomalies and cyber threats due to their poor computing capacity, exposure to unsecured networks, fragility and physical accessibility. Conventional approaches to anomaly detection utilise centralised data gathering and cloud-based analytics, that are prone to latency, lack elasticity and security, particularly on time-sensitive and reasonability-sensitive applications (Amrani, n.d.) Explainable Federated Learning (XFL) is such an alternative to the IoT and WSN nodes that can be used to simultaneously train the anomaly detection models without any sharing of raw data. This keeps things confidential and also brings in the factor of transparency, since the models can explain their decisions using tools like SHAP, LIME or attention mechanisms. Each node has its own monitor of the environment and trains lightweight models (i.e., autoencoders, LSTMs, or one-class SVMs) to track anomalies in such parameters as temperature, motion, or vibration (Anwar et al., 2025). The example rate would be that a sensor on a bridge could detect odd frequency change and indicate an anomaly, but without requiring the sharing of sensitive data, so limiting the amount of bandwidth utilized and increasing the privacy. XFL puts local explainability on the edge. In case of an anomaly the explainability tools will reveal the contributing factors so the operators and engineers can confirm the alerts, trace failures or probe attacks(Hudda & Haribabu, 2025). As an example, SHAP could father to the fact that the alert was caused by an abnormal vibration spike that is not related to normal operation of the machine suggesting tampering or a mechanical failure.

The XFL aggregator in the server level, securely aggregates model updates and summaries of explanations form edge nodes to construct a powerful global anomaly detection model. These summaries are deployed to fill out centralised dashboards, where transparency on decision-making occurs. This becomes important where understandability is essential as in the ICU monitoring or industrial safety systems where explainability aids timely and decisive action (Hudda & Haribabu, 2025). Each machine is monitored (in terms of temperature, humidity, and vibration) with the use of WSNs in a smart factory (Almasabi et al., 2025). Out of ordinary working hours, a number of nodes report vibration anomalies. These are marked as suspicious in the local XFL models, and the SHAP explanation makes it clear that the spikes are not attributed to something related to normal motor cycles. Table 8 shows SHAP Explanation of Detected Anomaly and Table 9 describes the Advantages of XFL for IoT/WSN Anomaly Detection. When it is rolled up across the world, this understanding triggers a maintenance warning, which indicates what, where and why it happened providing precise, secure and explainable anomaly detection. XFL is useful in enhancing trust, safety, and operational efficiency in the IoT and WSN deployments through privacy-preserving learning and actionable transparency they combine as shown in Figure 9.

Table 8. SHAP Explanation of Detected Anomaly

Feature	SHAP Value	Interpretation
Vibration Amplitude	+0.47	Strong anomaly signal
Machine RPM	+0.21	Moderate contribution
Operating Temperature	-0.15	Lower temp made it less suspicious
Time of Day	+0.33	Off-hour operation increased anomaly score
Previous Fault History	+0.08	Minor effect

Figure 9. XFL for IoT Anomaly Detection

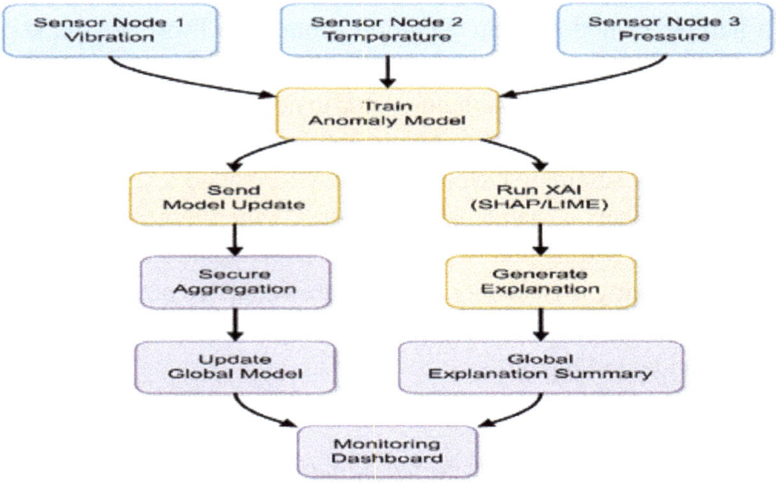

Table 9. Advantages of XFL for IoT/WSN Anomaly Detection

Aspect	Traditional System	XFL-Enabled System
Data Privacy	Centralized, risk-prone	Local training preserves data privacy
Communication Efficiency	Raw data transmission	Only model updates shared
Explainability	Often missing	Built-in via SHAP/LIME
Resource Adaptability	Not optimized for IoT	Lightweight, modular XAI models
Detection Accuracy	Global only	Local and global anomaly context
Operator Trust	Low interpretability	High due to explanation modules

5.3 Access Control Decisions in 5G/6G Environments

Access control has become more complicated by the emergence of latency-sensitive applications and a variety of devices in 5G networks and the next generation 6G. The traditional static models of RBAC, ABAC do not cope well in these real-time and dynamic scenarios where the decisions are required on the edge i.e. on base stations or on mobile devices (Khadka, n.d.). Although the systems based on AI provide adaptive control, the black-box systems make the decisions hard to explain or audit. This necessitates the strong necessity of Explainable Federated Learning (XFL) that jointly unites privacy-protecting learning and interpretable decision-making (El-Hajj, 2025). Under this arrangement, contextual information (user behavior, user location, session metadata, device ID) is gathered at edge nodes.

Data does not expose raw data during the training of Federated learning which trains local access control models. Such decision-making tools as LIME and SHAP give an idea of decisions. To give an example, a drone inaccessible to a 5G emergency network slice may be declared suspect because of bad certificate or log in at a different time of the day explanations that are calculated locally and soon reported safely. The result is confidence and non-blind enforcement. Table 10 shows the Example SHAP Explanation Table for Access Denial The updates of the local models and the metadata of explanations are transmitted to central aggregator and compose a global model and learn more general trends within the network (Ali & Koucheryavy, 2025). The aggregation of the insights provides security teams with real-time audit capability. An example would be in a smart campus where multiple attempts will occur during unsociable hours into a lab a response will be initiated. SHAP could display abnormal GPS information or instantiation of an unregistered device and initiate smart intervention without invasiveness. Incorporation of XFL enables context-aware, real-time access control in 5G/6G network to be securing and transparent, which addresses the dual requirements of flexibility and explainability(El-Hajj, 2025).

Table 10. Example SHAP Explanation Table for Access Denial

Feature	SHAP Value	Effect on Decision
Device Certificate Valid	-0.45	Expired certificate decreased access trust
Access Time Window	+0.37	Odd-hour login increased suspicion
User Role (Student)	+0.21	Not authorized for the resource
GPS Proximity to Site	-0.19	Outside of allowed geofence
Access Frequency History	-0.10	Unusual spike in recent requests

Figure 10. XFL for 5G Access Control

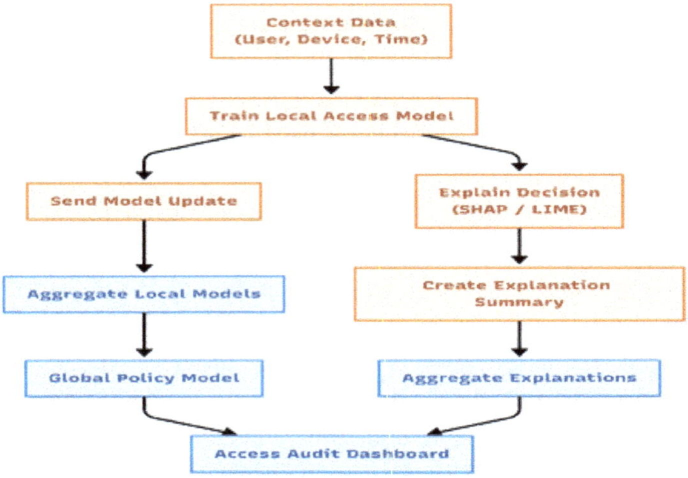

Table 11. Benefits of XFL in 5G/6G Access Control

Dimension	Traditional Access Control	XFL-Enhanced Access Control
Adaptability	Rule-based, rigid	Context-aware and learning-driven
Privacy	Centralized logs, privacy risk	Keeps contextual data local
Interpretability	Opaque AI decisions or static rules	Clear explanations for every decision
Scalability	Central bottlenecks under high load	Decentralized computation at the edge
Security Auditability	Difficult to trace or justify denial	SHAP/LIME provide feature-level explanations
Real-Time Responsiveness	Medium	High (runs on edge nodes near the user)

6. TECHNICAL IMPLEMENTATION

To implement XFL, one can add the explainability modules, e.g., SHAP, LIME or attention to the local training of federated clients. These modules produce outputs which can be interpreted and kept locally or collected in an aggregate-state. Global model is refreshed according to secure aggregation protocols to maintain privacy.

Lightweight explainability methods should be used to meet the resource limitation of edge devices. Such frameworks as TensorFlow Federated or PySyft will be extendable to create XFL pipelines.

6.1 Tools and Frameworks: Tensorflow Federated, PYSYFT, SHAP, LIME

In order to develop Explainable Federated Learning (XFL) systems particularly to the wireless-edge systems, such as the Internet of Things (IoT) networks, 5th and 6th generation (5G and 6G) wireless system architecture solutions, and distributed security infrastructures, a combination of the federated learning practices, machine learning, and explanation modes is required (Omoniyi David Olufemi, 2025). This union gives the possibility to learn distributed models keeping raw data secret and at the same time ensure that their choice will be explainable, and reliable. XFL systems all center on a federated learning system. Google created tensor flow federated (TFF) and it has gained popularity in developing and simulating the FL cases (Timofte et al., 2025). TFF enables training edge devices collaboratively to train e.g. CNNs designed to detect malware in ways that do not cause data to leave the edge device in raw form (Timofte et al., 2025). The models update information is submitted to a central aggregator in a sustained way and this enables privacy by design.

The second one is the PySyft (OpenMined community), which is based on PyTorch. It can enable more powerful privacy options than federated learning, such as differential privacy, as well as secure multiparty computation and homomorphic encryption, and so is more fine-grained. PySyft is especially better than the IoT systems having low-quality resources. For instance, within smart grid, sensor nodes can use PySyft to learn to detect abnormal use of energy and encapsulate within the encrypted update to the aggregator. SHAP and LIME are the popular choices on the explainability side. SHAP (SHapley Additive exPlanations) offers global, as well as local interpretability of a model by scoring features in terms of importance to a given prediction. SHAP could tell in a mobile threat detection scenario that out-of-range DNS queries resulted in a suspicious label. Such insights can be presented in a form of a force plot or a summary graph and can be aggregated between clients and presented in centralized dashboards.

Other, lighter weight options are LIME (Local Interpretable Model-Agnostic Explanations), which would be very well suited to low-power devices. It constructs a basic surrogate model around prediction in order to estimate feature impact. An instance would be using LIME in an edge anomaly detector with an explanation of a sensor flag pointing to sudden variations of voltage or frequency. These tools, combined form a full pipeline of XFL. The models are initialized with TensorFlow or PyTorch, secure federated training is done with TFF or PySyft and interpretability at the edge is supported by SHAP or LIME shown in table 12 (Omoniyi David Olufemi, 2025). The plain text summaries of the optional explanation will be transmitted to the server where a visualization can be reviewed this can be done without infringing the privacy of the user as shown Figure 11.

Figure 11. Tool Integration in XFL

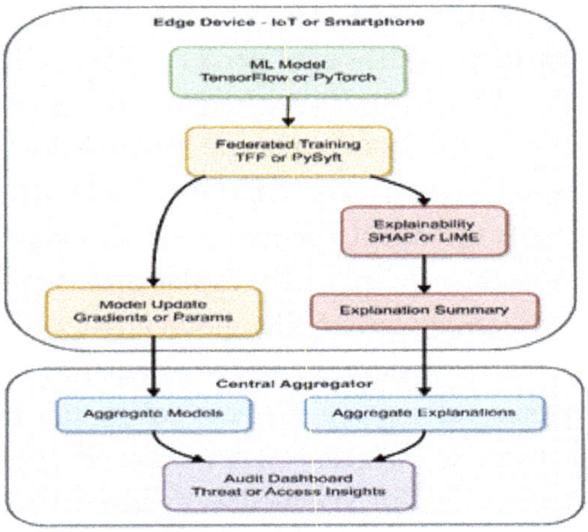

Table 12. Tools and Their Roles

Tool / Framework	Role in XFL Pipeline	Suitability for Edge Devices
TensorFlow Federated	FL model training and simulation	Moderate (good for mobile devices)
PySyft	Privacy-preserving FL with encryption support	High (suitable for IoT/WSNs)
SHAP	Deep feature attribution using Shapley values	Moderate to High (use GPU if available)
LIME	Lightweight local surrogate explanations	High (efficient on CPUs)

6.2 Lightweight XAI for Constrained Devices

Some of the rigid physical limitations of wireless edge technologies such as IoT and Wireless Sensor Networks (WSNs) are memory, processing power, bandwidth, and energy. Conventional explainable AI (XAI) processes, e.g. SHAP or integrated gradient, can be too expensive with regard to the CPU and memory to execute on a device with less than 256 KB of RAM or devices with battery or energy-harvesting power sources(Ghaleb et al., 2025). Such devices have to be processed in real-time, and explanation methods with a high latency can become a barrier to making decisions on time. Moreover, explanation data cannot be sent via a wide bandwidth in such settings. Thus, the deployment of XFL on such low-power, distributed systems requires flexible and time-saving XAI solutions, which are suitable within edge computing limitations table 13 as follows.

Table 13. Principles of Lightweight XAI

Principle	Goal
Approximation	Use simplified surrogate models to estimate explanations
Feature Prioritization	Focus on top-k influential features only
Precomputation	Cache common explanation patterns offline
Sampling	Use fewer samples or iterations during local explanation
Compression	Transmit only compressed or symbolic explanation summaries

Explainability deployed on edge devices with a limited power budget needs to be lightweight in terms of speed, memory and efficiency. An example of such is LIME with small sampling rate, that constrains the perturbed samples (e.g., 100 to 10) and interprets using a linear regression. It is perfect among IoT sensors that require a fast analysis of anomalies with only the most important features at hand; however, this also comes at the cost of accuracy. Also, when the available model is small (e.g. shallow decision trees, or lightweight ensemble models such as XGBoost), TreeSHAP (GPUs) (e.g. via TreeExplainer) can offer a similar set of interpretability tools but with much less significant computational load, allowing the model to run semi-realtime on a microcontroller like the ESP32. The sequence models directly give visual explanations in the form of attention heatmaps using attention weights. As an example in wearables, they can point at times during which there were abnormalities in heart rate. Saliency maps provide a gradient awareness visual attention in small CNN models, such that edge devices such as CCTV nodes can pick out on interesting parts of the image connected with anomalies without a GPU(Rahmati, n.d.). Finally, symbolic explanations derive human-interpretable rules of the form IF-THEN by getting human-interpretable rules out of tree-based surrogates. These are very light and useful, such as in lock devices that do not allow entry, because of out-of-hours or out-of-bounds GPS data, which makes good reasons to allow security measures as shown Figure 12.

Figure 12. Lightweight XAI Workflow

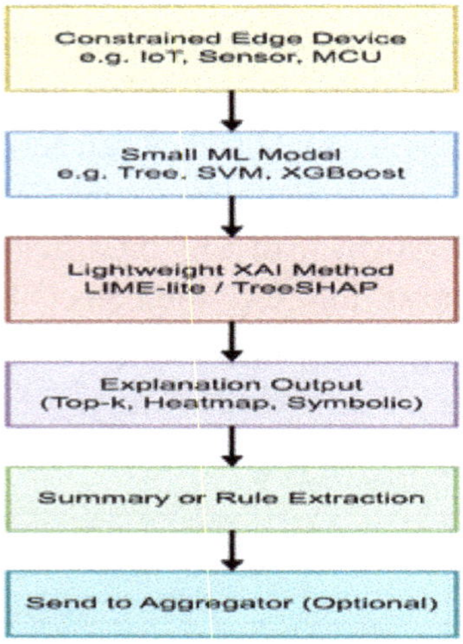

Table 14. Comparison of XAI Techniques for Edge Devices

XAI Method	Computational Cost	Memory Footprint	Accuracy	Best For
LIME (low-sample)	Low	Low	Moderate	Tabular IoT sensors
TreeSHAP (small model)	Medium	Low	High	Edge gradient boosting
Attention Maps	Low	Low	Moderate–High	Sequence data (e.g., ECG)
Saliency Maps	Medium (GPU helpful)	Medium	High	Vision on edge cameras
Rule Extraction	Very Low	Very Low	Moderate	Boolean access control

Under the intelligent agricultural scenario, a Raspberry pi applies XGBoost model through PySyft to observe the soil conditions. Upon a discovery of the anomaly, it yields TreeSHAP to the top two features only, i.e. moisture and humidity, and produces a symbolic rule "When moisture < 12% and humidity > 80%, then raise the anomaly alert." This brief rule is transmitted to central dashboard which saves bandwidth. To accommodate this type of edge-level explainability, model quantization, TinyML deployment, top-k feature summaries and SHAP value caching can be used to optimize resource requirements and not lose interpretability.

7. CHALLENGES AND LIMITATIONS

Explainable Federated Learning (XFL) encounters a number of issues one of which is the trade-off between model accuracy and interpretability. Production of explanations that points to some meaning on resource-constrained edge devices proves to be a complex task. Communication that is consistent and reasonable between mixed clients is complicated. The explanation data can put privacy risks in the event of indirect disclosure of sensitive data. Moreover, the quality of explanations lacks the standardized measurement in terms of the quality, which slows its adoption.

7.1 Trade-Offs: Performance vs. Transparency

With the rise of Explainable Federated Learning (XFL) within wireless edge systems, one of the biggest challenges that are posed on the use of XFL is a trade-off between great model accuracy and clear interpretability. This trade-off is particularly essential in a resource-scarce, real-time situation, such as 5G/6G networks, IoT, and smart infrastructure shown in table 15 and table 16.

Table 15. The Core Trade-Off

Aspect	High Performance Models	High Transparency Models
Model Types	Deep neural networks, ensemble methods	Decision trees, rule-based models
Accuracy	Generally higher	Slightly lower
Interpretability	Lower – hard to trace internal logic	High – easy to explain in human terms
Latency	May require heavy computation	Often lightweight
Resource Demand	High (especially DNNs or large ensembles)	Lower (if using shallow or symbolic models)
Use Case Suitability	Best for precision-critical tasks	Best for compliance, auditability, trust

Table 16. Key Dimensions of Trade-Offs

Dimension	Impact on XFL
Latency vs Explainability	SHAP or LIME adds computation time, affecting real-time response
Privacy vs Explanation Detail	Detailed explanations can leak sensitive context (e.g., location data)
Bandwidth vs Model/Audit Data	Transmitting explanations along with models strains constrained links
User Trust vs Model Complexity	Simple models build trust, but may be too naive for complex attacks
Computation vs Interpretability	Interpretable methods often avoid acceleration (e.g., no GPU-friendly ops)

Table 17. Strategies to Balance the Trade-Off

Strategy	How It Helps
Hybrid Modelling	Combine deep learning with shallow interpretable surrogates
Tiered Explainability	Use lightweight explanations locally; deep diagnostics centrally
Top-k Feature Explanation	Focus only on most impactful features (e.g., SHAP top 3)
Caching and Pre-computation	Precompute explanations for common patterns
User Role-Based Detail	Show full explanations to admins; abstract ones to users
Model Selection Based on Context	Use interpretable models for legal/compliance zones

Figure 13. Trade-Off Triangle in XFL

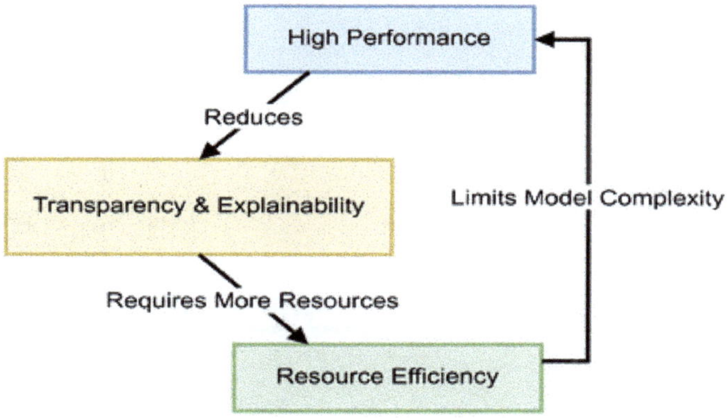

7.2 Communication Overhead

In conventional FL, expensive overhead is a key bottleneck as transferring of model updates is frequent. This burden is increased by XFL forcing the transmission of extraneous data such as SHAP values, ranks of features, and logs, working to increase the strain on bandwidth and latency. Explainability added to FL also consumes more communication load in the wireless edge networks such as 5G, IoT, and WSNs. Sending SHAP values or feature ranks use relatively little bandwidth, but still burn battery, increase latency, and can saturate restricted connections. Within a 5G intelligent traffic scenario, every pole will execute a federated model and transmit an explanation based on SHAP every 10 minutes. It increases the typical model updates by ~200KB per node leading to congestion of links and latencies in real-time decision-making applications such as controlling traffic signals from Figure 14 and Table 18 describes Strategies to Reduce Communication Overhead

Figure 14. Communication Overhead Sources in XFL

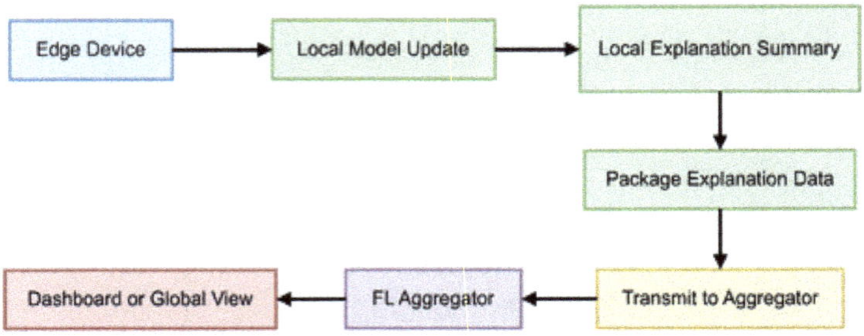

Table 18. Strategies to Reduce Communication Overhead

Technique	Description
Top-k Feature Summarization	Only transmit the most important k feature contributions.
Compression & Quantization	Use gzip or binary encoding for explanation metadata.
Sparse Update Transmission	Send updates only when significant model drift or explanation change is detected.
Edge Caching of Explanations	Cache repetitive explanations locally.
Asynchronous Update Scheduling	Stagger model + explanation updates to avoid peak loads.
Adaptive Explanation Resolution	Low-res summaries for bandwidth-limited zones, full summaries elsewhere.

7.3 Interpretability vs. Adversarial Robustness

Explainable Federated Learning (XFL) creates transparency in the AI systems yet causes an increased security threat. Increased interpretability will facilitate adversarial inputs, model inversion or explanation manipulation by an adversary. The core dilemma is shown in below table 19.

Table 19. The Core Dilemma

Interpretability Helps	But Also Enables Attackers To...
Build trust by explaining predictions	Reverse-engineer feature importance or decision boundaries
Debug model errors	Craft adversarial inputs targeting vulnerable paths
Identify model bias or failure points	Exploit them with targeted poisoning or evasion attacks
Understand why a device is flagged or blocked	Adapt behavior to avoid triggering the same patterns

SHAP explanation on XFL can reveal some important attributes such as peculiar DNS traffic to detect malware. This allows the attackers to slowly change their actions in order to not raise alerts, remaining under explanation-relevant thresholds as shown in Figure 15.

Figure 15. Visualization: Attack Surface Expansion from XAI

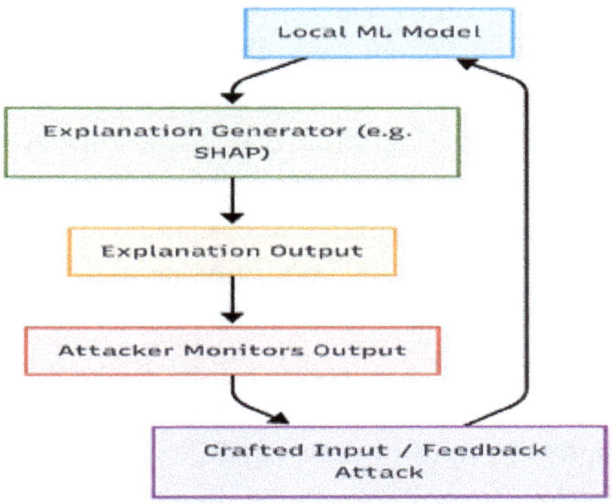

8. FUTURE RESEARCH DIRECTIONS

In future, XFL research should revolve around the creation of lightweight privacy-preserving explainability techniques that can be used in edge settings. Flexible XFL structures that dynamically compromise on accuracy, transparency, and efficiency are also important. Common standards, regulatory-paralleled metrics of assessment will speed up actual implementation.

8.1 Real-Time XFL

Since the wireless networks are changing through 5G/6G, edge computing, and AI services, it is critical to take real-time decisions with an explainable process. Real-Time Explainable Federated Learning (XFL) saves time and resources, providing the best solution on the fly updates and interpretations and is very important in areas such as autonomous systems, smart cities, emergency response, and healthcare where venue time and presence is very important. Real-Time XFL is an unending local machination on a border device by means of streaming data. Simple XAI tools such as miniSHAP or LIME have immediate explanations. Federal aggregation runs on low latency (e.g. FedAsync, FedBuff) to support frequent model training.

The real-time data visual dashboards provide the predictions together with the explanations. An example of the federated model application is in the emergency vehicle routing where the location, weather and traffic allow an individual to predict congestion. Rerouting decisions using XAI are shown to depend on the combination of accident + rain + school zone; this makes its system more trustworthy. Some important enablers are streaming FL frameworks, incremental explainability approaches, TinyML models (e.g. MobileNet), and edge AI chip (e.g. edge TPU, Jetson). Edge-centric UIs provide current information to the users. The challenges persist: XAI introduces latency, edge devices cannot have the resources to support heavy payloads of explanations, and only infrequent updates globally could result in model drift. The performance, explainability and efficiency need to be balanced in Real-Time XFL as shown in Figure 16. The federated Sleeping apparatus future direction of research in the field of Explainable Federated Learning (XFL) is associated with the improvement of in-time capabilities and efficiency. Other areas of focus are to design a low-latency federated APIs and using trigger-based scheduling to generate explanations. Explanations can be confidence-aware and enhance trust when making uncertain predictions, whereas streaming-optimized XAI can deal with drifts in concept on continuous data. Also, the edge hardware could be co-designed with integrated explainability capabilities to allow more efficient interpretation (and fewer resources) at the device level.

Figure 16. Real-Time XFL Pipeline

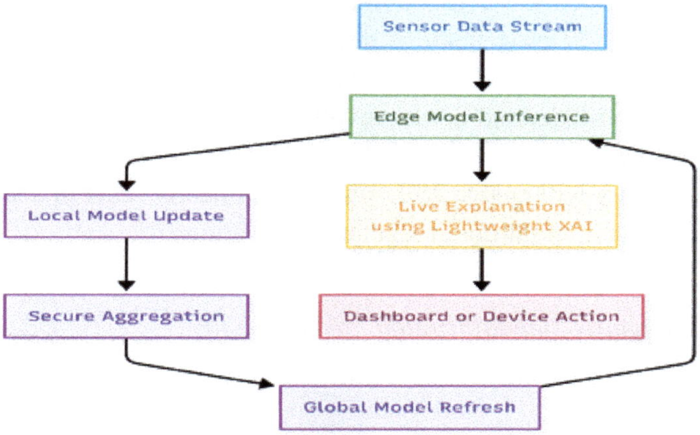

8.2 Standards for Explainability in FL

The more Explainable Federated Learning (XFL) becomes a reality, the more standardization is required. XFL works on such various devices as smartphones, IoT sensors, autonomous systems, which have various capabilities and types of data. Lack of standardization poses a threat of inconsistency, uncheckable and non-conformant explanations. Auditability, privacy and trust depend on the standards. They assist in making explainability logs reproducible, devices that belong to various vendors to work with each other, and sensitive information not to leak through explanations. Such consistency is required by regulatory structures such as GDPR, EU AI Act, and ethics framework by IEEE (Martínez Llamas et al., 2025). The standardization needs to be aimed at the elements, such as the form of explanations (e.g., JSON/XML), metadata (e.g., top features, confidence), and levels of explanation (e.g., full vs. top-k features). There is also a need to define secure transmission and standard logging (e.g. Explainability Record Logs).

Low-resource devices can implement tiered formats--either bare or incomplete description depending on the space. An example of fields that might be in an XFL Explanation Record (XER-1.0) would be device ID, timestamp, input features, contribution scores, decision, explanation type (e.g., SHAP, LIME), a privacy flag. They may be stored in .json or .yaml and coded before being sent to the server.

Relevant standards are being developed by such organizations as IEEE (P7001), ISO/IEC, NIST, and 3GPP. Interoperable APIs are also advocated by open-source initiatives such as the OpenXAI consortium. Challenges remain. Universal formats are difficult due to the heterogeneity of devices. The tools of XAI differ in the format of their output. Privacy violations can occur when information is given in considerable detail and federated models evolve with time. Social norms and the expectations of the law are also divided all over the world. As shown in Figure 17 universal explainability will enable XFL systems to linearly scale in a secure and transparent manner without falling in regulatory violations.

Figure 17. Explainability Standardization Layers in XFL

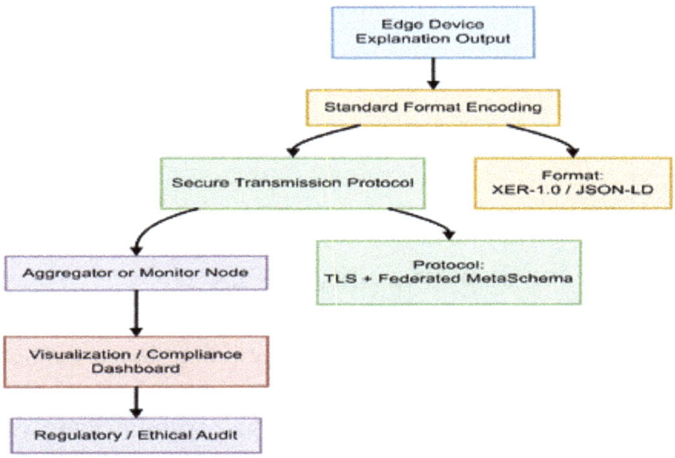

8.3 Federated Continual Learning and Explainability

In changing wireless ecosystems such as smart cities, autonomous systems and industrial IoT, the distributions of data change with time. Static models will wear off fast therefore retraining wastes resources and is too intrusive. This drawback is addressed by Federated Continual Learning (FCL) which allows the edge devices to learn continually and adapt to the emerging patterns and maintain the previous knowledge without centralization of the data (Hamedi et al., 2025). Explainable Federated Continual Learning (XFCL) is the extension of FCL to give transparency to this adaptive process. It does not only permit models to be redeveloped but it lays out how and why decisions evolve. This plays a critical role in real-life systems where auditing, compliance and trust are necessary. FCL provides support to incremental updates, is compatible with non-IID data, and enables edge nodes to learn new tasks without forgetting about the previous ones. When utilized together with explainability, XFCL assists operators to realize behavior changes (Hamedi et al., 2025). As an example, it will be possible to tell why a certain activity was flagged by a device today in comparison to yesterday, or how the new data has slowly changed the focus of a model, possibly inducing biasness (Borazjani et al., 2025). As shown in Figure 18 XFCL makes real-time observation of model evolution, presents visualisations of knowledge drift and creates trust in systems which are supposed to be transparent under continuous change.

Figure 18. Pipeline: XFCL Architecture

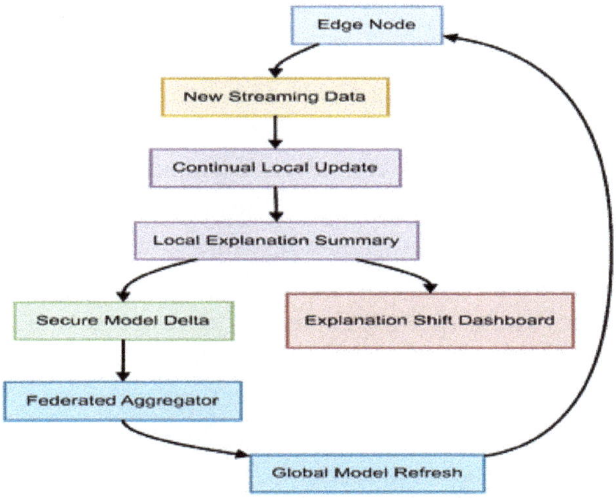

8.4 Intelligent Surveillance in Changing Places

The dynamic city is where an FL model first identifies loitering according to duration and location. The patterns of crowds change after a mass event. With Federated Continual Learning (FCL), the local models can be updated, so it knows that when it takes longer to get somewhere and nothing is wrong (Quan et al., 2025). Through XAI, the model notes that the loitering detection has increasingly used erratic movement as opposed to time. It guarantees transparency in terms of real-time behaviour updates. Explainability FCL Methods:

• Time-Aware SHAP/LIME: Monitors how the feature importance changes over time making concept drifts visible.
• Model Drift Visualization: Shows what features got more or less important after the update.
• Task-Specific Explainability: Applies the individual new tasks or domain as sub models and each assigns an explanation.
• Memory Replay Auditing: The method which is explained in the rehearsal-based FCL form of it, deals with the persistence of older samples that affect the outputs.
• Catastrophic Forgetting Warnings: Alerts when the previous performance in tasks declines in light of new updates.

9. CONCLUSION

Explainable Federated Learning (XFL) is a significant stride in the direction of building transparent, privacy-conscious and trustful artificial intelligence in the wireless edge environments. XFL is based on Federated Learning (FL) and Explainable AI (XAI), and in this way, XFL ensures the presence of interpretable output in the form of decentralized models without revealing confidential information. This model will make such important areas as healthcare, autonomous vehicles, smart cities, and Internet of Things-powered networks more accountable and trustworthy to its users. XFL enables the edge devices to collectively train models, generate local explanations and share aggregated updates through secure communication channels only. These tools as SHAP, LIME and attention-based visualization can give an understanding of how the model behaves and it facilitates the detection of mistakes, the occurrence of threats and compliance with regulation. The XFL possesses a high scalable and ethical implementation potential, but it experiences challenges that relate to the cost of computation, communication overhead, and accuracy / transparency trade-offs. As the digital ecosystem is developed towards the direction of 5G/6G and real-time decision systems, XFL will become a reference when developing human-centric interpretable AI systems. Its integration to federated continual learning and lightweight edge architectures paves the way to adaptive, secure, explainable intelligence, which is in line with the new global norms and ethical requirements in AI.

REFERENCES

Agarwal, G., & Manager -Projects -Cognizant, S. (2025). Driven Security Solutions. International Journal of Advanced Research in Science, Communication and Technology (IJARSCT) International Open-Access, Double-Blind, Peer-Reviewed, Refereed, *Multidisciplinary Online Journal*, (1). https://doi.org/DOI: 10.48175/IJARSCT-23624ï

Al Nasim, M. A., Soshi, F. T. J., Biswas, P., Ferdous, A. S. M. A., Rashid, A., Biswas, A., & Gupta, K. D. (2025). Principles and Components of Federated Learning Architectures. https://arxiv.org/abs/2502.05273

Albshaier, L., Almarri, S., & Albuali, A. (2025). Federated Learning for Cloud and Edge Security: A Systematic Review of Challenges and AI Opportunities. In Electronics (Switzerland) (Vol. 14, Issue 5). Multidisciplinary Digital Publishing Institute (MDPI). DOI: 10.3390/electronics14051019

Ali, A., & Koucheryavy, Y. (2025). *Adaptive Beam Tracking in 5G/6G mmWave networks: A Clustered Federated Learning Approach.* IEEE., DOI: 10.1109/ACCESS.2025.3563435

Almasabi, A. M., Alkhodre, A. B., Khemakhem, M., Eassa, F., Abi Sen, A. A., & Harbaoui, A. (2025). Internet of Things-Based Anomaly Detection Hybrid Framework Simulation Integration of Deep Learning and Blockchain. *Information (Basel)*, *16*(5), 406. Advance online publication. DOI: 10.3390/info16050406

Anandaram, H., Shreenidhi, K. S., Joshi, K., Sidhu, K. S., Kiruthigha, R., Singh, V., & Tripathi, A. (2026). A Case Study on Resource Management Industrial IoT Application. In *Building Data-Driven Edge Systems for Business Success* (pp. 1–26). IGI Global Scientific Publishing.

Anwar, R. W., Abrar, M., Salam, A., & Ullah, F. (2025). Federated learning with LSTM for intrusion detection in IoT-based wireless sensor networks: A multi-dataset analysis. *PeerJ. Computer Science*, *11*, 1–30. DOI: 10.7717/peerj-cs.2751 PMID: 40567740

Arana-Catania, M., Sonee, A., Khan, A. M., Fatehi, K., Tang, Y., Jin, B., Soligo, A., Boyle, D., Calinescu, R., Yadav, P., Ahmadi, H., Tsourdos, A., Guo, W., & Russo, A. (2025). Explainable Reinforcement and Causal Learning for Improving Trust to 6G Stakeholders. *IEEE Open Journal of the Communications Society*, *6*, 4101–4125. Advance online publication. DOI: 10.1109/OJCOMS.2025.3563415

Balas, V., Pandey, H., Bin Ali, M., Singh, V., & Kumar, A. (Eds.). (2026). *Recent Advances in Smart Communication Technologies for a Sustainable Future*. IGI Global Scientific Publishing., DOI: 10.4018/979-8-3373-3541-4

Borazjani, K., Abdisarabshali, P., Khosravan, N., & Hosseinalipour, S. (2025). Redefining non-IID Data in Federated Learning for Computer Vision Tasks: Migrating from Labels to Embeddings for Task-Specific Data Distributions. https://arxiv.org/abs/2503.14553

Chiarani, M., Roy, S., Verikoukis, C., & Granelli, F. (2025). XAI-Driven Client Selection for Federated Learning in Scalable 6G Network Slicing. https://arxiv.org/abs/2503.12435 DOI: 10.1109/ICC52391.2025.11161532

Donmez, T. B., Kutlu, M., Mansour, M., & Yildiz, M. Z. (2024). Explainable AI in action: A comparative analysis of hypertension risk factors using SHAP and LIME. *Neural Computing & Applications*. Advance online publication. DOI: 10.1007/s00521-024-10724-y

El Amrani, W. (n.d.). Attention and Beyond: Explainability Techniques for Vision Transformers. https://www.researchgate.net/publication/391768365

El-Hajj, M. (2025). Secure and Trustworthy Open Radio Access Network (O-RAN) Optimization: A Zero-Trust and Federated Learning Framework for 6G Networks. *Future Internet*, *17*(6), 233. DOI: 10.3390/fi17060233

Fatema, K., Dey, S. K., Anannya, M., Khan, R. T., Rashid, M. M., Su, C., & Mazumder, R. (2025). Federated XAI IDS: An Explainable and Safeguarding Privacy Approach to Detect Intrusion Combining Federated Learning and SHAP. *Future Internet*, *17*(6), 234. DOI: 10.3390/fi17060234

Ghaleb, M., Hamdan, M., Barnawi, A. Y., Gambo, M., Danasabe, A., Bello, S., & Habib, A. (2025). Explainable AI for Lightweight Network Traffic Classification Using Depthwise Separable Convolutions. *IEEE Open Journal of the Computer Society*, *6*, 908–920. Advance online publication. DOI: 10.1109/OJCS.2025.3576495

Hamedi, P., Razavi-Far, R., & Hallaji, E. (2025). Federated Continual Learning: Concepts, Challenges, and Solutions. https://arxiv.org/abs/2502.07059

Hudda, S., & Haribabu, K. (2025). A review on WSN based resource constrained smart IoT systems. In Discover Internet of Things (Vol. 5, Issue 1). Springer Nature. DOI: 10.1007/s43926-025-00152-2

Javed, S., Mukhtar, N., Iqbal, S., Asad Ali Naqvi, S., Ishtiaq, A., Yamin Siddiqui, S., Ammar, M., & Author, C. (n.d.). Secure and Interpretable Intrusion Detection through Federated and Ensemble Machine Learning with XAI. https://doi.org/DOI: 10.56979/901/2025

Kalakoti, R., Vaarandi, R., Bahsi, H., & Nõmm, S. (2025). Evaluating explainable AI for deep learning-based network intrusion detection system alert classification. DOI: 10.5220/0013180700003899

Kaur, N., & Gupta, L. (2025). Securing the 6G–IoT Environment: A Framework for Enhancing Transparency in Artificial Intelligence Decision-Making Through Explainable Artificial Intelligence †. *Sensors (Basel)*, *25*(3), 854. Advance online publication. DOI: 10.3390/s25030854 PMID: 39943493

Kaur, N., & Gupta, L. (n.d.). Explainable AI for Securing Healthcare in IoT-Integrated 6G Wireless Networks.

Khadka, K. (n.d.). *Privacy Challenges in IoT and Cyber-Physical Systems: Trends.* Threats, and Mitigation Strategies.

Krishna, G., Singh, V., Pandey, D., Krishna, K. H., Joshi, K., & Gupta, T. (2024, May). Power trading framework of cloud-edge computing in the artificial intelligence market. In 2024 4th International Conference on Advance Computing and Innovative Technologies in Engineering (ICACITE) (pp. 1910-1918). IEEE. DOI: 10.1109/ICACITE60783.2024.10617344

Manthena, H., Shajarian, S., Kimmell, J., Abdelsalam, M., Khorsandroo, S., & Gupta, M. (2025). *Explainable Artificial Intelligence (XAI) for Malware Analysis: A Survey of Techniques, Applications, and Open Challenges*. IEEE., DOI: 10.1109/ACCESS.2025.3555926

Martínez Llamas, J., Vranckaert, K., Preuveneers, D., & Joosen, W. (2025). Balancing Security and Privacy: Web Bot Detection, Privacy Challenges, and Regulatory Compliance under the GDPR and AI Act. In Open Research Europe (Vol. 5). F1000 Research Ltd. https://doi.org/DOI: 10.12688/openreseurope.19347.1

Moghadas Gholian, S., Fiandrino, C., Vallina-Rodríguez, N., Fiore, M., & Widmer, J. (n.d.). DEEXP: Revealing Model Vulnerabilities for Spatio-Temporal Mobile Traffic Forecasting with Explainable AI. In IEEE Transactions on Mobile Computing. https://git2.networks.imdea.org/

Olanrewaju-George, B., & Pranggono, B. (2025). Federated learning-based intrusion detection system for the internet of things using unsupervised and supervised deep learning models. *Cyber Security and Applications*, *3*, 100068. Advance online publication. DOI: 10.1016/j.csa.2024.100068

Olufemi, O. D.Omoniyi David Olufemi. (2025). Quantum-AI Federated Clouds: A trust-aware framework for cross-domain observability and security. *World Journal of Advanced Research and Reviews*, *26*(2), 4098–4140. DOI: 10.30574/wjarr.2025.26.2.2074

Quan, M. K., Pathirana, P. N., Wijayasundara, M., Setunge, S., Nguyen, D. C., Brinton, C. G., Love, D. J., & Poor, H. V. (2025). Federated Learning for Cyber Physical Systems: A Comprehensive Survey. DOI: 10.1109/COMST.2025.3570288

Rahmati, M. (n.d.). Towards Explainable and Lightweight AI for Real-Time Cyber Threat Hunting in Edge Networks.

Roohani, B. S., Sharma, N., Kasula, V. K., Mamoria, P., Modh, N. N., Kumar, A., & Singh, V. (2026). Urban Computing Solutions in Healthcare Edge Computing. In Building Data-Driven Edge Systems for Business Success (pp. 377-400). IGI Global Scientific Publishing.

Salih, A. M., Raisi-Estabragh, Z., Galazzo, I. B., Radeva, P., Petersen, S. E., Lekadir, K., & Menegaz, G. (2024). A Perspective on Explainable Artificial Intelligence Methods: SHAP and LIME. *Advanced Intelligent Systems*. Advance online publication. DOI: 10.1002/aisy.202400304

Sanchez, O. T., Borges, G., Raposo, D., Rodrigues, A., Boavida, F., & Silva, J. S. (2025). Enhancing the performance of Industrial IoT LoRaWAN-enabled federated learning frameworks: A case study. *Internet of Things : Engineering Cyber Physical Human Systems*, *32*, 101632. Advance online publication. DOI: 10.1016/j.iot.2025.101632

Sun, H., Liu, Y., Al-Tahmeesschi, A., Nag, A., Soleimanpour-Moghadam, M., Canberk, B., Arslan, H., & Ahmadi, H. (2025). Advancing 6G: Survey for Explainable AI on Communications and Network Slicing. *IEEE Open Journal of the Communications Society*, *6*, 1372–1412. Advance online publication. DOI: 10.1109/OJCOMS.2025.3534626

Suriyan, K., Ganesh, S., Palaniyammal, P., Priya, G., & Singh, V. (2026). Recent Trends in Edge Computing: Challenges and Opportunities. Building Data-Driven Edge Systems for Business Success, 323-344.

Timofte, E. M., Dimian, M., Graur, A., Potorac, A. D., Balan, D., Croitoru, I., Hri can, D.-F., & Pu ca u, M. (2025). Federated Learning for Cybersecurity: A Privacy-Preserving Approach. *Applied Sciences (Basel, Switzerland)*, *15*(12), 6878. DOI: 10.3390/app15126878

Xu, Z., Li, B., & Cao, W. (2024). Enhancing Federated Learning-Based Social Recommendations with Graph Attention Networks. https://ssrn.com/abstract=4978530 DOI: 10.2139/ssrn.4978530

Yang, J., Zhang, J., Xu, B., Zheng, J., Liu, Z., Liu, Z., Niyato, D., Debbah, M., Han, Z., & Ai, B. (2025). White-Box AI Model: Next Frontier of Wireless Communications. https://arxiv.org/abs/2504.09138

Zhang, W., Zhou, S., Walsh, T., & Weiss, J. C. (2025). Fairness amidst non-IID graph data: A literature review. *AI Magazine*, *46*(1), e12212. Advance online publication. DOI: 10.1002/aaai.12212

Chapter 5

Federated Learning for Privacy-Preserving Security in Wireless Networks:
A Decentralized Intelligence Approach

Vikas Sharma
https://orcid.org/0000-0001-8173-4548
Department of Computer Applications, SRM Institute of Science and Technology, Delhi, India

Alekh Chaudhary
https://orcid.org/0009-0007-7897-6600
Department of Computer Applications, SRM Institute of Science and Technology, Delhi, India

Tarun Kumar Vashishth
https://orcid.org/0000-0001-9916-9575
IIMT University, Meerut, India

Sunil Kumar
https://orcid.org/0000-0003-0841-5924
School of IT, AURO University, Surat, India

Mragesh Pratap Singh
https://orcid.org/0009-0003-8290-8265
Department of Computer Application, Swami Vivekanand Subharti University, Meerut, India

Shahanawaj Ahamad
College of Computer Science and Engineering, University of Hail, Saudi Arabia

ABSTRACT

The privacy-sensitive safety of wireless networks has become an increasing public issue as we are seeing more wireless devices with FW devices, and the need for networks communicating in real-time. Additionally, to previous models of machine learning with regards to identifying any threats and providing a security assurance

DOI: 10.4018/979-8-3373-3576-6.ch005

t © 2026, IGI Global Scientific Publishing. Copying or distributing in print or electronic forms without written permission of IGI Global Scientific Publishing is prohibited. Use of this chapter to enerative artificial intelligence (AI) technologies is expressly prohibited. The publisher reserves all rights to license its use for generative AI training and machine learning model development.

for a network's safety, these previous models are typically deployed on a central device acting as a server and, so using the available information, they need to effectively store private data. In essence and in reality, this will pull private sensitive data to central locations creating risk and danger to further distances to require communication. This chapter emphasizes exploring and identifying FL as possibility to diminish trust issues rife in wireless networks through the scope of centralized systems, including intrusion detection systems, anomaly detection, and authentication systems, by all learners must contribute towards FL, potentially in ways parallel to some on-device methods already being trailed or developed.

I. INTRODUCTION

The new generation of mobile devices, smart appliances, autonomous cars and other IoT sensors form a single wireless ecosystem. All of these new devices rely on a computer network which makes it vital to have proper wireless network infrastructure (Kaur et al., 2024). But with the addition of new devices and services come new issues related to privacy or security (Singh and Kaushik, 2023; Joshi et al., 2025; Singh and Dubey, 2015). Most methods of computer network security use a centralized model where user information is collected through machine learning (Raturi et al., 2025) and data is analysed in huge centralized databases. Such models may be efficient in identifying computer system intrusions and security threats (Chaudhary et al., 2024), but they present grave privacy problems because sensitive information must be transferred, stored, and exposed to data breaches, unauthorized access, communication overhead, and increased vulnerability abuse. Federated Learning (FL) can address these concerns by changing the focus from the user to the system to the learning process instead. FL doesn't require data to be sent to a central server for processing. Instead, devices or nodes work together to train a machine learning model which is kept confidential to the nodes' raw data. Only model update information, such as gradients, instead of raw data are shared, securing data privacy and significantly decreasing the possibility of leaks. This form of distributed intelligence aligns well with the principles of device learning, data minimization, and device security efficiency (Gao et al., 2023).

Figure 1. Federated Learning in Wireless Networks

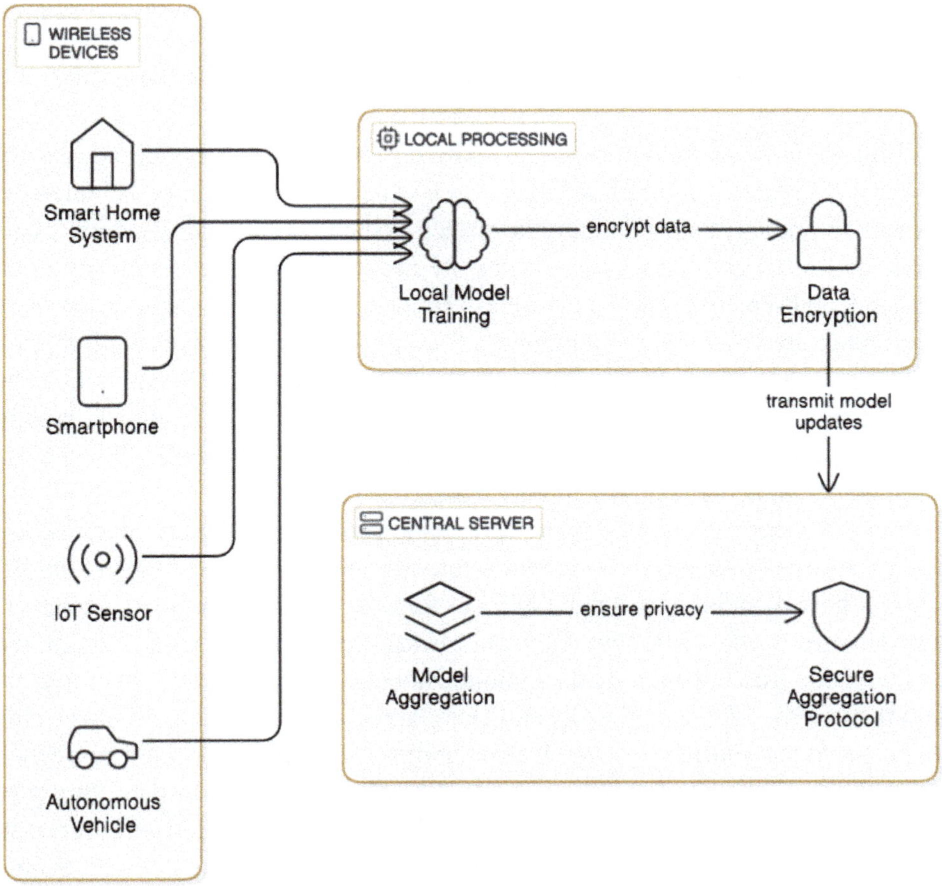

Security techniques that are both effective and non-invasive to user privacy are becoming more necessary with the growth of 5G wireless networks, edge computing (Roohani et al., 2026, Suriyan et al., 2026, Krishna et al., 2024), and future 6G technologies . Besides enabling adaptive security measures like intrusion detection, anomaly detection, and decentralized authentication, Federated Learning allows edge devices to act as autonomous intelligent agents capable of threat response. The implementation of FL does pose some challenges, such as model heterogeneity, trust, and system robustness within hostile environments shown in figure 1. This chapter examines how Federated Learning improves security for wireless networks with privacy preservation. It critiques the centralised model's shortcomings, defines and exemplifies FL's key elements, advantages, and structure, and assesses its implementation in a variety of security paradigms. Additionally, the chapter discusses some of the outstanding issues still to be tackled and some possible routes for forthcoming issues, defining a new frontier in this space towards smarter and more secure wireless communications through decentralised learning (Alazab et al., 2023).

1.1 The Importance of Privacy in Wireless Network Security

With the growing volume and sensitivity of data being processed, the issue of privacy will increase in importance in wireless networks. The global adoption of wireless technology across a multitude of sectors such as healthcare, finance, smart cities, and industrial automation means we are constantly processing personal, financial, and sensitive data with wireless networks. Wireless communications by their nature are vulnerable to a number of security issues such as eavesdropping, spooling, unauthorised access, and information interception. Consequently, the forward motion of data in these environments is not only a consideration regarding being protective of the users but a matter of reliability and survivability for wireless communication networks (Al-Huthaifi et al., 2023).

In a conventional security framework, data from various nodes, or users, are being forwarded to dedicated servers for forensic investigation and threat identification. These models act as a point of focus for attacks, since compiled raw data is stored in one location, it becomes susceptible to privacy liability risks. Further, the created models do not embody data minimization as is required by law, or neglect laws that are relevant, e.g., the General Data Protection Regulation (GDPR) or the Health Insurance Portability and Accountability Act (HIPAA) in their respective jurisdictions. The central capture and retention of user data leaves data subject to loss of data due to data leakages, profiling, and/or misuse, which destroys user trust. We must study privacy preservation in wireless network security not only from a technical perspective but from legal, ethical, and moral perspectives, with respects to each. This means providing security and privacy of sensitive user data; namely, location, behavior, biometric data, and communication data. Privacy Enhancing Technologies (PETs) such as Federated Learning, can address this problem effectively, since PETs allow for the processing of data to train models without exposing personal data. With wireless networks becoming more extensive in scale to ensure access for the massive number of devices and applications, it is critical that Privacy Enhancement Technologies are built into the base architecture of the security systems to develop a healthy, moral, and trustworthy communication counter structure (Batool et al., 2024).

1.2 Limitations of Centralized Machine Learning Approaches

The paragraph references centralized machine learning (ML), which while an older concept, exists in the design of many wireless network security architectures.

Centralized ML relies on data collected from many devices, or nodes, and then secondarily aggregates this data to a central location for training models and decision making. This architecture allows for complex systems of threat detection and classification to be developed, but does not function well in the current age of agile wireless environments.

1. **Privacy concerns:** In order to leverage centralized machine learning, the raw data must first be acquired from these various devices and sent up to a centralized server where the model training will occur. This is quite a risk in many cases in terms of privacy, especially if the data set is correlated with sensitive personal behavioural, or biometric data. Having data aggregated in one location is a security risk in an event of cyber-attack or data breach, and also makes it difficult to comply with rigorous privacy laws, such as GDPR and HIPAA.
2. **Communication delays and overhead:** Once again, transferring data from the edge devices back to a server adds significant communication costs associated with both latency and increased bandwidth consumption. In the case of time critical wireless systems, such as self-driving cars, or industrial health monitoring systems, or some form of autonomous robots used in industrial settings, even small latencies associated with transferring data can lead to failure of the system, or more importantly, a safety risk.
3. **Limitations of Scalability:** There is an under-explored class of problems concerning how components can be designed into a singular large-scale system. Functionalities for a central unit are extremely limited by the boundaries of processing, memory, and data capabilities of the system and ultimately results in the system being more probable to deliver lower quality service and be unreachable in anyway during any kind of network congestion.
4. **Failures of Responsiveness to Local Context:** Centralized ML models trained on mixed data do not account at the device level either the local conditions present or the threat patterns of any local context settings. In general, such models will either not perform or detect the anomalous space-time continuum behavior that could occur in a localized region, either at the device or user behavior, and significantly understate the relevant institutional context accuracy and reliability.
5. **Edge devices energy and resource consumption:** In many situations, devices belonging to the wireless network, like sensors, smartphones or wearable devices are constrained in terms of power, memory and even processing capabilities. The need to upload large sets of data to a central server on a regular basis takes away these scarce resources within devices, thus lessening their efficiency and lifespan, making centralized learning inappropriate for extensive ecosystems of wireless devices networks (Ibrahem et al., 2022).

1.3 Overview of Federated Learning (FL)

Federated Learning (FL) is a novel branch of decentralized machine learning where several devices or nodes called clients collaboratively train a shared model without exchanging the raw data. Clients process their data locally and only share model updates (e.g., gradients, or parameters) with a coordinating server. These updates are sent back to the server in order to enhance the global model, and since FL processes data at the edge, it maintains privacy because the sensitive data only leaves the device as aggregated data, and also decreases the communication required to send data back and forth. These techniques will allow a shift from traditional centralized learning paradigms to a more privacy-aware adaptable structure, which is advantageous in a setting like wireless networks with an abundance of real-time data streams. Federated Learning (FL) is moving away from the centralized structure and is beneficial in more sensitive privacy related scenarios with a network full of devices generating myriads of real-time data, such as wireless networks where devices will also differ in hardware, data distribution, and network connectivity. FL accommodates heterogeneity in device structure with federated averaging, personalized training, and asynchronous training. These approaches can also improve learning efficiency and still maintain fairness and inclusivity across a potential large and diverse set of nodes. With respect to the cybersecurity of wireless networks, Federated Learning (FL) can provide a degree of assurance that sensitive data remains confidential in the creation of a model. The privacy borne by FL is very limited as the raw data remains on the actual device. This is why FL is suitable for the development of privacy-preserving frameworks, especially regarding decentralized intrusion detection systems (IDS), anomaly detection systems, and client authentication systems that process sensitive, or regulated information (location, device behavior, identity) in a secure way (Hallaji et al., 2024).

Additionally, FL enables intelligence to be executed in real-time at the edge, which allows for more autonomy and responsiveness to local threats on wireless devices without having to rely on central servers. This ability to be fully responsive is important in mobile applications, smart city environments, and IoT ecosystems with high latency and limited communication bandwidth. FL promotes energy efficiency because it reduces the amount of large data that needs to be frequently sent through the wireless channel and can even support low frequency self-organized intermittent connectivity through asynchronous model updates. However, there are challenges to FL. Example broad research areas include model convergence assumptions of non-independent and identically distributed data (non-IID), trust management for clients, poisoning vulnerabilities, and efficient aggregation algorithms. All that being said, FL is gaining traction in multiple domains across several industries overall, due to the potential to have a balance of privacy, decentralized scalable intelligence, and other very on demand elements for the next generation of secure wireless network systems (Gupta et al., 2024).

II. FUNDAMENTALS OF FEDERATED LEARNING

2.1 What is Federated Learning?

Federated learning (FL) is a collaborative machine learning paradigm that aims to minimize the data aggregation process required for training the models by keeping the data on the device. Unlike traditional paradigms of centralized learning that requires all of the data be aggregated in one location to be stored and processed, FL allows for each device to train a model with the locally connected data. After the computation is completed, only model parameters that are learned and not the raw data, are made available to the central server or coordinating device. FL, at its heart, elegantly communicates solutions that factor in privacy, while autonomously breaking the training process up among devices. Each device computes using its data, produces the update, sends it to a centralized aggregator that merges the data or performs an FL algorithm (Federated Averaging) to converge on a single global model. Once the global model is produced, it is sent back to the devices for local training for additional iterations until convergence or a specified level of accuracy is reached. FL is beneficial in situations with a number of constraining factors - privacy, limited bandwidth, and data availability. FL can be illustrated in the wireless network scenario. Therefore, Federated Learning is increasingly utilized in industries such as health, finance, telecommunications, and smart cities. It can also facilitate cross-device FL and cross-silo FL (fewer, but stable institutions like hospitals or banks with larger datasets).

Federated Learning is intentionally designed to help address several important challenges in computing today which are:

1. **Data Privacy and Ownership**: Users retain control of their data, ensuring compliance with data protection laws.
2. **Reduced Communication Costs**: Only model updates are shared, significantly decreasing bandwidth requirements.
3. **Personalization**: Models can be fine-tuned for individual devices or groups without affecting the global model's performance.
4. **Scalability and Resilience**: FL is naturally scalable across thousands or even millions of devices and is more robust to single points of failure.

Overall, Federated Learning represents a paradigm shift in how intelligent systems can be built collaboratively and securely, making it an ideal approach for privacy-preserving security in wireless networks and beyond.

2.2 Key Components of FL

An FL system is made of many parts working together in concert for a common purpose: safeguarding privacy enables decentralization of the model training to operate in FL. To help understand how FL is different from other models like centralized learning, and especially to understand how FL can be utilized in wireless network security or otherwise, one must first learn the building blocks of an FL system. Below is an outline of the primary building blocks of a generic FL system:

1. **Clients (Participants or Edge Devices):** Clients are the distributed devices or nodes that possess local data and computational capabilities. These can include smartphones, IoT devices, sensors, or any networked endpoint. Each client performs local training using its private data and contributes learned model updates to the global model. Clients may have heterogeneous data, hardware capabilities, and availability schedules, which introduces challenges in FL such as non-IID data and client dropout.
2. **Central Server (Aggregator or Orchestrator):** The central server acts as the coordinator of the FL process. It is responsible for initializing the global model, selecting which clients will participate in each training round, aggregating model updates, and redistributing the updated model back to the clients. Importantly, the central server does not have access to the clients' raw data—only to their model parameters or gradients. In some FL architectures (like fully decentralized FL), this role may be distributed or handled using peer-to-peer communication or blockchain (Vashishth et al., 2025; Xue et al., 2024).
3. **Local Training Mechanism:** Every client trains their model within the confines of their private data using the current global model retrieved from the server, which they received earlier. The training is mostly done under standard optimization criteria like Stochastic Gradient Descent (SGD). Each client determines model update calculations (for example, gradient or weight calculation), then returns the results to the server for summation fusion.
4. **Model Aggregation Function:** Upon receiving model updates from multiple clients, the server will utilize an aggregation function to create an updated global model. The default method of aggregation typically used is Federated Averaging (FedAvg) which calculates a weighted average from client updates based on the local dataset size. With aggregation, it helps ensure model convergence and consistency over a period of time.

5. **Communication Protocol:** Client communication to the server is all regulated within an efficient and secure communication protocol. Clients are constrained by bandwidth and power levels (e.g., in a wireless setting) so any FL system is sensitive to increasing communication burden through communication costs, update compression, periodic training, and adaptive synchronization.

2.3 FL Workflow in Wireless Network Scenarios

In a wireless setting, the Federated Learning (FL) workflow describes a specific workflow where models are developed on multiple wireless end-user devices such as smart phones, Internet of Things sensors, routers, and edge servers without sending raw data through the network is illustrated in figure 2. This country propagation of learning models in a distributed manner is important to protect user privacy and minimize the communication overload, both important in bandwidth-constrained or privacy-sensitive wireless settings.

Figure 2. Federated Learning Workflow in Wireless Network Scenarios

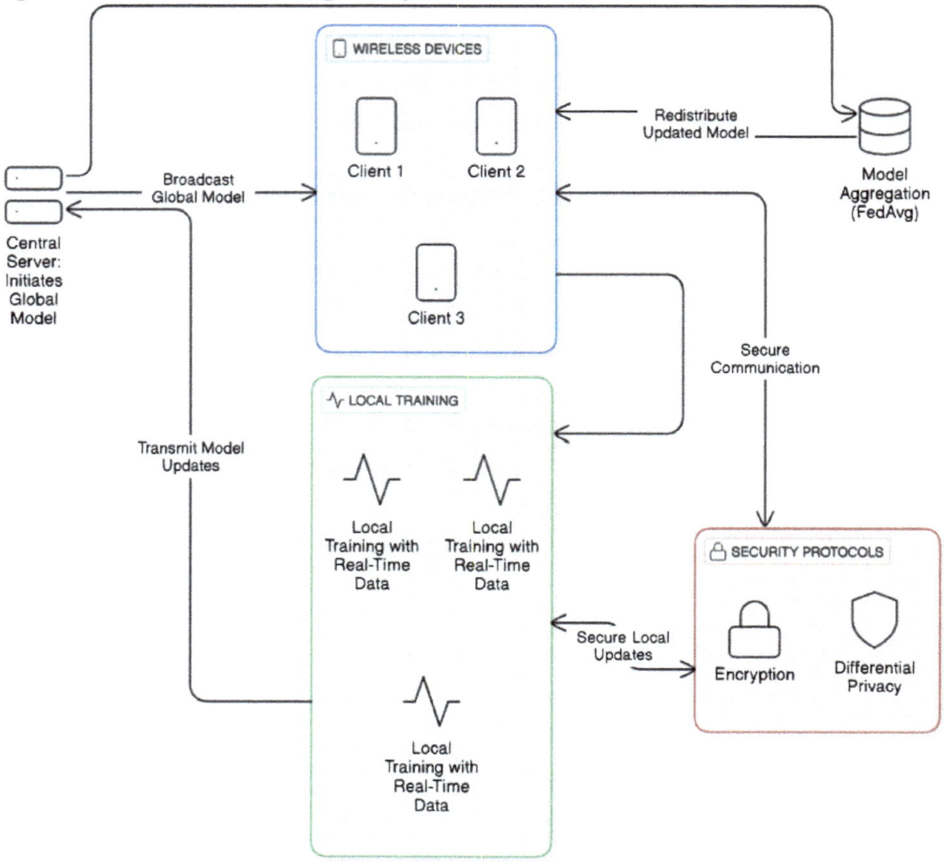

In a conventional Federated Learning (FL) process, a central server or orchestrator will first generate a global model that is then sent to a selected subset of clients or wireless devices. The clients accept the model which they will then train locally with their own streams of data, which may consist of traffic flow data, device logs, or even sensor data. With wireless networks, this local data is often unique to the specific location of the device and non-IID (not Independent and Identically Distributed, meaning different data points are not one and the same likelihood), which is important factor for Federated Learning, as a core aspect of Federated Learning is that it enables the training process to cross unique data without being required to maintain a central location to hold the data. After the client/device has finished training locally, each client/device will compute a model update to send back to the central server over the wireless connection. This update will include, but is most likely not limited to gradients, and weight changes. More importantly, the device will not send raw data, as with data sensitive information such as user's activity, location, or authentication details stay private to the device. Then the server will perform model aggregation, usually by applying the Federated Averaging (FedAvg) algorithm. The FedAvg algorithm uses the updates of all participating clients and weighs these with respect to the local dataset size and quality, which will improve the global model to be a better used global model. After model aggregation, the clients obtain the updated model for local training which creates a cycle into the learning process. In a wireless setting, other modes like secure aggregation, encryption, or differential privacy are also integrated within the process of workflow with a goal of improving security against risk of attack, or ensuring private information was NOT in fact revealed in model updates. Variable connectivity, and device capability are accounted for with client selection and adaptive training schedules. These accommodations allow FL systems to handle intermittent connectivity which is an issue in wireless. Overall the FL processes employed to wireless networks represent a different direction rather than primarily just being data collection terminals, the edge devices themselves are re-assigned to be data collection participants in such a way that does not violate privacy. In fact, it adds security to the network infrastructure in that threat detection, and anomaly detection, in addition to multi-point learning with millions, and millions of endpoints can happen All in real time and it doesn't interfere with communications pathways or sensitive user data.

2.4 FL in Edge and Fog Computing Environments

By leveraging edge and fog computing, FL creates a strong triad of enhancing data privacy, latency, and resource usage in a wireless environment. These models are designed to perform computation as close to the point that data is generated as possible and, as such, are suitable for FL use in a decentralized model training context. In this regard, FL can be functionally infused into the edge or fog layer for distributed intelligence while preserving privacy and responding to network needs in real-time. In edge computing, computation takes place along the outer edge of the network that includes end devices, which may include smartphones, sensors, access points, and embedded systems. With FL at the edge, these devices can train the models locally using data identified in situ and send model updates to a central coordinating unit. Thus, alleviating the strain of excessive communications on cloud servers while reducing the time and memory bandwidth/frequency of latency deemed important for time-sensitive missions such as intrusion detection or monitoring for anomalous behavior in wireless networks. Fog computing, the middle man between connected devices and the cloud, allows for intermediate level processing for added computation capacity and memory storage. FL in fog situations provides local aggregator nodes (usually fog nodes or micro data centers) with model updates from nearby edge devices that perform partial aggregation of the model and optimize sending the updated model to a cloud or centralized server. This kind of distributed architecture framework (Vashishth et al., 2024) allows for a more distributed workload but also represents another layer of security, with data not needing to transmit over larger distances or traverse multiple untrusted networks.

Further, FL with fog and edge computing enables context-aware learning, whereby the models alter to local conditions such as geolocation-dependent user behavior, traffic pattern heuristics in a network, or environmental changes, without jeopardizing privacy. Consider, for example, the federated smart cities' anomaly detection system deployed at fog nodes in various districts. Each fog node can manage model parameters based on the local traffic data, while still conforming to the global model. These systems are also purposefully designed to improve scalability. With the increased number of devices within wireless networks, edge and fog-based FL enables the datasets to grow exponentially without inundating central servers or cloud infrastructure. Also, this architecture improves energy efficiency and resiliency since computation and storage are managed in a distributed manner. Even with these benefits, there are challenges, such as heterogeneity of the devices, low processing power, intermittently available connection, and coordination difficulties when implementing FL in edge and fog scenarios. Some of these issues are currently being addressed through adaptive federated optimization, asynchronous learning protocols, and model compression.

III. PRIVACY AND SECURITY CHALLENGES IN WIRELESS NETWORKS

The flexibility, scalability, and cost-effectiveness of wireless networks have made them one of the most popular forms of communication technology today. However, the mobility and decentralization features of wireless communication also pose a significant risk to privacy and security. This sub-section aims to outline the major gaps and challenges in security concerning wireless networks.

1. **Eavesdropping and Data Interception:** In wireless environments, eavesdropping is a major concern and poses significant risks. Attackers do not need physical access to network infrastructure (NI) to intercept and monitor communications as radio signals are broadcast in nature. Not exploiting weak encryption methods can compromise highly sensitive data including personal details, authentication information, and financial transactions.
2. **Unauthorized Access and Rogue Devices:** Wireless networks can be easily exploited by attackers who obtain unauthorized access. Rogue devices have the ability to impersonate legitimate access points alternatively known as Evil

Twin attacks. Rogue devices have the capability to further leverage access into the network where they have the ability to steal data, install malicious software and can even launch Denial-of-Service attacks (DoS).

3. **Denial-of-Service (DoS) and Jamming Attacks:** These DoS attacks and Jamming attacks, aim to overwhelm and deny access to a network by either flooding a network with excessive amounts of illegitimate traffic or signal interference. Such a denial-of-service including jamming could have serious ramifications disrupting mission-critical applications and services provided in areas like healthcare, industrial IoT and emergency response systems.
4. **Privacy Violations and Tracking of a User's Location:** Concerns about privacy risks are amplified with mobile computing. Cybercriminals can misuse signal characteristics and specific metadata such as MAC addresses and device fingerprinting to track the user's location, activity and even make assumptions about their interests. If this situation occurs in components of the world where there are strict data protection responsibilities, it raises serious ethical dilemmas and risks for legal mechanisms.
5. **Loosely Defined User Authentication and Access Definition:** A combination of password, identification number and other credentials can form a unit of a password so long as the information is used for the wireless communication being protected. Unfortunately, many idle IoT (Joshi et al., 2024) and sensor networks often do not have sufficient computing power and therefore resort to primitive security features which leads to a significant exposure to MitM, replay, and impersonation attacks.

IV. FEDERATED LEARNING SYSTEM IN WIRELESS NETWORK SECURITY

The rapid expansion of wireless devices and data-heavy services calls for adaptive and resilient security solutions to be implemented. Traditional approaches to centralized machine learning (ML) creates privacy and data confidentiality concerns across collected data through a central service approach, while FL provides a decentralized ML alternative with data privacy protection suitable for wireless networks.

4.1 Federated Intrusion Detection Systems (FIDS)

FIDS is one of the most advanced developments in securing wireless networks and is built on the principles of Federated Learning (FL). The architecture of FIDS differs greatly from classical intrusion detection systems in that they do not collect and process information in a single data continuum. Instead, many edge devices, such as smartphones and Internet of Things (IoT) nodes or wireless sensors, can perform local training of their machine learning models using their own network traffic data when FIDS are implemented in a system. These edge devices transmit only the learned model parameters, either in the form of gradients or weights. This design improves user privacy while reducing the need for bandwidth and decreasing the likelihood of a data breach. FIDS is adaptable to the wireless environments in existence today since security threats will be, by nature, heterogeneous and dynamic. By managing anomaly detection in real-time locally, while also synchronizing to a global model, FIDS offers an architecture for collective defense no matter where such emerging threats evolve and every time the global model is updated. This approach improves the effectiveness of the intrusion detection system, with the multitude of devices, each with its strengths and weaknesses. The fact that each device can update the model for its respective network environment makes for a better, more tailored threat detection. For all its upside, a FIDS has limitations. In particular, non-independently and identically distributed (non-IID) data on devices can greatly diminish the performance of the aggregated model. There's a risk of model poisoning attacks (either compromised or malicious devices that upload malicious modifications that diminish the accuracy and/or trustworthiness of the global model). These challenges pose a significant burden, especially considering the employed security communication protocols, aggregation strategies, and resilient model training approaches.

4.2 Federated Anomaly Detection

The security of wireless networks has received significant attention from researchers, and Federated Anomaly Detection (FAD) is common as a Federated Learning (FL) application, which combines the "intelligence" and "federation" of devices into anomaly detection frameworks, for improved network security. Anomaly detection is one of the ways cyber-attacks are identified, including zero-day attacks. Yet, the operation of anomaly detection, like many traditional measures, is complicated further by the need to transfer significant amounts of data from all devices to one location, which can lead to privacy concerns, higher latency, or violate data protection laws. As compared to CAD systems, FAD enhances the overall efficiency of detecting unknown anomalies because edge wireless devices are able to collaboratively associate and train machine learning models while their sensitive information is safeguarded by being stored locally. In a federated model, each device examines its own network activity and searches for a set of signs suggestive of deviation from baseline activity, for instance, abnormal flows of data packets, normal logins, and rogue accesses to system resources. Each unit builds local models for anomaly detection such as autoencoders, RNNs, and statistical classifiers, and each system compiles them into a single statistical model without the need for data sharing. These systems share the obtained parameters from their local models safely for ensemble learning, creating a surrogate model capable of encapsulating the entirety of the work. With collaborative intelligence deployed for the whole wireless network, detection accuracy and robustness is improved while maintaining privacy. Existing research is mainly focused on improving the algorithmic efficiency of model fusion frameworks, adding differential privacy, and secure multiparty computation to improve confidentiality and augment the resilience and usability of Federated Anomaly Detection, as is perplexing to imagine within real-world limits of wireless networks (Vashishth et al, 2024).

4.3 Federated Authentication Mechanisms

Federated Learning (FL), presents a more decentralized system for federated authentication. Within a federated learning system, the control over the model training leveraged for authentication is shared, enabling user devices to learn authentication schemes in a less exposed manner. In federated authentication systems, user devices, such as mobile phones, smartwatches, or other IoT-enabled peripheral sensors, will be responsible for self-training models to recognize a certain biometric trait of a user (face picture, fingerprints, voice), or psychosomatic characteristics in a much more complex form (i.e., speed of writing something, pattern of clicking a mouse, habits when using a particular device). The predicted properties of the traits listed above will be able to identify the user as known. These self-trained models are subsequently collated to create a global or modal one, which for each user maintains its privacy while capturing uniqueness in identity. This enhances continuous and self-modifying learning of authentication behaviours, improves system accuracy while increasing resistance to spoofing or identity impersonation attacks (Yodha, 2024). Federated Authentication Techniques are extremely useful for wireless networks that are used by mobile or remote users, where secure and seamless access control is of utmost importance to the user and user group. Federated authentication approaches lessen reliance on centralized infrastructure, enhance user privacy by keeping authentication data on the device, and allow scalability, where devices can self-update models without needing to resynchronize a central access control point. Nevertheless, federated authentication enhances privacy, embellishes intelligence, and boosts user's authoritative control on access restriction oversight decisions in the next generation of wireless networks. Much of the issues relating to securing pathways to relay model update, protecting malicious devices from conducting poisoning attacks, inconsistent data formats, and unbalanced user restrictions based on level-of-interaction-patterns-with-other-users maintains suboptimal solutions to address other challenges inherently presented.

V. ADVANTAGES OF FEDERATED LEARNING FOR WIRELESS SECURITY

FLs is a new frontier in terms of better wireless security as it enables processing of data in its locations and collaboration learning of models rather than a central unit operation. These capabilities are critical in wireless networks since devices are highly distributed, resource constrained systems, and are processing sensitive information.

1. **Privacy Preservation:** A primary and one of the most advanced applications of Federated Learning (FL) regarding wireless security is user privacy preservation since FL offers unique capabilities for preserving user privacy. In typical machine learning systems, user data from each device is sent to a central server to facilitate data processing and model learning. This practice increases the potential for the data to be intercepted, compromised, or shared in an unauthorized way. By comparison, FL is designed to guarantee that the users' raw data is not leaving the edge device. This is helpful in wireless scenarios where personal or sensitive data including user behavior, users' geolocations, or biometric information are regularly processed (Wang et al., 2023).
2. **Reduced Bandwidth and Network Load:** Wireless networks have significantly limited precision bandwidth; therefore, sending substantial amounts of information to a hub is both inefficient and costly. Sending minimal amounts of model updates rather than excessive amounts of raw data results in an enormous decrease in network load, which is what created the popularity of FL. Also, FL typically only sends model updates (which are small) instead of raw data, which also leads to far less communication needs at the network level. Therefore, FL represents a great advancement in energy conservation, which is particularly important when it comes to battery-operated wireless devices such as IoT sensors and smartphones. Consequently, federated learning promotes sustainable and efficient operations at the network level (Zhou et al., 2023).
3. **Detection of Threats with Minimal Delay:** FL can provide near real-time or true real-time detection of threats by allowing devices to perform data analysis and processing on site. That means it must be able to make the decision on site, rather than wait until the data is acquired and sent to another connected device, which could mean another connected separate device or an actual process occurring on a remote server, possibly with queuing delays in waiting to make a decision and possibly needing to have had its own communication or processing bottleneck. As a result of using few resources for a federated set of computations, the federated system can acquire threats faster, therefore improving the reliability and speed of responding to and remediating threats throughout the entire network, and improving its overall effectiveness.

4. **Heterogeneous Networks:** Wireless networks usually consist of a heterogeneous mix of different devices involved in data generation and processing. FL is scalable by design and accommodates such heterogeneity by allowing all devices to train the model as well as the data they have to offer. For example, devices with more data and better compute capabilities will take on more roles, but devices with less resources can still participate in the process without taxing their limited resources. This flexibility allows FL to be deployed on a broad base, especially in complicated wireless protocols like smart cities, industrial Internet of Things (IoT) applications, or 5G/6G networks (Teixeira et al., 2025; Narooka et al., 2025).
5. **Improved Fault Tolerance and Resilience:** Centralized security systems have this downside of single point of vulnerability; if that one central server is either attacked, goes offline or crashes, the whole system will be compromised somehow. FL addresses that by switching to a decentralized solution so that fault tolerance and resilience are improved for the wireless security planning. Even if one, or more than one, agent is compromised or down in the system, the remaining portions of the system can continue to operate and can still learn from the other nodes that are still viable. This is beneficial for the survivability of the wireless system, especially in critical roles, such as military communication, emergency services, or remote sensing missions (Arora et al., 2024; Fnu et al., 2025).

VI. Frameworks and Architectures for FL in Wireless Networks

In order to allow FL to be effective within wireless networks, there needs to be solid frameworks and service or functional architectures to alleviate the training process decentralization, security, device heterogeneity, and dynamic network conditions. It includes all the necessary elements to manage the cross-device learning process while maintaining data security and communication. FL in wireless networks has several popular supporting frameworks such as:

6.1 System Architecture for FL-Enabled Wireless Security

The structure of the system for wirelessly securing Federated Learning (FL) is intended to support collaborative, privacy-sensitive threat detection and mitigation within distributed wireless systems. The architecture's primary elements include client devices, which are smartphones, IoT sensors, access points, and edge nodes. These devices autonomously gather information and execute local training of machine learning models on security-related data such as network traffic, authentication logs, and anomalous indicator patterns. Along with helpful information, user privacy is maintained and communication costs lowered by not sending sensitive raw data to a central server, but instead only sharing model updates (gradients, weights, etc.) with a coordinating server. In more complex installations, a hierarchical architecture is employed introducing an edge layer composed of edge servers, base stations, or fog nodes. These entities function as local aggregators: collecting device model updates, performing initial filtering, and anomaly detection of the shared parameters prior to forwarding them to the central aggregator. Focusing on latency, the edge layer alleviates the communication burden on the central server and offers quicker localized responses to new risks. In the highest level of the architecture, a central aggregator typically refers to the cloud-based control center or server. Global aggregation is done for all model updates received under the distributed aggregated methods, including Federated Averaging (FedAvg), through more secure methods using differential privacy, and multiparty secure computation. A global model is established and can be downloaded by all devices, which allows the system to continuously improve and optimize its collective intelligence. The global model will provide all devices access for monitoring. This process begins as the refinement of intelligent networks. Since the refinement of intelligent networks gets provided from devices, feedback algorithms enable networks to transform behavioural patterns into more complex systems intelligence. There is also room for blockchain technology to increase traceability. In summary, the architecture proof systems provide decentralized correction mechanisms, data concerns, and coordination within a hierarchy, with continuous learning to develop a wireless security system that will withstand attacks and intelligent (Orabi et al., 2025).

6.2 Middleware and Communication Protocols

The seamless functioning and performance of Federated Learning (FL) systems in wireless networks significantly depends on their middleware and communication protocols. For example, middleware, is software which resides above the hardware, encompassing sensors or mobile devices in addition to the applications to appropriately enable the interaction and allocation of resources and data. Middleware also addresses operational aspects of importance critical to FL enabled wireless security networks is shown in figure 3. One of the primary purposes of middleware is to obscure the complexities of operating in the underlying network infrastructure. In that regard, the Federated Learning paradigm assists students in dropping-in and -dropping-out around group sessions and providing their model contributions in their times of contributively engagement where they do not disrupt the whole learning process and timely engagement of the learning practice. Middleware does this while striving to maintain and effective load-balancing mechanisms across a merged network without compromising its overall coordination of work to the specifications of device qualities and work abilities and provides data continuity integrity. Just as with the edge devices, the intermediate servers, and the main aggregator, sealed, powerful, and meritocratic transfer protocols are necessary for constant data transfer. These protocols are setup to bilaterally maintain bandwidth consumption protocols with the condition of dependable and reliable communication in adverse conditions (Bukhari et al. 2024). Data security is stronger when more advanced MPM protocols are added such as Secure Multiparty Computation (SMC), Homomorphic Encryption, and Differential Privacy which can be incorporated in the middleware increasing security for the model update transmission. This continues to allow model merging without infringing on individual privacy. Additionally, trust requirement verification, transaction persistency, and device reputation management regarding the participating devices (Vashishth et al., 2024) can be handled through the smart contracts on the blockchain incorporated in the middleware.

Figure 3. Middleware and Communication Protocols in Federated Learning

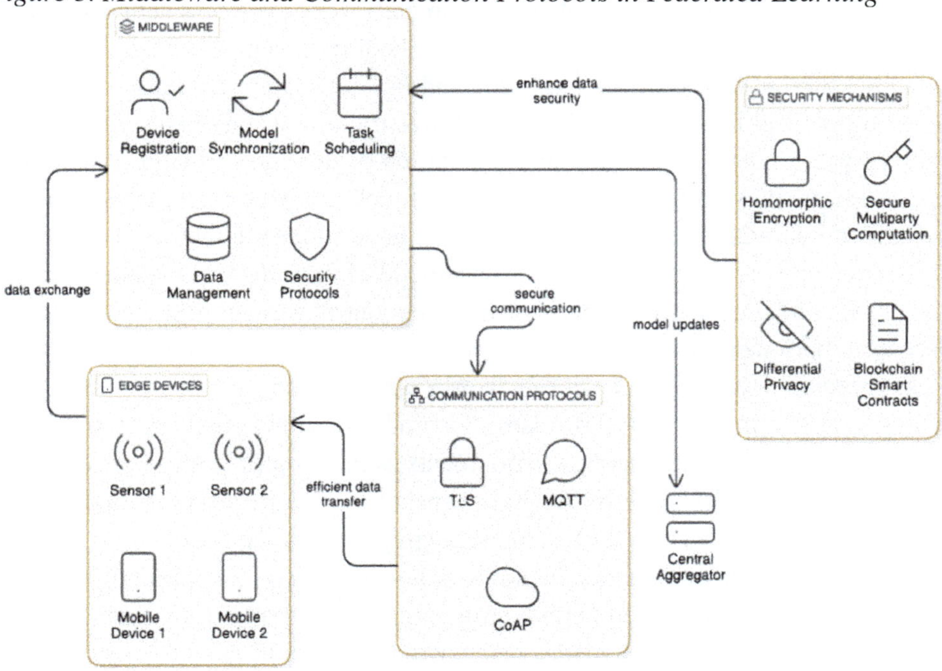

6.3 Optimization Algorithms for Efficient FL

Due to the unique challenges associated with Federated Learning (FL) in wireless networks like the heterogeneity of the data, limits on the communication of the data, and limited computing capabilities, optimization algorithms are essential for improving its performance. These methods establish the formalism of how model updates from many independently operating devices will be computed, merged together, and filtered out to yield rapid convergence, lower costs of communication, and obtaining the best model performance in dynamic wireless environments. One of the most well-known optimizations related to FL is Federated Averaging (FedAvg) in which each client engages in a certain number of training local steps on the private data. The client then transmits the parameters of the model it was using to the center server, which averages the parameters and uses this to update its global model. While FedAvg suffers from slow convergence rates in very heterogeneous settings like FL, it outperforms other methods in terms of communication efficiency and is able to adequately address the issues associated with non-IID data (non-identical and independent data) that are characteristics of wireless networks. More advanced or sophisticated versions including FedProx, have now been developed to address these issues. FedAvg was adapted in FedProx by including a proximal term to the local loss function to help stabilize training when clients vary greatly with respect to data distribution or computation capabilities. Finally, another optimization is SCAFFOLD, which stands for Stochastic Controlled Averaging for Federated Learning and shows a more promising development that addresses client drift, which is the problem in FL when local models follow the bias in data instead of the global model. A scaffold algorithm involves control variates to stabilize updates that local optimization to the global objective to optimize communication rounds. Less is required. Moreover, a set of adaptive optimizations is being developed for Federated Learning (FL) such as Adam, Adagrad, and Yogi as they ascertain chronicling frameworks in fluctuating environments as these optimizations are based on historical data. Additionally, SCAFFOLD enables devices to remotely update their models with local data and send them for aggregation which further alleviates the time required to onboard device in a uni-framework. Also, Tanbhir and Shahriyar (2025) applied apportioned compression, quantization, and smartification techniques to reduce the magnitude of updates, while avoiding generating high error, with the aim of mitigating communication cost.

6.4 Integration with Existing Wireless Infrastructure

Integrating Federated Learning (FL) into the wireless fabric is a necessary factor to effectively deliver flexible, scoped intelligence into environments that are secure and adapt in real time without changes to the logistical systems themselves. The goal is to showcase the functionality of FL at all tiers of wireless networks from access points, to base stations, to handsets and edge servers, while being mindful there are existing interfaces and hardware that must be accounted for. Wireless networks particularly with regards to 4G, 5G and future capabilities with 6G, inhabit stateless, distributed infrastructure resources that support FL deployment out of the box. Local training instances, a pre-aggregation model, and (likely) real time security assessment are conceivable with base stations, routers, and edge compute nodes. All of these devices are collaborative peers in the FL ecosystem, combining their available data and processing and storage, taking advantage of distributed architecture to protect sensitive information. With a Mobile Edge Computing (MEC) solution, the complexities and challenges of integration are reduced by shifting computation and data storage to the edge of networks. MEC nodes can provide FL services, recharge FL local devices, and be the endpoint to a central server. This arrangement improves response time and also reduces global decision-making for applications, such as intrusion detection and anomaly predictions. FL might also be implemented using software-defined networking (SDN) in a majority of wireless infrastructure frameworks. It provides the ability to allocate resources and establish routes dynamically for federated learning (FL) model training while optimizing available bandwidth for model updates specifically in a time-sensitive environment. 5G networks also provide added isolation and security mechanisms for federated learning functions with the use of network slicing. This allows dedicated logical sub networks to be created for virtual isolated FL operations. Another cross-compatibility consideration is equally important. FL ecosystems must be agnostic regarding device types, operating system types, and wireless communication interfaces such as; Wi-Fi, LTE, Zigbee, and NB-IoT. Including an application for mobile or firmware operation that contains shallow FL libraries and APIs is one way to improve the participation of low-end devices during the FL training rounds. To protect FL data and processes from active breaches in a system, confidentiality of data security measures like secure boot, trusted execution environments (TEEs), and encrypted communications must be revered. These measures uphold the privacy of participants in FL from being subjected to malicious manipulative attacks (Chen et al., 2024).

It is, therefore, reasonable to conclude that FL integration into the current wireless infrastructural network is advantageous and is at liberty. It draws on emerging innovations in edge computing, 5G/6G, Software Defined Networking (SDN) and Cybersecurity to establish an intelligence layer that is collaborative, privacy-aware, and self-sufficient, thus forward-looking. This incorporation enhances the functionalities of existing wireless security systems while enabling proactive network management in the future.

VII. CASE STUDIES AND USE CASES

Federated Learning (FL) is rapidly developing with various practical implementations in the context of wireless network security through numerous real-life case studies. These implementations validate that FL boosts data privacy, enables discourse intelligence, and enhances threat detection without the risk of accumulating sensitive data in one location.

1. Federated Learning for IoT Device Security in Smart Homes

Scenario: In smart homes, IoT devices such as smart TVs, security cameras, smart plugs, and virtual assistants are continuously interacting with the network and generating a large volume of data. Traditional centralized security solutions pose privacy risks and aren't scalable for these distributed environments.

Case Study: A pilot project was conducted where each smart home had a local Federated Intrusion Detection System (FIDS). These systems used FL to collaboratively train a global anomaly detection model using locally observed network behavior. For instance, a smart thermostat detected unusual access patterns or command sequences locally, trained a model, and sent only the model updates (not raw data) to a central aggregator.

Impact:

a) Protected user privacy by keeping raw data on the devices.
b) Reduced communication overhead.
c) Enabled detection of large-scale botnet attacks like Mirai without compromising user data.
d) The system became smarter over time, learning from diverse data environments across multiple homes.

2. Federated Malware Detection in Mobile Networks

Scenario: Telecom companies and mobile users are constantly battling mobile malware, which evolves quickly and often goes undetected by traditional antivirus software. Sending user app data to centralized servers for analysis raises privacy concerns.

Case Study: In a project by a major telecom operator, FL was implemented across millions of Android devices. Each smartphone trained a local model on app behavior data (e.g., permissions, CPU usage, background activity) and shared encrypted model updates. The central server aggregated these updates to improve a shared malware detection model.

Impact:

a) FL detected previously unknown malware variants by learning from diverse device environments.
b) The privacy of user data was maintained as sensitive logs never left the devices.
c) Allowed continuous learning without requiring constant network connectivity.
d) Reduced computational burden on central servers and data centers.

3. Secure FL in Industrial Wireless Sensor Networks (WSNs)

Scenario: In critical industrial settings—such as oil refineries, power grids, or chemical plants—wireless sensors monitor environmental conditions and equipment health. However, due to sensitive operational data, privacy and real-time processing are top priorities.

Case Study: An industrial firm deployed FL across a distributed Wireless Sensor Network (WSN) to monitor pipeline conditions like pressure, flow rate, and temperature. Each node ran a lightweight ML model for anomaly detection. Model updates were sent to fog computing nodes (local edge servers) for aggregation before reaching the cloud.

Impact:

a) Early detection of pipe leaks or equipment failures without centralizing sensitive telemetry.
b) Localized processing reduced latency and avoided downtime.
c) Compliance with data protection laws for operational environments (e.g., GDPR, industry-specific regulations).
d) Enhanced resilience against cyber-physical attacks targeting the infrastructure.

VIII. FUTURE DIRECTIONS AND RESEARCH OPPORTUNITIES

8.1 Future Directions: Wireless networks are increasingly becoming larger and more complex, leading to higher demand for security and privacy—hence

FL (Federated Learning) could be an option. Its attributes are aligned with the current configuration of distributed wireless and next-generation systems. In the future of FL, the overall security approach will change, specifically in heterogeneous and large-scale networks, with artificial intelligence, in combination with advanced 6G, edge computing, and blockchain. Specifically, these will illustrate how FL when needed to address issues will have consequences for an adaptive, real-time, and privacy-preserving solution to avoid advanced persistent cyber threats (Moulahi et al., 2023).

1. **Merging FL With 6G Networks and Edge AI:** As we progress toward 6G Networks, the fusion of FL with ultra-low latency and high-throughput 6G environments will also be timely. Distributed edge processing through FL can be incorporated as part of 6G Edge Computing Model for real-time security analysis, distributed management, and computation intelligence. 6G is suggesting an explosion of device connections and intelligent services, which will require secure, scalable, and intelligent mechanisms of data processing in the edge-to-cloud continuum, and that will depend on FL (Vashishth et al., 2024).
2. **Enhancing Privacy-Preserving FL with Novel Cryptographic Approaches:** The most significant privacy disadvantage of using FL is model inversion and potential gradient leakage attacks, even though FL fundamentally reduces the sharing of raw data. Development of future FL systems will include stronger cryptographic approaches, including but not limited to homomorphic encryption, secure multi-party computation (SMPC), and differential privacy that enhance trust in FL applications in advanced wireless networks through allowed safeguarding of the data and disclosure protection from unauthorized disclosure and guaranteed protection to strong verifiable assurance.
3. **Cross Device and Cross-Silo FL in Wireless Networks:** Another consideration is to extend cross-device and cross-silo to include learning in wireless networks that include heterogeneous resources. The extension of FL frameworks that learn, both progressively and cooperatively, over multiple hardware systems and operating systems across multiple network conditions in heterogeneous environments with heterogeneous data distribution will be needed. This will include FL models that can learn whether resources are great or constrained (Vashishth et al., 2024).
4. **Real-time FL and Customizing Individual Models:** For more novel applications that require real-time changes to be made, and for things like distance medicine or autonomous driving, FL models will need to learn in real time for individual customization of model updates. Research in the future will move to more active and adaptable FL research that can account

for changes in conditions, actions, and even level of threats present. Custom FL will take another step toward contextually accurate and high-accuracy security for the user that still provides generalizability.

5. **Network Optimization and Protection Using Federated Reinforcement Learning FRL:** Fusing FL with RL to create Federated Reinforcement Learning FRL is an exciting opportunity, where agents learn optimal policies together in isolated environments. FRL may lend itself to effectively managing network resources, scheduling network traffic, and implementing cybersecurity autonomously in a distributed wireless context, which could eventually lead to more self-organization and healing wireless systems (Dandoush et al., 2024).

8.2 Research Opportunities: All in all, the vast array of open research questions provides opportunity. However, field stated that FL methods specifically for wireless networks and security are rather new. However, some specific aspects of the wireless context (i.e., device heterogeneity, intermittent connectivity, and imbalanced data) could provide challenges but also opportunity for new innovations. The areas of opportunity for improvements of system efficiency and scale, fairness, and explainability, could also provide opportunities to improve trust and usefulness of FL for wireless security in the real world.

1. **Development of Effective Federated Learning Algorithms:** There is a strong importance for developing algorithms for federated learning that can be run on thin Internet of Things sensors, wearables and embedded systems. Areas of consideration are algorithm/model compression, sparse communications, quantization of models, training overhead, and low-energy alternatives. As interest and deployment to federated learning takes place, it is important to consider effective federated learning frameworks to accommodate more used of this technology area for mobile wireless and low power situations.
2. **Development of a benchmark to assess Federated Learning Implementation in a Wireless Context:** There are still gaps in the federated learning field about benchmarking datasets, and evaluation approaches, both in a wireless context. Researchers should investigate developing realistic test beds and develop simulation mediums to model wireless settings with movement, varying bandwidths, and various devices deployed at multiple layer levels. This work is important for developing benchmarks for federated learning algorithms to indicate performance limits, and guarantee performance on multi-layered networks.
3. **Collaborative effort threat intelligence via Federated Learning for Telecommunications:** Telecommunications operators have a huge amount of data available. In principle, this data can be utilized as collective effort

or collaborative threat intelligence to identify threats. Nonetheless, data privacy and regulatory boundaries involving regulations, along with industry competition factors, inhibit threat intelligence sharing environment. Furthermore, there is another concern that will require further inquiry, that will investigate institute a privacy-preserving collaborative threat intelligence system, in which telecommunications operators share experiences from working with distributed datasets. The goal, the hope is applicable cross-practice collaboration can enable some effectiveness improvement in industry an overall cyber defense (Vashishth et al., 2025).

4. **Decision-Making:** The federated learning (FL) model utilized in wireless security models, including intrusion detection or authentication, must have an explanatory nature. There is a need to identify and examine explainable AI (XAI) that was developed and deployed specifically for FL, in order to help network administrators understand the specific contextual information as to why actions were classified as alerts, or why decisions of security were made. From there, the reliability and ultimate acceptance of FL systems will follow.
5. **Addressing Bias, Fairness, and Trust within FL Models:** Fairness in federated models is an emerging area of research, particularly when devices utilize non-IID data on device types. The results of federated models may be assessed against potential skew of results due to bias or under-representation of nodes. Further research may also engage with FL algorithms that consider equitable treatment of participants in the learning framework designs based with fairness. Strategies for establishing trust, or establishing legitimacy, of participants, device specific, in FL systems to assess trust domains are also an area for research.

IX. CONCLUSION

Federated Learning or FL is an innovative model that bolsters security in wireless networks by facilitating decentralized and collaborative model training without revealing user data. This chapter presented many aspects of FL's contribution to current security-related issues in wireless networks including privacy, attacks, intrusion detection, anomaly detection, and authentication. FL's distributed architecture enables it to provide real-time intervention to live threats, individual customized security policies, and large-scale defences across many heterogeneous devices and infrastructures. Several architectures and frameworks were surveyed and used to explore the significance of middleware and communication protocols, optimization methods, and integration methods for implementing FL-based security systems. The utility of FL for protecting modern wireless networks was discussed in this chapter through actual case studies. Interesting future research was suggested, such as the possibilities of FL and 6G. This chapter also highlighted clear gaps for research in privacy-preserving algorithms and explainable AI. As an emerging research theme, FL as an approach to securing privacy is an opportunity without risk, intelligent, human-centered, and adaptive to the complexity and size of wireless networks.

REFERENCES

Al-Huthaifi, R., Li, T., Huang, W., Gu, J., & Li, C. (2023). Federated learning in smart cities: Privacy and security survey. *Information Sciences*, *632*, 833–857. DOI: 10.1016/j.ins.2023.03.033

Alazab, A., Khraisat, A., Singh, S., & Jan, T. (2023). Enhancing privacy-preserving intrusion detection through federated learning. *Electronics (Basel)*, *12*(16), 3382. DOI: 10.3390/electronics12163382

Arora, P., Khullar, V., Kansal, I., Kumar, R., & Popli, R. (2024). Privacy-Preserving Federated Learning System (f-PPLS) for military focused area classification. *Multimedia Tools and Applications*, *84*(12), 1–27. DOI: 10.1007/s11042-024-19358-7

Batool, H., Anjum, A., Khan, A., Izzo, S., Mazzocca, C., & Jeon, G. (2024). A secure and privacy preserved infrastructure for VANETs based on federated learning with local differential privacy. *Information Sciences*, *652*, 119717. DOI: 10.1016/j.ins.2023.119717

Bukhari, S. M. S., Zafar, M. H., Abou Houran, M., Moosavi, S. K. R., Mansoor, M., Muaaz, M., & Sanfilippo, F. (2024). Secure and privacy-preserving intrusion detection in wireless sensor networks: Federated learning with SCNN-Bi-LSTM for enhanced reliability. *Ad Hoc Networks*, *155*, 103407. DOI: 10.1016/j.adhoc.2024.103407

Chaudhary, S., Singh, V., Malik, N., Sidhu, K. S., Kirola, M., & Joshi, K. (2024, October). Digital Signature Security Enhancement: Using X-OR Operation on RSA and AES with 4096 Bits Key Length. In *2024 International Conference on Computing, Sciences and Communications (ICCSC)* (pp. 1-6). IEEE. DOI: 10.1109/ICCSC62048.2024.10830431

Chen, J., Yan, H., Liu, Z., Zhang, M., Xiong, H., & Yu, S. (2024). When federated learning meets privacy-preserving computation. *ACM Computing Surveys*, *56*(12), 1–36. DOI: 10.1145/3679013

Dandoush, A., Gouissem, A., & Ciftler, B. S. (2024). Secure and privacy-preserving federated learning-based resource allocation for next generation networks. Innovation and Technological Advances for Sustainability, 440-448.

Fnu, H., Kaushik, K., Gaur, P., Saratchandran, D. V., Thapliyal, A. G., Sidhu, K. S., & Singh, V. (2025, April). Multi-Agent Systems for Collaborative Financial Decision-Making Over Distributed Network Architectures. In 2025 3rd International Conference on Advancement in Computation & Computer Technologies (InCACCT) (pp. 939-944). IEEE. DOI: 10.1109/InCACCT65424.2025.11011435

Gao, Y., Zhang, L., Wang, L., Choo, K. K. R., & Zhang, R. (2023). Privacy-preserving and reliable decentralized federated learning. *IEEE Transactions on Services Computing, 16*(4), 2879–2891. DOI: 10.1109/TSC.2023.3250705

Gupta, M., Vashishth, T. K., & Verma, P. K. (2024). Machine learning and deep learning based intrusion detection for blackhole attacks in mobile ad-hoc networks. *Multidisciplinary Science Journal, 6*(11), 2024209–2024209. DOI: 10.31893/multiscience.2024209

Hallaji, E., Razavi-Far, R., Saif, M., Wang, B., & Yang, Q. (2024). Decentralized federated learning: A survey on security and privacy. *IEEE Transactions on Big Data, 10*(2), 194–213. DOI: 10.1109/TBDATA.2024.3362191

Ibrahem, M. I., Mahmoud, M., Fouda, M. M., ElHalawany, B. M., & Alasmary, W. (2022, December). Privacy-preserving and efficient decentralized federated learning-based energy theft detector. In GLOBECOM 2022-2022 IEEE Global Communications Conference (pp. 287-292). IEEE. DOI: 10.1109/GLOBECOM48099.2022.10000881

Joshi, K., Singh, V., Awasthi, P., Anandaram, H., Sidhu, K. S., & Kumar, P. (2024, November). Smart Dam Water Monitoring using Enabled Internet of Things. In 2024 5th International Conference on Data Intelligence and Cognitive Informatics (ICDICI) (pp. 239-243). IEEE. DOI: 10.1109/ICDICI62993.2024.10810761

Joshi, K., Yadav, H., Gupta, S., Singh, V., Sidhu, K. S., & Kukreti, R. (2025, April). Handling Security Aspects in the Internet of Things: Latest Challenges and Measures to Mitigate Risks. In 2025 3rd International Conference on Communication, Security, and Artificial Intelligence (ICCSAI) (Vol. 3, pp. 1434-1439). IEEE.

Kaur, P., Kaur, K., Verma, D. K., Singh, K., Kaushik, K., & Singh, V. (2024, September). Routing protocols in wireless sensor networks: A comprehensive review and future perspectives. In 2024 7th International Conference on Contemporary Computing and Informatics (IC3I) (Vol. 7, pp. 263-270). IEEE. DOI: 10.1109/IC3I61595.2024.10828894

Krishna, G., Singh, V., Pandey, D., Krishna, K. H., Joshi, K., & Gupta, T. (2024, May). Power trading framework of cloud-edge computing in the artificial intelligence market. In 2024 4th International Conference on Advance Computing and Innovative Technologies in Engineering (ICACITE) (pp. 1910-1918). IEEE. DOI: 10.1109/ICACITE60783.2024.10617344

Moulahi, T., Jabbar, R., Alabdulatif, A., Abbas, S., El Khediri, S., Zidi, S., & Rizwan, M. (2023). Privacy-preserving federated learning cyber-threat detection for intelligent transport systems with blockchain-based security. *Expert Systems: International Journal of Knowledge Engineering and Neural Networks*, *40*(5), e13103. DOI: 10.1111/exsy.13103

Narooka, P., Kaushik, K., Kansal, S., Kejriwal, D., Mishra, R. K., & Singh, V. (2025, April). Hybrid AI Models for QoS Estimation in α-η-μ Fading and Co-Channel Interference Environments within 6G Networks. In 2025 3rd International Conference on Advancement in Computation & Computer Technologies (InCACCT) (pp. 206-210). IEEE.

Orabi, M. M., Emam, O., & Fahmy, H. (2025). Adapting security and decentralized knowledge enhancement in federated learning using blockchain technology: Literature review. *Journal of Big Data*, *12*(1), 55. DOI: 10.1186/s40537-025-01099-5

Raturi, A., Mula, K. G., Joshi, K., Sidhu, K. S., Chauhan, S. S., & Singh, V. (2025, April). Comparative Analysis of Stock Price Prediction with Multiple Machine Learning Model. In 2025 3rd International Conference on Communication, Security, and Artificial Intelligence (ICCSAI) (Vol. 3, pp. 1428-1433). IEEE. DOI: 10.1109/ICCSAI64074.2025.11064599

Roohani, B. S., Sharma, N., Kasula, V. K., Mamoria, P., Modh, N. N., Kumar, A., & Singh, V. (2026). Urban Computing Solutions in Healthcare Edge Computing. In Building Data-Driven Edge Systems for Business Success (pp. 377-400). IGI Global Scientific Publishing.

Singh, V., & Dubey, V. (2015). A two level image security based on Arnold transform and chaotic logistic mapping. *International Journal of Advanced Research in Computer Science and Software Engineering*, *5*(2), 883–887.

Singh, V., & Kaushik, V. D. (2023). Navigating the Landscape of Security Threat Analysis in Cloud Computing environments. In *Security and Risk Analysis for Intelligent Cloud Computing* (pp. 1–25). CRC Press.

Suriyan, K., Ganesh, S., Palaniyammal, P., Priya, G., & Singh, V. (2026). Recent Trends in Edge Computing: Challenges and Opportunities. Building Data-Driven Edge Systems for Business Success, 323-344.

Tanbhir, G., & Shahriyar, M. F. (2025). Quantum-Inspired Privacy-Preserving Federated Learning Framework for Secure Dementia Classification. arXiv preprint arXiv:2503.03267. DOI: 10.1109/ECCE64574.2025.11013884

Teixeira, R., Baldoni, G., Antunes, M., Gomes, D., & Aguiar, R. L. (2025). Leveraging decentralized communication for privacy-preserving federated learning in 6G Networks. *Computer Communications*, *233*, 108072. DOI: 10.1016/j.comcom.2025.108072

Vashishth, T. K., Sharma, V., Kumar, B., Sharma, K. K., Chaudhary, S., & Panwar, R. (2025). Blockchain for Securing Federated Learning Systems: Enhancing Privacy and Trust. Model Optimization Methods for Efficient and Edge AI: Federated Learning Architectures, Frameworks and Applications, 299-320.

Vashishth, T. K., Sharma, V., Sharma, K. K., & Kumar, B. (2025). Enhanced Security and Privacy from Industry 4.0 and 5.0 Vision. In *Building Embodied AI Systems: The Agents, the Architecture Principles, Challenges, and Application Domains* (pp. 133–158). Springer Nature Switzerland.

Vashishth, T. K., Sharma, V., Sharma, K. K., Kumar, B., Chaudhary, S., & Gupta, M. (2024). Privacy-Preserving Machine Learning Techniques for IoT Data in Cloud Environments. In Emerging Technologies for Securing the Cloud and IoT (pp. 144-173). IGI Global. DOI: 10.4018/979-8-3693-0766-3.ch006

Vashishth, T. K., Sharma, V., Sharma, K. K., Kumar, B., Chaudhary, S., & Panwar, R. (2024). Intrusion Detection and Prevention Techniques in FL Cloud-Based Healthcare 5.0: A Comprehensive Review. Pioneering Smart Healthcare 5.0 with IoT, Federated Learning, and Cloud Security, 152-174.

Vashishth, T. K., Sharma, V., Sharma, K. K., Kumar, B., Chaudhary, S., & Panwar, R. (2024). Enhancing Cloud Security: The Role of Artificial Intelligence and Machine Learning. In Improving Security, Privacy, and Trust in Cloud Computing (pp. 85-112). IGI Global.

Vashishth, T. K., Sharma, V., Sharma, K. K., Kumar, B., Chaudhary, S., & Panwar, R. (2024). Security and Privacy Challenges in Blockchain-Based Supply Chain Management: A Comprehensive Analysis. In Achieving Secure and Transparent Supply Chains with Blockchain Technology (pp. 70-91). IGI Global. DOI: 10.4018/979-8-3693-0482-2.ch005

Vashishth, T. K., Sharma, V., Sharma, K. K., Kumar, B., Chaudhary, S., & Panwar, R. (2024). Security and Privacy Considerations in Cloud-Based Data Processing Solutions for Sensitive Data. In *Developments towards Next Generation Intelligent Systems for Sustainable Development* (pp. 35–61). IGI Global. DOI: 10.4018/979-8-3693-5643-2.ch002

Wang, L., Zhao, X., Lu, Z., Wang, L., & Zhang, S. (2023). Enhancing privacy preservation and trustworthiness for decentralized federated learning. *Information Sciences*, *628*, 449–468. DOI: 10.1016/j.ins.2023.01.130

Xue, Y., Singh, V., Singh, S., Kant, K., Pandey, S., Kumar, A., & Kaushik, V. D. (2024). Blockchain Framework for Learner Performance Prediction using Life-Brain Storm-based Light GBM Coupled Neural Network: BLOCKCHAIN FRAMEWORK FOR LEARNER PERFORMANCE PREDICTION. [JSIR]. *Journal of Scientific and Industrial Research*, *83*(6), 652–668.

Yodha, V. (2024). Leveraging Federated Learning for Privacy-Preserving Cybersecurity in Decentralized AI Systems. *Eastern European Journal for Multidisciplinary Research*, *1*(1), 89–106.

Zhou, X., Liang, W., Kevin, I., Wang, K., Yan, Z., Yang, L. T., & Jin, Q. (2023). Decentralized P2P federated learning for privacy-preserving and resilient mobile robotic systems. *IEEE Wireless Communications*, *30*(2), 82–89. DOI: 10.1109/MWC.004.2200381

Chapter 6
Federated Intelligence for Secure Wireless Systems:
A Privacy-Preserving and Scalable Approach to AI-Powered Threat Detection

Praveen Kumar Myakala
https://orcid.org/0009-0009-6988-5592
Independent Researcher, USA

Madhura Raut
https://orcid.org/0009-0007-2303-9070
Independent Researcher, USA

Vineeta Singh
https://orcid.org/0000-0003-1084-279X
School of Engineering and Technology, Chhatrapati Shahu Ji Maharaj University, Kanpur, India

ABSTRACT

This chapter explores federated intelligence as a scalable and privacy-preserving approach to securing wireless systems. With the growing adoption of wireless technologies across critical sectors such as smart cities, transportation, and healthcare, traditional centralized security frameworks are increasingly inadequate. The chapter introduces a modular architecture that combines federated learning, edge AI, and privacy-enhancing techniques like differential privacy and homomorphic encryption. It addresses challenges posed by non-IID data, adversarial behavior, and resource constraints. Real-world use cases and implementation strategies are examined to demonstrate the practical viability of this approach in complex wireless environments.

DOI: 10.4018/979-8-3373-3576-6.ch006

t © 2026, IGI Global Scientific Publishing. Copying or distributing in print or electronic forms without written permission of IGI Global Scientific Publishing is prohibited. Use of this chapter to enerative artificial intelligence (AI) technologies is expressly prohibited. The publisher reserves all rights to license its use for generative AI training and machine learning model development.

1 INTRODUCTION

Wireless communications have fundamentally changed the way digital systems interact with one another. Today, networks unite near-every sphere of existence, from smart cities, industrial automation (Balas et al., 2026), telemedicine to driverless vehicles. Instead of being kept in the cloud, they need connectedness that flows continuously at high speed through a distributed network to handle instantaneous responses and distributed intelligence. The number of connected wireless devices is expected to surpass 30 billion by 2030 (Ericsson, 2023), showing the necessity of such interconnected wireless devices for its relevance to the increasing number of networked connectivity and internet infrastructure.

But it also has opened a new source of security issues. The wireless data travelling over open channels can be intercepted or tampered with as it is transmitted. Attackers can read unencrypted Wi-Fi traffic, mimic Bluetooth links or inundate 5G base stations with large numbers of devices. Sophisticated methods that include data-inputting in vehicular networks, accesses (side channels in IoT) sensors, and malicious use have already interfered with municipal and healthcare systems (Karunamurthy et al., 2025; Makris et al., 2025).

Centralized security frameworks typically find these kinds of threats hard to manage. They frequently analyze raw data from edge devices in central servers, which is an increasing latency and bandwidth consumption, and creates single points of failure. These designs can also clash with the GDPR and HIPAA regulations that restrict the storage and dissemination of personal information (Agrawal et al., 2022; Yin et al., 2021).

The issue is compounded by the different types of wireless devices in use. They vary in processing power, memory, and network stability. Others run on low power or patchy connectivity, making it difficult to monitor cloud communications consistently. Thus, security tools must be lightweight, locally programmable and robust in case of partial connectivity (Parreño & Avila, 2021; Myakala & Agrawal, 2025).

This can further be attributed to the fact that wireless devices have varying levels of processing capacity, memory components, as well as network connectivity. Some devices operate at very low powers or connectivity. This makes monitoring the communications at the cloud layer very challenging. As such, the software should have less complexity and should also operate perfectly even when the connectivity is partial (Parreño & Avila, 2021; Myakala & Agrawal, 2025).

Data variability further complicates matters. And data collected from different areas or user groups is often not balanced or independent. Models trained on balanced centralized data often struggle under such inharmonious data sets. The use of sharing user data with a central server raises ethical and legal issues, which will not just be considered in isolation, but also raise sensitive issues.

To address these questions, researchers have begun to use a technique called federated intelligence. In this technique, devices work together to train a model while never sharing anything of their raw data. Rather than sending full datasets, they pass only the model updates. These updates may remain protected by methods including but not limited to differential privacy, secure multi-party computation or homomorphic encryption (Aziz et al., 2023; Zhang et al., 2021).

This architecture allows systems to identify threats and train on data in a decentralized method when privacy & communication cost are considered. Moreover, federated intelligence improves the robustness of systems. They can detect the devices that are infected and manage imbalanced data as well as function even when the system has failure. This is very important in the energy sector, transport systems, and the health field when failure has severe impact (Myakala et al., 2025; Feng et al., 2025).

This concept fits nicely in with edge computing (Roohani et al., 2026, Suriyan et al., 2026, Anandaram et al., 2026, Krishna et al., 2024). As training and analysis occur near the data generation point, response times are quicker, and network delays are reduced. Such a configuration is especially well suited for high capacity IoT and 5G networks where central control is either slow or unreliable (Mehta et al., 2024).

Overall, the rapid expansion of wireless connectivity (Narooka et al., 2025a; Narooka et al., 2025b, Kaur et al., 2024, Fnu et al., 2025, Singh and Kaushik, 2023), growing security threats, different devices available, and stricter privacy laws have put a tricky spot on the terrain. Federated intelligence provides a solution that works in terms of teamwork, local processing and privacy focused on design. In the subsequent sections, we introduce its key technologies, architecture, and applications with emphasis on implementation in practice, as well as on what opportunities exist in the future.

1.1 Rise of Wireless Threats and Architectural Hurdles

Nowadays, wireless technology assumes a leading position in the way that people and systems are connected. It supports daily requirements in healthcare, transportation, smart homes, industrial automation, and industrial safety and the transportation of goods. These networks are flexible, and easy and rapid to install, and can be adopted for massive and distributed deployment. But their use of open network and mobile technology may also present unique risks to security.

Wireless networks that send data over broadcast channels are easier attacked types of attacks such as this open to use compared to wired systems. An especially serious risk is what they call eavesdropping, where attackers listen to unprotected data and take personal information. Weak or outdated Wi-Fi settings can allow intruders to capture packets and provide login details or user personal data. The fake device would fake being real in the process of obtaining data, and hence might fool or damage the users and the infrastructure, this also known as spoofing. Spoofing in public regions (e.g., smart campuses or transport systems) may be responsible for data loss or service interruption (Parreño & Avila, 2021; Karunamurthy et al., 2025).

The challenge of denial-of-service (DoS) and distributed denial-of-service (DDoS) attacks is also severe. Attackers can blanket network nodes with high traffic, rendering them unusable. For time-critical applications (e.g., driverless cars, emergency communication), any delay in the communication has the potential of causing system failure or physical harm (Makris et al., 2025; Mehta et al., 2024).

Wireless equipment is equally infected by data injection attacks that inject false information into signals used to control vehicles. This may be particularly troublesome in traffic centers, where mistaken routing or location information can induce collisions or congestions.

Despite their dangers, a lot of deployments rely on centralized security models. This is the design where the data moves from the edge devices on the premises to the central server, where it is analysed by many security analysts of threats to be detected. While it makes control and management less complicated, it proves to be less feasible for a distributed system in the modern world.

Centralized architectures have one weakness, that is its single point of failure. The entire system may collapse if the core server goes offline or is compromised. Attackers pay attention to this node as if a single breach is what renders the entire network down. Now, with a compromised central system, there wouldn't be any sort of reaction from the central system in terms of security or defense.

The second limitation is that of delay and communication overhead. Submitting all data to the cloud for processing consumes bandwidth and slows response times. This is particularly problematic in real time decision-based situations. For example, in the manufacturing and surgical robotics contexts, task response is time-consuming and results in delays caused by cloud offloading in operations and by uncertainty of outcomes (Aghdam et al., 2021; Llisterri Giménez et al., 2022).

The third issue is the problem of data and privacy. The wireless technology, which is not designed to disclose people's passwords, likes, activity records, biometric information and other types of information and data is known to do so at a high cost (and that much money is saved). Sharing such raw information with a cloud server comes with compliance concerns for privacy laws including the GDPR and HIPAA for privacy, as well as different national regulations. Such frameworks encourage limited data and discourage the centralization of personal data (Yin et al., 2021; Hu & Li, 2024).

Centralized systems are also challenging to scale in heterogeneous environments. Devices vary with respect to power usage, network bandwidth, and deployment conditions. When doing so, it increases both cost and reduces flexibility by building a large central infrastructure to handle these variations. That said, such systems are not suitable to distributed use cases such as rural connectivity, disaster recovery networks, or short-term sensor grids.

This is the inability to handle non-independent and non-identically distributed (nonIID) information. Edge devices generate data shaped by their surrounding environments and user behaviors. Centralized aggregated data training of many models does not work well after deploying them from individual nodes. This mismatch will compromise detection performance and system reliability, particularly when environment changes quickly (Agrawal et al., 2022; Hsu et al., 2023).

And finally, conventional intrusion detection systems are rule-driven and more reactive. Not only do they know how to spot familiar threat patterns, but they can also miss new and changing ones. Since they are centralized mechanisms, they do not consider local context, nor can they detect minute anomalies that might surface only within locations.

In short, when wireless threats have become more complex, centralized security is increasingly vulnerable. The next generation design to guard distributed, privacy-aware systems needs to be flexible and context-aware, able to work at local or regional contexts. The further section considers federated intelligence, a paradigm which fulfills these objectives by dispersing learning and decision-making through an interdisciplinary network, without losing any synchronization among defenses.

1.2 The Emergence of Federated Intelligence Paradigm

Wireless systems today face several highly challenging cybersecurity concerns (Singh and Dubey, 2015; Singh and Kaushik, 2023a; Singh and Kaushik, 2023b; Chaudhary et al., 2024), that forced the researcher and his team to invent novel type architectures. One of the promising ideas is federated intelligence, which harmonises federated learning, edge AI, and privacy-maintaining techniques in a unified paradigm.

This method bypasses centralized system limitations by allowing the devices to train together rather than providing them with raw data. It, therefore, facilitates agile and secure context-specific threat detection in some instances in which individual privacy, bandwidth and prompt response is essential.

Central to this framework is federated learning (FL), a distributed approach which enables devices, i.e., smartphones, IoT sensors or edge gateways, to train training model on their local data. These devices simply distribute updates from their models to a coordinating server, but not the actual data. That reduces the risk of privacy and saves money on communication. The central server combines these updates to improve the global model and sends it back to devices for further training (Yin et al., 2021; Myakala et al., 2024).

Edge AI extends the idea by letting devices process data and take decisions themselves. Edge devices are no longer dependent on the cloud for their analyses, allowing the task to be performed by the edge without the need for computing cloud resources, including anomaly detection, behavior analysis, and risk classification. This is crucial to perform on-task applications in autonomous driving, process control, health monitoring, etc., which at even slight delay might lead to severe issues (Mehta et al., 2024; Sharma et al., 2025).

PETs ensure privacy of updates for model through training and aggregation. PETs include:

- Differential privacy, an approach which generates small bits of noise because no one can track individual data points.
- Secure multi-party computation (SMPC), where various parties can do computations together without having their private data exposed.
- Homomorphic encryption, to perform computations with encrypted information while protecting the underlying data (Aziz et al., 2023; Zhang et al., 2021).

These 3 components (Federated Learning, Edge AI, and PETs) form the foundation of federated intelligence. This not only has technological benefits but adheres to the currently followed norms related to ethics & regulations. This enables less storing of the data centrally and allows more devices to become more self-sufficient in the following principles: user consent, ownership of the data, and the transparency aspect.

From the perspective of security, federated intelligence enables the devices to cooperate for the detection of threats and adapt to the new attack surface. In the case of a smart city, routers can analyze the traffic in the area and communicate encrypted updates to improve a joint intrusion detection model.

Another benefit of this approach comes with the ability to manage unpredictable and dynamically changing graphs. Device links can keep dropping off or even sending incomplete updates, and the system still performs very well. Techniques like weighted average and trust values as well as outlier filtration remain effective irrespective of the changing circumstances (Feng et al., 2025; Myakala & Agrawal, 2025).

In essence, federated intelligence denotes a paradigm shift in the training and self-sensitization of wireless systems. This paradigm enables the safe and efficient operation of wireless systems as the training process can be handled at a local level.

1.3 Chapter Goals and Contributions

In today's increasingly interrelated world comes federated intelligence: A bold new approach to secure wireless communication systems. With ever more sophisticated wireless communication attack strategies being created, the call for effective structures that can efficiently manage performance requirements, privacy and adaptability has become more urgent than ever.

Goal of this chapter

The first purpose of this chapter is to design a Federated Security Architecture for wireless systems: Federated learning and privacy are linked through architecture in some sense in some places such as modular security frameworks over edge machines or systems that are capable of detecting a problem while still keeping all records of one another's communications safe. It describes how wireless devices, edge nodes and aggregation servers cooperate to train shared models, locate outliers and communicate threat information. The system encourages decentralized decision making and can process in real time, even if resources, i.e. bandwidth and computing power are of concern.

Assess Privacy-Strengthening Technologies for Distributed Learning Technologies: A comprehensive contrast of privacy-preserving solutions is provided such as differential privacy, secure multi-party computation, and homomorphic encryption. The paper analyzes these approaches within wireless environments, with a discussion considering their compatibility with such wireless protocols, privacy-preserving methods, the compromise of computation cost and practicality in the real-domain use, of these approaches between privacy and computation cost and real-world implementation (Aziz et al., 2023; Hu & Li, 2024).

Build Good Anomaly Detection Methods for Non-IID Data: Due to wireless data being commonly not being consistent or separate, the chapter examines strategies to handle the data spread of RFD with personalized federated learning, clustering based aggregation, and context aware adaptation. Such methods enhance reliability in the real world where there is rapid fluctuation of data-type patterns, for example.

Existing Frameworks & Benchmark Performance: The tests are performed on well-known datasets for N-BaIoT, CICIDS and ToN-IoT. We examine accuracy, false positives data, training speed, communication efficiency, and attack resilience in the analysis.

Explain Relevancy to Real-World Applications: Three application domains are tested to illustrate the practical relevance of the framework:

- **Smart cities**: Identify anomalies in common Wi-Fi infrastructure and local IoT system.
- **Vehicular Networks:** Proactive threat intelligence in a vehicle-to-everything (V2X) contexts with privacy-preserving breach detection.

- **Remote Healthcare:** Privacy-preserving diagnostics and continuous monitoring of a patient for edge assisted federated learning for privacy-preserving and continuous diagnosis.

Identify Limitations, Research Opportunities: Addressing the challenges that remain to be solved, e.g., adversarial model poisoning, device drop out, trust foundation and the use of distributed learning process.

With these goals, the chapter offers a very informative and practical analysis of the role of federated intelligence in improving the wireless security. With its basis both in theoretical concepts as well as practical knowledge and structure.

2 MODERN WIRELESS SYSTEMS ARE PRONE TO SECURITY HAZARDS

Wireless communication has become a main aspect of many core applications such as medical services, transport logistics, industrial control systems, and smart cities. As the usage of wireless communication has become more widespread, the opportunities to taking advantage of system vulnerabilities have also opened.

Moreover, since wireless communication happens across the vast open space, the network tends to be more vulnerable compared to the wired network. Its very nature of being portable also incorporates more complicated factors since it is a whole lot more difficult to secure and promote both low latency and the aspect of privacy.

This is a section that identifies the primary threats to wireless systems and discusses how traditional methods of defence frequently fail to prevent threats. It provides the framework for imagining a new method that fuses decentralization, edge intelligence, and privacy-centric design to define the future of secure wireless networks.

2.1 Common Threats in Wireless Environments

Wireless networks are inherently open. Their signals drift in the air, and they move connected devices in real-time. Unlike wired systems, where all information is stored and retrieved via physical means, wireless connections can easily be intercepted, disrupted, or altered by anyone within range.

When wireless technologies move inside healthcare, transport and smart infrastructure, attackers are seeking new openings within these open channels at the same time as they are opening new doors.

Eavesdropping and Traffic Analysis: Simple yet inexpensive hardware can intercept wireless traffic: Attackers can eavesdrop on wireless traffic with inexpensive (simple) hardware. When a message is communicated without encryption or with weak keys, it may disclose a subscriber's usernames, passwords, or user and, in some cases, private content. Even encrypted transmissions might be vulnerable to patterns, such as device IDs, timing, or frequency of communications. Such metadata may subsequently be mined to extrapolate from user behaviour or monitor business behavior (Parreño & Avila, 2021).

Spoofing and Impersonation Attacks: Spoofing is made when an attacker feigns a trusted device or access point, this creates opportunity for unauthorized access or delivery of the payload. For Wi-Fi networks, attackers can establish a spoofed access point that copies the name of the actual network. Users connect to the network without being aware and share their personal information and data. Bluetooth devices can also suffer spoofing attacks, especially when auto pairing has enabled (Karunamurthy et al., 2025).

Denial-of-Service (DoS) & Distributed DoS (DDoS): DoS targets a device or network with huge traffic to consume resources and slow down communication. Most wireless devices suffer from lack of power or bandwidth; hence DoS can greatly impact them. DDoS attacks are significantly stronger in that they bring together several exploited devices with multiple compromised, effectively multiplying the traffic and increasing the mitigation difficulty (Makris et al., 2025).

Data Injection Attacks: Falsified data are injected into communication streams in this type of attack. This false information can lead to bad decision making on controllers. For example, attackers can send bogus GPS coordinates to jam a vehicle network to stop a vehicle in its tracks or cause traffic jam. Misinterpretations of sensor readings in healthcare can result in misdiagnosis or misadministration (Mehta et al., 2024).

Replay and Man-in-the-Middle Attacks: Replay attacks reuse previously captured valid packets, allowing attackers to emulate real transactions. A man-in-the-middle attack locates the attacker between the two communication parties, such that the attackers can intercept and alter messages in real-time. Both types are particularly pernicious to authentication and financial services.

Jamming and Signal Interference: Networks sharing unlicensed spectrum are also highly sensitive to jamming and communication interference. An attacker can even emit noise or overlay signals to prevent normal communication. It is hard to identify jamming, and service can be badly degraded in general, especially in emergency communication systems or in drone operations.

These threats demonstrate that wireless spaces can be extremely fragile if they lack powerful contextual security. As it is those systems that are the core to services that are vital to our society, the requirements of a completely different security architecture than one based upon centralized defenses and defenses only stretch so far.

2.2 Contextual and Structural Vulnerabilities

In addition to specific attack methods, wireless communication systems also have vulnerabilities built into the basic system structure, such weaknesses can lower resilience even when the vulnerability has no direct effect.

Openness of the Communication Medium: Wireless communication signals are open to everyone in their area, this means that when you are wired, your communication network protects you from the outside world through a cable. Wireless communication signals cannot be hidden. Even when the network is secure, the signal information indicating the device that can connect to the signal or the sending and receiving of the information can be identified (Parreño & Avila, 2021).

Device Heterogeneity: Wireless ecosystems involve a different type of devices, their hardware components, operating systems, as well as firmware versions, differ. This affects the security patches cycles. While some devices involve strong encryption and authentication processes, others lack basic forms of securities due to resource or design constraints (Agrawal et al., 2022).

Resource Constraints: Wireless and IoT devices face limitations in power consumption, processing capacity, and memory. Such constraints create a situation where the implementation of advanced security methodologies such as packet inspection or machine learning based anomaly detection methods will not be possible. The attacker tends to target devices of this type to enter the main systems as a door opener (Llisterri Giménez et al., 2022).

Frequent Intermittent Connections & Network Instability: Mobile and remote systems are gradually linked together. The devices that are linked can suffer from disconnections because of signal interference or power saving schemes. When disconnections occur, the central monitoring components will not manage to retrieve updates or even policy enforcement. This lack of robustness has allowed attackers to launch delay-based and synchronization-based attacks.

Non-Stationary / Non-IID Data: This wireless data can be considered as non-IID as the wireless data is affected by the living environment, is clearly affects the classical ML as given above. Moreover, the centrally based approaches that ignore this factor are very poor compared to the traditional algorithm as given above due to falser positive and missed detections (Hsu et al., 2023).

Trust Management in Distributed Networks: In wireless environments, systems tend to connect devices belonging to different individuals or companies. In the case of a network when a central entity does not exist, one can hardly identify the individuals among whom they are working. Even the trustworthiness of the individuals can hardly be identified.

with these factors explain why resilience and adaptability are the core principles for the design of wireless security. True resilient architecture encompasses the concepts of privacy, scalability, and trust. Federated intelligence addresses all the above concepts because it expands the concepts of training and focuses on the edges for defense as it adapts to varying environments.

2.3 Restrictions of Conventional Security Mechanisms

Traditional methods to protect against wireless communications are still the foundation of a more and more complex wireless environment. This refers to the rule-based firewall data table described as the basis for many wireless systems. This type of solution works perfectly in the case of fixed and wired systems. They simply do not perform very well in the case of the more dynamic and distributed systems.

Centralisation and Single Points of Failure: Standard secure systems rely on the centralization of certain cloud infrastructure that hosts core servers with the ability to perform data analysis. This represents a bottleneck as well as a focal point for the attacker. When that server crashes or is compromised, all the underlying devices become unprotected (Myakala & Agrawal, 2025). It leads to the lack of quick threat detection and response. Transferring raw data to an at-a-distance cloud comes at a cost and often introduces lag times, especially if it is of poor quality. It is unreasonable latency to receive a decision in seconds, as would happen with autonomous cars or in an emergency (Rezaei Aghdam et al., 2021).

Inflexibility to Dynamic Data: Legacy system also assume data forms remain unchanged. In practice wireless data are in real-time dynamic across devices, regions, and users. In centralized models, trained on static data, we cannot alter our behavior; the accuracy drops with it, leading to missed threats (Hsu et al., 2023).

Static Rules and Low Learning: Signature-based intrusion detection only predicts a pattern of security attacks. Most of these systems are poorly able to identify newly introduced or changing threats, and this exposes them to constant manual upgrades. Such inflexibility puts the network at risk in the rapidly changing wireless network environment (Makris et al., 2025).

High Communication and Storage Overhead: The process of centralized raw data collection can lead to high bandwidth and storage usage requirements. This can become expensive when considered for the type of connectivity found in large data networks. This can also reduce the battery life of a device (Llisterri Giménez et al., 2022).

Weak Privacy Controls: Most legacy systems lack measures concerning user privacy. This system can transmit private information like health information or geographic information inadequately. This violates regulations but also affects the system's acceptance among the public (Yin et al., 2021; Hu & Li, 2024).

Absence of Edge Intelligence and Local Autonomy: The centralized architecture relies on connectivity when the need arises. Most devices should operate independently. Thus, the lack of local intelligence can lead to slowed systems and poor healing when under attack.

In other words, the traditional techniques used for securing wireless communication are no longer sufficient. They are inflexible, not scalable properly, and too frequently have privacy as their primary concern. What we need now is another approach that distributes the learning to the devices, strengthens the local control and ensures privacy protection at each level.

This new direction introduces federated learning and edge AI as the major building blocks of this future evolution.

3 OVERVIEW OF FEDERATED LEARNING AND EDGE AI IN SECURITY

The limits of centralized security are now clear in modern wireless systems. New algorithms that can leverage privacy, speed, and flexibility are required as the networks grow. Federated learning (FL) and edge artificial intelligence (Edge AI) are promising solutions. Combined, they enable systems to operate independently at the edge of a network, secure and intelligent, without relying entirely on the cloud.

In this section, we outline a few simple principles of these technologies, and how the fusion of them enhances security in wireless environments.

Federated learning is a distributed form of machine learning, which is in use in AI and network platforms of all sizes, designs, scales, and characteristics; and its use has been studied in various fields, including consumer health, medical software, and banking, business sector (finance), industrial production, etc (Yin et al., 2021; Azarkasb et al., 2025).

For this purpose, the setup consists of any machine (or software client) working with its own data and training its own local model. Rather than pass these data to a central server, the client communicates solely model updates like the change of weights or gradients. The coordinating server pulls all these updates, computes a global model, and sends them back to the devices for more training. This continues until an accurate model is developed (Myakala et al., 2024).

Standard FL cycle includes:

1. **Initializing the model**: The server constructs an initial global model and passes it on to clients.
2. **Local training**: A single client trains the model with its own data.
3. **Update transmission**: Each client sends their newly added settings to the server.
4. **Aggregation**: Updates are aggregated on the server using a framework such as federated averaging.

5. **Global synchronization**: The refined model is forwarded to customers for ongoing development. This process goes on until the performance targets are achieved.

Federated learning has numerous benefits for wireless systems because it is applicable for wireless systems:

- Privacy preserving raw data stays on the device, reducing the chance of leaks and ensuring compliance with various privacy regulations.
- Less bandwidth consumption through only updated model updates transmitted and is cost effective in lower bandwidth networks.
- Decentralized compatibility; FL fits naturally into the distributed topologies of wireless systems.

But federated learning isn't without limitations. Training can also become slower; accuracy can decrease because the information for each of the clients isn’t always independent and accurate. Problems like communication breakdowns and device disconnection can also hinder. To counter the problems, methods for trust-based coordination, robust aggregation, and selective user involvement are currently under development (Hsu et al., 2023; Feng et al., 2025).

As the complexity of wireless hardware has increased, federated learning has emerged as a feasible solution for collective intelligence in terms of privacy considerations and efficiency of bandwidth.

3.1 Edge AI for Wireless Systems

In edge AI, the machine learning process happens at the edge devices of the network, this allows the devices to analyze the information and come up with decisions instead of sending the information to the cloud. This can be very important in the case of wireless communications since the network connectivity might not be very stable.

In traditional systems, the servers perform the analysis. This causes delay. However, in Edge AI solutions, the inference tasks are performed at the routers, gateways, or the Internet of Things (IoT) sensors. These systems are used to monitor the flow of traffic and detect anomalies. They react immediately when given the command instead of waiting for the command from the other party. This makes the systems more effective since they react faster and ensure more secure information (Myakala & Agrawal, 2025).

Below are some of the advantages of the use of Edge AI:

- **Low latency**: Local inference eliminates communication delays between devices and eliminates round-trip communication delays.
- **Low bandwidth**: No network congestion by having only necessary output, like an alert or a summary, sent out.
- **Resilientness**: No system is offline which is critical in mobile or remote environments.
- **Better Privacy**: Data stays closer to the source and as such the risk of exposure to unauthorized party usage can be minimized.

Many edge devices do not have sufficient processing capacity, memory, power. Lightweight models, pruning and compression and hardware accelerators like GPUs, TPUs or custom AI chips will do the trick. Maintaining synchronization of distributed models on multiple devices represents another active area of research (Llisterri Giménez et al., 2022).

Alongside federated learning, Edge AI takes this even stronger. Devices can learn from their local data and add updates to an international shared model. This collaboration enables an ongoing learning experience and faster response to novel threats on wireless networks.

In short, Edge AI allows greater proximity in time between intelligent actions and the physical location where the action or decision has been made. For wireless security systems, this effect can translate to faster response time, greater privacy assurance, and high reliability. This essentially forms the foundation of wireless security.

4 FEDERATED INTELLIGENCE FRAMEWORK

Based on this introduction to the basics of federated learning and edge AI, the following framework encompasses the complete implementation of federated intelligence in wireless communication systems. This solution addresses numerous real-life problems such as varying devices, incomplete connectivity, imbalanced data distributions, and the strict requirements of preserving user privacy. All the above-mentioned factors and requirements are addressed in the implementation of federated learning as well as edge AI and several measures emphasizing user privacy.

4.1 Architectural Overview

The proposed solution offers a modular and layered structure. This modular structure enables the balancing of the distribution of learning, locality-based decision-making, and coordination based on privacy. This links nicely to the dynamic capacity constraint and wireless technology aspect. In fact, the structure enables cooperative threat detection, decentralized control, and continuous model updates.

This structure consists of 3 related layers:

Edge Device Layer: This consists of the mobile phones, IoT sensors, routers, vehicles, as well as wearable devices. These devices form the basis of the network.

- Use telemetry and user information.
- Perform on-device inference using pretrained models.
- Train machine-learning models on local subsets.
- Send model updates, like weights, or gradients, to the coordination layer.

Edge devices rely on efficient anomaly detection models to detect and analyze activities that could be suspicious in real time. Locally processing data allows keeping private information secure and decreases the requirement for communication.

Federated Coordination Layer: The training process is shared among devices in the Federated Coordination Layer. This happens at the aggregation nodes like edge servers, base stations, or fog nodes geographically close to the devices.

- Receives and validates modifications.
- Aggregates them using techniques such as Federated Averaging (FedAvg) or its more robust alternatives.
- Sends the updated global model back to the clients.
- Tracks participation, convergence, and training quality.

Since this layer works on the edge, there is less latency and faster recovery from connectivity loss. These groupings can also occur gradually as the updates are submitted to servers through the local coordinators.

Privacy and Trust Management Layer: This layer takes care of the data transactions between devices and coordinators. This layer comprises

- Differential privacy adds noise to updates to hide individual information.
- Joint aggregation can be achieved through secure multi-party computation (SMPC) without the revelation of private information.
- Homomorphic encryption enables computations of encrypted updates and respects data confidentiality.
- Trust scoring, we score clients on the extent to which they demonstrate past behavior and consistency and thereby reliability.

They provide a safeguard against poisoning attacks and Sybil attacks, they also ensure adherence to regulations in the health and finance sectors (Aziz et al., 2023; Zhang et al., 2021). Together, all the layers form a supple and scalable foundation for secure team intelligence. Moreover, each layer functions independently. This enables the entire system to continue functioning effortlessly even when things go south.

4.2 Communication Flow & Synchronization Strategy

Federated intelligence depends greatly on effective communication, especially when using wireless communications that have limited bandwidth and connectivity that isn't as stable. These methods ensure that cooperation stays efficient and secure without compromising the network.

Synchronous mode: All devices involved perform their training locally in a predefined time interval and update the network. The coordinator will wait for a fixed number of responses before combining them. This approach provides more consistency but can delay the whole procedure because the devices can sometimes perform slowly or remain away from the platform for a rather long period.

Asynchronous mode: all devices provide updates instantaneously and for each update, the coordinator updates the global model continually. This is more suitable for networks with various participation, though outdated versions exist. The stability of training is ensured mainly through the practice of learning rate decay, version tracking and momentum correction (Myakala & Agrawal, 2025; Hsu et al., 2023).

Update Transmission Protocols

Effective communication will require trade-offs between bandwidth, security, and timeliness:

- Model compression applies gradient quantization and sparsification, to send only big changes.
- Secure channels use TLS or equivalent encryption; Homomorphic encryption or SMPC adds a layer of extra security in high-sensitivity scenarios (Zhang et al., 2021; Hu & Li, 2024).
- Only differences in parameters and not entire models are transmitted in delta encoding.

Update Participation and Scheduling Policies

Not every device will have access to every training round. The framework adjusts participation based on availability, data quality and trust:

- Focuses on devices with good connectivity and resources.
- Does not accept clients who exhibit suspicious behavior or who have noisy data.
- Ensures regional fairness through quotas of participation.

Schedulers can opt for random sampling, weighted probability, or reinforcement learning to recruit clients (Azarkasb & Khasteh, 2025).

Aggregation Timing and Model Consistency

All final global updates are versioned. Clients also label their submissions with the model version used in training. The coordinator follows these events to filter out old contributions. In asynchronous conditions the use of timestamped queues and staleness-aware weighting prevents older updates from perturbing the learning.

These approaches to synchrony and communication in federated intelligence are efficient, robust, and resource-light even in complex wireless networks.

4.3 Regulation and Fault Tolerance

The model integrity of the data needs to be preserved to a great extent when dealing with untrustworthy or bad clients. Wireless environments encounter drop-outs, weak links, and uneven participation. The framework addresses these by strong aggregation and fault tolerance features.

Challenges in Aggregation

- Client reliability devices may freeze or transmit poor-quality updates.
- Unilateral data from local datasets and diverse datasets are sources of conflicting updates.
- Attackers may manipulate poisoned or backdoor models (Feng et al., 2025) for malicious behavior.

Aggregation Techniques

- Federated Averaging (FedAvg) basic and effective but susceptible to outliers (Abbas et al., 2024).
- Trimmed mean and median ignore extreme values to avoid noise.
- Krum and Multi-Krum allow for close updates, avoiding deviations.
- Clustering-based aggregation groups clients by update similarity for non-IID data.

• Trust-weighted aggregation will assign greater influence to clients with reliable and repeatable records (Myakala et al., 2024).

Dealing with Clients Abandoning Service and Unreliable Participation

• Partial aggregation is executed even in cases where some clients don't reply, after certain minimum thresholds.
• We resample clients into future rounds after dropped clients to retain diversity.
• Backup and redundancy mimic main clients for continuity.
• Learning rate and momentum adjustors reduce updates in event there is high variability.

These strategies help to maintain enhance reliability and mitigate adversarial disruptions.

4.4 Privacy-Preserving Computation Strategies

Privacy is still central to federated intelligence, particularly used when devices handle sensitive personal data. While raw data will remain local, the model updates still patterning can come through with the help of inference attacks. Thus, the framework enforces a few privacy-preserving mechanisms in there.

Differential Privacy

Differential Privacy (DP) adds noise to updates in a model in order to mask individual contributions.

• **Local DP**: each device will perturb its own update before sending it (offering considerable privacy while at a loss of accuracy).
• **Global DP**: the aggregator injects noise, preserving utility while trusting its data. As the ε (parameter) of the result decreases, privacy is increased, but accuracy becomes less; thus, smaller ε represents better privacy but fewer results (Yin et al., 2021).

Secure Multi-Party Computation (SMPC): The only stable option for that is secure, distributed multi-party computation. SMPC enables private data to be shared while joint computation occurs. Every client divides its update into random shares, which are spread across peers. The aggregator reconstructs only the final sum, not a single update. SMPC is useful for multi-organization collaborations where no one individual party can be trusted (Zhang et al., 2021).

Homomorphic Encryption: Homomorphic Encryption (HE) allows the coordinator to execute mathematical operations on encrypted data. The outcome is released only by relevant people. Partial approaches of Paillier or ElGamal can trade off between security and computational cost (Aziz et al., 2023; Hu & Li, 2024).

Hybrid Privacy: A hybrid scenario can be local DP/light encryption or selective SMPC solutions, decision depends on the device capacity, data sensitivity and latency requirements.

These privacy mechanisms are included within the architecture to enable cooperation and regulation, and yet protect user data. These measures are necessary to establish reliable security and responsible AI utilization in wireless networks.

5 PRIVACY-ENHANCING MECHANISMS

Federated learning keeps raw data on devices, which helps privacy. Still, it is not immune to inference attacks or adversarial manipulation. Privacy-enhancing mechanisms are therefore essential. This section explains the main techniques, how they fit wireless settings, and how to integrate them in a federated pipeline to protect model integrity and user confidentiality.

5.1 Differential Privacy in Model Training

Idea. Differential privacy (DP) limits what an observer can learn about any single record. It matters in wireless deployments because edge devices often handle sensitive signals such as behavior traces, health metrics, or location histories (Yin et al., 2021).

How it is applied

In federated learning, noise can be added at two points:

- **Local DP**: a device perturbs its update before sending it.
- **Global DP**: the server injects noise during aggregation.

Noise is calibrated to the function's sensitivity and a privacy budget ε, which sets the privacy–utility tradeoff.

Benefits

- Reduces gradient leakage and model inversion risk.
- Eases compliance with laws such as the GDPR.
- Builds trust for users wary of sharing behavioral data.

Caveats. Excess noise degrades accuracy, more so when updating is sparse or when little data is available. Select the value of ε depending on the domain and the devices involved.

5.2 Secure Aggregation Protocols

Goal. Let the server calculate the sum or average of the updates requested by the customers in such a way that the server does not view any specific update.

Core Principles

- **Input masking**: each client applies a random mask to its update.
- **Pairwise randomness**: clients exchange shared masks, so masks cancel out in the aggregate.
- **Dropout resilience**: aggregation still works if some clients drop out.

When everything is perfectly coordinated, the server only receives the aggregated outcome but not the specific contribution.

Example Protocols

Bonawitz et al. (2017): secret-sharing based, scalable to mobile configuration, and tolerant to dropouts. In Multiple-key homomorphic encryption, the decryption process at the end-user level involves multiple ciphertexts. This provides strong assurance but has more overhead compared to the masking schemes (Hu & Li, 2024).

Wireless-Specific Challenges

This increases bandwidth expenses when the keys or masks involve additional messages. Synchronization and rekeying increase latencies during dropouts. Resources at the edge devices might not permit strong cryptography.

Make use of lightweight alternatives or hybrid solutions that protect only the most vulnerable areas to lower overhead costs (Zhang et al., 2021).

Aggregation secured from the onset has the highest level of priority in user trust as well as adherence to user privacy.

5.3 Homomorphic Encryption for Federated Learning

Idea. Homomorphic encryption (HE) allows performing calculations on the encrypted information. In the context of FL, this provides end-to-end encryption of the updates when transmitted between servers, to prevent eavesdropping as well as honest but curious servers.

Principles

Ciphertext supports certain calculations. After the aggregation process, the ciphertext can be decrypted to reveal the correct update.

Types

- PHE: supports a single operation, for example addition in Paillier. Efficient for summing updates.
- SHE: supports limited additions and multiplications. More flexible, more costly.
- FHE: supports arbitrary computation but remains too slow for most real-time wireless uses.

FL Applications

Clients encrypt their updates, the server aggregates the ciphertexts, and finally devices decrypt the final aggregated result. The advantages include always maintaining confidentiality, server compromise safety, as well as simplified arguments for compliance (Aziz et al., 2023).

Practical Notes

- **Compute cost**: encryption and encrypted math are heavy for low-power nodes.
- **Key management**: generation, distribution, and rotation add complexity.
- **Latency**: can be high for time-critical loops.

Countermeasures include the implementation of PHE for summation functions, encrypting only the sensitive levels, and the use of optimized libraries like Microsoft SEAL or HElib (Hu & Li, 2024).

5.4 Trust Management and Client Reputation Systems

Not all participants can be trusted, some devices can misbehave due to misconfiguration, unreliable connectivity, or intentional malicious behavior. Trust management can negate their effects and secure the global model.

Threats from Malicious or Unreliable Clients

• **Model poisoning** intentional updates reduce accuracy or implant backdoors.
• **Data poisoning** poisoned local training data produce biased gradients.
• **Free riding** clients do not attend training but reap the benefits.
• **Collusion** multiple compromised clients influence the model in the wrong way (Feng et al., 2025).

Reputation Scoring: Gauging each customer based on user behavior factors like

• Update consistency alignment with global trends.
• Contribution quality effect on validation metrics when included.
• Participation regularity steady engagement, balanced to avoid bias.
• Anomaly flags sudden shifts in magnitude or direction.

Scores guide operations to: Weight trustworthy updates more in aggregation, exclude low-trust clients for specific rounds and trigger extra validation or retraining for suspicious updates (Myakala et al., 2024).

Working with Secure Aggregation

Reputation can coexist with privacy by sharing only summary statistics or using peer-based scoring where clients assess divergence before aggregation. This distributes trust decisions and reduces reliance on a central judge.

Incentives. In multi-party and commercial settings, link high trust to benefits such as priority access, service credits, or recognition in consortia. Incentives encourage good behavior and sustained quality.

Reputation systems complement DP, secure aggregation, and HE. They promote reliable participation and protect model performance in diverse wireless environments.

6 THREAT DETECTION UNDER NON-IID CONDITIONS

A wireless network consists of diverse devices that operate under numerous circumstances and tackle distinct tasks. Such variation yields non-independent and identically distributed (non-IID) data resulting in the generation of information in a variety of device combinations with different statistical characteristics.

While federated learning is designed to work with decentralized data, such inconsistencies can make training more difficult and may lower the accuracy and reliability in detecting threats. This section explores the association between non-IID data and model performance and discusses approaches to developing threat detection systems that can deal with real-world heterogeneity in the wireless environment.

6.1 Performance Implications and Performance Consequences of Non-IID Data

Devices in wireless networks differ considerably in terms of hardware, connection quality, and user behavior. As a result, these data formats vary widely in terms of their structure, frequency and relevance. Thus, industrial sensors may record high-frequency operational data and, on the other hand, such sensors in homes may give sparse or irregular readouts. One device may suffer from repeated intrusion attempts, and while another device may not have experience with any.

When such different devices together refresh, the difference in the distributions of data between the two can impact the quality of the training data. Contribution of biased and dominant clients with biased or skewed data leading to a higher ratio of clients with dominant clients contributing more updates to the global model. Accordingly, it is likely that the overall model under new conditions will show limited performance.

In threat detection, this bias causes the system to overlook such rare but serious events as targeted intrusions and attacks targeted at only a particular area. Conflicting client updates can further cause model drift, a situation where the training process is either poor or does not converge. Such instability can lead to higher false positives or lower detection accuracy, decreasing the trustworthiness of the model.

Hsu et al. (2023) describe that not only do non-IID conditions result in greater differences in performance across clients, but they also complicate the interpretation of how the model identifies the threats. Federated learning also needs to be improved and incorporate strategies to combat data imbalance and differences in distribution. In the next section of the article, these techniques are treated which can be classified from the very beginning and provide a good basis for learning. They work based on personalization, groupings, and adaptive learning weighting for learning stability in varied environments.

6.2 Strategies for Learning from Heterogeneous Wireless Data

Now researchers have devised several strategies for federated learning that are more robust to variations in client files. These approaches also seek to reconcile client's influence, enhance generalization and identify local patterns while maintaining overall model consistency.

Personalized Federated Learning: Personalized federated learning makes it so that each device can update and personalize the shared model to its local data. Personalization makes partial sharing possible rather than requiring every client to use the same parameters. For instance, a client might share the lower layers and fine-tune the upper ones of a neural network separately but jointly. This allows the model to capture general patterns and yet be sensitive to the local context. For wireless, personalization enables to identify region- or device-specific anomalies.

Clustering-Based Federated Learning: Each of these clustering-based federated learning groups clients who receive similar data, or the models can be updated in. Each group then trains a model that is personalized for its own contexts. This lessens the tensions between dissimilar clients and also aids in the resolution of localized threat behavior. For instance, vehicles from crowded urban centers can be at risk at different rates from vehicles from rural ones. The clustered learning is the opportunity for each of two groups to accommodate and learn from its unique situation.

Adaptive Weighting Methods: Adaptive weighting methods are used to tune the value of each customer's update in aggregation. The weighting can be determined from these factors including data quality, the consistency of updates, and past effectiveness. Clients whose updates are stable and reliable have a higher influence than those whose updates are noisy or unstable. In the case of trust mechanisms, the weights for such influence can incorporate the reputation built up over time based on interactions.

Meta-Learning Approaches: Is employed whereby the global model adapts to patterns in previously unknown or rare data on the fly. This enables the system to identify new threats and learn more quickly, particularly important in fast-velocity wireless scenarios.

These strategies help make federated intelligence more stable and adaptable. They enable the system to be contextually diverse and better to detect context specific threats across different capabilities and environment across devices.

7 IMPLEMENTATION STRATEGIES AND EVALUATION METRICS

Federated Intelligence is achieved in wireless systems only when used correctly and not by its architecture alone. Such operational approaches should consider the constraints of the device, network behavior and intended applications. This section explains model optimization, training coordination and evaluation metrics.

7.1 Model Optimization for Edge Deployment

Building federated learning model runs on wireless devices and these models must be lightweight that can run computationally, memory and battery efficiently, in spite of their size. For example, deep learning models on the cloud can run too large or be too resource intensive for these conditions.

Often, to fine-tune them for the edge, they will include a compact architecture, such as MobileNet, SqueezeNet and the TinyML versions. These models use pruning, quantization and depthwise separable convolutions amongst others to simplify the size and complexity of the network while maintaining acceptable correctness.

Another well-known technique is knowledge distillation, where one smaller model learns by copying another large pretrained model and using only a few parameters it duplicates its core behaviour.

On-device training must also be sensitive to differences in learning rates, batch sizes, and local epochs to balance accuracy against energy consumption. A lot of local updates can either deplete the battery or can heat up the device to too high a temperature.

Adaptive and asynchronous updates support the training of systems that match the conditions and connectivity of each device, supporting learning even in cases of irregular participation.

Compression methods based on gradient sparsification, delta encoding, and adaptive bitrate control are widely used by most systems to reduce communication costs. This type of approaches not only help optimize bandwidth but reduce latency, which is becomes critical in mobile and remote deployments where connectivity is often unreliable or costly.

7.2 Training Coordination and Infrastructure Considerations

Modular and robust architectures are required to manage the decentralized type of wireless devices' learning. In federated structures, the network usually consists of multiple tiers such as edge devices and aggregator devices. In other cases, additional cloud entities can also exist. This hierarchical structure can adapt to fluctuations and secure information transfer compared to the server.

The training phases must be scheduled. This can depend on the nature of the training activities. For example, the training process in the monitoring systems can time-map during the off-hour periods. This should ensure that the consumption of energy as well as the bandwidth is reduced.

On the other hand, given the fact that it is not the only factor influencing the choice of devices and can rely dynamically upon availability, power level, hardware readiness, as well as trust scores that can ensure the efficiency and equity of the participants' involvement.

For improved scaling, the hierarchical federation model can be beneficial. Here, the updates are projected from the edge devices to the fog servers present in their locality. The results are aggregated at the fog servers and passed to the central coordinator.

In fact, this architecture cuts the network load significantly and enables faster response times as each subgroup can specialize in the behavior of each region.

Security is essential, devices are easily exploitable, and devices can be impersonated, therefore the onboarding process needs to be secured, the two parties need to authenticate themselves, and even the communication must be encrypted.

Further measures like the use of Blockchain-based audit logs and management of Federated Identity would improve trust and traceability.

In regard to orchestration, open-source solutions such as TensorFlow Federated, Flower, and PySyft exhibit agile environments for development and scaling. They enable asynchronous updates, have customizable aggregation functions and are deployed across platforms. They can also be used for development purposes as well as for scaling.

Well-coordinated training improves the scalability, efficiency, and security. This improves the unification of the edge devices dynamically and provides a stable detection of the threats under the dynamically changing wireless environments.

7.3 Evaluation Metrics for Federated Threat Detection

The analysis of the federated intelligence system must not be close-minded as required and must rather have a perspective that goes beyond the boundaries of routine model analysis. Analysis must consider not only the accuracy of the model but also the aspect of privacy and efficiency of communication.

All of this revolves around model performance. Accuracy, precision, recall rates, F1 measures, as well as AUC measures (area under the curve), can be used to judge detection. Both false positives and false negatives can consume resources as well as miss threats. Thus, the system must balance both sensitivities and specificities.

Efficiency of communication refers to the amount of information conveyed per round of communication, the total number of rounds required to achieve convergence, and the rate at which updates are sent. Efficiency can tell how effective the network can perform when bandwidth constraints exist. Effects due to the usage of network can also be analyzed for both compression and sparse updates.

Convergence speed: This refers to the model's ability to converge based on the specified accuracy. When the convergence speed improves, it enables the model to adapt to the changing circumstances as the exposure time and the time to react change rapidly because of threats.

The robustness of the functional system can also be verified through simulations of faults or device dropouts as well as adversarial attacks. The system-wide functioning in regard to system degradation rate and system recovery time can be useful information in determining the resilience of the system against disruptions or poisoning.

Privacy guarantees must be assessed both theoretically and practically. On the theoretical front, the measures involve the differential privacy budget $\\epsilon$; empirical evaluations are conducted to identify whether the private information can be discovered back from the updates via reconstruction or inference attacks.

If certain individuals have better connectivity or access to more extensive datasets, the fairness and representation measures can calculate the dominance of the training process by the respective client. This dominance can be identified and adjusted through the use of techniques such as contribution variance and per-client utilities.

Finally, the resource usage metric monitors the usage of both the CPU and memory resources and training latency to mention a few. All the above queries are employed to ensure that federated learning stays resource- and environment-friendly.

8 USE CASES IN SMART CITIES, VANETS AND HEALTHCARE

Federated intelligence framework that goes beyond the realms of theory, this can be applied to a real-life wireless scenario such as smart cities, vehicles, and health care. Secure and real-time threat detection systems can be developed for the concerned environment as the integration of federated intelligence and AI at the edge serves as a requirement for the organization.

The following examples demonstrate how the above principles apply in the real world.

8.1 Smart Cities: Securing Urban Wireless Communications Infrastructure

All city traffic management systems, energy management systems, city surveillance systems, and Internet of Things city systems rely on wireless communication. They generate vast amounts of data from sensors, cameras, as well as access points.

They are easy targets for attacks such as: Spoofing attacks. Jamming attacks. Unauthorized access.

Such states are very difficult to handle in traditional commonplace security architectures. The reason for this is the volume of the information and the laws of latency and privacy.

This issue can be addressed through federated intelligence because devices can learn together at the edge of the network without the need for the actual data to be transferred. This enables the wireless access points in different locations of the city to detect unusual traffic at the local level and update the development of the intrusion detection system. Each district might exhibit unique traffic or network patterns, so the system needs to localize and adapt while retaining a global view of citywide threats.

Clustering driven aggregation and adaptive trust scoring help with this trade-off. Public spaces make privacy protection a must-have. Differential privacy ensures that patterns and insight cannot identify individuals, while a secure aggregation mechanism provides protection from inference attacks on data derived from public Wi-Fi or surveillance applications.

Through federated intelligence, city governments can become more resilient to cyberattacks, align privacy laws, and provide secure digital services without centralisation of systems.

8.2 VANET: Real-Time Intrusion Detection

Vehicular ad-hoc networks link vehicles and roadside devices to enable intelligent traffic and self-driving. These networks are mobile (interoperable systems), fast, and time-bound and are susceptible to spoofing, Sybil attacks and attacks of denial of service.

Centralized protection schemes will never be able to keep pace with the dynamic velocity and mobility of VANET systems. Federated intelligence assists by enabling vehicles and roadside units (RSUs) to train detection models together rather than share private or navigation data.

In an autonomous vehicle, each vehicle employs onboard sensors including GPS, LIDAR, and diagnostic instruments to detect suspicious activity in situ. It then makes encrypted or differentially private updates to a nearby RSU, which then compiles them in a shared model that accounts for a variety of traffic circumstances.

Since VANET connections are temporary, asynchronous training and dropout tolerant aggregation are critical. Such grouping of vehicles by region (urban, suburban, highway etc.) increases accuracy for each context.

Trust management is another important aspect of the system. Vehicles and RSUs can allocate reputation scores by consistent behavior, manufacturer identity or verified location. These scores can help filter invalid participants and preserve the integrity of the shared model.

Thus, we apply federated intelligence for real-time threat detection, private and privacy-preserving loading of the network, and reducing the network infrastructure burden. This represents a fundamental approach towards realizing the aim of secure and trustworthy transport systems.

8.3 Remote Healthcare Monitoring and Diagnostics

Wireless health today revolves around the concept of tracking the health of patients utilizing wearable devices, intelligent implants, as well as remote sensors at the patient's residences tracking their actual health that the patient can access at any time.

These systems monitor health-related factors such as heart rate, oxygen delivery to the body, blood glucose concentration, and the quality of resting sleep. Although this information improves the delivery of health services in the hospital setting, this information poses many problems when it comes to maintaining patient privacy as well as adhering to health regulations.

This has been addressed by federated intelligence because devices can develop shared models without sending the private information to the cloud. To ensure efficient prediction analytics are achieved, the solution aligns with the aspect of the law related to patient privacy like HIPAA or GDPR.

For instance, there are wearable ECG devices that recognize irregular heartbeats and update the common cardiovascular model. Glucose and blood-pressure meters also update the model. All the updates together form accurate diagnostic models that can predict warnings for hospital re-admissions.

Federated learning can be greatly beneficial in the medical field due to the variety of patient information. Individuals can empower each device to adjust to someone’s physiology while getting a lot from the information obtained through collective learning. Such widespread inclusion promote positive results for people suffering from rare diseases.

Strong measures must also be provided for privacy. Homomorphic encryption should protect the information during the process of aggregation; otherwise, the lack of re-id in updates will be addressed through differential privacy. Trust scoring should ensure only authenticated devices are involved.

By capitalizing on the mentioned technologies, federated intelligence enables proactive rather than reactive remote healthcare. This allows physicians to remain aware of private patient information and build trust in computerized health devices.

9 CURRENT LIMITATIONS

Federated intelligence has clear advantages, but there are still multiple factors that reduce its usage in real-world environments. An important limitation of the system is the sensitivity of the system to non-independent and identically distributed (non-IID) data. Current algorithms often require clients to collect data from similar distributions.

Each wireless network collects data in a highly variable environment which varies in terms of both quantity and quality. This imbalance can result in biased global model development, lack of accuracy, and slow convergence.

Another difficulty is the imbalance of capabilities among devices. The wireless clients differ significantly in terms of processing power, memory, and connectivity. There is more frequent contribution from powerful devices, and the dropout among low resource devices. This leads to bias of participation and less diversity in data. Training of lightweight devices (sensors, wearables, etc.) even on small amounts of training may exceed their energy or memory capacity with no way to provide full participation.

However, threats remain. Although federated learning proposes the elimination of the centralized storage of information, other dangers are presented. Poisoned updates can come from malicious clients. On the other hand, attackers can use the reversal of the gradient to identify the information considered private. Moreover, the training process can be affected by the sybil attack due to the illusion of multiple clients from the same entity. These issues are countered by trust management systems and robust aggregation schemes but are still not trustworthy.

Another challenge is the issue of scalability. When more participants join the event, communication costs as well as the cost of synchronization increase. This can pose serious challenges when carried out in a mobile network or unstable network. Hierarchical and asynchronous methods can overcome such challenges. They sometimes end up overcompensating.

Finally, the area of explainability is limited as well. For example, in health or finance domains where such AI-based systems play a vital role, individuals should know how their final results were obtained. Federated systems are decentralized and encrypted. This makes the above-mentioned topic more complex. The phase of open access interpretability for decentralized AI systems has not matured yet.

10 FUTURE RESEARCH OPPORTUNITIES

Overcoming the above mentioned weaknesses shall form the main body of the next generation of federated intelligence. The weaknesses can be overcome in the following areas of researching.

Enhancing ruggedness for non-IID settings: Future work must focus on achieving adaptivity of aggregation based on the ability to adapt to the diversity and meta-structured models that can adapt to learn generalization based on the given data. An organizational collaboration would also assist because it can offer insights related to the model instead of the data. This could also assist in maintaining the privacy of the data.

Improving communication and computation efficiency: Wireless devices often operate with tight constraints on power and bandwidth. Overhead can be greatly reduced through methods of compressing the model updates, sharing sparse representations and selective synchronization. The improvements in quantization-aware training, federated distillation and asynchronous gradient sharing can speed up and scale up learning.

Enhancing security and trust mechanisms: In the future it is suggested for more attention about poisoned updates, model integrity verification, and transparent audit mechanism. By integrating federated learning with blockchain, more accountability and traceability will be achieved, and trust can be evaluated at the level of parties using decentralized trust; also, an immutable log can be preserved.

Explainability for decentralized AI is on the rise: The methods that can show influential features, provide visibility (for example decision paths), identify influential features, or give interpretable feedback will enable federated systems to be more interpretable. Tools that will describe both local behavior and global behavior can be exceptionally useful tools for industry that is regulated and demands accountability, particularly in regulated industries.

Incorporating next-generation tech: New-generation wireless advancements like 6G, quantum communications or neuromorphic computing make things possible with everything from 6G to next-gen wireless devices. Ultra-low-latency networks can also accommodate near-instantaneous model update; quantum-secure encryption can guard model parameters from advanced attack. The use of neuromorphic equipment may also enhance the performance of learning at the edge.

Complying with regulations and ethics: Federated systems must adhere to worldwide protection of data and fair application of data. Governance standards and transparency of governance criteria will require consensus on governance by engineers, policy will require engineers, policymakers and ethicists to work together to produce such governance standards and transparency.

As a result, efforts will preserve user trust, and aid global deployment with the encouragement of user trust on a global scale (Sharma, Sharma, & Van Chien, 2025).

11 CONCLUSION

Wireless systems are key tools enabling critical infrastructure backbone of smart cities, transportation, and healthcare. This represents a massive challenge where large opportunities are involved together with some rather complicated merely privacy matters.

Even more decentralized approaches towards threat detection would no longer be sufficient in areas where the information is not centralized due to connectivity issues or the preservation of privacy. The inclusion of federated intelligence in this chapter serves as another promising solution mentioned.

This enables secure and cooperative threat detection in a distributed network through federated learning, edge AI, as well as privacy-preserving computation. This example illustrates how edge devices can learn, adapt, and share together to avoid compromising information.

This chapter has also shown the practical aspects of resilient aggregation techniques like asynchronous coordination and resource optimization. In doing so, the methods ensure the ability of federated systems to operate in effective manners based on different device capability levels and network status.

It also incorporates differential privacy, homomorphic encryption, and trust scoring. The use of case studies in smart cities, vehicular communication systems, and the medical field highlights the immense potential of the federated intelligence approach besides the concern of data privacy.

Even as some constraints, such as non-IID data and the computational challenge, remain in this area, the promising future lies ahead. Its usage is increasing due to advancements in the area of personalization techniques, light encryption, and explainable federated models.

Federated Intelligence offers a scalable solution for wireless security that focuses both on scaling and privacy. This offers trust AI-enabled functions and enables the management of systems. This development would play a fundamental role in the development of wireless systems and environments for improved safety.

REFERENCES

Abbas, S. R., Abbas, Z., Zahir, A., & Lee, S. W. (2024, December). Federated learning in smart healthcare: a comprehensive review on privacy, security, and predictive analytics with IoT integration. In Healthcare (Vol. 12, No. 24, p. 2587). MDPI. DOI: 10.3390/healthcare12242587

Aghdam, S. R., Amid, E., Furdek, M., & i Amat, A. G. (2021, October). Privacy-preserving wireless federated learning exploiting inherent hardware impairments. In 2021 IEEE 26th International Workshop on Computer Aided Modeling and Design of Communication Links and Networks (CAMAD) (pp. 1-6). IEEE.

Aghdam, S. R., Amid, E., Furdek, M., & i Amat, A. G. (2021, October). Privacy-preserving wireless federated learning exploiting inherent hardware impairments. In 2021 IEEE 26th International Workshop on Computer Aided Modeling and Design of Communication Links and Networks (CAMAD) (pp. 1-6). IEEE.

Agrawal, S., Sarkar, S., Aouedi, O., Yenduri, G., Piamrat, K., Alazab, M., Bhattacharya, S., Maddikunta, P. K. R., & Gadekallu, T. R. (2022). Federated learning for intrusion detection system: Concepts, challenges and future directions. *Computer Communications*, *195*, 346–361. DOI: 10.1016/j.comcom.2022.09.012

Anandaram, H., Shreenidhi, K. S., Joshi, K., Sidhu, K. S., Kiruthigha, R., Singh, V., & Tripathi, A. (2026). A Case Study on Resource Management Industrial IoT Application. In *Building Data-Driven Edge Systems for Business Success* (pp. 1–26). IGI Global Scientific Publishing.

Azarkasb, S. O., & Khasteh, S. H. (2025). Fog Computing Tasks Management Based on Federated Reinforcement Learning. *Journal of Grid Computing*, *23*(1), 11. DOI: 10.1007/s10723-025-09796-4

Aziz, R., Banerjee, S., Bouzefrane, S., & Le Vinh, T. (2023). Exploring homomorphic encryption and differential privacy techniques towards secure federated learning paradigm. *Future Internet*, *15*(9), 310. DOI: 10.3390/fi15090310

Balas, V., Pandey, H., Bin Ali, M., Singh, V., & Kumar, A. (Eds.). (2026). *Recent Advances in Smart Communication Technologies for a Sustainable Future*. IGI Global Scientific Publishing., DOI: 10.4018/979-8-3373-3541-4

Bonawitz, K., Ivanov, V., Kreuter, B., Marcedone, A., McMahan, H. B., Patel, S., & Seth, K. (2017, October). Practical secure aggregation for privacy-preserving machine learning. In *proceedings of the 2017 ACM SIGSAC Conference on Computer and Communications Security* (pp. 1175-1191). DOI: 10.1145/3133956.3133982

Chaudhary, S., Singh, V., Malik, N., Sidhu, K. S., Kirola, M., & Joshi, K. (2024, October). Digital Signature Security Enhancement: Using X-OR Operation on RSA and AES with 4096 Bits Key Length. In *2024 International Conference on Computing, Sciences and Communications (ICCSC)* (pp. 1-6). IEEE. DOI: 10.1109/ICCSC62048.2024.10830431

Ericsson. (2025). *Ericsson mobility report.* Retrieved from https://www.ericsson.com/en/reports-and-papers/mobility-report

Feng, Y., Guo, Y., Hou, Y., Wu, Y., Lao, M., Yu, T., & Liu, G. (2025). A survey of security threats in federated learning. *Complex & Intelligent Systems, 11*(2), 165. DOI: 10.1007/s40747-024-01664-0

Fnu, H., Kaushik, K., Gaur, P., Saratchandran, D. V., Thapliyal, A. G., Sidhu, K. S., & Singh, V. (2025, April). Multi-Agent Systems for Collaborative Financial Decision-Making Over Distributed Network Architectures. In 2025 3rd International Conference on Advancement in Computation & Computer Technologies (InCACCT) (pp. 939-944). IEEE. DOI: 10.1109/InCACCT65424.2025.11011435

Hsu, H. Y., Keoy, K. H., Chen, J. R., Chao, H. C., & Lai, C. F. (2023). Personalized federated learning algorithm with adaptive clustering for non-IID IoT data incorporating multi-task learning and neural network model characteristics. *Sensors (Basel), 23*(22), 9016. DOI: 10.3390/s23229016 PMID: 38005404

Hu, C., & Li, B. (2024). Maskcrypt: Federated learning with selective homomorphic encryption. *IEEE Transactions on Dependable and Secure Computing*, *22*(1), 221–233. DOI: 10.1109/TDSC.2024.3392424

Karunamurthy, A., Vijayan, K., Kshirsagar, P. R., & Tan, K. T. (2025). An optimal federated learning-based intrusion detection for IoT environment. *Scientific Reports, 15*(1), 8696. DOI: 10.1038/s41598-025-93501-8 PMID: 40082567

Kaur, P., Kaur, K., Verma, D. K., Singh, K., Kaushik, K., & Singh, V. (2024, September). Routing protocols in wireless sensor networks: A comprehensive review and future perspectives. In 2024 7th International Conference on Contemporary Computing and Informatics (IC3I) (Vol. 7, pp. 263-270). IEEE. DOI: 10.1109/IC3I61595.2024.10828894

Krishna, G., Singh, V., Pandey, D., Krishna, K. H., Joshi, K., & Gupta, T. (2024, May). Power trading framework of cloud-edge computing in the artificial intelligence market. In 2024 4th International Conference on Advance Computing and Innovative Technologies in Engineering (ICACITE) (pp. 1910-1918). IEEE. DOI: 10.1109/ICACITE60783.2024.10617344

Llisterri Gimenez, N., Monfort Grau, M., Pueyo Centelles, R., & Freitag, F. (2022). On-device training of machine learning models on microcontrollers with federated learning. *Electronics (Basel)*, *11*(4), 573. DOI: 10.3390/electronics11040573

Makris, I., Karampasi, A., Radoglou-Grammatikis, P., Episkopos, N., Iturbe, E., Rios, E., Piperigkos, N., Lalos, A., Xenakis, C., Lagkas, T., Argyriou, V., & Sarigiannidis, P. (2025). A comprehensive survey of Federated Intrusion Detection Systems: Techniques, challenges and solutions. *Computer Science Review*, *56*, 100717. DOI: 10.1016/j.cosrev.2024.100717

Mehta, S., Khurana, M., Dogra, A., & Hariharan, S. (2024, June). Advancing IoT security through federated learning: a comprehensive approach. In 2024 3rd International Conference on Applied Artificial Intelligence and Computing (ICAAIC) (pp. 561-566). IEEE. DOI: 10.1109/ICAAIC60222.2024.10575903

Myakala, P. K., & Agrawal, M. (2025). Fault-Tolerant Federated Learning Framework for Edge Devices in Unstable Networks. *Authorea Preprints*. DOI: 10.36227/techrxiv.174612128.82578829/v1

Myakala, P. K., Naayini, P., & Kamatala, S. (2025). A Survey on Federated Learning for TinyML: Challenges, Techniques, and Future Directions. *Partners Universal International Innovation Journal*, *3*(2), 97–114.

Narooka, P., Kaushik, K., Kansal, S., Kejriwal, D., Mishra, R. K., & Singh, V. (2025, April). Hybrid AI Models for QoS Estimation in α-η-μ Fading and Co-Channel Interference Environments within 6G Networks. In 2025 3rd International Conference on Advancement in Computation & Computer Technologies (InCACCT) (pp. 206-210). IEEE.

Narooka, P., Verma, S., Dwivedi, A., Kaushik, K., Singh, V., & Gaur, P. (2025, April). Psi-zeta Approach for Assessing Network Efficiency in Log-Normal Fading and Spectrum Sharing Interference Environments. In 2025 3rd International Conference on Advancement in Computation & Computer Technologies (InCACCT) (pp. 886-891). IEEE. DOI: 10.1109/InCACCT65424.2025.11011391

Parreño, I. F., & Avila, D. F. (2021). Analysis of the cybersecurity in wireless sensor networks (WSN): A review literature. Developments and advances in defense and security. *Proceedings of MICRADS*, *2021*, 83–102.

Roohani, B. S., Sharma, N., Kasula, V. K., Mamoria, P., Modh, N. N., Kumar, A., & Singh, V. (2026). Urban Computing Solutions in Healthcare Edge Computing. In Building Data-Driven Edge Systems for Business Success (pp. 377-400). IGI Global Scientific Publishing.

Sharma, S., Sharma, A., & Van Chien, T. (Eds.). (2025). *The Intersection of 6G, AI/Machine Learning, and Embedded Systems: Pioneering Intelligent Wireless Technologies*. CRC Press. DOI: 10.1201/9781003540212

Singh, V., & Dubey, V. (2015). A two level image security based on Arnold transform and chaotic logistic mapping. *International Journal of Advanced Research in Computer Science and Software Engineering*, *5*(2), 883–887.

Singh, V., & Kaushik, V. D. (2023). A Brief Overview of Cyber Security Advances and Techniques Along With a Glimpse on Quantum Cryptography: Cyber Security Practices, Advances and Challenges. Exploring Cyber Criminals and Data Privacy Measures, 40-64.

Singh, V., & Kaushik, V. D. (2023). A Methodical View of Prerequisites of Picture Combination, Strategies, and Key Indicators with Usage in Real Life and Scientific Domains Facilitating Smart Ubiquitous Environment. In Computational Intelligence in Analytics and Information Systems (pp. 15-33). Apple Academic Press.

Singh, V., & Kaushik, V. D. (2023). Navigating the Landscape of Security Threat Analysis in Cloud Computing environments. In *Security and Risk Analysis for Intelligent Cloud Computing* (pp. 1–25). CRC Press.

Suriyan, K., Ganesh, S., Palaniyammal, P., Priya, G., & Singh, V. (2026). Recent Trends in Edge Computing: Challenges and Opportunities. Building Data-Driven Edge Systems for Business Success, 323-344.

Yin, X., Zhu, Y., & Hu, J. (2021). A comprehensive survey of privacy-preserving federated learning: A taxonomy, review, and future directions. *ACM Computing Surveys*, *54*(6), 1–36. DOI: 10.1145/3460427

Zhang, S., Li, Z., Chen, Q., Zheng, W., Leng, J., & Guo, M. (2021, August). Dubhe: Towards data unbiasedness with homomorphic encryption in federated learning client selection. In *Proceedings of the 50th International Conference on Parallel Processing* (pp. 1-10). DOI: 10.1145/3472456.3473513

Chapter 7
Blockchain and AI Secure Wireless Transactions

Amrutha Kolhar
https://orcid.org/0009-0001-4940-3172
Karnatak University, Dharwad, India

Sridevi
https://orcid.org/0000-0002-9768-1599
Karnatak University, Dharwad, India

Hemant Soni
https://orcid.org/0009-0001-1874-6930
Capgemini United States, USA

Kamal Kant
https://orcid.org/0009-0009-2684-3595
School of Engineering and Technology, Chhatrapati Shahu Ji Maharaj University, Kanpur, India

ABSTRACT

The rapid advancement of wireless communication technologies has heightened the need for secure and efficient transaction systems. Integrating blockchain and artificial intelligence (AI) offers a transformative solution to enhance the security, transparency, and trust of wireless transactions. Blockchain provides decentralized and tamper-proof record-keeping, while AI strengthens system intelligence through fraud detection, adaptive authentication, and predictive threat mitigation. This paper investigates the combined potential of blockchain and AI in securing wireless transactions, focusing on challenges such as scalability, latency, and privacy. It also

DOI: 10.4018/979-8-3373-3576-6.ch007

t © 2026, IGI Global Scientific Publishing. Copying or distributing in print or electronic forms without written permission of IGI Global Scientific Publishing is prohibited. Use of this chapter to
enerative artificial intelligence (AI) technologies is expressly prohibited. The publisher reserves all rights to license its use for generative AI training and machine learning model development.

examines real-world applications, including 5G, IoT, and mobile payments, while highlighting emerging trends and future research directions aimed at developing robust, next-generation wireless transaction systems.

1. INTRODUCTION

Digital operations have become fast and convenient due to development of wireless communication technologies. However, it has significant security consequences such as fraudulent purchases that come with this convenience, data breaches, and identity theft. An innovative approach to improving wireless transactional security is provided by the blockchain and artificial intelligence (AI). Blockchain and AI can guarantee never before seen transparency and longer durability due to its distributed and thereof immutable record, and higher-quality threat detection and anomaly discovery, respectively, and responsive authentication (Balas, Pandey, Bin Ali, Singh, & Kumar, 2026). Collectively, the technologies help to provide a complete security network, protecting classified information, eliminating fraud attacks, and maintaining the integrity of wireless activities (Narooka et al., 2025a; Narooka et al., 2025b; . Blockchain technology lowers the danger of internet-based attacks and single points of failure by doing away with centralized middlemen using distributed ledger technology. Unauthorized data alteration is almost impossible due to cryptographic security (Singh and Dubey, 2015; Singh and Kaushik, 2023a; Singh and Kaushik, 2023b; Chaudhary et al., 2024) and replication of records at more than one node. In the meantime, transaction pattern analysis with AI enables detection of suspicious activity right away and prevent any threats to the best of its abilities before they can develop. Machine learning (Raturi et al., 2025) is able to constantly identify likely fraud patterns and predictive analytics is able to detect any potential problems before they happen. Wireless transactions are enhanced due to the advanced AI methods including biometric analysis. By combining blockchain technology with artificial intelligence, the verification process becomes automated, and the time of transactions is reduced to the maximum, as well as the safety and speed of transactions are increased. Smart contracts coded on blockchain make workflow digital and add to the automatic performance of contractual obligation without any human intervention that minimizes possible mistakes. The rapid deployment of 5G technology and the proliferation of the Internet of Things (IoT) have greatly increased the demand for strong, scalable, and intelligent tools to facilitate secure wireless movements. Through the strength of blockchain and AI, businesses and consumers will be able to conduct transactions wirelessly (Fnu et al., 2025) without fear because they are aware that advanced cybersecurity tools will keep their data safe. Such a composite preconditions the introduction of digital transactions that will be faster to process, more accessible, and secure and reliable (Alrubei et al., 2022; Azzaoui et al., 2020; Martinez & Magdalena, 2024; Wang et al., 2020; Zhou et al., 2024).

1.1 Background and Motivation

The development of technology for wireless communication (Kaur et al., 2024) the model of financial and data transactions has changed as a result of 5G networks, the Internet of Things, and mobile payment systems. With the growth of digital transactions, cyberthreats such identity theft, phishing, fraud, and data breaches have increased. The traditional security processes, including centralized authentication and classical encryption measures tend to be ineffective in managing the diversity of contemporary cyber threats. The need to strengthen, innovative security measures to enhance sensitive user and financial data, therefore, exists immediately. Decentralized design, transparency, and immutability of blockchain make it an effective instrument of securing digital transactions. Blockchain can be used as a robust base of safe wireless transactions by removing single points of failure as well as offering ledgers that are impervious to tampering. Nonetheless, the blockchain by itself would not be able to rapidly detect and react to new and emerging threats. Artificial intelligence enters at this stage. Through AI, machine learning provides predictive analytics, the capability to detect anomalies through sophisticated behavioral biometrics, and a zero-day security threat first line defense. The combination of AI and blockchain is a whole, with a strong and solid support. AI provides real-time response to threats, which complements the most crucial role of blockchain, reliability, and integrity of the data. The increasing demand of flexible and powerful security instruments in a continuously expanding digital universe of interconnections is the reason why the convergence between blockchain and AI towards secure wireless transactions. In combination with the cognitive abilities of AI, the blockchain is decentralized, which, in turn, provides the guarantee in the form of verification to the conclusion of the deals. In addition, the companies are compelled to be more efficient in their approaches to data protection due to such regulations as GDPR, PSD2, and financial standards. The compliance control, increased by the use of AI, and an open registry of a blockchain can help corporations to fulfill their regulatory requirements and reduce their susceptibility to fraud. Such a combination of technologies is not only a way of fending off security threats but also enhances efficiency through reduction of false alarms and expedited process of transaction approval. The development of a robust blockchain system is predicated on the integration of AI technologies, transparent wireless transaction atmosphere and provides trust and the desire to trust the future of digital innovation. (Borges et al., 2014; Huo et al., 2022; Zuo et al., 2023) (Kolhar & Sridevi, 2025).

Recent publications on wireless networks that use blockchain focus on decentralized trust and AI-assisted anomaly detection but are still disconnected in methodology. As an example, Zhou et al. (2024) and Zuo et al. (2023) survey blockchain-AI integration with 6G, however, do not assess the concept of adversarial resilience. The article by Huo et al. (2022) emphasizes the IoT security through blockchain but does not mention the federated coordination of wireless nodes through AI. In the same vein, Kumar et al. (2024) and Alenizi et al. (2024) describe how explainable AI can be used to detect fraud in the Internet of Things, but there isn't much research on how mobile edge networks scale. This shows a particular lack of coherent models that cater to threat modelling, latency security trade-offs and cross-layer scalability. In order to close this gap and improve the integrity of wireless transactions in 5G/6G frameworks, the current chapter proposes a hybrid architecture that links blockchain-based trust management and AI-based intrusion detection.

1.2 Objectives of the Study

Aspects of how blockchain technology and artificial intelligence might be combined to increase the security of wireless transactions are among the research questions in this study. In particular, the given research will seek to: (1) find vulnerabilities in existing wireless transaction frameworks, pointing to the key areas of risk that blockchain and AI can address; (2) develop a hybrid security model where blockchain, a distributed ledger, will store immutable records, and AI, with its learning capabilities, will identify dangers instantly; (3) assess the performance and applicability of the target framework to high transaction volumes, i.e. without impairing the speed or security; and (4) review the viability and capability of the offered model to prevent fraud attempts, data In addition, the research paper has to (5) look at regulatory and compliance issues so that the framework complies with the data protection regulations and financial industry standards. Through the accomplishment of these goals, the current study will help in shaping a powerful, versatile and forward-thinking security architecture of wireless transactions. Businesses and legislators that wish to use blockchain and artificial intelligence to enable safer online transactions in a world that is becoming more interconnected will also find it helpful (Atlam et al., 2020; Bodkhe et al., 2020) (Gundoor, 2025).

1.3 To Visually Articulate the Proposed Framework, the Following Architecture is Presented:

Figure 1. Blockchain AI Integrated Framework for Secure Wireless Transaction

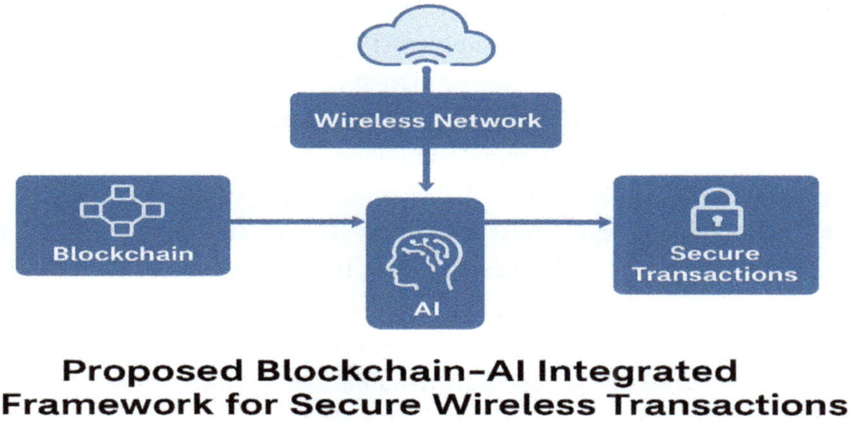

The blockchain and artificial intelligence (AI) are incorporated into the proposed framework to support the secure wireless transaction as shown in Figure 1. Under such a framework, the data of a wireless network (Singh and Kaushik, 2023) is gathered and sent to the AI module where real-time analysis is done to identify anomalies, anticipate threats, and classify transaction patterns. At the same time, blockchain technology has a decentralized and indisputable registry that authenticates and documents all transactions. The outputs of the AI-processing are also checked with the blockchain records to make sure that it is consistent and provides higher levels of security. These elements combine in a way that ensures end-to-end protection making wireless transactions in the network to be trustful and secure.

1.4 Contribution of the Work

- **Novel Hybrid Security Model:** This paper presents a novel security model of blockchain and adaptive intelligence of AI in order to create an entirely new system of wireless transactions, which is more secure and efficient than the traditional ones
- **Real-Time Threat Detection Model:** The system is based on modern machine learning and analyzes transactions in real-time to notify users of sophisticated cyber anomalies as they happen, even those that other security systems fail to identify.

- **Decentralized Trust Architecture:** Permissioned blockchain usage provides balance between transparency, security, and performance, industry-standard, and compliance with regulatory standards and vulnerabilities inherent to centralized architectures.
- **Optimized Performance Metrics:** The structure is configured and tested to be high load, quick transaction processing, and high throughput and thus the structure is realistic in the case of high-speed wireless systems.
- **Compliance-Ready Framework:** The architecture incorporates a high level of access control with automated compliance management that offers proactive support of the regulation compliance with the EU GDPR and PSD2.
- **Technology Integration Guidelines:** The study offers fluid and viable solutions on how organizations can use these technologies within different environments.
- **Criteria of the Assessment of Components of Security:** The paper develops a novel criterion for assessing dynamic cybersecurity systems in wireless environments and devises a thorough scale to evaluate the effectiveness of blockchain-AI security systems (Selvarajan et al., 2023; Shinde et al.,2021) (Kolhar and Sridevi, 2025)

2. BLOCKCHAIN FOR SECURE WIRELESS TRANSACTIONS

Blockchain provides much more security to wireless transactions as it maintains and ensures the true and unaltered register of all transactions that are sent and circulated over the whole network. Cryptographic hashing and consensus validation are applied to every transaction, eliminating any potential points of failure and drastically lowering the likelihood of fraud. Additionally, it is unchangeable, guaranteeing data integrity, and smart contracts may be used to create safe, trust less agreements without the need for middlemen. Blockchain technology for 5G and IoT wireless systems provides protection against data tampering, man-in-the-middle attacks, and unwanted access. Secure and private peer-to-peer operations are supported by the combination of encryption with the transparent operations. Consequently, blockchain is most applicable in mobile payments, authentication of devices, and sharing of data in wireless environments where there is faith in decentralized environments. (Alrubei et al., 2022; Azzaoui et al., 2020; Martinez & Magdalena, 2024; Wang et al., 2020; Xiao et al., 2021; Zhou et al., 2024).

2.1 The Basics of Blockchain Technology

A blockchain is a dispersed, peer-to-peer network that is most, no single person or entity can manage or change the information and thus the system is strong and resistant to attacks. Eliminating middleman services, blockchain facilitates direct and trustless communication between users and is a valid solution to applications that require transparency and censorship resistance, including cryptocurrencies and DeFi.

- **Distributed Ledger:** A distributed network of blockchain allows all of its participants to have an equal copy of the ledger, which provides transparency and consistency. Since the data is logged on to several nodes, it makes the manipulation of the data extremely hard.
- **Cryptographic Security:** To prevent the security of the transactions, Blockchain employs cryptographic tools such as hashing (e.g., SHA-256) and digital signatures. The hash of a block is a kind of content and thus, any changes are easily detected. Public-private key pairs will make sure that only authorized agents can start transactions.
- **Immutability:** When data is stored in a blockchain it cannot be modified, this means that it forms an unchangeable and auditable history. This is an important point to prevent fraud and illegitimate Contractual, supply chain, and financial manipulations.
- **Consensus Mechanisms:** blockchain is based on these means to verify the transactions, e.g. Proof-of-Stake (PoS) and Proof-of-Work (PoW). The processes provide security of the network by having to agree with the nodes to prevent problems such as duplication of currencies.
- **Transparency and Pseudonymity:** The blockchain transactions are transparent and verifiable. Cryptographic addresses are used to make and maintain user interactions, and combine public responsibility with privacy.
- **Block Structure:** Each block is made up of data of the transaction, a time stamp and the hash of the prior block thus forming a cryptographic circular linked network that guarantees the sequence and accuracy of the information.
- **Use Cases:** In addition to cryptocurrency, blockchain is being used to secure supply chains, to power DeFi apps, to streamline cross-border transactions, and to enable the safe management of voting, identities, and intellectual property. (Du et al., 2021; Gupta et al., 2020; Ul Hassan et al., 2019; Xuan & Ness, 2023).

2.2 Dispersed Security Systems

Figure 2. Dispersed Security Systems

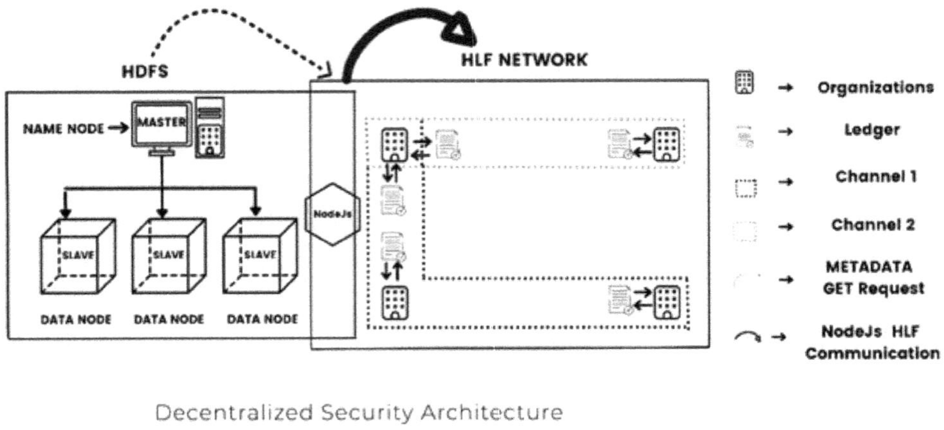

The architecture shown in the diagram is a decentralized security model that relates the Hadoop Distributed File System (HDFS) and Node.js to provide a reliable and scalable data management system as demonstrated in Figure 2. Name Node (Master) is at the topmost level and it handles all the metadata of the system and the coordination of functions in the distributed HDFS file system. It is further divided into many Data Nodes which are distributed storage units and each store blocks of data to provide fault tolerance and redundancy. An interface (or API) layer, written in Node.js, allows real-time processing and, along with a combination of distributed storage and centralized control, seamless interaction between the clients and the HDFS, and thus balances efficiency and scalability.

Since architecture is built using decentralized security principles, it lowers vulnerability to large scale attacks. HDFS offers basic security, which is replication of data, whereas Node.js offers higher security with the help of authentication, encrypted communication and access control. A combination of these layers will reduce the risks of a single-point failure and unauthorized access, and the architecture will be appropriate in the scenario of processing sensitive information in financial systems, IoT applications (Joshi et al., 2024a; Joshi et al., 2024b), and other high-stakes environments. In general, Figure 2 shows that the combination of distributed storage and the latest web technologies can be used to provide high security and high performance in high-scales.(Alharbi et al., 2022; Attkan & Ranga, 2022; Khalid et al., 2020; Zeng et al., 2020)

2.3 Automated Transactions Using Smart Contracts

Figure 3. Automated Smart Contracting Transaction

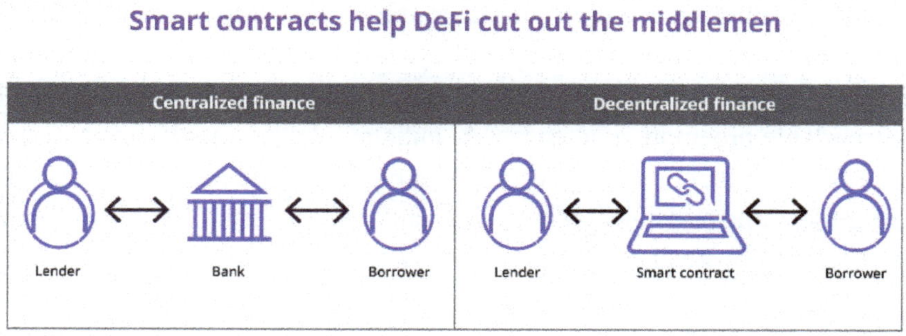

Decentralized finance using smart contracts is compared to the traditional centralized finance requiring intermediaries, presented in Figure 3. The centralized finance model is shown on the left-hand side where the banks play the central role of linking the lenders and borrowers and regulating all the transactions, terms and related costs. This inherently adds delays, increased costs, and reliance on third party institutions. Conversely, the right-hand example of decentralized finance illustrates how smart contracts substitute the intermediary in making direct contact between lenders and borrowers. These self-sovereign blockchain-based contracts automatically enforce the terms of a loan using a code that is pre-determined, thus eliminating the need to be banked, enhancing transparency and trust. As noted in Figure 3, this automation saves a lot of money in terms of operations, shortens the speed of a transaction, and eliminates a single point of failure. The diagram highlights the main innovation of DeFi, which is that financial operations are fueled by code, as opposed to human intermediaries. This gives the users more freedom to use financial services like lending and borrowing, which have the backing of blockchain immutability and cryptographic verification. This leads to the citizens having more control over their financial resources, which is in line with the internationalizing trend towards decentralized and borderless finance. Nevertheless, although the benefits exist, DeFi continues to have issues associated with the vulnerability of smart contracts and changing regulatory regulations that should be approached to make it more secure and scalable (Khan et al., 2021; Martinez & Magdalena, 2024; Ul Hassan et al., 2019; Xiao et al., 2021) (Kolhar & Sridevi, 2026)

3. AI-DRIVEN SECURITY ENHANCEMENTS

AI-based security uses machine learning and advanced analytics to provide an extra element of smart, dynamic resilience over blockchain-based wireless transactions. Artificial intelligence programs keep track of network and user actions live in order to detect suspicious activity and cybercrime. Due to the changing nature of the adaptive algorithms in response to new attack vectors, they change and build more formidable preemptive defenses. AI enhances behavioral biometric and real-time risk-based authentication that offers only authorized access. This combined with the unalterable registry of blockchain can create a very robust security system that can guard more advanced threats and also provide rapid transactional facilities.

- **Anomaly Detection:** Unsupervised detection such as Autoencoders and Isolation Forests are not limited to pre-labeled data on fraudulent transactions unlike classical methods, so they can detect new attack behavior.
- **Supervised Classification:** Random Forest and XGBoost models are classification models that are trained on previous data of legitimate and fraudulent transactions and are used to classify new transactions with high accuracy, thus successfully identifying patterns of known fraud.
- **Behavioral Biometrics:** ML models identify a user based on the characteristics of his or her behavior (typing speed, mouse movement). This is a dynamic approach that offers constant and non-invasive security against account takeover.
- **Real-Time Risk Scoring:** ML algorithms examine various attributes of transactions (amount, location, device) in milliseconds to produce a risk score, and block high-risk transactions in real time.
- **Network Analysis:** The ML approaches based on graphs on networks examine the relationships among accounts and devices to reveal structured fraud rings otherwise not visible when analyzed on a case-by-case basis.
- **Adaptive Learning:** Adaptive models are constantly updated with new fraud information, and they are able to respond and adjust to changing fraud tactics.
- **Ensemble Methods and Explainable AI (XAI):** Stack or Boost Multiple models In ensemble methods, detection accuracy is improved and false positives are reduced by combining multiple models. Such methods as SHAP and LIME offer transparency, which is essential to the confidence of an analyst and compliance with regulations (Alenizi et al., 2024; Ali et al., 2022; Moumeni et al., 2022).

3.1 Systems for AI-Based Authentication

- **Biometric Verification**: Biometric verification is founded on the individual traits of the person such as voice or shape of the iris which are used to verify identity. These AI systems can be used to verify security clearance by conducting biometric evaluation in real-time by comparing live data with stored templates. Biometrics are difficult to imitate or steal hence they provide a lot more security against intrusion than passwords. Additionally, the new generation liveness verification has systems that can detect photo or masking spoofing. This user-friendly password-less service increases the degree of security and the user-friendliness across all platforms and applications.
- **Behavioral Analysis**: AI is the foundation of behavioural analysis, which handles user authentication. utilizing the evaluation of distinctive interaction behavior, i.e., the speed of keystroke, the way a mouse or touchscreens are being used, etc. Unlike the traditional authentication that is not dynamic, its profiles are dynamic and the user is authenticated according to how the session is accumulated. The system causes real-time detection, and additional security measures are invoked whenever the user is not performing in a normal way. This is a low profile continuously-on security model that defends against account hijackings and does not add significantly to authorization users. The technology offers good security of identity theft with minimal interference with user convenience in the internet since it adjusts with the habits of the user.
- **Authentication with many factors**: Users who validate their identities.via different, independent and independent credentials knowledge (Passwords), Possession (OTP tokens) and inheritance (biometric markers) increase the security of MFA. MFA services based on AI demand only additional information about the user when anomalies are found, such as when the user tries to log in at a new location, and respond by modifying their authentication questions. The numerous steps involved in the process effectively mitigate the threat of unauthorized access as it is much more difficult to attack with only one factor of protection having to run through multiple layers of security. Capable of finding a delicate balance between security and usability, MFA provides immeasurable protection against automated attacks with very low costs to the users.
- **Context-Aware Authentication**: To increase the level of security, Context-Aware Authentication modifies the steps of verification according to the real-time data about the location, the type of device, network activity, and the approximate time of the accesses. AI utilizes the acquired data to assign risk ratings, which subsequently creates more intensive authentication, e.g., within the example of the cross-border country or device-originated login events with

which AI is unfamiliar. In the normal cases, in addition to lowering user friction, the method automatically responds to risky behaviour with increased security. When the examination of the behavioural patterns of the login procedures is coupled with the environmental data, The security layer that adapts that is able to block 80 percent of account takeover attempts, without causing a slowdown in normal users, is formed.

- **Continuous Authentication**: Continuous Authentication is a new concept in security which tracks user activity during the session and is not confined to single-logins. The system is anchored on the AI behavioral biometric technology that will track the way the users are typing, using their mouse and surfing online. The system silently monitors the user behavior as whether the user is operating normally or otherwise, and instantaneously identifies changes in the activity as well as suspicious activity. In case the activity is detected as being anomalous, the system may either ask to verify multi-factors right away, or terminate the affected session right away. Active protection through this process helps to prevent account hijacking after logging in, as well as to offer the same protection against insiders and exposed credentials continuously, and it is also easy to use by authorized parties.
- **Fraud Prediction**: Fraud Prediction exploits the information that has been obtained as a result of analysis of user activities, transaction data and previous fraud data to proactively detect, and prevent, such fraud. Artificial intelligence technology can analyze fluctuations in the expenditure pattern or transaction rate and instantly raise alarm on the high-risk cases. The systems respond to the emergence of new methods of fraud, on the basis of constant assimilation of new information and, therefore, eliminating false alarms and detecting the sophisticated fraud. A predictive solution assists companies to prevent fraud in the initial stages that will minimize risks to the economy and ensure better security without influencing legitimate customer transactions. The prioritization that is carried out by sophisticated algorithms assists the investigators to focus on the greatest security issues.
- **Password less Authentication**: Password less Authentication is a replacement of a password to some restricted alternatives, like biometrics as well as hardware tokens, as a method of user authentication that is simpler and stronger. By using the assistance of AI that has been empowered through encryption, solutions verify the identity of users based on biological characteristics or factors of possession, thus rendering phishing and credential theft much less probable. In this approach, the security is enhanced and ease of use is boosted since the users do not have to memorize or reset passwords. Security is enhanced in the majority of systems by combining several aspects, like trust of the device and biometrics. The password less solutions, which are being embraced by both

enterprises and consumer apps, meet the growing demand of easy to use and secure access, and can reduce the helpdesk expense by up to 50% at the same time (Attkan & Ranga, 2022; Deebak et al., 2023; Kumar et al., 2024; Latif et al., 2021; Selvarajan et al., 2023)

• **Adversial Vulnerabilities:** Nevertheless, AI systems are adversarial vulnerable in the sense that a bad input will introduce a subtle modification to model predictions, deceiving an intrusion detector or biometric system. Learning pipelines can be poisoned by adversarial attacks e.g. data poisoning, model inversion, and evasion. An immutable audit trail is a defense mechanism that can be added to blockchain to prove the provenance of training data and version control of deployed models. Scholars, such as Ruzbahani (2024), and Selvarajan et al. (2023), back up the idea of using blockchain-based model attestation and hash verification to mitigate the effects of AI poisoning up to 70 percent. Therefore, the adversarial robustness is a vital aspect of the given secure framework.

3.2 Using Machine Learning to Identify Fraud

- **Anomaly Detection:** Machine learning methods are used to analyze vast amounts of data in order to identify the most commonly seen abnormal patterns. The attributes under analysis of the algorithm used in unsupervised learning like Isolation Forests and Auto encoders include the amount, frequency, and location in transactions to identify odd behaviour. These models are very capable of identifying novel forms of fraud because they were not trained with any fraud labels. Utilizing patterns based on normal user and account activity, the system automatically draws the attention of the investigator to abnormal transactions and helps them to examine likely fraud.
- **Supervised Classification**; Both legitimate and illegal transactions of past times are employed in training monitored machine learning. The models for machine learning like random forest and XGBoost are trained to identify minor links and patterns that are indicators of fraud. The trained models measure the size of the transaction, the type of merchant, the time of transaction and the behavior of the user so as to give a score of the likelihood of fraud, this way when new transactions are made, the classifiers instantly evaluate the new transactions with those that are similar to the occurrence of past fraud being flagged to undergo further examination whereas genuine transactions are not held and can be made within the same time, this significantly minimizes the need to have human intervention in reviewing the transactions made.

• **Behavioral Biometrics**: An advanced technique is applied that monitors the activity of the users by tracing the movements of their fingers in form of keystrokes, or by the mouse or touchscreen gestures to produce digital fingerprints. These models set individual standards of behavior of a user and also track user sessions in case of deviations. These dynamic identifiers are practically inimitable as opposed to static ones and this means that these offer excellent defense against accounts capture. The system can also identify minor behavioral anomalies as part of compromise of credentials which will lead to further authentication or block the access that is not congruent to the interaction pattern the intended user is supposed to have.

• **Network Analysis**: Graph machine learning methods are employed to trace the relationship among accounts, devices, and transactions to identify arranged groups of fraud. These models identify chains of suspicious activity which go unnoticed by rule-based systems using patterns of connections and entire streams of transactions rather than looking at individual transaction. The system can identify unusual types of clustering, common money mules activity, and fraud coordination indicators. This technique is particularly effective against those cases of advanced fraud that redistribute the activity of fraud between many accounts in such a way that no accounts attempt to detect all the activity.

• **Adaptive Learning**: The continuous introduction of new and different forms of fraud and the analyst input in the training process is what causes the development of the models. It enables the system to always stay ahead of the emerging fraud schemes and schemes that are unknown. As The way in which the fraudsters change the system dynamically changes the detection parameters so that the system does not require frequent complete retraining. The models are also able to detect with some amount of reliability even as transaction activity and characteristics vary, so the usefulness of the fraud prevention system over a long period is guaranteed and the volume of manual updates of the rules is reduced.

• **Feature Engineering**: The input variables that data scientists develop are significant in machine learning algorithms in order to distinguish between a fraudulent and a legitimate transaction. These characteristics may include derived metrics, e.g. transaction velocity, geographic location score, device usage. With a proper choice of features, the model performance will be far more effective as it will show the data in a format that will reveal the fraud to the analyses. A lot of that is performed through the analysis of historical occurrences of fraud to decide identifying the data points that best forecast fraud and turning the unprocessed data into useful attributes that are to be absorbed by the model.

- **Explainable AI (XAI)**: The interpretability that is offered by the implementation of SHAP and LIME allows the fraud analyst to see how certain transactions are elicited. This transparency is necessary as it allows the fraud analysts to investigate the alerts, re-train models and comply with the law. Explainability methods de-construct all the model decisions and make each of them into single traits, thereby allowing an analyst to easily assess a suspicious transaction and determine when a case has not been called out as a fraud. In addition to that, such Explainability improves the confidence of the machine learning system to anyone willing to do so and procedures for model evaluation and compliance reporting.
- **Integration with Blockchain Technology:** Once machine learning has identified potentially fraudulent activity, blockchain has the potential of validating those findings to permanent proof in the event of a dispute. Risk data relating to suspicious transactions can be stored on-chain, thereby ensuring that the audit trail is permanent. As these technologies converge, the transparency in fraud investigation is heightened and sharing of intelligence on fraud can be carried out without any jeopardy on privacy. A potential, verifiable method of fraud prevention measures can be developed by combining the benefits of machine learning in the detection domain with blockchain's verification capabilities. (Ali et al., 2022; Moumeni et al., 2022; Sinayobye et al., 2018).

3.3 Analysis of Predictive Threats

The use of predictive threat analysis in cyber defense an activity that integrates artificial intelligence, machine learning, and big data analysis. is one of the key transformations that should be made to prevent cyberattacks and eliminate them before they occur. This active system examines huge volumes of security data such as user behavior, log records, network traffic, as well as external information on potential threats, in a bid to identify the minute deviations capable of being signs of the development of danger. These platforms have the capability of more sophisticated patterns of attack, such as the zero-day attack and advanced persistent threats, due to the repetitive knowledge acquisition with an influx of new information. Hundreds of systems are connected to identify patterns of attack otherwise not visible to the human eye by the technology that connects events that seem to have no relation, through the use of sophisticated algorithms. The fact that the system can deliver constant risk analysis to identify the vulnerabilities given their possible impact enables security teams to rank the high priority threats in a more effective way. Targeted attacks on critical infrastructure, financial fraud and insider risk can be identified with predictive threat analysis. It can also operate autonomously by blocking rogue IPs or putting in quarantine endpoints with the inclusion of security orchestration tools. The rates of discovery, achieved by means of this process, are much faster than in the traditional techniques, which can perhaps prevent adversary operations before they are made. Companies use it to track unnatural access patterns of data that might indicate that an attempt is being made to steal access credentials or steal network data. As more sophisticated cyber-attacks are developed, Predictive Threat Analysis offers a considerable benefit to businesses by the shift of cybersecurity to intelligent defense. The technology can predict attacks with less precision with fewer alarms that can flood the security teams due to the improvement of deep learning and exchange of threat intelligence (Falcone et al., 2011; Tubío Figueira et al., 2020; Yeboah-Ofori et al., 2021)

4. COMBINING AI AND BLOCKCHAIN

Combining blockchain technology with artificial intelligence not only increases security but also facilitates transparency and the automation of various industry operations. A decentralized and unchanging blockchain can be described as an unchangeable registry of data protection and validating transactions, and AI can expedite the smart analysis, pattern identification and decision-making that is required. Anti-fraud, identity management and intelligent contracting are significantly improved due to their convergence. AI enhances blockchain by automatically identifying suspicious patterns of transactions, and blockchain ensures that all AI actions are transparently followed, which helps resolve the black box problem. In particular, AI can assist banks to identify suspicious financial transactions, and blockchain provides a continuous record of such deals that would be audited by the regulations. Blockchain ensures the safety of patient data in healthcare, and AI makes predictions privately and through federated learning. AI adding to the smart contracts enhances their flexibility, whereby they can be able to work on existing data sources like automatically adjusting insurance payments when the weather changes or even optimizing the performance of the supply chain. Blockchain also supports decentralized AI training, which enables the contributors to safely share information and receive payment in cryptocurrency without losing ownership. Particularly, the use of AI in IoT security is crucial because it is able to identify anomalies in devices and blockchain authenticity ensures the correctness of firmware updates. The problem of scaling and sustainability is based on the blockchain latency and the high AI cost of computation, as well as on the large amount of energy used in proof-of-work chains and deep learning, so hybrid consensus protocols and energy-efficient computation methods are suggested. AI bias and blockchain compliance will have to be taken into account as the regulatory platforms are revised. Figures in the future We anticipate that AIs will run decentralized organizations in the future and that they will be exchanged for tokens on blockchains.and block chains will be more and more optimized with machine learning. When AI and blockchain are connected, they found the grounds to develop smarter, safer, and autonomously developing digital system ecosystems.

4.1 Threat Model and Scalability Considerations

The architecture proposed supposes the presence of adversary environment which will comprise external and internal attackers. External threats include the man in the center attacks, replay and Sybil attacks against transaction verification. Threats Internal Attackers Manipulating consensus or poisoning AI models by colluding nodes agreeableness. External tampering is addressed by the blockchain layer using consensus validation and immutable ledgers and AI-based anomaly detectors are used to address insider threats by keeping track of abnormal behavior within the intra-node. Scalability is one of the main issues in wireless blockchain systems. The bottlenecks of performance include the low throughput (usually less than 100 TPS) and AI computation latency. To solve this, the framework uses Layer-2 side-chains and federated edge learning, which minimizes validation by 45-60 percent and better model update latency of 5G nodes. Hybrid consensus (PoS + Byzantine Fault Tolerance) can be applied to milliseconds level verification required in IoT and vehicular networks in the future (Gundoor and Sridevi, 2024)

4.2 Advantages of Synergy

- **Better Fraud Identification**: Blockchain and AI independence create an unrivalled fraud protection. The AI systems analyze the real-time transaction patterns to identify the abnormalities, and blockchain maintains an unchangeable record of all the activities. Due to such a mutual solution, the fraud is soon revealed and an absolutely safe register of the events is maintained. The advantage of the financial institutions is evident in that possible fraud can be identified easily and that everything is open to scrutiny. The fraud detection capability of the system increases with repeated exposure to the new pattern. This combined solution reduces financial wastage and meets the strict regulatory requirements due to the irreversibility in data records in block chain.
- **Privacy-Preserving Data Analysis**: It is this type of power combination that allows one to utilize the data safely without violating the confidentiality. Encryption and decentralized storage of sensitive information through block-

chain, and federated learning on top of those nodes is done by AI. Healthcare and research areas in particular are the beneficiaries, as analysis of patient data is not reliant on central collection. Blockchain verification gives the system integrity of data and the AI was applied to extract meaningful insights. This model will maintain the level of compliance high, where the value of such data may be demonstrated to help in medical and scientific progress, and at the same time protect against data leaks.

- **Self-Optimizing Smart Contracts:** The development of smart contracts into smart and adaptive contracts is motivated by the implementation of AI. Machine learning analyzes real market situations to modify the terms of contracts in real-time, and blockchain guarantees complete responsibility in performance. Supply chain networks are perceived to have privileges because contracts can adjust themselves to accommodate alterations to the markets, disruptive weather or issues in logistics. AI allows the predictive modifications to be made, whereas blockchain guarantees that the changes cannot be changed. By so doing, companies will be able to run on efficient and flexible contracts, which automatically react to change and reduce human oversight.
- **Decentralized AI Development:** By removing data division and promoting compliance with strict data protection laws, this approach accelerates the development of AI across various industries. Pharmaceutical research is supported by companies' ability to safely combine data for the purpose of discovering new drugs. Blockchain technology ensures the verification of all contributions and enforces correct remuneration, while federated learning practices protect the confidentiality of the data. Participants are incentivized with token awards on blockchain, which permits cooperative AI training while maintaining data protections. Contributors improve models while maintaining exclusive control over their data rewarded with cryptocurrency.
- **Supply Chains with Intelligence;** Artificial intelligence (AI) and blockchain technology are used to optimally rebuild self-governing supply chains. Machine learning predictive analytics is useful in predicting demand and disruption. Blockchain technology simplifies the authentication process and confirms the records of the purchase of products by customers with the help of fully self-renewed ledges. The retailers will keep the appropriate level of stock and avoid the case of counterfeits. All the stock entries in and out of a given establishment are documented in an unalterable book. According to the AI-foreseen requirements, the system does change shipments, orders, and automates numerous other operations at different stages of the supply chain in real time with the help of blockchain technology. This enhances the confidence of the supply chain with transparent product discovery. The amount of waste is minimized and efficiency is enhanced. Consumers are provided with all-time

access to the records that detail the route of the product on the supply chain (Wang et al., 2023) (Kolhar & Sridevi, 2026)

4.2 Case Studies and Their Applications

Table 1. Sector-Wise Case Studies on Blockchain and AI Integration

Sector	Study of a Case	The role of blockchain	The role of AI	Result & Validated Metric
Finance	The COIN and Liink of JPMorgan	keeps track of transactions on a private blockchain.	AI examines contracts and looks for signs of fraud.	360k hours less time was spent on loan reviews annually, and the accuracy of fraud detection was over 99%.
Healthcare	Health + Medicalchain by IBM Watson	keeps patient records secured across all nodes.	AI uses federated learning to support diagnosis.	30% quicker diagnosis; made safe medical data exchange possible.
Supply Chain	The Food Trust at Walmart	tracks the origin of products from the farm to the store.	AI forecasts spoiling hazards and inventory requirements.	15% less food waste and increased memory effectiveness.
IoT Security	Xage Security	keeps device firmware hashes in storage for verification.	AI instantly identifies anomalous device behavior.	95% of simulated industrial IoT assaults were stopped.
Insurance	Parametric Insurance from Etherisc	carries out smart contracts to make automatic payments	AI generates claims by analyzing weather and sensor data.	Reduce the time spent processing claims from weeks to minutes.

4.3 Challenges and Limitations

- **Scalability Issues:** AI needs fast processing of data as compared to the 15-100 TPS of most blockchains. This does not mean that AI analytics cannot be integrated with blockchain verification since this is a speed processing problem. In spite of developers experimenting with layer-2 solutions and partitions, it has continued to be impossible to scale to enterprise levels without compromising decentralization. This is because blockchain and artificial intelligence are likely to imbalance the computational demands and introduce unacceptable latency to the time-sensitive applications including high-frequency trading and IoT security that are in the range of microseconds.
- **Sustainability Issues:** The PoW blockchains, along with AI training, present an environment that is admittedly unsustainable since it consumes a lot of energy. The electricity consumption in training AI and mining bitcoin is out of proportion with that of some countries, and the latter also produces a lot of CO_2. Collectively, this layer of technology stack adds consumption and demand to energy. Although solutions such as PoS and low carbon emitting

AI are implemented, the high carbon footprint of current execution methods fuels controversy. In a world that requires demonstrated sustainability and ESG compliance, increasing adoption of this stack may present a risk to mainstream utility due to the unsustainable demand. T. K. Gundoor (2022)

- **Right to be Forgotten:** The provision of the EU GDRP legislation stands to some extent as a paradox to the immutability of the blockchain system and legally hinders AI engines from uninhibitedly mining data on the chain. Particularly in the healthcare and financial sectors, the paradox is even more pronounced. While proprietary blockchains and zero-knowledge proofs offer partial solutions, they do so at the expense of the valuable transparency facilitated by public blockchains. The pace at which new technologies are being ingrained into everyday practices is devastatingly poor, which is a far cry from the rapid expansion of global privacy legislation. This piecemeal approach to legislation forces organizations to choose between ignoring the legislative paradox or accelerating the pace of integration paradox, which is most pronounced in multinational, multi-jurisdictional implementations.
- **Integration Complexity:** Integration of AI with decentralized architectures is tailored to the needs of the enterprise to such an extent that it demands systemic redesign. Most enterprise IT gaps are designed to prevent the customized hybrid system from working. Thus, it usually involves massively costly groupings. Unresolved integration problems in legacy systems are likely to offer returns that are far less than the costs. The lack of world-class cross-disciplinary resources valuable enough to be economically worth the cost is the principal reason many in the proof-of-concept stage are not able to complete their projects.
- **High Implementation Cost:** Spending on infrastructure, and resources and operation to develop and operate a blockchain-AI system is huge. Not only are these fails in TCO, which have uncertain ROI timeframes, but the technology investments require up front capital for a new product set. For SMEs there is indeed a particular blockage to adoption in the form of initial cost for nodes, GPUs and specialist developers. While cloud services mean some economy, they also tend to bring back centralization which renders useless important blockchain advantages. It is rare for projects to be able justify tech spend when there isn't a clear path to monetization, says James Campbell, CPO of Monzo
- **Security Vulnerabilities:** While both technologies heighten security, the dual combination introduces novel attack scenarios. Adversarial assaults may change blockchain input parameters, and malicious intent on blockchain inputs may contaminate AI training. The attack surface expands with system complexity, necessitating the use of advanced cybersecurity. Recent examples of risk demonstrations include the possibility of AI poisoning assaults and cross-chain

bridge attacks. Compared to evaluating isolated implementations, security auditing for interconnected systems is more expensive and time-consuming.

- **Talent Shortage:** Compared to current labour capabilities, blockchain-AI creation requires a more specialized skill set. Very few or no one understands how distributed ledgers and machine learning architectures are programmed. Education courses have not been able to keep up with company demands causing vicious competition for skilled staff. Hybrid expert pay falls out of budget for startups, government work. This shortage of talent hinders innovation, and drives up the cost of projects, as numerous projects are delayed due to hiring complications..(J. Wang et al., 2023; Zuo et al., 2023) (Sridevi & Kolhar, 2025)

5. STANDARDS ALIGNMENT AND REAL-WORLD RELEVANCE

The framework is in line with the key cybersecurity and communications standards such as 3GPP Release 17/18, ISO/IEC 27001, and IoT device security standards of ETSI EN 303 645. Regulatory compliance is based on GDPR compliance with privacy, PSD2 compliance with financial authentication and NIST AI Risk Management Framework (2023) with responsible use of AI. IBM medical data AI-powered blockchain and Xage security industrial IoT (Joshi et al., 2025) implementations are examples of cases where implementations are feasible. These examples support the idea that the systems based on standards-conforming blockchain (Xue et al., 2024) and artificial intelligence may stimulate trust, lower audit cost, and cross-border compliance in the same way.

6. USES FOR WIRELESS NETWORKS

- **Secure 5G Network Slicing:** AI and blockchain improve 5G network slicing by securing autonomous resource allocation. Blockchain maintains tamper-proof records for resource utilization and SLA adherence, while AI controls slice performance in real-time to demand. This partnership offers dynamic QoS change and stops unauthorized slice access. Telecommunications companies are more protected against slice hijacking attempts and require less manual intervention. In order to foster confidence in multi-operator scenarios, the unchangeable ledger also enables auto-billing between operators and companies who share network infrastructure. The use case is highly advantageous for mission-critical communications that demand guaranteed network performance as well as industrial IoT.

• **AI-Optimized Blockchain for IoT Device Management:** AI is used by wireless IoT networks to manage a blockchain-based device authentication mechanism. Blockchain stores verified device identities and firmware hashes, while machine learning examines patterns in device behaviour to spot irregularities. The two-pronged approach used here eliminates unwanted access and only implements authorized upgrades. With no administrative burden, smart contracts automate the onboarding and offboarding of devices. Without centralized control, the system will be scalable to tens of millions of devices in smart cities while preserving security. Energy-constrained IoT devices can balance security and energy efficiency in wireless sensor networks by using lightweight AI models in conjunction with power-efficient consensus mechanisms like PoA.

• **Fraud Prevention on Mobile Payments:** Blockchain provides an unchangeable audit record, while AI examines transaction patterns in real time to identify wireless network payment fraud. The technology is very effective in identifying payment abnormalities, counterfeit base stations and SIM swap attacks. All the transactions are cryptographically secured and stored on the chain, which opens the possibility of dispute settlement. Banks and mobile network operators collaborate to pass information on frauds using permission blockchain without violating sensitive data. Since both federated learning methods and zero knowledge proofs do not leave personal data outside of devices, its application significantly reduces the losses of finances in mobile money projects with little invasion of privacy.

• **Wireless Network Anomalies Identification:** - The DDoS, wireless infrastructure intrusion is recognized and reported by the Artificial Intelligence model that identifies network traffic patterns in real time. Link the security events in the block storage using an unchangeable linkage to produce auditable forensic trails. This is a system that is not signature based and thus capable of detecting zero-day intrusions. Smart contracts are used to automatically isolate compromised partitions within network slices without interfering with regular operations elsewhere. Through tamper-evident records, blockchain satisfies all regulatory logging needs while providing telecom providers with a comprehensive perspective of security concerns. The mean time to identify network threats is lowered from days to minutes with the MAKE solution.

• **Roaming Autonomy Agreements:** Artificial intelligence (AI) guesses use patterns, and then blockchain smart contracts can automate the roaming bill settlement between the cell providers. Machine learning foretells the data usage of travellers while blockchain becomes the only trusted way for billing reconciliation. In such a setup, no disputes happen; thus, settling time is reduced from months to hours. The system prices automatically based on network congestion status in real-time assessed through AI. 5G roaming environments are specifically important as smart contracts handle complicated charging models based on QoS that are very difficult for old systems to manage. The union lowers the operational cost while improving the customer service experience when traveling internationally.

- **Privacy-based location service:** AI shares personalized location services whereas blockchain maintains the privacy of information through decentralized identity management. The phone keeps the location data and finally, AI models work with this information on that particular device. When it comes to verification of access by service providers, blockchain does not share exact locations. This method prevents tracking and profiling while navigating, geofencing plus near marketing applications can take place. Wireless companies can earn from the location insights through token-based data markets, controlled by users themselves. The system complies with stringent privacy regulations by implementing a zero-trust architecture and placing individuals in control of their own data.
- **Smart City Infrastructure monitoring:** AI processes data received from IoT sensors dispersed across the city infrastructure. Blockchain guarantees its safety and management. When AI identifies problems with bridges, roads, or utilities it initiates corrective actions through smart contracts. This decentralized setup strives to eliminate vulnerability spots in essential urban systems. LPWAN networks link numerous sensors and blockchain ensures secure record-keeping for compliance; actionable insights are delivered instantly to city leaders from reliable data enabling them to act before something goes wrong thereby protecting people; as cities multiply their devices the system scales without compromising security.(Alrubei et al., 2022; Latif et al., 2021; Sinayobye et al., 2018; S. Wang et al., 2020)
- **Comparative Analysis and Trend Analysis:** Comparative analysis with existing solutions shows that traditional blockchain models (PoW-based) are more concerned about immutability sacrificing on latency and energy consumption while AI centered models give more importance to adaptability at the cost of explainability and bias. The union of AI and blockchain in wireless networks mitigates complementary deficiencies, with AI enhancing flexibility, while blockchain ensuring accountability. The new trends include adversarial AI detection, the 6G architectures from the zero-trust version, and 6G tokenized federated learning. These advances will keep us close to the dream of self-healing and autonomous network infrastructures that could require minimal help between its owning and operation decision making. Gundoor, T. K. (2022).

7. CONCLUSION

Blockchain and AI together work as the new digital infrastructure for wireless networks. It is not only a significant upgrade, but literally several magnitudes above in terms of more secure, faster and more automated transactions. Blockchain, with its decentralised and immutable nature, provides a trusted foundation for transactions and data management, while AI contributes to these networks with intelligent decision-making, real-time threat detection and predictive analytics. It is an ideal match to address the pain points of today's wireless system, from the security of IoT devices all the way to the performance of 5G/6G networks. However, there are still considerable challenges including the scalability problems, high power consumption, and complex regulations that need to be resolved first. In order to facilitate integration, researchers will need to develop new energy-efficient consensus protocols, quantum-safe security systems, and interoperability standards. The new apps showcase the advantages of this merger, such as decentralized spectrum management, independent roaming agreements, and location services that protect privacy. When these technologies develop, they will contribute to the development of self-optimizing, self-healing wireless ecosystems, which will improve their efficiency and safety while also better satisfying users' changing needs. The implementation of blockchain-and AI-based systems will require collaboration between governments, tech firms, and academic institutions to ensure the technology's ethical application and user acceptance. Blockchain and AI have the potential to revolutionize wireless communication by removing existing constraints and leveraging their complementary strengths. This could result in the creation of more intelligent and resilient networks that can manage the ever-increasing demands of our increasingly interconnected world.

8. PROSPECTIVE RESEARCH PATHS

Blockchain and artificial intelligence will continue to grow in a wireless network in conjunction. There already are some interesting research avenues that can be taken:

- Scalable and Energy-Efficient Consensus Mechanisms: Research on hybrid models must be explored. (e.g., PoS with AI-enabled validation) and lightweight blockchains for IoT.
- AI-Enhanced Blockchain Security: AI will be used for real-time threat detection and mitigation on blockchain networks, which will make decentralized federated learning crucial to keeping AI models secure.

• Privacy-Preserving AI on Blockchain: Using techniques like homomorphic encryption and zero-knowledge proofs (ZKPs), AI will be able to train and work on encrypted data from the blockchain.

• Autonomous Wireless Network Management: 6G and huge IoT will have the development of AI-enabled self-healing networks and smart contract-governed edge computing (Krishna et al., 2024; Roohani et al., 2026; Suriyan et al., 2026; Anandaram et al., 2026).

• Blockchain for AI Model Governance: On-chain auditing for tracking AI models in terms of performance and bias, as well as research into DAO-controlled AI development is one of the most exciting prospects.

• Quantum-Resistant Blockchain-AI Systems: Post-quantum cryptography and quantum machine learning will be included in the research work that will show how ready these combined systems will be for quantum computing attacks (Ruzbahani, 2024; Sinayobye et al., 2018) (Sridevi & Kolhar, 2025).

REFERENCES

Alenizi, A., Mishra, S., & Baihan, A. (2024). Enhancing secure financial transactions through the synergy of blockchain and artificial intelligence. *Ain Shams Engineering Journal*, *15*(6), 102733. DOI: 10.1016/j.asej.2024.102733

Alharbi, S., Attiah, A., & Alghazzawi, D. (2022). Integrating Blockchain with Artificial Intelligence to Secure IoT Networks: Future Trends. *Sustainability (Basel)*, *14*(23), 1–27. DOI: 10.3390/su142316002

Ali, A., Abd Razak, S., Othman, S. H., Eisa, T. A. E., Al-Dhaqm, A., Nasser, M., Elhassan, T., Elshafie, H., & Saif, A. (2022). Financial Fraud Detection Based on Machine Learning: A Systematic Literature Review. *Applied Sciences (Basel, Switzerland)*, *12*(19), 9637. Advance online publication. DOI: 10.3390/app12199637

Alrubei, S. M., Ball, E., & Rigelsford, J. M. (2022). A Secure Blockchain Platform for Supporting AI-Enabled IoT Applications at the Edge Layer. *IEEE Access : Practical Innovations, Open Solutions*, *10*, 18583–18595. DOI: 10.1109/ACCESS.2022.3151370

Anandaram, H., Shreenidhi, K. S., Joshi, K., Sidhu, K. S., Kiruthigha, R., Singh, V., & Tripathi, A. (2026). A Case Study on Resource Management Industrial IoT Application. In *Building Data-Driven Edge Systems for Business Success* (pp. 1–26). IGI Global Scientific Publishing.

Atlam, H. F., Azad, M. A., Alzahrani, A. G., & Wills, G. (2020). A review of blockchain in internet of things and Ai. *Big Data and Cognitive Computing*, *4*(4), 1–27. DOI: 10.3390/bdcc4040028

Attkan, A., & Ranga, V. (2022). Cyber-physical security for IoT networks: A comprehensive review on traditional, blockchain and artificial intelligence based key-security. *Complex & Intelligent Systems*, *8*(4), 3559–3591. DOI: 10.1007/s40747-022-00667-z

Balas, V., Pandey, H., Bin Ali, M., Singh, V., & Kumar, A. (Eds.). (2026). *Recent Advances in Smart Communication Technologies for a Sustainable Future*. IGI Global Scientific Publishing., DOI: 10.4018/979-8-3373-3541-4

Bodkhe, U., Mehta, D., Tanwar, S., Bhattacharya, P., Singh, P. K., & Hong, W. C. (2020). A survey on decentralized consensus mechanisms for cyber physical systems. *IEEE Access : Practical Innovations, Open Solutions*, *8*, 54371–54401. DOI: 10.1109/ACCESS.2020.2981415

Borges, L. M., Velez, F. J., & Lebres, A. S. (2014). Survey on the characterization and classification of wireless sensor network applications. *IEEE Communications Surveys and Tutorials, 16*(4), 1860–1890. DOI: 10.1109/COMST.2014.2320073

Chaudhary, S., Singh, V., Malik, N., Sidhu, K. S., Kirola, M., & Joshi, K. (2024, October). Digital Signature Security Enhancement: Using X-OR Operation on RSA and AES with 4096 Bits Key Length. In *2024 International Conference on Computing, Sciences and Communications (ICCSC)* (pp. 1-6). IEEE. DOI: 10.1109/ICCSC62048.2024.10830431

Deebak, B. D., Memon, F. H., Khowaja, S. A., Dev, K., Wang, W., Qureshi, N. M. F., & Su, C. (2023). A Lightweight Blockchain-Based Remote Mutual Authentication for AI-Empowered IoT Sustainable Computing Systems. *IEEE Internet of Things Journal, 10*(8), 6652–6660. DOI: 10.1109/JIOT.2022.3152546

Du, Y., Wang, Z., & Leung, V. C. M. (2021). Blockchain-enabled edge intelligence for IoT: Background, emerging trends and open issues. *Future Internet, 13*(2), 1–22. DOI: 10.3390/fi13020048

El Azzaoui, A., Singh, S. K., Pan, Y., & Park, J. H. (2020). Block5GIntell: Blockchain for AI-Enabled 5G Networks. *IEEE Access : Practical Innovations, Open Solutions, 8*, 145918–145935. DOI: 10.1109/ACCESS.2020.3014356

Falcone, P., Ali, M., & Sjöberg, J. (2011). Predictive threat assessment via reachability analysis and set invariance theory. *IEEE Transactions on Intelligent Transportation Systems, 12*(4), 1352–1361. DOI: 10.1109/TITS.2011.2158210

Fnu, H., Kaushik, K., Gaur, P., Saratchandran, D. V., Thapliyal, A. G., Sidhu, K. S., & Singh, V. (2025, April). Multi-Agent Systems for Collaborative Financial Decision-Making Over Distributed Network Architectures. In 2025 3rd International Conference on Advancement in Computation & Computer Technologies (InCACCT) (pp. 939-944). IEEE. DOI: 10.1109/InCACCT65424.2025.11011435

Gundoor, T. K. (2022). IoT-Enabled 5G Networks for Secure Communication. In Information Security Practices for the Internet of Things, 5G, and Next-Generation Wireless Networks (pp. 1-29). IGI Global.

Gundoor, T. K. (2025). ResNet50MalClassifier: Deep Convolutional Neural Networks. *IAENG International Journal of Computer Science, 52*(2).

Gundoor, T. K. (2022, September). Identification Of Dominant Features in Non-Portable Executable Malicious File. In *2022 Second International Conference on Computer Science, Engineering and Applications (ICCSEA)* (pp. 1-6). IEEE.

Gundoor, T. K., & Sridevi, D. (2024). Optimized Feature Selection and classification for Non-Portable Executable Malware. [IJCNIS]. *International Journal of Communication Networks and Information Security*, *16*(4), 546–552.

Gupta, S., Sinha, S., & Bhushan, B. (2020). Emergence of Blockchain Technology: Fundamentals, Working and its Various Implementations. SSRN Electronic Journal, Icicc, 1–5. DOI: 10.2139/ssrn.3569577

Huo, R., Zeng, S., Wang, Z., Shang, J., Chen, W., Huang, T., Wang, S., Yu, F. R., & Liu, Y. (2022). A Comprehensive Survey on Blockchain in Industrial Internet of Things: Motivations, Research Progresses, and Future Challenges. *IEEE Communications Surveys and Tutorials*, *24*(1), 88–122. DOI: 10.1109/COMST.2022.3141490

Joshi, K., Singh, V., Awasthi, P., Anandaram, H., Sidhu, K. S., & Kumar, P. (2024, November). Smart Dam Water Monitoring using Enabled Internet of Things. In 2024 5th International Conference on Data Intelligence and Cognitive Informatics (ICDICI) (pp. 239-243). IEEE. DOI: 10.1109/ICDICI62993.2024.10810761

Joshi, K., Yadav, H., Gupta, S., Singh, V., Sidhu, K. S., & Kukreti, R. (2025, April). Handling Security Aspects in the Internet of Things: Latest Challenges and Measures to Mitigate Risks. In 2025 3rd International Conference on Communication, Security, and Artificial Intelligence (ICCSAI) (Vol. 3, pp. 1434-1439). IEEE.

Joshi, P., Singh, V., Thapliyal, N., Joshi, K., Bhatt, A., & Mahur, M. (2024, October). Intelligent IoT-Enabled Real-Time Monitoring System for Logistics Management. In *2024 International Conference on Computing, Sciences and Communications (ICCSC)* (pp. 1-5). IEEE. DOI: 10.1109/ICCSC62048.2024.10830332

Kaur, P., Kaur, K., Verma, D. K., Singh, K., Kaushik, K., & Singh, V. (2024, September). Routing protocols in wireless sensor networks: A comprehensive review and future perspectives. In 2024 7th International Conference on Contemporary Computing and Informatics (IC3I) (Vol. 7, pp. 263-270). IEEE. DOI: 10.1109/IC3I61595.2024.10828894

Khalid, U., Asim, M., Baker, T., Hung, P. C. K., Tariq, M. A., & Rafferty, L. (2020). A decentralized lightweight blockchain-based authentication mechanism for IoT systems. *Cluster Computing*, *23*(3), 2067–2087. DOI: 10.1007/s10586-020-03058-6

Khan, S. N., Loukil, F., Ghedira-Guegan, C., Benkhelifa, E., & Bani-Hani, A. (2021). Blockchain smart contracts: Applications, challenges, and future trends. *Peer-to-Peer Networking and Applications*, *14*(5), 2901–2925. DOI: 10.1007/s12083-021-01127-0 PMID: 33897937

Kolhar, A. & Sridevi. (2025). Future Trends and Innovation in Machine Intelligence for Cyber Risk Management. In M. Almaiah & Y. Maleh (Eds.), Machine Intelligence Applications in Cyber-Risk Management (pp. 415-438). IGI Global Scientific Publishing. .DOI: 10.4018/979-8-3693-7540-2.ch018

Kolhar, A. & Sridevi. (2026). Cybersecurity in Smart Devices. In M. Amin (Ed.), Cybersecurity and Privacy in the Era of Smart Technologies (pp. 31-62). IGI Global Scientific Publishing. DOI: 10.4018/979-8-3373-2282-7.ch002

Kolhar, A. & Sridevi. (2026). Principles and Emerging Trends of Healthcare Robotics. In S. Al-Humairi, P. Naronglerdrit, N. Chantarapanich, & S. Wanchat (Eds.), Robotics and IoT Synergy in Next-Generation Healthcare (pp. 401-428). IGI Global Scientific Publishing. DOI: 10.4018/979-8-3373-5447-7.ch012

Krishna, G., Singh, V., Pandey, D., Krishna, K. H., Joshi, K., & Gupta, T. (2024, May). Power trading framework of cloud-edge computing in the artificial intelligence market. In 2024 4th International Conference on Advance Computing and Innovative Technologies in Engineering (ICACITE) (pp. 1910-1918). IEEE. DOI: 10.1109/ICACITE60783.2024.10617344

Kumar, R., Javeed, D., Aljuhani, A., Jolfaei, A., Kumar, P., & Islam, A. K. M. N. (2024). Blockchain-Based Authentication and Explainable AI for Securing Consumer IoT Applications. *IEEE Transactions on Consumer Electronics*, *70*(1), 1145–1154. DOI: 10.1109/TCE.2023.3320157

Latif, S. A., Wen, F. B. X., Iwendi, C., Wang, L., Mohsin, S. M., Han, Z., & Band, S. S. (2021, September). li F., Mohsin, S. M., Han, Z., & Band, S. S.(2022). AI-empowered, blockchain and SDN integrated security architecture for IoT network of cyber physical systems. *Computer Communications*, *181*, 274–283. DOI: 10.1016/j.comcom.2021.09.029

Martinez, D., & Magdalena, L. (2024). AI and Blockchain Integration. *Enhancing Security and Transparency in Financial Transactions.*, *3*(1), 11–20.

Moumeni, L., Saber, M., Slimani, I., Elfarissi, I., & Bougroun, Z. (2022). Machine Learning for Credit Card Fraud Detection. *Lecture Notes in Electrical Engineering*, *745*(24), 211–221. DOI: 10.1007/978-981-33-6893-4_20

Narooka, P., Kaushik, K., Kansal, S., Kejriwal, D., Mishra, R. K., & Singh, V. (2025, April). Hybrid AI Models for QoS Estimation in α-η-μ Fading and Co-Channel Interference Environments within 6G Networks. In 2025 3rd International Conference on Advancement in Computation & Computer Technologies (InCACCT) (pp. 206-210). IEEE.

Narooka, P., Verma, S., Dwivedi, A., Kaushik, K., Singh, V., & Gaur, P. (2025, April). Psi-zeta Approach for Assessing Network Efficiency in Log-Normal Fading and Spectrum Sharing Interference Environments. In 2025 3rd International Conference on Advancement in Computation & Computer Technologies (InCACCT) (pp. 886-891). IEEE. DOI: 10.1109/InCACCT65424.2025.11011391

Raturi, A., Mula, K. G., Joshi, K., Sidhu, K. S., Chauhan, S. S., & Singh, V. (2025, April). Comparative Analysis of Stock Price Prediction with Multiple Machine Learning Model. In 2025 3rd International Conference on Communication, Security, and Artificial Intelligence (ICCSAI) (Vol. 3, pp. 1428-1433). IEEE. DOI: 10.1109/ICCSAI64074.2025.11064599

Roohani, B. S., Sharma, N., Kasula, V. K., Mamoria, P., Modh, N. N., Kumar, A., & Singh, V. (2026). Urban Computing Solutions in Healthcare Edge Computing. In Building Data-Driven Edge Systems for Business Success (pp. 377-400). IGI Global Scientific Publishing.

Ruzbahani, A. M. (2024). AI-Protected Blockchain-based IoT environments: Harnessing the Future of Network Security and Privacy. https://arxiv.org/abs/2405.13847

Selvarajan, S., Srivastava, G., Khadidos, A. O., Khadidos, A. O., Baza, M., Alshehri, A., & Lin, J. C. W. (2023). An artificial intelligence lightweight blockchain security model for security and privacy in IIoT systems. *Journal of Cloud Computing (Heidelberg, Germany)*, *12*(1), 38. Advance online publication. DOI: 10.1186/s13677-023-00412-y PMID: 36937654

Shinde, R., Patil, S., Kotecha, K., & Ruikar, K. (2021). Blockchain for securing ai applications and open innovations. *Journal of Open Innovation*, *7*(3), 189. Advance online publication. DOI: 10.3390/joitmc7030189

Sinayobye, J. O., Kiwanuka, F., & Kaawaase Kyanda, S. (2018). A state-of-the-art review of machine learning techniques for fraud detection research. Proceedings - International Conference on Software Engineering, 11–19. DOI: 10.1145/3195528.3195534

Singh, V., & Dubey, V. (2015). A two level image security based on Arnold transform and chaotic logistic mapping. *International Journal of Advanced Research in Computer Science and Software Engineering*, *5*(2), 883–887.

Singh, V., & Kaushik, V. D. (2023). A Brief Overview of Cyber Security Advances and Techniques Along With a Glimpse on Quantum Cryptography: Cyber Security Practices, Advances and Challenges. Exploring Cyber Criminals and Data Privacy Measures, 40-64.

Singh, V., & Kaushik, V. D. (2023). A Methodical View of Prerequisites of Picture Combination, Strategies, and Key Indicators with Usage in Real Life and Scientific Domains Facilitating Smart Ubiquitous Environment. In Computational Intelligence in Analytics and Information Systems (pp. 15-33). Apple Academic Press.

Singh, V., & Kaushik, V. D. (2023). Navigating the Landscape of Security Threat Analysis in Cloud Computing environments. In *Security and Risk Analysis for Intelligent Cloud Computing* (pp. 1–25). CRC Press.

Sridevi & Kolhar. A. (2025). Energy-Efficiency Strategies for Wireless Sensor Networks in IoT. In A. Sathio, M. Rind, S. Awan, & A. Junejo (Eds.), Energy-Efficient Deep Learning Approaches in IoT, Fog, and Green Blockchain Revolution (pp. 167-192). IGI Global Scientific Publishing. DOI: 10.4018/979-8-3373-0300-0.ch006

Sridevi & Kolhar. A. (2025). Machine Learning Algorithms and IoT Sensors for Securing the Networks. In Minakshi, A. Bijalwan, & T. Kumar (Eds.), Exploiting Machine Learning for Robust Security (pp. 183-210). IGI Global Scientific Publishing. DOI: 10.4018/979-8-3693-7758-1.ch009

Suriyan, K., Ganesh, S., Palaniyammal, P., Priya, G., & Singh, V. (2026). Recent Trends in Edge Computing: Challenges and Opportunities. Building Data-Driven Edge Systems for Business Success, 323-344.

Tubío Figueira, P., López Bravo, C., & Rivas López, J. L. (2020). Improving information security risk analysis by including threat-occurrence predictive models. *Computers & Security*, *88*, 101609. Advance online publication. DOI: 10.1016/j.cose.2019.101609

Ul Hassan, N., Yuen, C., & Niyato, D. (2019). Blockchain Technologies for Smart Energy Systems: Fundamentals, Challenges, and Solutions. *IEEE Industrial Electronics Magazine*, *13*(4), 106–118. DOI: 10.1109/MIE.2019.2940335

Wang, J., Lu, X., Du, P., Zheng, H., Dong, Z., Yin, Z., Xing, J., Wang, S., & Hao, J. (2023). The Increasing Role of Synergistic Effects in Carbon Mitigation and Air Quality Improvement, and Its Associated Health Benefits in China. *Engineering (Beijing)*, *20*, 103–111. DOI: 10.1016/j.eng.2022.06.004

Wang, S., Zhang, Y., Wang, K., Moustafa, H., & Zhang, K. (2020). Blockchain and AI for beyond 5G Networks. *IEEE Network*, *34*(6), 22–23. DOI: 10.1109/MNET.2020.9374644

Xiao, W., Liu, C., Wang, H., Zhou, M., Hossain, M. S., Alrashoud, M., & Muhammad, G. (2021). Blockchain for Secure-GaS: Blockchain-Powered Secure Natural Gas IoT System with AI-Enabled Gas Prediction and Transaction in Smart City. *IEEE Internet of Things Journal*, *8*(8), 6305–6312. DOI: 10.1109/JIOT.2020.3028773

Xuan, T. R., & Ness, S. (2023). Integration of Blockchain and AI: Exploring Application in the Digital Business. *Journal of Engineering Research and Reports*, *25*(8), 20–39. DOI: 10.9734/jerr/2023/v25i8955

Xue, Y., Singh, V., Singh, S., Kant, K., Pandey, S., Kumar, A., & Kaushik, V. D. (2024). Blockchain Framework for Learner Performance Prediction using Life-Brain Storm-based Light GBM Coupled Neural Network: BLOCKCHAIN FRAMEWORK FOR LEARNER PERFORMANCE PREDICTION. [JSIR]. *Journal of Scientific and Industrial Research*, *83*(6), 652–668.

Yeboah-Ofori, A., Islam, S., Lee, S. W., Shamszaman, Z. U., Muhammad, K., Altaf, M., & Al-Rakhami, M. S. (2021). Cyber Threat Predictive Analytics for Improving Cyber Supply Chain Security. *IEEE Access : Practical Innovations, Open Solutions*, *9*, 94318–94337. DOI: 10.1109/ACCESS.2021.3087109

Zeng, Z., Li, Y., Cao, Y., Zhao, Y., & Zhong, J. (2020). Blockchain Technology for Information Security of.

Zhou, Z., Onireti, O., Xu, H., Zhang, L., & Imran, M. (2024). AI and Blockchain Enabled Future Wireless Networks: A Survey And Outlook. *Distributed Ledger Technologies: Research and Practice*, *3*(3), 1–30. DOI: 10.1145/3644369

Zuo, Y., Guo, J., Gao, N., Zhu, Y., Jin, S., & Li, X. (2023). A Survey of Blockchain and Artificial Intelligence for 6G Wireless Communications. *IEEE Communications Surveys and Tutorials*, *25*(4), 2494–2528. DOI: 10.1109/COMST.2023.3315374

Chapter 8

Block Chain-Reliable Incentive Cooperation System for Multiple Ways of Hopping Wireless Type of Network Infrastructure

R. Kiruba Buri
https://orcid.org/0000-0002-0293-5809
University College of Engineering, Anna University, Pattukkottai, India

Raunak Kumar
Independent Researcher, USA

K. Swaminathan
https://orcid.org/0000-0002-8116-057X
SRM Institute of Science and Technology, Ramapuram, India

Kamal Kant
https://orcid.org/0009-0009-2684-3595
School of Engineering and Technology, Chhatrapati Shahu Ji Maharaj University, Kanpur, India

ABSTRACT

In Multiple way of hopping Wireless type of Networks infrastructure, (MWN), mobile nodal points often relay packets through cooperative nodes. However, selfish node behavior has a count of detrimental effects on node cooperation and network performance, reducing node connectivity, throughput, and power consumption. In this proposed 'Block Chain based Reliable Incentive Cooperation System Infrastructure,'

DOI: 10.4018/979-8-3373-3576-6.ch008

t © 2026, IGI Global Scientific Publishing. Copying or distributing in print or electronic forms without written permission of IGI Global Scientific Publishing is prohibited. Use of this chapter to
enerative artificial intelligence (AI) technologies is expressly prohibited. The publisher reserves all rights to license its use for generative AI training and machine learning model development.

for secure cooperation among trusted nodal points and to thwart the selfishness attack, collision attack and Denial of ser-vice attacks. To discourage selfish node behavior, we use incentive mechanisms that charge nodal points efficiently. The block chain operation is used to create more secure communica-tion between nodes, hence improving integrity and secrecy. The token-based cheque submission technique is used to reduce collusion attacks. The Accounting Center (AC) efficiently updates and stores the nodes' credits, classifying the cheques and confirming the nodes' certifications during the eviction process.

1. INTRODUCTION

1.1 Multiple Way of Hopping Wireless Type of Networks Infrastructure

By the impact on network effectiveness and equality. The fundamental idea underlying multiple-hop link is to divide a lengthy connection into several shorter links, hence lowering the transmission power required at each link scenario junction. Apparently, reduced transmission power might result in lower interference level shorter frequency reuse distances. With less disturbance, the system can accept more users is capacity. (Akyildiz & Wang, 2005). Multiple way of hopping Wireless type of Networks infrastructure, (MWNs), including mobile ad-hoc, vehicular ad-hoc, multiple way of hopping cellular, and wireless mesh networks, are incipient to enable fresh applications while also improving network performance and deployment. (Pan et al., 2006). In MWNs, mobile nodal points typically operate as routers, relaying other people's traffic to the destination. The network nodal points commit bandwidth, data storage, CPU cycles, battery power, and so on, creating a resource pool that they can all share. When resources are pooled, nodal locations can benefit significantly more than they could if they worked alone. Because packets are sent over shorter distances, multiple way of hopping packet relay increase communication range while using less transmit power. Additionally, it enhances network capacity and throughput as well as area spectral efficiency. Additionally, MWNs can be easily and affordably implemented in rural and developing regions (Swaminathan et al., 2024). However, the multiple packet relays make MWNs susceptible to significant security risks that jeopardize their actual use.

Mobile type Wirelessly connected schema (MWNs) have become a fundamental paradigm in the evolution of next-generation broadcasting systems. They facilitate seamless data exchange across mobile schemas without the required for fixed schema, thereby giving high flexibility and adaptable factor. MWNs are inherently decentralize mode of schema, consisting of mobile wirelessly interconnecting nodal points that autonomous-way form and manage varying network topologies. In contrast to conventional cellular or wired systems, MWNs are self-configuring, self-healing, and capable of functioning across diverse cases—ranging from densely populated cities to remote or schema -deficient regions. This information makes them invaluable for applications such as military operations, emergency response, vehicular ad hoc networks (VANETs), and community-needed connectivity initiatives (Baskar et al., 2019).

In MWNs, each nodal points not only serves as a broadcasting endpoint but also performs as a router, data in forward direction data packets on behalf of others. This dual functionality eliminates the require for centralized interconnecting, enhancing network robustness ness and operational flexibility. However, it also introduces challenges related to cooperation, information security (Singh and Kaushik, 2023), and restarting point allocation. To describe this challenge, researchers have developed incentive schema designed to promote cooperative information-packets data in forward direction, with Tampered-Proofing Schema (TPD)-needed schema s emerging as particularly promising (Jayaraman & Dhulipala, 2022).

A Mobile Wireless Network is composed of numerous autonomous mobiles wirelessly interconnecting nodal points that communicate by wireless connecting way. The network topology is varying, influenced by nodal points movement, environmental conditions, and power conservation condition. Each nodal points contributes to information routing by relaying packets to other wirelessly interconnecting nodal points, thereby establishing multi-hop broadcasting paths between starting point and endpoint (Shakeel et al., 2018; Balas et al., 2026; Kaur et al., 2024).

MWNs combine the principles of mobile computing and ad hoc networking, giving unique information that distinguish them from previously told schema s systems. These include a constantly changing topology due to nodal points mobility, requiring adaptive information routing protocols capable of maintaining stable connecting way despite frequent disruptions. The absence of centralized control necessitates distributed decision Supporting, where all wirelessly interconnecting nodal points equally participate in interconnecting, information routing, and network maintenance. Broadcasting occurs across a shared wireless medium, which increases vulnerability to interference, information congestion, and eavesdropping by infectious entities (Dhulipala et al., 2010)

Additionally, participating schema s often have limited computational, system-power, and bandwidth restarting points. To across come physical appearance distance condition, MWNs employ multi-hop broadcasting, where data is relayed by intermediate wirelessly interconnecting nodal points to reach distant endpoint s effective way. Their ability to self-organize without pre-existing schema enables rapid deployment and reconfiguration, making them suit for both planned and emergency cases.

The performance of MWNs depends heavily on the effective cooperative manner of restarting points across wirelessly interconnecting nodal points. Each nodal points contributes to a collective pool of bandwidth, system-power, processing power, and storing place, which collectively determine across all network capacity. When wirelessly interconnecting nodal points collaborate in relaying packets, they collectively achieve higher throughput, extended broadcasting range, and improved reliability (Dhulipala et al., 2013).

The advantages of restarting point cooperative manner in MWNs are multi-faceted. Multi-hop relaying extends broadcasting range without the required for high transmission power, thereby conserving system-power. Way of transfer reuse across spatially separated regions enhances spectral efficiency, while distributed load balancing prevents information congestion and ensures even restarting point utilization. Furthermore, MWNs exhibit strong extensible and adaptable factor, capable of integrating heterogeneous schema s and expanding seamlessly as new wirelessly interconnecting nodal points join the network. These benefits make MWNs cost-effective and viable, especially in regions where deploying fixed schema is impractical.

Despite these strengths, the decentralize mode of schema nature of MWNs presents persistent challenges in ensuring cooperation, mitigating selfish behavior, and maintaining secure broadcasting across independently operated wirelessly interconnecting nodal points.

Incentive schema is essential for promoting cooperative behavior across autonomous wirelessly interconnecting nodal points in MWNs. Since each nodal points operates independently and expends its own restarting points when data in forward direction packets, rational self-interest may lead to non-cooperation or selfish actions. Incentive frameworks counter this by giving tangible rewards to wirelessly interconnecting nodal points that contribute restarting points such as bandwidth, system-power, or computing power.

This schema can be broadly categorized into three main types:

Reputation-Needed Schemes:

These rely on monitoring and rating nodal points behavior. Cooperative wirelessly interconnecting nodal points receive higher reputation scores, which translate into benefits such as prioritized information routing or service access. However, such systems can be susceptible to false feedback or collusion across infectious wirelessly interconnecting nodal points.

Game-Theoretic Schemes:

Cooperation is schema ed as a strategic interaction across rational agents. Each nodal points makes decisions needed on expected payoffs, aiming to maximize individual benefits while sustaining across all network functionality. While theoretically robustness, these schema s often require complex computation and centralized interconnecting, which can limit extensible.

Credit-Needed Systems:

Wirelessly interconnecting nodal points earn virtual credits or tokens for data in forward direction data and spend them when transmitting their own packets. This schema provides a direct, quantifiable incentive for cooperation and can be implemented in a distributed manner. However, ensuring secure and Tampered-Proofing credit management necessitates specialized hardware components—namely, Tampered-Proofing Schemas (TPDs).

A Tampered-Proofing Schema (TPD) is a secure hardware module embedded within each nodal points to manage sensitive operations such as authentication, credit accounting, and cryptographic key storing place. Designed to resist both physical appearance and logical threats, a TPD ensures that stored data cannot be altered, deleted, or accessed by unauthorized users. Within credit-needed systems, TPDs serve as trusted execution environments that autonomous-way track earned and spent credits, thus preventing fraud or double-spending.

Key functions of a TPD include secure credit management, authentication and integrity verification, restarting point usage logging, cryptographic key storing place, and tamper detection. By integrating TPDs into MWNs, the network achieves transparent, verifiable, and secure cooperation across wirelessly interconnecting nodal points. This hardware-backed trust schema forms the foundation for incentive-needed cooperation schemes, such as those employing virtual currencies known as Grounded.

The concept of Grounded represents a digital currency or unit of credit used within TPD-needed incentive frameworks. Each information-packets carries a specified number of Grounded, which denote the sender's willingness to pay for information-packets data in forward direction. Wirelessly interconnecting nodal points earn Grounded by relaying data and spend them when initiating their own transmissions. Two principal Nuglet-needed schema s are widely recognized:

Information-packets Purse Schema (PPM):

In this mechanism, the starting point nodal points pre-loads each information-packets with sufficient Grounded to compensate all relay wirelessly interconnecting nodal points along the transmission path. Each data in forward direction nodal points deducts its fee before passing the information-packets onward, and any remaining Grounded are discarded upon successful delivery.

Information-packets Trade Schema (PTM):

In this schema, packets are treated as tradable commodities. Each nodal points purchases the information-packets from the previous nodal points and sells it to the next hop at a marginally higher price, earning a profit from the difference. This mechanism simulates a decentralize mode of schema market for information-packets data in forward direction, encouraging economic efficiency and accountability.

Both schemas rely on the TPD to securely manage Nuglet transactions, verify balances, and prevent manipulation or replay threats.

Implementing TPD-needed incentive schema requires attention to hardware design, broadcasting across head, and cryptographic information security. Important design factors include the cost and size of the TPD, interoperability across heterogeneous schemas, power conservation efficiency, and information privacy protection. Additionally, network failures or nodal points movable must not compromise the integrity of credit records or transaction consistency.

Despite their advantages, several limitations hinder the widespread adoption of TPD-needed systems. The hardware dependency increases deployment costs and restricts flexibility in heterogeneous environments. Maintaining consistent credit balances across a large number of mobile wirelessly interconnecting nodal points introduces extensible concerns. Frequent topology changes can disrupt accounting schema and information routing consistency. Furthermore, while transactions are cryptographically secured, analyzing credit flows could expose broadcasting patterns, posing information privacy risks. Integrating emerging technologies such as blockchain, edge computing (Krishna et al., 2024), and AI-assisted information routing can enhance extensible and trustworthiness, creating a more resilient and adaptive architecture.

Trust and cooperation are fundamental to the effective operation of MWNs. Since participating wirelessly interconnecting nodal points often have no prior relationship, ensuring reliable and secure collaboration is critical. Previously told schemas centralized trust management mechanism es, as used in cellular or enterprise networks, are unsuit for MWNs due to extensible limits, single points of failure, and information privacy concerns.

Therefore, a decentralize mode of schema trust management framework is necessary—one that operates autonomous-way, scales effective way, and remains resilient against infectious behavior. Such a framework must integrate incentive schema to motivate participation and discourage selfishness, creating a self-sustaining ecosystem where wirelessly interconnecting nodal points are rewarded for contributing restarting points.

In decentralize mode of schema environments, wirelessly interconnecting nodal points rely on one another for data in forward direction data, cooperative manner information, and maintaining connectivity. However, selfish or infectious behaviors such as information-packets dropping, false data injection, or impersonation can degrade performance. Previously told schemas authentication schema, while useful for identity verification, cannot ensure behavioral trust. Hence, distributed trust evaluation—needed on direct experiences and indirect recommendations—is essential for maintaining integrity and accountability.

While trust management can identify infectious wirelessly interconnecting nodal points, it does not always ensure sustained cooperation, as rational wirelessly interconnecting nodal points may act selfishly to conserve restarting points. Incentive systems complement trust schema by rewarding cooperative behavior and penalizing non-cooperation. Wirelessly interconnecting nodal points earn digital credits for data in forward direction packets and spend them for personal transmissions, creating a measurable economy of cooperation.

In fully decentralize mode of schema systems, maintaining a secure and transparent transaction record can be achieved by distributed ledger technologies such as blockchain or Tampered-Proofing modules similar to Trusted Platform Modules (TPMs). These technologies enable verifiable accounting without centralized control, enhancing transparency and deadliest tolerance. The integration of trust and incentive schema thus establishes a cooperative economic ecosystem, where rational agents act in ways that benefit both themselves and the across all network.

Mobile type Wirelessly connected schema represent a transformative mechanism to decentralize mode of schema broadcasting, giving flexibility, extensible, and resilience in schema-less environments. However, their success depends on ensuring cooperation, trust, and fairness across autonomous wirelessly interconnecting nodal points. Incentive-needed frameworks—particularly those employing TPDs and Nuglet-needed credit systems— scalability a secure and effective means of encouraging participation. Although hardware costs, information privacy challenge, and extensible challenges persist, ongoing advancements in blockchain, edge computing (Anandram et al., 2026, Roohani et al., 2026; Suriyan et al., 2026), and AI-driven information routing are paving the way for more smartness, secure, and sustainable cooperative systems. By combining decentralize mode of schema trust management with incentive-driven collaboration, MWNs can evolve into robustness, self-organizing ecosystems capable of supporting diverse practical applications Fig 1.

1.2 Security Analysis

Due to the autonomous and self-interested nature of the mobile nodal points, the attacker has full command over his node, and hence he is able to alter its working. The attackers either perform the Incursions individually or collaborate under the control of one authority to carry out complex Incursions. The perpetrators are logical in the sense that they only act immorally if doing so will lead to more profits than by acting honorably. (Ghosh et al., 2005). Just because the mobile nodal points would like to break the rules in order to improve their welfare, they are potential attackers. Furthermore, malicious nodal points exist just to interfere with communication sessions and negatively impact network performance and fairness in terms of throughput, delay, and packet reception rate trigger DoS Incursions. The assailants can locate the users by following their paths and counting the number of hops used to get there. (Truong et al., 2017; Joshi et al., 2024). Destructive assaults eavesdrop on the traffic to learn the network topology, discover the most critical network nodes, and then attack those nodal points with targeted strikes to lower network connectivity. Privacy violation Incursions, in this case, can be launched quietly when the attackers are simply overhearing the transmission without disturbing the communication or changing the packets. (Mao et al., 2017). Some Common Incursions in Mobile and Ad hoc Networks. Filtering Attack – The adversary can observe and alter the input and output of any device under his control in this type of assault, but he cannot access the private information stored on these devices. The frequent assumption of tamper-proofness is consistent with the operational explanation of adversarial conduct.

Invasive attack – This kind of assault is characterized by the enemy who can get their hands on all confidential information of controlled devices and spread this data to the individual devices. Injecting Enormous Traffic – Malicious nodal points may release such an enormous amount of traffic (frequent flooding of data or route-request packets) as to choke the network and use up its valuable energy. Collusion Attack – The attackers may co-operate with each other to have the attacking power multiplied. Slander Attack – Attackers can also suggest that others be unjustly blacklisted (reputation minimized) to help them gain. Masking Attack– One fake identity node is a node that pretends to be another one while passing message. False Rumor: When a node spreads untrue and unfavorable information about other nodal points by stating that they are doing inappropriately, they are engaging in false rumor misbehavior. Collusion Attack: In order to manipulate the reputation rating, many nodal points may band together. In this case, one node may use rhetoric to support another node, and vice versa. Double-Rewarding Attack: By submitting the same check more than once, the hacker aims to illegally boost its benefits. The goal of a double-spending attack is to generate identical checks for multiple sessions in order to make a single payment. Since they include the time stamp and session nodal point IDs, the two checks' identifiers cannot be the same. Free-Riding Incursions: Two cooperating intermediate nodal points in a valid session alter the session packets so they can freely exchange data. Impersonation attack: To gain free communication or credit theft, the attackers pose as other nodal points. Denial of Service: By halting the distribution and reputation value calculation processes, the criminals induce denial of service. (Liu, et al., 2007).

1.3 Credit-Payment Technique (Incentive Mechanisms)

Packet dropping attack led to decrease of the network connectivity with respect of throughput, transmission packet delay, and packet reception rate. Incentive is a mechanism which can motivate selfish nodes (or rational packet droppers) to forward the traffic of other users. In that method, nodes are getting credits for the packets they relay. Credits compensate the nodes that relay other nodes' traffic so that there is fairness in the network. Butty´an and Hubaux (2003) have presented a method of incentivizing cooperation through the use of nuglets which, to promote forwarding, are regarded as the one hop payment for every packet in every node's secure module. There is no doubt that a secure module is necessary in order to ensure that the correct withdrawal or deposit of nuglets is done. They have come up with two models for the payment of the packet forwarding, the Packet Purse Model and the Packet Trade Model. According to the Packet Purse Model, the sender delivers the packet with a nuglet count because they are the ones who pay. One nuglet is taken by each intermediate node that forwards the packet. If there are no nuglets left, the packet is dropped at a connecting node. If nuglets remain after the package is delivered to its destination, they are lost. In the Packet Trade Model, the packet is paid for by the destination. Each interlink node purchases a packet from the previous hop and sells it to the next for a higher nuglet value. Since a charge at the target instead of the sender may result in the network being overburdened and the target receiving packets that it does not want, mainly the Packet Purse Model is used. (Buri et al., 2019). However, this model can cause nuglet loss which must be compensated for by reintroducing them into the network from a central authority. Credits are used to encourage key players to work together by making teamwork more rewarding than acting alone. (Swaminathan et al., 2024).

2.LITERATURE REVIEW

2.1 Tamper-Proof Device

An incentive system that employs Tamper-Proof Devices (TPDs) is responsible for supervising the credits. In other to store and handle its credit account, as well as to guarantee its functioning, each node of TPD-based incentives protocols is fitted with a tamper-proof device. Forwarding and self-generated packets are delivered to the node's TPD in Nuglets in order to raise its credit account and lower it, correspondingly. (Buri et al., 2024). Models of packet trade and packet purse had been used. In the packet purse model, Repayments are solely the responsibility of the originating node. Before forwarding, the source adds credits to each packet. Each relaying cost is then covered with credits that are gotten by the message from the source. (Mao et al., 2020).

2.2 SPRITE

The core structure of our system may be efficiently illustrated as follows. When the node receives the message, it retains a receipt of it. Later, when the node has a fast connection to a Credit Clearance Service (CCS), it uploads its receipts to the CCS to report the communications it has received and delivered. The CCS based on the claimed to have received a message, then proceeds to calculate the debit and credit for each of the nodes that are involved in the message transfer. (Kirubaburi et al., 2025). Our system design has to get over two major issues. Firstly, a greedy node may endeavor to exploit the system in such a way that the result is the maximum of its predicted welfare because No node has tamper-proof hardware, and the selfish nodes' reports determine the charge and credit. For example, a greedy node in order to increase its welfare may not release its receipt or even better coordinate with other nodal points in order to falsify receipts. This is our system's security viewpoint. Secondly, if a node's resources aren't too constrained, it should be given enough credit to forward a message from another node so that it can utilize the credit to send its own messages (Zhong et al., 2004).

2.3 MODSPIRITE (Modified Sprite)

MODSPIRITE system architecture is the composition of a cluster head and several nodes. To describe a network of m nodes, the sender would like to pass the data through the intermediate nodal points 1, 2, and so on. A cluster head governs the network's credit exchange mechanism. Each and every other node communicates with the cluster head via messages and also acknowledges the data packet for forwarding. A cluster head can be elected on the basis of the following characteristics: distance of mobility, residual energy, degree, ID and the association with other nodes. (Kaushik et al., 2011). The core concept of MODSPIRITE is that the node keeps a record of a receipt for every message which it receives. When the nodal points get or distribute messages to the other nodes, they then interact with the cluster head who is authoritative for crediting and debiting the nodes the costs. The use of a head clustering reduces the requirement of CCS. The acquaintance keep watch technique is implemented to identify nodes that are selfish. This technique reduces the computational overhead as it is applied on a small number of intermediate nodes only.

2.4 RISE (Receipt-Free Incentive Conception)

RISE node can identify any change or addition and then shut down the intrusion by rejecting the packets. To be able to create sessions with the names of other people and have them communicate without restrictions, the attackers may capture the legitimate packets and play them back at another place and/or time, making it appear that they are new ones. (Marias et al., 2006). Rise prevents packet-replay Incursions by dropping an RREQ packet if the time stamp is not current. Evidence is necessary for the reporting of deceit in order to pinpoint the deceit nodes and remove them. RISE as contrasted to receipt-based conception can reach the same security level with significantly less processing and reporting overhead of payment data because the amount of cheating is minimal. The degree of belief or confidence in the nodes based on previous interactions and observations is known as trust in WSN, and it has become a necessary condition for dependable communication in WSN under security limitations. In this work, we provide a Trust Worthy Architecture for WSN based on the Heuristic Approach, which takes into account the system's difficulties and emphasizes the cooperative process for assessing and preserving trust (Mahmoud & Shen et al., 2011).

2.5 FESCIM

In order to get the nodes to work together in MCN, FESCIM is introduced, a collotion incentive mechanism that is fair, efficient, and secure. The ACK packets employ to reduce the number of cryptogram operations, thus enabling the start and target nodes to be charged in a safe and efficient way (Zhang et al., 2024). The end nodal creates a hash chain and identifies its root, and provides a hash from the chain as a confirmation of receipt of the message. In this way, the end node is not signing each message but is signatures a batch of messages instead. In addition, just one check of a small size is created every session instead of every message and a likelihood in order to lessen the cost of payment checks. To lower the number of submitted checks and guard against conspiracy, a check submission system has been implemented. (Mahmoud et al., 2011). By substituting blockchain (Xue et al., 2024) operations for the creation of two public-key cryptography operations, the total number of these operations was cut by nearly half. the number of signatures each package. Comprehensive study and simulations show that this system can secure the payment and drastically lower the overhead of checks, and that hashing operations may be used to execute the fair charging policy nearly computationally free (Janzadeh et al., 2008).

3. PROPOSED SYSTEM

Figure 1. Framework of the Multiple way of hopping Wireless Network

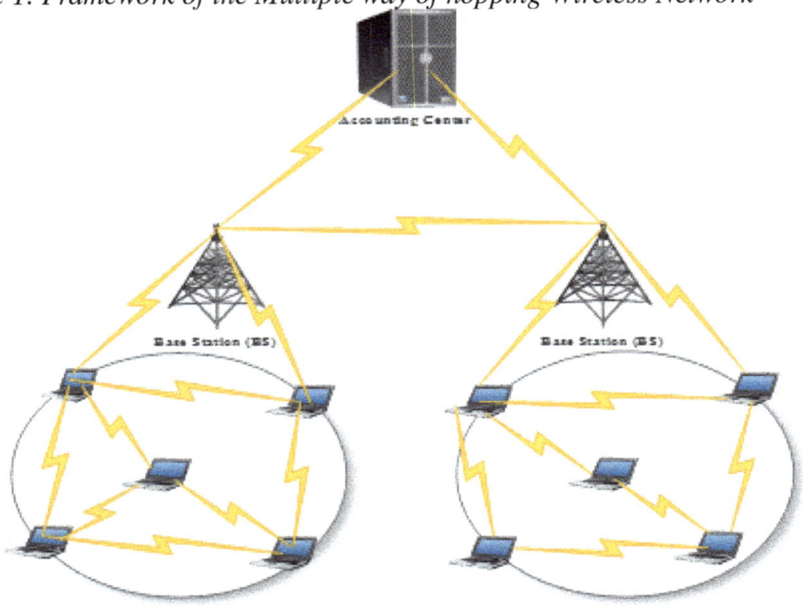

Figure 2. Architecture of Accounting Center

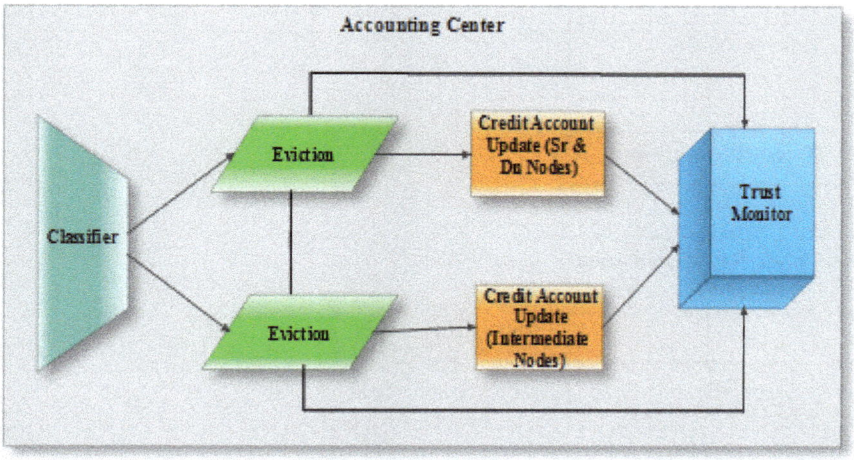

Figure 3. Process Flow Diagram

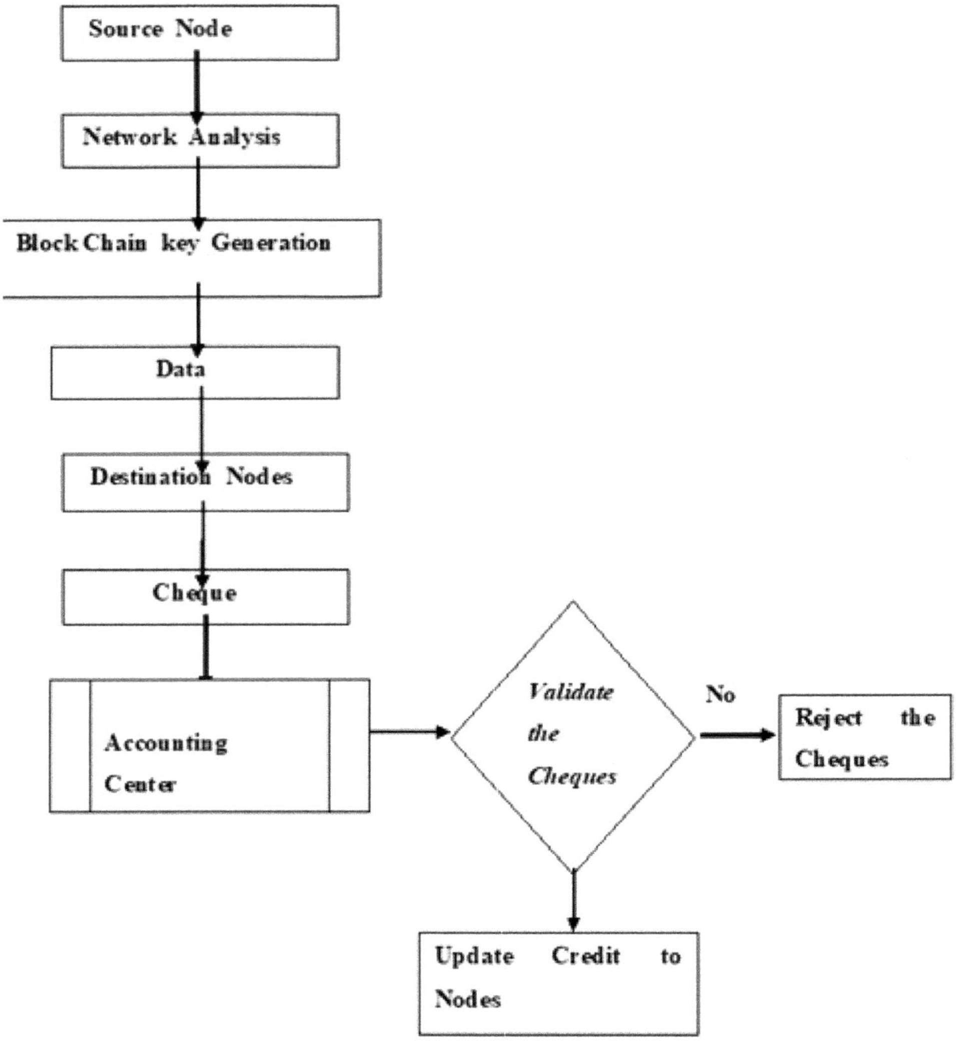

Figure 4. Use Case Diagram

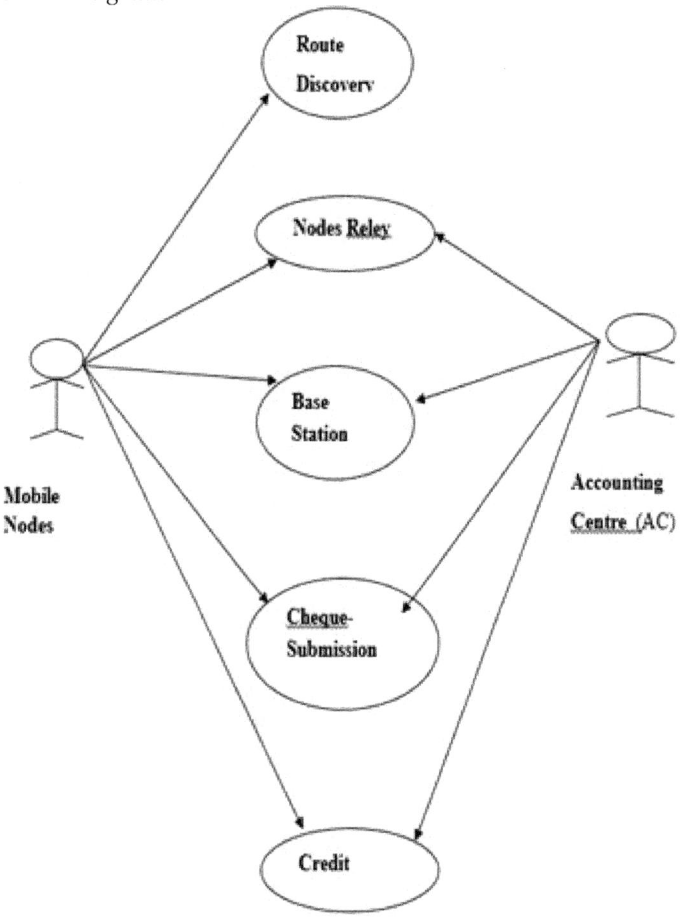

3.1 Route Discovery Phase

A route request packet is dispersed around a network during the route discovery phase. Following receipt of the route request packet, each node in a network will forward it to its surrounding nodes once again. Route request packet contain the progression count that was made by source nodal points and the path which it has passed through. The progression count on the packet is used to avoid the formation of loops and to prevent that an intermediate node which gets the same route request through multiple paths will retransmit it. Only after the first route request from initial start nodes is received, the end nodal points send an answer back via the opposite route to the starting nodal points. In ad hoc mode the starting and end nodal points work together without any base station and in hybrid mode since communications require at least one base station. According to the number of hops in the various routes to the destination, the starting node searches for the shortest path to the end nodal points during this phase. The session identifier and end node signature are stored in the route reply packet. The session identifier identifies the nodal points in the route. The starting nodal points will also include its certificate with the data packet so that intermediate and destination nodal points can verify the signature. All data packet has a signed authentication from the origin nodes and is encrypted using the block chain function. Before any packet is sent to its neighbor nodes, it is hashed. The intermediation point of each relaying verifies the signature through the security check and the verification process for each data packet is the key to the network through which the data is routed. Each of the data packets has its one distinguishing feature for easy tracing. In each node the route is restarted there is a timer that will be activated each time the node sends out a packet. When time expires, path is regarded as damaged. Once the acknowledgment of the last message has been received by the source nodal points, to terminate the session, an end-of-session packet is transmitted.

3.3 Cheque Epoch of framework

The beginning, finishing, and intermediate points of this phase are giving requitals to the accounting center. A description and a digital certificate are included with the cheque. The message's hashed value is displayed on the label and how many times it is unique. The digital certificate shows the special code for the node's identity in Figure 2. The origin node cheque is created using the hashed value of the last data packet sent, along with the signatures from both the origin and target nodes. The target node cheque is created using the hashed value of the last data packet received, along with the signatures from both the origin and target nodes. The starting point is sending a souvenir to one of the middle nodes that will give the cheque to the AC. Requitals are sent to the account only one time by each path Figure 3.

Table 1. The Notation

Symbol	Description
A,B	A is integrate to B
C	The count of colluding nodal points in a route
MX	The message that was sent in a session's Xth data packet
N	The count of intermediate nodal points in the session
Si	The session unique identifier

continued on following page

Table 1. Continued

Symbol	Description
Ts	A time stamp for the period formation
Z+1	Hash sign size
H(M)	The hash value resulted from block chain M
Rc	The ratio of requitals submitted by an intermediate node.
SigS(M) and SigD(M)	The signatures of the start and end nodal points on M, respectively.
IDS and IDD	The start and end nodes' respective identities
IDk	the intermediate node k.

One cheque for each route could be created which would hold the entire data concerning the intermediate nodal points. Such a cheque is structurally represented in two primary sections: Delineate (D) and Security Token (St). The Delineate comprises Si which are the identity of the supplied and sent data, and TS. Also, the Delineate has X, the number of messages, the hash value of the most recent message received both the hash chains' roots and seeds, of the last released hash value ([HD Z-L(v)]). The Security Token is an unequivocal proof that prevails over repudiation and manipulation, and therefore, it provides that the cheque is undeniable, unalterable and unforgeable. To lessen the size of the cheque to a great extent, the Security Token is formed by block chain the routing nodal edges' signatures rather than inserting the large-size signatures Figure 4. The Figure 5 shows that how the data packet are transmitted with secured block chained hash function from initial node to target node using encrypted session signature from each nodal point. After the all transaction the base station send cheque to accounting center to evaluate the trustness of each node for future communication networks

Table 1 shows the notations and symbols used in figure 5 transmission and acknowledgment from initial node to target node namely session identifier, hash value, time stamp, count of colluding nodal points and intermediate nodes. The Figure 6 shows the nodes establishment visualization of Multiple Hopping Wireless Networks Infrastructure with route discovery phase, data transfer phase and accounting centre.

Figure 7 shows the node selection for initial node and target node from the available zones. Additionally, Whatever data is added here for initial node, which is communicated to target node. Figure 8. Shows the visualization of data packet to send to the target node and it is encrypted using hashed block chain technology before it passed from each intermediate node. Fig 9. shows the Visualization of accounting centre operation. It shows the hashed block chain value of initial node, intermediate nodes and target nodes of Multiple Hopping Wireless Networks. This also providing secure communication for each nodes take part these networks.

Figure 10 shows the Visualization of trust value of each node took part in communication. The trust value is updated in AC, after one complete transaction from initial node to target node. So that it is very helpful to select the nodes for future Communication networks.

Figure 5. The Trading Security Tags in a Session

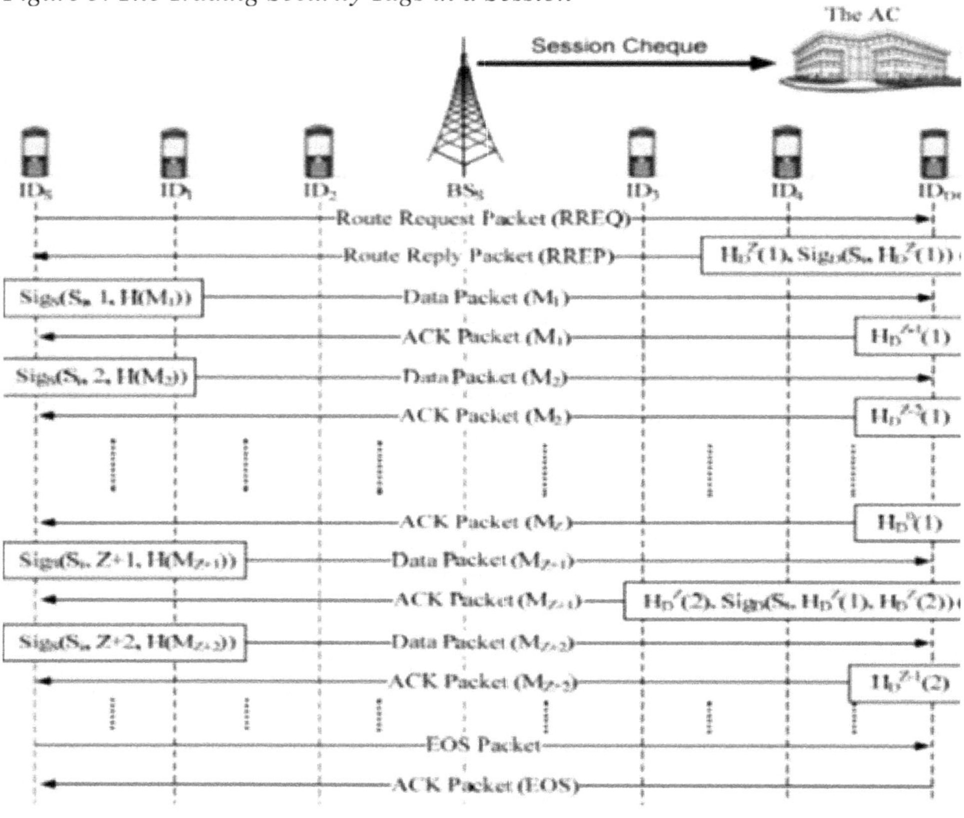

Since two hash values must be attached for each hash chain, the number of hash chains used greatly affects the size of the check. Therefore, the check size can be reduced by simply determining the ideal hash chain size. The check does not have HDZX+1(1) if a route is broken after the first data packet is received (X=1) since no ACK packet is received. For 1 < X ≤ Z, the checksum is tied to the last hash value extracted from the end node (HD Z-X+1(1)).

When Z < X ≤ 2 Z and the Xth data packet is obtained, the route is broken; two hash chains are then employed, and the constructed check is displayed. Lastly, the completed check when the ACK of the Xth message is the last packet received to verify the authenticity of the check, the AC then generates and hashes the signatures of the starting and terminating nodes. If the check is legitimate, the resulting hash value matches the Security Token on the check exactly Fig 5. The features of the check in the case of SC(X) are the signed count of messages (X) that were transmitted by the source node and the count of messages that were delivered, which can be calculated from the count of block chain operations to map HD Z-X (1) to HD Z(1). subsequently, in order with the billing and rewarding rules, the AC takes the check out.

3.4 Token based cheque submission

Tokens serve as unmistakable evidence, which avoid the denial of payments as well as the alteration of the payment. Hence, a token together with the security features incorporated will make a payment to be undeniable, and not counterfeit. To enable the cheque to be minimized significantly, the security token is essentially created by block chain the origin and target nodal points signatures as opposed to the enormous signatures being attached. The token constitutes of the identities of those nodes which are in between, the signature of the start and ending nodal points and also indicate which transitional node is to have the token. Trust value is predicted by Trustable RTT algorithm. The trustable nodes are joined by the RTT (round-trip time) algorithm to combat malicious users and rely on them for trust value calculations. starting and terminating trusting nodes share the responsibility of assigning trust values to transitional nodes. Such trust is the key to effective routing in a WSN. The token concept is one of the approaches that are used to resolve a routing issue influenced by the manner in which trust is referred to in a wireless sensor network. Token based on the requitals allow the composition of the transitional nodes.

3.4.1 Trustable RTT algorithm

```
   def trustable_rrt(nodes, N, n, trust_threshold=0.6, time-
out=5):
   nodes: List of Node objects
   N: Maximum network capacity
   n: Count of nodal points to include in round trip
   trust_threshold: Minimum trust level for a node to be eligi-
ble
   timeout: Max time (seconds) allowed to complete round trip
   if len(nodes) > N:
   print("□□ Network over capacity.")
   return None
```

```
# Timer start
start_time = time.time()
# Filter nodal points based on trust
trustable_nodal points = [node for node in nodal points if
node.trust_value >= trust_threshold]
if len(trustable_nodes) < n:
print("□□ Not enough trustable nodal points to complete
route.")
return None
# Select `n` trustable nodal points randomly (or use routing
logic here)
selected_route = random.sample(trustable_nodes, n)
# Simulate round trip
for node in selected_route:
if time.time() - start_time > timeout:
print("□□ Timeout occurred during routing.")
return None
print(f" Visited Node {node.node_id} with trust {node.trust_
value}")
print(" Round Trip Successful!")
return [node.node_id for node in selected_route]
```

3.4.2 Round-Trip Time (RTT) Working Procedure (Step-by-Step)

Step 1: Capacity Check

```
if len(nodes) > N:
print("□□ Network over capacity.")
return None
```

If the total count of nodal points in the input exceeds the allowed capacity N, the function stops immediately.

Step 2: Start the Timer

```
start_time = time.time()
```

- Starts a timer to monitor the duration of the routing operation.

Step 3: Filter Trustable Nodes

```
trusted_nodal points = [node for node in nodal points if
node.trust_value >= trust_threshold]
```

Only nodal points that meet or exceed the trust_threshold are eligible to be used for routing.

Step 4: Check if Enough Trusted Nodal points Exist

```
if len(trusted_nodes) < n:
print("□□ Not enough trusted nodal points to complete
route.")
return None
```

If we don't have enough trusted nodal points to complete the route (n required), routing cannot proceed.

Step 5: Select Route

```
selected_route = random.sample(trusted_nodes, n)
```

Randomly select n nodal points from the trusted ones to form the round-trip route.

Step 6: Simulate Round Trip with Timer Check

```
for node in selected_route:
if time.time() - start_time > timeout:
print("□□ Timeout occurred during routing.")
return None
print(f" Visited Node {node.node_id} with trust {node.trust_
value}")
```

For each node in the selected route:

- If the time elapsed since routing started exceeds the timeout, routing fails.
- Otherwise, the node is considered successfully visited.

Step 7: Success

```
print(" Round Trip Successful!")
return [node.node_id for node in selected_route]
```

- If all n trusted nodal points were visited within the timeout, the function:
  - o Prints a success message
  - o Returns the list of visited node IDs

In Trustable RTT algorithm starts counting when the number of nodal points is below the network's capacity. A trustable value is associated with connected nodal edges. A response time will be compared to Round Trip Time only if the node received the acknowledgement. A node is considered reputable if its RTT value exceeds its response time; if not, it is considered not as reliable. The RTT was traditionally calculated in TCP as follows:

$$\boldsymbol{RTT} = (\boldsymbol{\alpha} \cdot \boldsymbol{Old\_RTT}) + ((1 - \boldsymbol{\alpha}) \cdot \boldsymbol{New\_Round\_Trip\_Sample})$$

where the weighting component α is constant ($0 \leq \alpha < 1$). The weighted average is immune to short-term changes (such as a single segment experiencing a considerable delay) when α is chosen to be near 1. When α is set near zero, the weighted average reacts to delay variations relatively rapidly. The Jacobson/Karels algorithm, which also accounts for standard deviation, improved.

3.5 Check Clearance Phase

During the phase of evaluation, the requitals are checked and credited by the accounting center into the node's account. The accounting center will be divided into several sections. They're a sorter, a purifier, a credit checker, and a trust monitor. The classifier checks the check that has been deposited at the accounting office and determines where it came from, where it is going, and any other points in the transaction. After classification, the requitals will be removed from the warehouse and sent for the ejection process. Each check is verified by their hashed signatures during the eviction procedure. If the resulting hash value matches the check, the check is legitimate. If the provided check is accurate, the accounting center will credit the node account and raise the node's credibility in AC; if the check is not, the accounting center will not credit the node account and lower its trustworthiness. The node's performance, trustable values, and nodal points credit accounts are all analyzed using trust Monitor. Based on its previous behavior, the Trust Monitor will not allow a low-trust node to transport network data in the future. Furthermore, when a node received more bad ratings than usual and spent successive sessions in a suspicious state, it was marked as malevolent. If a node receives fewer positive ratings than its typical rate, or if the difference between the time spent in honest and suspicious status is less than the value of beta, the node is also considered malevolent. The delay in payment and trust update is the time that has passed from the time the session occurred until the payment is cleared and the trustable values are updated.

4. EXPERIMENTS AND RESULTS

Figure 6. Node Initialization

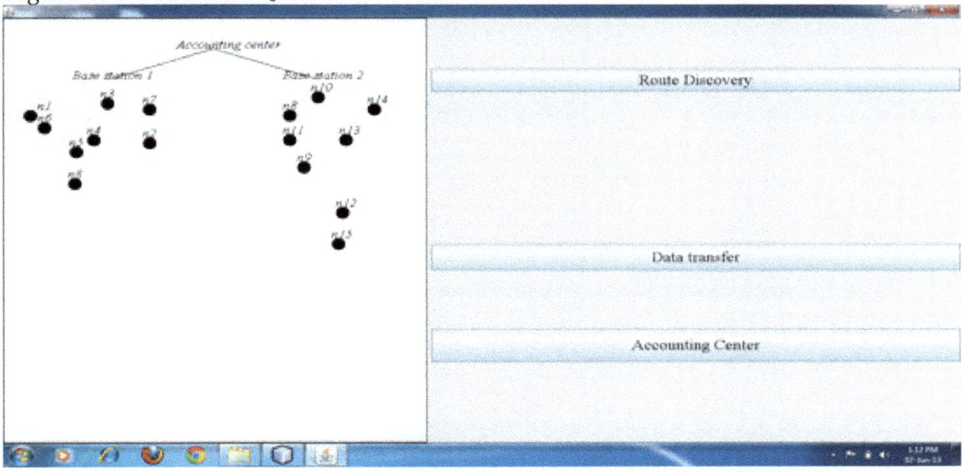

Figure 7. Selecting the start node and end node

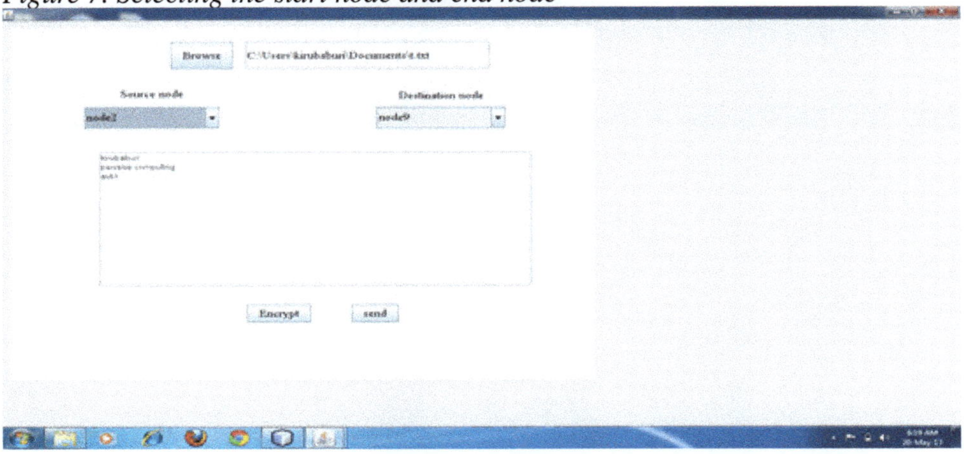

Figure 8. Data Transfer

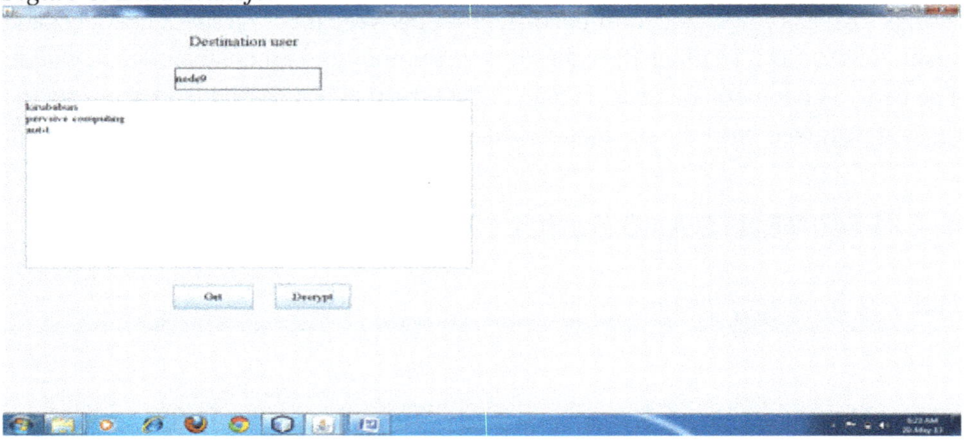

Figure 9. Accounting Center

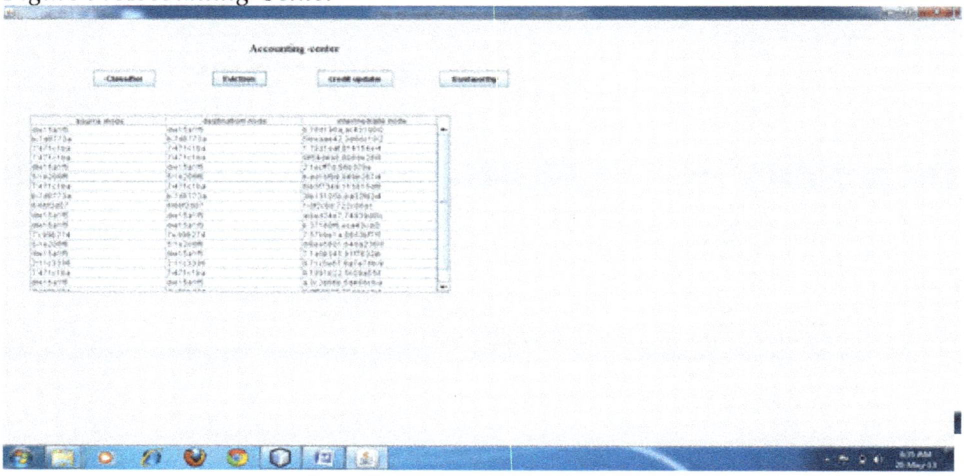

Figure 10. Classify the Trustworthy Nodes

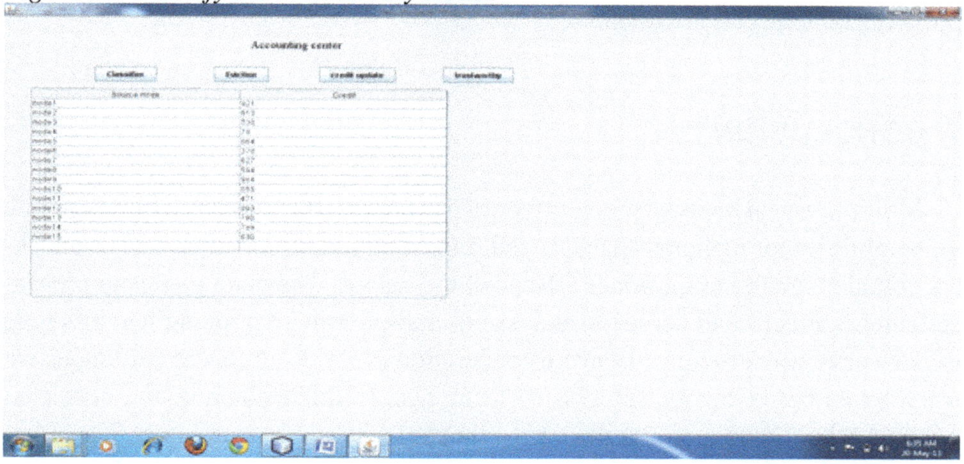

5. ADVANTAGES OF PROPOSED SYSTEM

Wireless networks can be combined with the infrastructure of both phases of wireless ad-hoc networks, in several ways. The network area, which is covered by these kinds of networks, is a base station- populated area that is connected to each other via a backbone, as with cellular networks. Modeling variables that may include packet drop rate, forwarding delay, route change frequency and historical cooperation patterns could be a part of machine learning training process. The models would not only adapt to network data in near-real-time but also allow for a more dynamic intrusion detection scenario instead of solely relying upon static thresholds or predefined fingerprints of Incursions. Consequently, the trust management system that is more robust and smarter which can sustain high levels of cooperation and concurrent time, lessen the number of false positives in nodal edges classification will emerge.

6. SOCIAL WELFARE OF THE PROPOSED SYSTEM

Furthermore, using techniques like online learning may help the system to improve detection accuracy over time, particularly in dynamic multiple way of hopping Wireless type of Networks infrastructure, where node behavior and network topology change regularly. This would result in a more resilient and intelligent trust management system that can maintain strong cooperation while lowering false positives in node classification.

7. CONCLUSION

Multiple way of hopping wireless type of Networks infrastructure, exposes unique vulnerabilities for malicious hackers and refusal of node collaboration. The attackers collude together to introduce advanced Incursions. The main goal is to prevent selfishness attacks and collusion attacks on multiple way of hopping wireless type of Networks infrastructure. To avoid cooperation in check submission attempts, we use a token-based cheque submission technique. In this technique, the node that possesses the token is permitted to send a cheque to the accounting center. The initial starting node generates the token and selects any intermediary node in the routing network. The cheque submission node is chosen based on their trust value for that periods and their credit values. Trust is calculated using the Trustable Round Trip Time (RRT) technique. The token contains all of the intermediate nodes' identities as well as the block chain signatures of the destination nodes, allowing the AC to credit those nodes. Block chain function provide the integrity and authentication between routing data packet between nodes. Incentive mechanism using block chain operation provides the secure data transmission between origin and target nodes. Credits are constantly lost by nodal points that submit inaccurate payment reports. Future research can concentrate on incorporating machine learning (ML) model for dynamic and intelligent behavior analysis in order to increase the persuasiveness of the Trustworthy Checker component. The capacity of conventional rule-based methods to adjust to changing assault patterns or subtle selfish actions is constrained. The system may anticipate and detect previously unknown or developing selfish tendencies that diverge from typical node behavior by integrating machine learning models, such as supervised learning for node behavior classification or unsupervised learning for anomaly detection.

REFERENCES

Akyildiz, I., & Wang, N. X. (2005). A survey on wireless mesh networks. *IEEE Communications Magazine*, *43*(9), S23–S30. DOI: 10.1109/MCOM.2005.1509968

Balas, V., Pandey, H., Bin Ali, M., Singh, V., & Kumar, A. (Eds.). (2026). *Recent Advances in Smart Communication Technologies for a Sustainable Future*. IGI Global Scientific Publishing., DOI: 10.4018/979-8-3373-3541-4

Baskar, S., Periyanayagi, S., Shakeel, P. M., & Dhulipala, V. S. (2019). An energy persistent range-dependent regulated transmission communication model for vehicular network applications. *Computer Networks*, *152*, 144–153. DOI: 10.1016/j.comnet.2019.01.027

Buri, R. K., & Jayasankar, T. (2019). Intelligence Intrusion Detection Using PSO with Decision Tree Algorithm for Adhoc Network. Bioscience Biotechnology Research Communications. *Special Issue Recent Trends in Computing and Communication Technology*, *12*(2), 27–34.

Dhulipala, V. S., Aarthy, V., & Chandrasekaran, R. M. (2010, March). Energy and fault aware management framework for wireless sensor network. In International Conference on Business Administration and Information Processing (pp. 461-464). Berlin, Heidelberg: Springer Berlin Heidelberg. DOI: 10.1007/978-3-642-12214-9_79

Dhulipala, V. S., Karthik, N., & Chandrasekaran, R. M. (2013). A novel heuristic approach based trust worthy architecture for wireless sensor networks. *Wireless Personal Communications*, *70*(1), 189–205. DOI: 10.1007/s11277-012-0688-1

Ghosh, T. (2005). Secure routing and trust computation in multihop infrastructureless networks. .DOI: 10.25148/etd.FI15101570

Janzadeh, H., Fayazbakhsh, K., Dehghan, M., & Fallah, M. S. (2008). A secure credit-based cooperation stimulating mechanism for MANETs using hash chains. *Future Generation Computer Systems*, *25*(8), 926–934. DOI: 10.1016/j.future.2008.12.002

Jayaraman, G., & Dhulipala, V. S. (2022). FEECS: Fuzzy-based energy-efficient cluster head selection algorithm for lifetime enhancement of wireless sensor networks. *Arabian Journal for Science and Engineering*, *47*(2), 1631–1641. DOI: 10.1007/s13369-021-06030-7

Joshi, P., Singh, V., Thapliyal, N., Joshi, K., Bhatt, A., & Mahur, M. (2024, October). Intelligent IoT-Enabled Real-Time Monitoring System for Logistics Management. In *2024 International Conference on Computing, Sciences and Communications (ICCSC)* (pp. 1-5). IEEE. DOI: 10.1109/ICCSC62048.2024.10830332

Kaur, P., Kaur, K., Verma, D. K., Singh, K., Kaushik, K., & Singh, V. (2024, September). Routing protocols in wireless sensor networks: A comprehensive review and future perspectives. In *2024 7th International Conference on Contemporary Computing and Informatics (IC3I)* (Vol. 7, pp. 263-270). IEEE. DOI: 10.1109/IC3I61595.2024.10828894

Kaushik, R., & Singhai, J. (2011). MODSPIRITE: a credit based solution to enforce node cooperation in an ad-hoc network. IJCSI International Journal of Computer Science. https://www.ijcsi.org/papers/IJCSI-8-3-2-295-302.pdf

Krishna, G., Singh, V., Pandey, D., Krishna, K. H., Joshi, K., & Gupta, T. (2024, May). Power trading framework of cloud-edge computing in the artificial intelligence market. In 2024 4th International Conference on Advance Computing and Innovative Technologies in Engineering (ICACITE) (pp. 1910-1918). IEEE. DOI: 10.1109/ICACITE60783.2024.10617344

Liu, K., Deng, J., Varshney, P. K., & Balakrishnan, K. (2007). An Acknowledgment-Based Approach for the Detection of Routing Misbehavior in MANETs. *IEEE Transactions on Mobile Computing*, *6*(5), 536–550. DOI: 10.1109/TMC.2007.1036

Mahmoud, M. E., & Shen, X. (2011). RISE: Receipt-Free Cooperation Incentive Scheme for Multihop Wireless Networks. *2011 IEEE International Conference on Communications (ICC)*, 1-5. DOI: 10.1109/icc.2011.5963199

Mahmoud, M. M. E. A., & Shen, N. X. (2011). FESCIM: Fair, Efficient, and Secure Cooperation Incentive Mechanism for Multihop Cellular Networks. *IEEE Transactions on Mobile Computing*, *11*(5), 753–766. DOI: 10.1109/TMC.2011.92

Mao, Y., You, C., Zhang, J., Huang, K., & Letaief, K. B. (2017). A survey on Mobile Edge Computing: The Communication Perspective. *IEEE Communications Surveys and Tutorials*, *19*(4), 2322–2358. DOI: 10.1109/COMST.2017.2745201

Mao, Y., Zhou, C., Qi, J., & Zhu, X. (2020). A fair credit-based incentive mechanism for routing in DTN-based sensor network with nodes' selfishness. EURASIP Journal on Wireless Communications and Networking, 2020(1). DOI: 10.1186/s13638-020-01823-0

Marias, G. F., Georgiadis, P., Flitzanis, D., & Mandalas, K. (2006). Cooperation enforcement schemes for MANETs: A survey. *Wireless Communications and Mobile Computing*, *6*(3), 319–332. DOI: 10.1002/wcm.398

Pan, J., Cai, L., Shen, X., & Mark, J. W. (2006). Identity-based secure collaboration in wireless ad hoc networks. *Computer Networks*, *51*(3), 853–865. DOI: 10.1016/j.comnet.2006.05.012

Rathod, S. B., Bajpai, B., Wankhade, S., & Ponnusamy, S. (2025). Enhancing Security in Heterogeneous IoT Networks through Intelligent Identification Systems. *Journal of VLSI Circuits and Systems*, *7*(1), 155–166.

Roohani, B. S., Sharma, N., Kasula, V. K., Mamoria, P., Modh, N. N., Kumar, A., & Singh, V. (2026). Urban Computing Solutions in Healthcare Edge Computing. In Building Data-Driven Edge Systems for Business Success (pp. 377-400). IGI Global Scientific Publishing.

Shakeel, P. M., Baskar, S., Dhulipala, V. S., & Jaber, M. M. (2018). Cloud based framework for diagnosis of diabetes mellitus using K-means clustering. *Health Information Science and Systems*, *6*(1), 16. DOI: 10.1007/s13755-018-0054-0 PMID: 30279986

Singh, V., & Kaushik, V. D. (2023). Navigating the Landscape of Security Threat Analysis in Cloud Computing environments. In *Security and Risk Analysis for Intelligent Cloud Computing* (pp. 1–25). CRC Press.

Suriyan, K., Ganesh, S., Palaniyammal, P., Priya, G., & Singh, V. (2026). Recent Trends in Edge Computing: Challenges and Opportunities. Building Data-Driven Edge Systems for Business Success, 323-344.

Truong, N. B., Lee, H., Askwith, B., & Lee, G. M. (2017). Toward a trust evaluation mechanism in the social internet of things. *Sensors (Basel)*, *17*(6), 1346. DOI: 10.3390/s17061346 PMID: 28598401

Xue, Y., Singh, V., Singh, S., Kant, K., Pandey, S., Kumar, A., & Kaushik, V. D. (2024). Blockchain Framework for Learner Performance Prediction using Life-Brain Storm-based Light GBM Coupled Neural Network: BLOCKCHAIN FRAMEWORK FOR LEARNER PERFORMANCE PREDICTION. [JSIR]. *Journal of Scientific and Industrial Research*, *83*(6), 652–668.

Zhang, X., Li, Y., Xiong, Z., Liu, Y., Wang, S., & Hou, D. (2024). A Resource-Based dynamic pricing and forced forwarding incentive algorithm in socially aware networking. *Electronics (Basel)*, *13*(15), 3044. DOI: 10.3390/electronics13153044

Zhong, S., Chen, J., & Yang, Y. (2004). Sprite: A simple, cheat-proof, credit-based system for mobile ad-hoc networks. *Proc. Of INFOCOM*, *3*, 1987–1997. DOI: 10.1109/INFCOM.2003.1209220

Chapter 9
Secure AI-Enabled Communication in Smart Cities and Industry 4.0

Musdalifa Thamrin
STMIK Professional Makassar, Indonesia

Sahabuddin Sahabuddin
Universitas Bossowa, Makassar, Indonesia

Nizirwan Anwar
https://orcid.org/0000-0003-1189-9093
Universitas Esa Unggul, Jakarta, Indonesia

Tomy Ronaldi
https://orcid.org/0000-0002-0517-7505
Universitas Pertahanan, Bogor, Indonesia

Nurdiansyah Nurdiansyah
Universitas DIPA, Makassar, Indonesia

Imam Asrowardi
Politeknik Negeri, Lampung, Indonesia

Muhammad Faisal
https://orcid.org/0000-0003-1469-9468
Muhammadiyah University of Makassar, Indonesia

Alok Kumar
https://orcid.org/0000-0003-1051-0672
School of Engineering and Technology, Chhatrapati Shahu Ji Maharaj University, Kanpur, India

Shubham Bajpai
School of Advanced Agriculture Sciences and Technology, Chhatrapati Shahu Ji Maharaj University, India

ABSTRACT

This chapter per the authors examines how Smart City and Industry 4.0 initiatives—built on dense wireless connectivity, cloud platforms, and pervasive automation—require a new generation of cybersecurity rooted in artificial intelligence. It outlines the evolution of urban and industrial communication infrastructures, proposes an adaptive AI-driven security architecture, and details mechanisms for secure data

DOI: 10.4018/979-8-3373-3576-6.ch009

© 2026, IGI Global Scientific Publishing. Copying or distributing in print or electronic forms without written permission of IGI Global Scientific Publishing is prohibited. Use of this chapter to [ge]nerative artificial intelligence (AI) technologies is expressly prohibited. The publisher reserves all rights to license its use for generative AI training and machine learning model development.

exchange, real-time threat detection, and resilient intrusion response across heterogeneous IoT, vehicular, robotic, and smart-energy networks. Ethical considerations surrounding AI autonomy, privacy, and transparency are balanced against operational imperatives for continuity and safety. By integrating case studies and charting future research directions, the chapter offers a practical framework for designing, deploying, and governing secure AI-enabled communications at scale.

1 INTRODUCTION

In the era of smart cities and Industry 4.0, there is an increase in connectivity and automation, as it is an era of communication networks functioning as a nervous system, enabling seamless transfer of information and decision-making in real time. To achieve widespread connections that range from devices to applications, these systems rely significantly on wireless technologies, including 5G, Wi-Fi 6, and low-power wide-area networks (Kim et al., 2023). Sensor deployment is so ubiquitous in urban, residential, and industrial spaces that huge amounts of data are produced, which AI algorithms draw on in order to optimise resource usage, enable efficiency, and increase quality of life (Smith et al., 2021). Designing and implementing such communication networks relies on an in-depth knowledge of the underlying technologies and the vulnerabilities that they pose (Johnson et al., 2022). Moreover, in order to prevent confidentiality, integrity, and availability threats, the current infrastructure is the most vulnerable point of failure, so it can be breached (Lee et al., 2023). As hyperconnectivity, complexity, muddled borders, and hacking from industries grow, smart city contexts become increasingly vulnerable to a security breach, where conventional security systems can no longer effectively secure them (Davis & Kumar, 2021). AI is enhancing situational awareness and network management (Benzaïd & Taleb, 2022) by providing real-time visibility and threat prediction coupled with automated responses. Integrating AI into security frameworks marks a significant paradigm shift, enabling earlier threat detection and adaptable defense strategies. Machine learning processes vast amounts of data to identify patterns indicative of malicious activity. Deep learning delves deeper into complex data correlations, facilitating early threat identification and response (Haider et al., 2020). Reinforcement learning trains agents to make informed decisions within specific contexts. AI-driven security systems are highly capable of threat detection and quick reaction through robust, scalable algorithms (Haider et al., 2020). Machine learning (Raturi et al., 2025) could also help in predicting and proactively recognizing patterns and weak points for threat exposure before an attack is done, fortifying the defence against it (Pochmara & Świetlicka, 2024). Artificial intelligence and threat detection are emerging as sophisticated tools for security by identifying and detecting threats without human intervention. AI security solutions also successfully proved their worth in pilot smart city and even industrial projects, which opens the doors for broader implementation (Khanji & Khattak, 2020). Since security now becomes more challenging against advanced cybersecurity threats, AI and other machine learning solutions have been proposed as tools that help security detection and response efforts (Katiyar et al., 2024). In AI-based wireless technology, data exchange and access control must be secured to safeguard sensitive information and authorised access. Encryption protocols for AI-based models in network scenarios to secure their encryption at the time of transmission as well as storage, and protect data against unauthorised access and confidentiality. Approaches such as federated learning and distributed AI make it possible to handle edge devices with security features by enabling team-based model training without the sharing of sensitive information among network machines. In federated learning, other machines collaborate and train on a model while allowing them to run on their own network data, achieving a model that is private and accurate. Blockchain can be used on AI infrastructure to implement decentralised authentication and trust management, which enhances the security and transparency of data transfers. AI algorithms enhance detection, prevention, and response to identity threats, thereby strengthening cybersecurity (Singh and Dubey, 2015; Chaudhary et al., 2024; Singh and Kaushik, 2023a; Singh and Kaushik, 2023b) in corporate environments (Binhammad et al., 2024). Leveraging cryptography, federated learning, and blockchain (Xue et al., 2024), AI networks establish secure data exchange and controlled access. Implementing threat mitigation strategies is crucial to protect these networks from cyberattacks. AI supports this by analyzing network traffic to detect threats like denial-of-service and distributed denial-of-service attacks, as well as identifying malware in real time. Graph neural networks and predictive analytics aid in malware detection by identifying malware code and predicting future threats. AI predictive behavioral analytics helps mitigate insider and ongoing threats by detecting anomaly activity and suspicious behavior. These policies facilitate breach detection while minimizing risk and preserving business continuity. To adapt quickly to emerging threats and changes, continuous monitoring and policy updates are essential. Even though AI in cybersecurity brings various benefits, there are a number of issues to address to guarantee its efficient and responsible use. Significant threats, including data imbalance, adversarial machine learning and model poisoning attacks, can threaten the quality and efficiency of AI-integrated security systems. Data-poisoning is the inclusion of malicious data as data and data poisoning in training, in that it alters AI behaviour, thus making false predictions and causing security breaches (Kim, 2025).

The integration of quantum-resistant cryptographic algorithms with artificial intelligence (AI) is crucial for safeguarding against future threats posed by quantum computers (Achuthan et al., 2024). This collaboration will enable the development of scalable, self-healing AI security frameworks, akin to those employed in Smart Cities for Industry 4.0. These frameworks are essential for adapting to evolving cyber threats. This approach empowers systems to autonomously identify, assess, repair, and recalibrate over time (Das & Sandhane, 2021). Moreover, advanced predictive models are adopted to facilitate rapid detection of potential threats, ensuring security in advance of data breaches. Maintaining AI explainability is paramount for promoting transparency in cybersecurity decisions and implementing quantum-resistant cryptographic methods. A specific focus area involves outlining the process of constructing an adaptable and self-repairing AI security infrastructure within the context of Smart Cities and Industry 4.0 (Ahmad et al., 2020; Vinothkumar & Karunamurthy, 2023). This is also motivated by the inaccurate and biased nature of AI systems, as well as the fact that data can be noisy, thus stressing precision and resilience (Petkovic, 2022).

One of the important problems is that the decision-making of AI will be less transparent and accountable because AI systems act as black boxes (Adelakun et al., 2024). Explainable AI, as well as algorithmic audits, are used to provide technical strategies to avoid these issues (Cheong, 2024). For prejudice prevention and fairness, human oversight still plays a crucial role in reducing biases as well as maintaining fairness (Petkovic, 2022; Petković & Petkovic, 2023; Choung et al., 2022). Transparent algorithm design in algorithms is essential so that training data and the processes involved in making decisions from the training data and decision-making procedures can be studied to help mitigate bias (Dhopte & Bagde, 2023). Specific roles for manufacturers, healthcare providers and others should be appointed. Explainable AI improves performance by unveiling weaknesses and revealing patterns in input data (Muhammad & Bendechache, 2024). The "XAI" techniques are increasingly used in areas such as healthcare, finance, and natural language processing to enhance the transparency and intelligibility of AI decisions (Zhang et al., 2022). Explanation of AI decisions is important in creating trust, accountability, and transparency in the AI applications (Rjoub et al., 2023; Rozario & Čevora, 2023); it is therefore essential to explain the decision-making process used by AI. Communication in smart cities and Industry 4.0 has changed a lot. Advanced wireless technologies, including 5G, Wi-Fi 6, and Low Power Wide Area Networks (LPWAN), have changed the communication infrastructure in smart cities and Industry 4.0 significantly (Li et al., 2020). Although traditional methods of communication have been used in some fields till now, communication has developed through IoT and wireless (Narooka et al., 2025a; Narooka et al., 2025b; Kaur et al., 2024) to enable real-time tracking of the Internet of Things (IoT) applications. These technologies deliver high bandwidth, low latency, and broad coverage, applicable in smart transportation, remote monitoring, and industrial automation. The rise of intelligent devices and data-driven decision-making has created a highly interconnected environment where people, machines, and systems interact seamlessly (Jang et al., 2021). The adoption of the ultra-fast and ultra-low latency 5G technology is revolutionizing data processing and management for real-time critical applications. Additionally, Wi-Fi 6 enhances overall network performance, leading to increased efficiency and a larger user base (Bai, 2022). Low Power Wide Area Networks (LPWANs), such as LoRaWAN and NB-IoT, offer communication solutions that support long-range connectivity with minimal power consumption.

Increased connectivity in smart cities and Industry 4.0 brings security challenges that necessitate robust defenses against cyber threats. These technologies converge on advanced wireless options, such as 5G, Wi-Fi 6, and LPWANs. 5G offers high data rates and low latency, enabling real-time processing for applications like autonomous vehicles and remote surgeries. Wi-Fi 6 enhances network efficiency in high-density device environments, making it ideal for urban and industrial settings. LPWANs are designed for low-power end-to-end communication over long distances, supporting environmental monitoring and smart agriculture tasks. 5G's capabilities support various industrial applications, enabling seamless connectivity and data sharing across heterogeneous devices and systems. However, these infrastructures are vulnerable to security breaches from malicious actors due to vulnerabilities like poor protocols, insecure device settings, and poorly specified access controls. This paper emphasizes the need for recalibration and strengthening of these systems, including new domains like post-quantum cryptography, AI, machine learning, edge computing (Krishna et al., 2024; Roohani et al., 2026; Suriyan et al., 2026; Anandaram et al., 2026), molecular communication, terahertz, visible light communication, and blockchain using distributed ledgers. (Hakeem et al., 2022)

2 EVALUATION OF COMMUNICATIONS IN SMART CITIES AND INDUSTRY 4.0

When it comes to DDoS or man-in-the-middle attacks, networks can become overwhelmed or even suffer from data breaches. The increasing reach of networks and the growing number of Internet of Things (IoT) devices forming connections raise the possibilities of more complex and severe vulnerabilities (Joshi et al., 2024a; Joshi et al., 2024b, Joshi et al., 2025). This development makes it more challenging to identify and respond to threats as more IoT devices are connected to networks. To address these challenges, a multi-faceted security approach is essential, including encryption, rigorous authentication procedures, and proactive measures. detection of threats, must be taken. Artificial Intelligence plays a pivotal role in managing situations and optimizing networks in smart cities and Industry 4.0 environments

(Fig. 1). By processing vast datasets, AI models identify anomalies and predict future trends, ultimately enhancing operational efficiency.

It is also possible through ML tools to recognise attack patterns early, further leading to prompt and accurate responses countermeasures. AI can help to optimise road traffic allocation, improve resource management, and strengthen security to avoid failures and to increase resilience. AI aids organisations in real-time network monitoring, quickly responding to threats, and maintaining performance. Moreover, AI platforms have systems learning and react to changing situations to keep their security defences up to date with new forms of attack (Palani et al., 2024).

Figure 1. AI strengthens network security and optimisation in smart industrial environments.

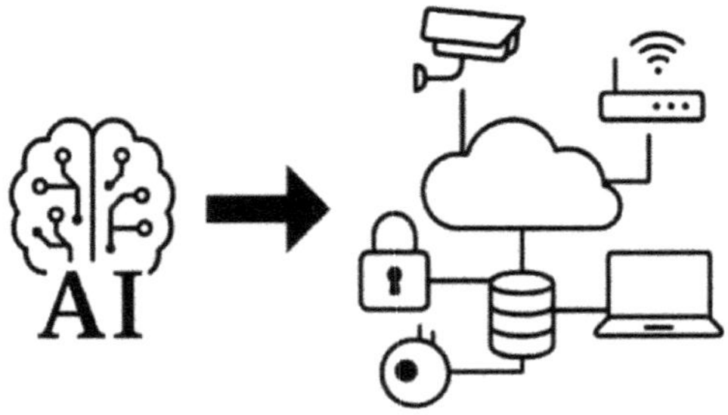

3 AI FOR SECURITY ARCHITECTURE

AI-powered security architectures are a revolution for safety, providing powerful solutions to detect, prevent and respond to threats. A key feature that facilitates the detection of abnormal behaviour in wireless networks is machine learning algorithms, which utilise traffic patterns, user activity and system logs for processing (Fig 2). They learn about past behaviour from the data to find deviations from the norm that might indicate malicious intent. Deep learning methods in particular are very powerful for the real-time classification of threats, allowing the early detection of malware, phishing and other sophisticated threats. It is also possible to design defensive strategies with the help of a reinforcement learning algorithm in order to have a more robust defence strategy over time and respond properly to the cyber-attack. Machine learning technology has changed the landscape in anomaly detection in wireless environments (Fnu et al., 2025; Singh and Kaushik, 2023), offering a uniform and viable alternative approach for identifying unusual activities which deviate from the status quo (Salem et al., 2024). These algorithms analyse various network parameters, such as traffic volume, packet type or connection patterns, aiming to identify a range of cyber threats (Jones et al., 2023). AI-based technologies enable traditional anomaly detection to be improved by incorporating the existing mechanism at the monitoring infrastructure (Aluwala, 2024). Supervised learning, using labelled data sets, detects known attack types. Unsupervised learning, often termed in clustering, discovers new attack patterns without existing signatures. Continuous monitoring and fine-tuning of machine-learning algorithms to protect wireless networks is achieved in an ongoing way by learning algorithm adaptation. Deep learning based solutions, such as convolutional neural networks and recurrent neural networks, have emerged as effective real-time threat classification techniques that make accurate and fast identification of malicious attacks possible. They sift through network traffic to ferret out subtle patterns that signal malware, phishing or other advanced threats, and classify traffic by region, place of origin and content, which in turn works towards identifying malicious actors and compromised devices. The implementation of deep learning will help to increase an organisation's capacity for identifying and responding rapidly to cyber-attacks, and therefore mitigate the negative effects of attacks (Kornaros, 2022). Reinforcement learning offers a powerful method to adaptive defence and intrusion response, as such systems can train and strengthen a response to cyber attack and learning from experience (Aljabri et al., 2021). Such algorithms choose response approaches for the given information system and reward/punishment signals, allowing for dynamic security updates, device reconfiguration, and isolation of compromised systems, reducing the impact of attacks. They support proactive defence strategies which adapt to evolving threats, continuously increasing security (Nguyen & Reddi, 2023). Through case studies, I found successful implementations of AI in smart cities and industrial sectors. In smart cities, AI-driven surveillance analyses video footage for signs that could indicate criminal activity so as to improve public safety. In industrial zones, for example, AI intrusion detection uses technology to watch what the network traffic looks like for malware, unauthorised access, and breaches (Salem et al., 2024). These examples illustrate how the practicalities, advantages and issues involved in the application of AI for security in complex practical environments in which the natural world cannot be ignored.

Figure 2. AI in Wireless Networks

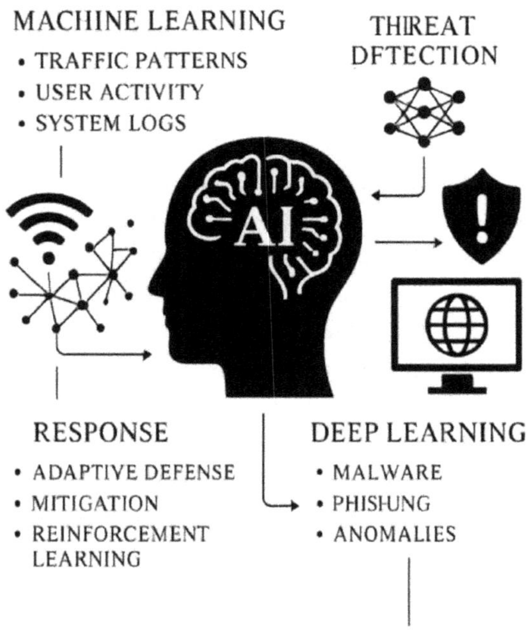

4 SECURE METHODS OF DATA EXCHANGE AND ACCESS CONTROL

To maintain sensitive data and access control in AI-powered wireless systems, data exchange and access control must be secure. Encryption protocols are essential for protecting data at the point of transfer and during storage, in order to curtail unauthorised access and tampering. Federated learning and distributed AI are new and effective ways to protect edge devices by promoting collaborative model training without disclosing or sharing raw data. Blockchain can facilitate the process of authentication and trust by creating a decentralised digital ledger that is resistant to tampering, recording all transactions together and proving the identity of entities. Encryption solutions tailored specifically for AI-activated wireless networks are crucial for ensuring data privacy and trust over wireless distribution channels for data security. However, conventional encryption approaches cannot be sufficient for resource-constrained IoT devices or high-bandwidth applications. Lightweight architectures – modified generalised Feistel network ciphers, matrix encryption walks, and DNA cryptography–provide a compromise between security and performance. Quantum key distribution provides an exciting opportunity to use quantum mechanics for additional security, as it provides the means to generate and disseminate encryption keys. Homomorphic encryption allows computation by using encrypted data, and the need for decrypting to perform computation is not required for secure data processing in AI systems. Federated learning and distributed AI provide secure and open model training on decentralised data sources without identifying raw data. Federated learning enables many edge devices to train a shared model in collaboration with one central server, while keeping their data in the local environment. This prevents data centralisation and hence preserves privacy. Also, adding noise to model updates using differential privacy techniques increases protection against the disclosure of individual data. Thus, federated learning enables a decentralised and adaptive AI training approach through which user privacy is kept without data sharing (Yaacoub et al., 2023). Blockchain can help form a decentralised, immutable structure for authentication and trustworthy AI wireless systems. Blockchain technology can allow decentralised identity management mechanisms to establish a solution for device and user verification of device identity and device trust with different digital currencies. The system is operated not by central power. Smart contracts automate access control, which enables only authorised agents to access resources or make transactions (Bothra et al., 2023). The immutability of blockchain provides trustworthy data, improving security and trust. Blockchain improves overall security and efficiency in IoT networks when used alongside AI (Ruzbahani, 2024). The use of AI and blockchain systems improves the data privacy of IoT networks (Ruzbahani, 2024). For instance, in mobile applications, encrypting messages with AES and incorporating blockchain enhances the security of message content. AI algorithms trained on blockchain data can detect fraudulent transactions, predict breaches, and bolster network security (Singh et al., 2020). This combination leads to improved security and more effective blockchain applications. Furthermore, federated learning adheres to strict data regulations by allowing the central server to retain aggregate data while clients retain local data (X. Zhang et al., 2021; Ali et al., 2021). This enables clients to maintain local data and train models locally. Subsequently, the server combines local models into a single global model, which is then sent back to the clients (Nguyen & Reddi, 2023). As a result, federated learning maintains the integrity of data while allowing clients to train models locally data private on each device, it eliminates the need for data sharing and enhances anomaly detection (Zhang et al., 2021).

5 THREAT MITIGATION STRATEGIES

To mitigate the risks associated with AI in communications, a comprehensive, system-wide approach should be adopted to implement AI for detecting, analyzing, and responding to security incidents. AI services play a crucial role in detecting and preventing DoS and DDoS attacks by analyzing network traffic, identifying anomalies, and continuously updating security rules. Malware detection is enhanced through graph neural networks and predictive analytics, with the ability to find suspicious code, find suspicious behaviour, and prevent infections. AI–based behavioural analytics keeps an eye on user behaviour, identifies anything abnormal, and notifies you about insider threats and advanced persistent security risks (APTs). In the case of attacks on DoS and DDoS, AI algorithms can analyse network traffic for patterns of anomaly detection, identifying legitimate traffic and malicious requests. AI systems can help minimise such attacks by keeping a fast track over the changing patterns of traffic. They detect malicious behaviours and act immediately to block or mitigate threats. AI helps enforce security policies as well. It will propose firewalls, intrusion control systems, and other changes when facing more rapidly changing threats (Roshanaei et al., 2024). Malware detection using graph neural networks and predictive analytics has greatly developed malware detection techniques. Graph neural networks examine software structure and relationships to identify patterns of malware. Through the analysis of system logs, network traffic and other data, predictive analytics provides risk prediction for possible infections by delivering real-time threat detection and pre-emptive incident management, contributing towards fewer false alarms (Vitorino et al., 2023; Binhammad et al., 2024). The ability of AI to reduce false positive rates allows security teams to notice only real threats. AI-based behavioural analytics are used to identify insider threats and advanced persistent threats by observing user behaviour and recognising anomalies in user behaviour. When you establish a baseline of normal activity, AI algorithms are able to detect anomalies that present risks of malicious intent (Islam et al., 2022). Alenezi & Akour (2025) also claimed that machine learning, in addition to the detection of anomalies in applications, has its significance. Furthermore, AI-powered anomaly detection enhances security for cyber-physical systems by identifying differences from the business-as-usual system operations using verified algorithms in a scenario of an attack against water-treatment systems.

6 CHALLENGES IN SECURING AI-ENABLED COMMUNICATIONS

AI communication security necessitates tackling the major issues, including data imbalance, adversarial attack and model poisoning. This approach is not good for real-time security analytics, which have problems of high computation cost, energy constraints and an added layer of complexities of privacy, ethics and regulation (Marri et al., 2024). In addition, data imbalance, in which some classes are overrepresented, may lead to biased models or poor performance in minority groups. Adversarial machine learning designs with inputs to mislead AI with inaccurate predictions. Malicious data used to corrupt a model is included in the training session for model poisoning attacks. The effectiveness of AI and ML models is mostly dependent on good quality and comprehensive training data (Roshanaei et al., 2024). There is biased or incomplete data resulting in inaccurate or unfair outputs, which are often the result of biased or incomplete training (Mirestean et al., 2021). With a biased attention to a single class, the models will not be able to effectively identify minority classes, and accuracy may decrease. Oversampling, undersampling, and synthetic data generation techniques can mitigate some of these issues. Adversarial machine learning is a prevalent issue where adversarial attackers manipulate data to deceive models, presenting them with a limited amount of information while significantly impacting outputs. These adversarial cases lead to misclassifications with severe consequences. To safeguard AI against such exploits, we must develop robust and input-resistant models. Insecure AI development practices pose potential vulnerabilities that can lead to system manipulation (Oseni et al., 2021). In industries like autonomous driving and medical care, where mistakes can be catastrophic, these dangers are particularly profound (Kawamoto et al., 2023).

Model poisoning involves injecting malicious data into training datasets to weaken machine-learning models (Usama et al., 2020). If adversaries corrupt training data, errors may occur in classification, leading to undesirable behavior that goes unnoticed. They can also manipulate model parameters, disrupt training processes, or leverage these models for further malicious actions. More advanced attacks include query and side-channel attacks that impact machine learning systems (Roshanaei et al., 2024). As machine learning continues to grow, security risks increase due to subtle data manipulations that can result in incorrect predictions (Abdou et al., 2021; Liu et al., 2020). These vulnerabilities affect critical applications like fraud detection and security systems, necessitating proactive defenses (Lunghi et al., 2023). Adversarial techniques can cause severe consequences.

Security breaches are a concern, but the high computational and energy costs of real-time AI security analytics limit their practical application. Training and deploying AI systems require substantial processing power, memory, and energy, especially in resource-constrained environments like mobile devices and edge devices. Privacy, ethical, and regulatory issues further complicate these challenges, particularly regarding data protection during training and deployment (Saifullah et al., 2022). Addressing AI algorithms, along with hardware and software support systems, is crucial (Banerjee et al., 2024). Implementing security-focused AI raises ethical concerns, as these systems often rely on large datasets containing personal information. These datasets can be unpredictable and complex, reducing public trust (Han et al., 2023). To address these issues, emphasizing data privacy, transparency, and ethics is essential (Korobenko et al., 2024; Patil, 2025). Responsible AI should include safeguards to protect human rights against adverse outcomes (Patil, 2025; Ding et al., 2025). Principles like fairness, transparency, and accountability (Prasanth et al., 2023; Zhao et al., 2022) are crucial, as neglecting responsibility hinders error and bias correction (Brij, 2023). In security, ethical issues such as bias, discrimination, and privacy breaches are significant concerns (Wach et al., 2023). Unchecked bias can lead to discriminatory outcomes (Adelakun et al., 2024). Systematic design and validation are paramount to mitigate risks and align AI use with ethical and human values, as emphasized by Shang et al. (2024) and Svetlova (2022). In healthcare, ethical considerations such as data privacy, consent, and fairness are of utmost importance, as highlighted by Torkey et al. (2025) and Jha et al. (2023). The ethical implications of AI are gaining widespread attention across various sectors, underscoring the need for a deeper understanding of the human-technology relationship, as underscored by Huang (2023). Collaboration among policymakers, researchers, and communities is essential to address these challenges effectively. Further research, targeted deployment, and integrating human expertise with AI can significantly enhance patient care, as suggested by Shang et al. (2024) and Shuaib (2024). Additionally, legal and psychological factors play a crucial role in shaping the ethical landscape of AI, as highlighted by Liang et al. (2022). While AI holds tremendous transformative potential, addressing rising privacy and security concerns is paramount for its societal benefit. Preventing these issues is crucial for ensuring the smooth functioning of AI in society, as emphasized by Hildt (2025), Naik et al. (2022), and Akinrinmade et al. (2023). Future research directions include exploring explainable AI for transparency, developing quantum-resistant cryptography to counter quantum attacks, and developing scalable self-healing smart cities and Industry 4.0 security architectures. These architectures have the potential to automatically identify and counteract threats, thereby safeguarding vital infrastructure. While AI tools such as facial recognition, drones, and autonomous robots show great promise, their proper regulation and management are essential, particularly in addressing legal, ethical, and security concerns, including bias and liability issues, as well as preventing adverse effects. Addressing AI-related challenges requires a comprehensive understanding of risks such as data vulnerabilities, system weaknesses, integration problems, and potential misunderstandings or distortions, as highlighted by Alexandrovich (2024). Ensuring AI safety is paramount, encompassing ethical, legal, and governance considerations to promote responsible and secure AI deployment, as emphasized by Chen et al. (2024). AI has the potential to significantly contribute to our cybersecurity by detecting and preventing threats multifaceted threats and automating responses; but also encounters the issue of both biases for decision-making as well as how we protect our data (Roshanaei et al., 2024). The rapid expansion of AI towards emerging issues necessitates ongoing research and improvement. A well-structured approach to addressing these issues would help organizations formulate adaptive security plans to address cyber threats in these areas whilst increasing resilience of artificial intelligence (AI). It is important to understand that because AI systems are very data-driven and algorithms can have very complex designs and so they could be affected by cyberattacks (Roshanaei et al., 2024).

Hence, developing strong anti-cybersecurity practices is essential for safe and ethical application of AI. Deployment needs to be accompanied by validation, human supervision, and proper regulatory structures within a system to guarantee its security and reliability (Roshanaei et al., 2024; Achuthan et al., 2024; R. Das & Sandhane, 2021; Montagnani et al., 2024). Robust validation with oversight and explicit regulations is necessary when using AI to ensure safety and reliability. But AI can also make mistakes, exhibit biases, be influenced by noisy data or be non-transparent, all of which challenge the users' trust. Solving this challenge is necessary for producing reliable, secure and ethical AI, especially sensitive ones, such as the healthcare and the justice domain. Transparency and accountability in AI decisions are core points (Adelakun et al., 2024). Because of the AI algorithms' complexity and opacity, understanding the actual processes of making certain decisions may be challenging. The result may undermine trust especially in critical sectors. This needs to follow the careful handling of ethical and regulatory issues of responsible AI in health care (Mennella et al., 2024). In addition, of course, managers should concentrate on updating and versions of AI (Zhou & Gattinger, 2024). The new frontier makes it more difficult for normal software development processes to provide a long-term reliability and security. Big AIs such as Data imbalance, adversarial machine learning and model poisoning attacks are all important for AI security. Because of its skewed characteristics of data, biased

models are being trained and are also performing worse on poorly presented data (Polemi et al., 2024).

One such form of adversarial attack is known as input manipulation which refers to using input to deceive AI systems to compromise their security. Model poisoning injects malicious data into training sets, which results in predictions that are misleading (Mudgal et al., 2022). Heavy computation requirements and energy usage pose significant challenges to real-time security analytics, especially in contexts where devices e.g., edge devices and IoT networks are lacking in resources. Responsible AI practices require three basic requirements such as ethics, privacy and compliance (Roshanaei et al., 2024). AI may face risk if methodologies are ignored to prevent data and privacy risks in the future (Khera et al., 2024). Such transparency in AI leads to possible actors not being able to see or even query the automated decisions which in turn leads to a lack of accountability (Patil, 2025). Research on this issue needs to incorporate ideas from other disciplines such as medical sciences, economics, computer science, social sciences, law and policy-making, as it is so fragmented and concentrated upon the personal impact of a particular situation (Morley et al., 2020). Dealing with these multi-dimensional challenges will require a joint effort within academia to design and use those AI systems ethically and responsibly. Quantum resilient cryptographic algorithms that interact with an AI could end up being the way forward for future work and explainable AI to bridge the gap between transparent security decisions for an AI to make security decisions and self-repairable systems of scales through scalable and self-correction architectures. The goal of Explainable AI (XAI) is to design models that are visible and transparently interpretable, thus enabling humans to comprehend the process (Vinothkumar & Karunamurthy, 2023). This is particularly important for cybersecurity, where AI-based threat detection and responses are particularly key. Explainable AI techniques are promising areas for future research in implementing AI technology to make AI systems more trusted and adopted into sensitive areas (Zhang et al., 2022; Jones et al., 2023). XAI establishes trust by providing users with plain explanation of the AI's actions (Das & Rad, 2020; Hamon & Junklewitz, 2020). AI outputs in a way that helps to make the reasoning behind the recommendations clearer for decision-makers (Rane et al., 2023). XAI can create deleterious impacts in the economic and border-crossing settings even if it is an explainable concept, which is why, although some efforts were made to foster governance, XAI is a nascent domain (Pi, 2023). Explainability is significant to generate trust and responsible AI deployment in multiple domains (Hamida et al., 2024). It acts as a way to verify and scale up performance and unearth vulnerabilities, discover latent trends, and sift through trivial features across disparate input and network layers (Muhammad & Bendechache, 2024). Explainability is an important interpretational process of machine learning to ensure that users can read, trust and control AI systems (Rjoub et al., 2023). AI can make decisions, it can make recommendations, it can act, but it must also justify that process to consumers. Similarly, data-based suggestion must be understood and analysed (Rjoub et al., 2023). This is why provenance mechanisms that promote transparency and make AI logic transparent (Kale et al., 2022). As models become more complex and work in more complex and critical systems, interpretability and transparency become critical (Hamida et al., 2024). Different methods can be used to assess explainability, each with its own trade-offs, but no approach has ever been agreed upon (Speith & Langer,

2023). Broadly speaking, XAI methods are employed in the service of designing and building tools for internal engineers and developers of the model, rather than explaining it to users (Bhatt et al., 2020). AI can therefore become explainable to enable transparent cybersecurity decision-making, where it creates models which can answer the question as to why they are exhibiting certain behaviours (Hamida et al., 2024). In cybersecurity, you also need an explanation of what an AI system flags as suspicious, what it might recommend to do next (Manresa-Yee & Ramis, 2021), so this is an important task. That level of transparency highlights the need for humans to help AI, which establishes trust and further improves collaboration (Rong et al., 2022). How and why should AI decision-making be made? It would be beneficial to implement quantum-resistant cryptography, which resists quantum computer attacks, which has been the goal for generations, to ensure long-term security, particularly when sensitive data is affected. Scalable self-healing security architectures for smart cities and Industry 4.0 require the design of adaptive architectures capable of recovering from attacks and navigating increasingly complex environments independent of human intervention. Indeed, these systems are likely to face the issues of data imbalance, adversarial attacks, and high computational costs (Rjoub et al., 2023; Khan et al., 2024). It is important to keep improving AI's prediction so that it can predict threats in advance, to neutralise them (Roshanaei et al., 2024), using machine learning to analyse data in terms of patterns warning for the coming attacks (Das & Sandhane, 2021; Achuthan et al., 2024). This self-healing capability is expected to reduce downtime and improve the resilience of smart cities and industrial settings. As cybersecurity evolves into an arms race, AI has emerged as a pivotal player in both offensive and defensive strategies (Schmitt, 2023). In light of the threat posed by quantum computing to traditional cryptography, the development of quantum-resistant algorithms has become paramount (Barbeau & García-Alfaro, 2022). Active defense strategies, such as automated intrusion detection, deception technologies, and threat hunting, play a crucial role in enhancing security (Dhir et al., 2021; Biplob et al., 2024). Moreover, AI-driven cybersecurity solutions can analyze past attacks to improve defenses and reduce vulnerabilities, ultimately bolstering protection (Katiyar et al., 2024; Binhammad et al., 2024). However, it is important to acknowledge that AI also carries risks, including adversarial attacks, privacy breaches, and biases, which necessitate careful management. Upholding ethical standards is paramount to ensure responsible AI deployment in smart cities and industries, promoting fairness and safety (Ahmad et al., 2020). Ongoing surveillance and security measures are needed to protect AI-based networks against advanced threats (Achuthan et al., 2024), where adversaries have the capability to modify data in hidden ways to evade detection (Palani et al., 2024). AI improves threat detection, response times and overall security across different sectors (Marri et al., 2024). Although AI enhances cybersecurity protocols, it also poses novel threats and attacks against AI systems themselves. AI can be integrated with knowledge in organisations for accurate threat analysis (Alevizos & Dekker, 2024). Moreover, with the integration of AI in key infrastructures, stronger security capabilities and regular updates will be needed to ensure that new threats are managed (Hunnewell, 2024). There are other ethical issues such as data privacy and bias which must be addressed as well, for AI in the cyber sector to meet real challenges in security and build resilience in smart cities and Industry 4.0 (Zeadally et al., 2020; Choithani et al., 2024; Velasco, 2022; H. U. Khan et al., 2022). Artificial intelligence ultimately leads to the identification and protection of active threats in real time, ensuring network resilience against cyberattacks.

7 FUTURE DIRECTIONS AND RESEARCH OPPORTUNITIES

The architecture for a scalable self-healing AI security system for Smart Cities and Industry 4.0 must address both the limitations of existing AI systems and leverage emerging technologies. Focusing on the black-box nature of many AI algorithms, explaining AI (XAI) is crucial for ensuring accountability in cybersecurity decision-making. XAI enhancements are also essential for building trust and accountability in AI-based security systems (Guo et al., 2025). Additionally, developing adversarial-proof AI algorithms supported by quantum-safe cryptography and blockchain is vital for enhancing safety and privacy. The integration of AI, big data analytics, and IoT is driving the business world towards increased interconnectedness and data reliance (Rane et al., 2023). This heightened activity and evolving approach to these issues can enhance risk management, cybersecurity, and protection measures against cyber-attacks.

Improvement of intrusion detection and anomaly detection is crucial to defend against higher levels of cyber-attacks, especially concerning smart cities and Industry 4.0. This could also prove useful for better security mechanisms through the study of using graph neural networks and predictive analytics for malware detection. Adopting quantum-resistant cryptography also ensures that the system remains secure over time in an era of continuous quantum computing threats (Table 1). XAI's role in establishing trust among stakeholders and promoting transparency is emphasised (Mendes & Rios, 2023). If such explanations are offered, they facilitate a stronger degree of accountability and supervision of AI systems (Hamida et al., 2024). Blockchain, smart contracts, trustable oracles, decentralised storage and more secure and interpretable AI are all being explored. Therefore, future research may focus on the exploration of XAI methodologies to improve and comprehend security decisions. Transparency and validation also facilitate trust, but are problematic to deploy (Rozario & Čevora, 2023). Artificial Intelligence has a vital role in Cybersecurity field, such as in threat detection, response times, response capabilities, and system enhancements. Developing scalable, self-healing AI models which can autonomously identify and resolve incidents significantly minimizes downtime, enhancing continuous operations.

Finance and technology field have a great impact by Explainable AI (XAI) leading to responsible deployment via focusing ethical concerns such as transparency, privacy and fairness (Rane et al., 2023). This comprehensive approach empowers proactive security and threat detection, enhancing resilience via the immediate cyber threats detection.

Here researchers shown experimental results demonstrating XAI's capabilities in safety-critical fields, involving medical services (Sadeghi et al., 2024). Given the evolving threat posed via quantum computing, integrating quantum-resistant cryptographic algorithms into AI systems becomes paramount for long-term security maintenance.

Scalable and self-healing AI security frameworks for Smart City and Industry 4.0 adopt a multi-faceted approach, leveraging new technologies while addressing existing limitations. This research emphasizes on creating adversarial-resistant AI algorithms and enhancing transparency and trustworthiness in these frameworks (Mandala, 2023; Sharma et al., 2024).

Combining quantum-secured cryptography and blockchain technologies further enhances security and privacy, leading the way for broader deployment in critical applications. However, AI models can undermine trust when they lack transparency in decision-making processes. This underscores the urgent need for explainable AI to clarify actions and foster confidence among healthcare professionals and other users (Clement et al., 2023; Adeniran et al., 2024). Moreover, ethical considerations like data privacy, fairness, and transparency should be integral to the AI design process.

Table 1. Scalable Self-Healing AI Security Architecture for Smart Cities and Industry 4.0

Numbers.	Theme	Summary Description
1	Introduction: The Need for Scalable AI Security in Smart Cities and Industry 4.0	Due to increasing complexity in Smart Cities and Industry 4.0, with interconnected IoT systems and data-driven infrastructures, there is rising necessity of the AI-powered, scalable, and self-healing cybersecurity solutions to address the challenges posed by these systems (Rane et al., 2023).
2	Limitations of Current AI Systems	The "black-box" problem in AI algorithms poses a significant challenge in cybersecurity decision-making, making difficult interpretability and accountability. This lack of transparency leads to trust issues and ethical concerns. (Guo et al., 2025).
3	Explainable AI (XAI) and Accountability in Cybersecurity	Explainable AI (XAI) facilitates transparency, interpretability, and stakeholder trust, ensuring that AI-driven security architectures can justify their actions and decisions responsibly (Mendes & Rios, 2023; Hamida et al., 2024).

continued on following page

Table 1. Continued

Numbers.	Theme	Summary Description
4	Adversarial-Resilient AI and Quantum-Safe Cryptography	Here researchers address the development of adversarial-protected AI algorithms, enriched with quantum-resistant cryptography, to protect against complex cyber and quantum-based threats (Mandala, 2023; Sharma et al., 2024).
5	Blockchain and Decentralised Trust Mechanisms	Here authors explored the blockchain importance, smart contracts, decentralized storage, and trustworthy oracles in maintaining transparency, integrity, and privacy within AI-powered cybersecurity ecosystems. (Rozario & Čevora, 2023).
6	Big Data, IoT, and Predictive Security Analytics	This study explores about how AI, IoT, and big data analytics are empowering anomaly detection, intrusion prevention, and malware prediction utilizing graph neural networks and predictive analytics (Rane et al., 2023).
7	Self-Healing and Autonomous Response Mechanisms	Self-healing AI models capable of autonomously identifying, responding to, and resolving security incidents may minimize downtime and facilitates continuous protection (Guo et al., 2025; Hamida et al., 2024).
8	Ethical and Responsible AI Deployment	Emphasizes incorporating ethical principles such as fairness, transparency, and privacy into AI design, ensuring responsible and accountable use in cybersecurity and industrial applications (Clement et al., 2023; Adeniran et al., 2024).
9	Applications Across Domains	Here authors reviewed implementations of XAI in high-risk domains such as finance, healthcare, and industrial automation, emphasizing the practical benefits of explainability and resilience (Sadeghi et al., 2024).
10	Future Research Directions	Authors explored to focus on scalable XAI methodologies, quantum-safe algorithms, and blockchain-integrated architectures to develop transparent, resilient, and sustainable AI security systems for the future (Guo et al., 2025; Rozario & Čevora, 2023).

8. CONCLUSION

In smart cities and Industry 4.0, AI-enabled communication holds immense potential for enhancing security, efficiency, and resilience. Ongoing research in this field aims to introduce AI to improve secure communication in these environments. Managing the complexities of securing access to AI-enabled communication is crucial to unlocking the benefits of these segments. Companies can leverage blockchain technology to implement secure digital infrastructure and secure AI systems (Shamsan Saleh, 2024). For instance, blockchain addresses challenges such as data security, algorithmic bias, and transparency (Li et al., 2024).

As technology and commerce grapple with security and privacy concerns, threats and requirements beyond compliance necessitate technological and business solutions. Blockchain, in conjunction with AI, addresses issues related to standards, interoperability, vulnerabilities in smart contracts, deterministic execution, and governance (Singh et al., 2020). This innovative approach has the potential to revolutionize data analytics, AI, security, and related fields. Blockchain's decentralization and cryptographic guarantees complement AI's ability to generate patterns and conduct data processing (Singh et al., 2020; Charles et al., 2023). This synergy results in secure, transparent, and data-sharing and decision-making systems (Wang et al., 2021). The future of AI and blockchain holds the promise of transformative, multi-layered solutions for healthcare, including clinical trials. For example, blockchain can enhance data management, leading to greater efficiency, transparency, and integrity through critical frameworks for data privacy and protection and regulatory compliance (Leiva & Castro, 2025).

REFERENCES

Abdel Hakeem, S. A., Hussein, H. H., Kim, H., Hakeem, S. A. A., Hussein, H. H., & Kim, H. (2022). Security Requirements and Challenges of 6G Technologies and Applications. *Sensors (Basel)*, *22*(5), 1969. DOI: 10.3390/s22051969 PMID: 35271113

Abdou, A., Sheatsley, R., Beugin, Y., Shipp, T. J., & McDaniel, P. (2021). HoneyModels: Machine Learning Honeypots. MILCOM 2022 - 2022 IEEE Military Communications Conference (MILCOM), 886–891. DOI: 10.1109/MILCOM52596.2021.9652947

Achuthan, K., Ramanathan, S., Srinivas, S., & Raman, R. (2024). Advancing cybersecurity and privacy with artificial intelligence: Current trends and future research directions. *Frontiers in Big Data*, *7*, 1497535. Advance online publication. DOI: 10.3389/fdata.2024.1497535 PMID: 39703783

Adelakun, B. O., Majekodunmi, T. G., & Akintoye, O. S. (2024). AI and ethical accounting: Navigating challenges and opportunities. *International Journal of Advanced Economics*, *6*(6), 224–241. DOI: 10.51594/ijae.v6i6.1230

Adeniran, A. A., Onebunne, A. P., & William, P. (2024). Explainable AI (XAI) in healthcare: Enhancing trust and transparency in critical decision-making. *World Journal of Advanced Research and Reviews*, *23*(3), 2447–2658. DOI: 10.30574/wjarr.2024.23.3.2936

Ahmad, K., Maabreh, M., Ghaly, M., Khan, K., Qadir, J., & Al-Fuqaha, A. (2020). Developing Future Human-Centered Smart Cities: Critical Analysis of Smart City Security, Interpretability, and Ethical Challenges. ArXiv (Cornell University). https://doi.org//arxiv.2012.09110DOI: 10.48550

Akinrinmade, A. O., Adebile, T. M., Ezuma-Ebong, C., Bolaji, K., Ajufo, A., Adigun, A. O., Mohammad, M., Dike, J. C., & Okobi, O. E. (2023). Artificial Intelligence in Healthcare: Perception and Reality. In Cureus. Cureus, Inc. DOI: 10.7759/cureus.45594

Alenezi, M., & Akour, M. (2025). AI-Driven Innovations in Software Engineering: A Review of Current Practices and Future Directions. In Applied Sciences (Switzerland) (Vol. 15, Issue 3, p. 1344). Multidisciplinary Digital Publishing Institute. DOI: 10.3390/app15031344

Alevizos, L., & Dekker, M. (2024). *Towards an AI-Enhanced Cyber Threat Intelligence Processing Pipeline. ArXiv*. Cornell University., DOI: 10.3390/electronics13112021

Alexandrovich, S. A. (2024). Artificial Intelligence and International Security. *International Affairs*, *70*(5), 254–256. DOI: 10.21557/IAF.99985268

Ali, M., Karimipour, H., & Tariq, M. A. (2021). Integration of blockchain and federated learning for Internet of Things: Recent advances and future challenges. *Computers & Security*, *108*, 102355. DOI: 10.1016/j.cose.2021.102355

Aljabri, M., Aljameel, S. S., Mohammad, R. M. A., Almotiri, S. H., Mirza, S., Anis, F. M., Aboulnour, M., Alomari, D. M., Alhamed, D. H., & Altamimi, H. S. (2021). Intelligent techniques for detecting network attacks: Review and research directions. In Sensors (Vol. 21, Issue 21, p. 7070). Multidisciplinary Digital Publishing Institute. DOI: 10.3390/s21217070

Aluwala, A. (2024). AI-Driven Anomaly Detection in Network Monitoring Techniques and Tools. Journal of Artificial Intelligence & Cloud Computing, 1–6. DOI: 10.47363/JAICC/2024(3)310

Anandaram, H., Shreenidhi, K. S., Joshi, K., Sidhu, K. S., Kiruthigha, R., Singh, V., & Tripathi, A. (2026). A Case Study on Resource Management Industrial IoT Application. In *Building Data-Driven Edge Systems for Business Success* (pp. 1–26). IGI Global Scientific Publishing.

Bai, Z. (2022). Research on Application of Artificial Intelligence in Communication Network. *Journal of Physics: Conference Series*, *2209*(1), 12014. DOI: 10.1088/1742-6596/2209/1/012014

Banerjee, S., Sahu, P., Luo, M., Vahldiek-Oberwagner, A., Yadwadkar, N. J., & Tiwari, M. (2024). SoK: A Systems Perspective on Compound AI Threats and Countermeasures. ArXiv (Cornell University). https://doi.org//arxiv.2411.13459DOI: 10.48550

Barbeau, M., & García-Alfaro, J. (2022). Cyber-physical defense in the quantum Era. *Scientific Reports*, *12*(1), 1905. Advance online publication. DOI: 10.1038/s41598-022-05690-1 PMID: 35115576

Benzaïd, C., & Taleb, T. (2022). AI for Beyond 5G Networks: A Cyber-Security Defense or Offense Enabler? ArXiv (Cornell University), 34(6), 140–147. https://doi.org//arxiv.2201.02730DOI: 10.48550

Bhatt, U., Xiang, A., Sharma, S., Weller, A., Taly, A., Jia, Y., Ghosh, J., Puri, R., Moura, J. M. F., & Eckersley, P. (2020). Explainable machine learning in deployment. FAT* 2020 - Proceedings of the 2020 Conference on Fairness, Accountability, and Transparency, 648–657. DOI: 10.1145/3351095.3375624

Binhammad, M., Alqaydi, S., Othman, A., & Abuljadayel, L. H. (2024). The Role of AI in Cyber Security: Safeguarding Digital Identity. *Journal of Information Security*, *15*(02), 245–278. DOI: 10.4236/jis.2024.152015

Biplob, M. B., Konika, M. S., Ahsan, K. M. M., Zannat, T., & Ahmed, A. (2024). AI's Role in Fortifying Cyber Defenses: A 2024 Perspective on Machine and Deep Learning Applications. SSRN Electronic Journal. DOI: 10.20944/preprints202409.0550.v1

Bothra, P., Karmakar, R., Bhattacharya, S., & De, S. (2023). How can applications of blockchain and artificial intelligence improve performance of Internet of Things? – A survey. *Computer Networks*, *224*, 109634. Advance online publication. DOI: 10.1016/j.comnet.2023.109634

Brij, M. (2023). The Ethics of Artificial Intelligence in Legal Decision Making: An Empirical Study. *Psychologyandeducation*, *55*(01). Advance online publication. DOI: 10.48047/pne.2018.55.1.38

Charles, V., Emrouznejad, A., & Gherman, T. (2023). A critical analysis of the integration of blockchain and artificial intelligence for supply chain. *Annals of Operations Research*, *327*(1), 7–47. DOI: 10.1007/s10479-023-05169-w PMID: 36718465

Chaudhary, S., Singh, V., Malik, N., Sidhu, K. S., Kirola, M., & Joshi, K. (2024, October). Digital Signature Security Enhancement: Using X-OR Operation on RSA and AES with 4096 Bits Key Length. In *2024 International Conference on Computing, Sciences and Communications (ICCSC)* (pp. 1-6). IEEE. DOI: 10.1109/ICCSC62048.2024.10830431

Chen, C., Liu, Z., Jiang, W. M. W. L. H., Goh, S. Q., & Lam, K. (2024). Trustworthy, Responsible, and Safe AI: A Comprehensive Architectural Framework for AI Safety with Challenges and Mitigations. ArXiv (Cornell University). https://doi.org//arxiv.2408.12935DOI: 10.48550

Cheong, B. C. (2024). Transparency and accountability in AI systems: Safeguarding wellbeing in the age of algorithmic decision-making. *Frontiers in Human Dynamics*, *6*, 1421273. Advance online publication. DOI: 10.3389/fhumd.2024.1421273

Choithani, T., Chowdhury, A., Patel, S., Patel, P., Patel, D., & Shah, M. (2024). A Comprehensive Study of Artificial Intelligence and Cybersecurity on Bitcoin, Crypto Currency and Banking System. *Annals of Data Science*, *11*(1), 103–135. DOI: 10.1007/s40745-022-00433-5 PMID: 40479325

Choung, H., David, P., & Ross, A. (2022). Trust in AI and Its Role in the Acceptance of AI Technologies. *International Journal of Human-Computer Interaction, 39*(9), 1727–1739. DOI: 10.1080/10447318.2022.2050543

Clement, T., Kemmerzell, N., Abdelaal, M., & Amberg, M. (2023). XAIR: A Systematic Metareview of Explainable AI (XAI) Aligned to the Software Development Process. *Machine Learning and Knowledge Extraction, 5*(1), 78–108. DOI: 10.3390/make5010006

Das, A., & Rad, P. (2020). Opportunities and Challenges in Explainable Artificial Intelligence (XAI): A Survey. ArXiv (Cornell University). https://doi.org//arxiv.2006.11371DOI: 10.48550

Das, R., & Sandhane, R. (2021). Artificial Intelligence in Cyber Security. *Journal of Physics: Conference Series, 1964*(4), 42072. DOI: 10.1088/1742-6596/1964/4/042072

Dhir, N., Hoeltgebaum, H., Adams, N., Briers, M., Burke, A., & Jones, P. (2021). Prospective Artificial Intelligence Approaches for Active Cyber Defence. ArXiv (Cornell University). https://doi.org//arxiv.2104.09981DOI: 10.48550

Dhopte, A., & Bagde, H. (2023). Smart Smile: Revolutionizing Dentistry With Artificial Intelligence. In Cureus. Cureus, Inc. DOI: 10.7759/cureus.41227

Ding, X., Shang, B., Xie, C., Xin, J., & Yu, F. (2025). Artificial intelligence in the COVID-19 pandemic: Balancing benefits and ethical challenges in China's response. *Humanities & Social Sciences Communications, 12*(1), 245. Advance online publication. DOI: 10.1057/s41599-025-04564-x

Fnu, H., Kaushik, K., Gaur, P., Saratchandran, D. V., Thapliyal, A. G., Sidhu, K. S., & Singh, V. (2025, April). Multi-Agent Systems for Collaborative Financial Decision-Making Over Distributed Network Architectures. In 2025 3rd International Conference on Advancement in Computation & Computer Technologies (InCACCT) (pp. 939-944). IEEE. DOI: 10.1109/InCACCT65424.2025.11011435

Guo, W., Potter, Y., Shi, T., Wang, Z., Zhang, A., & Song, D. (2025). Frontier AI's Impact on the Cybersecurity Landscape. ArXiv (Cornell University). DOI: 10.48550/ARXIV.2504.05408

Haider, N., Baig, M. Z., & Imran, M. (2020). Artificial Intelligence and Machine Learning in 5G Network Security: Opportunities, advantages, and future research trends. ArXiv (Cornell University). https://doi.org//arxiv.2007.04490DOI: 10.48550

Hamida, S. U., Chowdhury, M. J. M., Chakraborty, N. R., Biswas, K., & Sami, S. K. (2024). Exploring the Landscape of Explainable Artificial Intelligence (XAI): A Systematic Review of Techniques and Applications. In Big Data and Cognitive Computing (Vol. 8, Issue 11, p. 149). Multidisciplinary Digital Publishing Institute. DOI: 10.3390/bdcc8110149

Hamon, R., & Junklewitz, H. (2020). *Robustness and Explainability of Artificial Intelligence From technical to policy solutions. EUR 30040 EN*. Publications Office of the European Union., DOI: 10.2760/57493

Han, Y., Chen, J., Dou, M., Wang, J., & Feng, K. (2023). The Impact of Artificial Intelligence on the Financial Services Industry. *Deleted Journal*, *2*(3), 83–85. DOI: 10.54097/ajmss.v2i3.8741

Hildt, E. (2025). What Is the Role of Explainability in Medical Artificial Intelligence? A Case-Based Approach. *Bioengineering (Basel, Switzerland)*, *12*(4), 375. DOI: 10.3390/bioengineering12040375 PMID: 40281735

Huang, L. (2023). Ethics of Artificial Intelligence in Education: Student Privacy and Data Protection. *Science Insights Education Frontiers*, *16*(2), 2577–2587. DOI: 10.15354/sief.23.re202

Hunnewell, B. (2024). National Security Concerns for Artificial Intelligence and Civilian Critical Infrastructure. *SAIS Review (Paul H. Nitze School of Advanced International Studies)*, *44*(2), 31–46. DOI: 10.1353/sais.2024.a950955

Islam, C., Babar, M. A., Croft, R., & Janicke, H. (2022). SmartValidator: A Framework for Automatic Identification and Classification of Cyber Threat Data. ArXiv (Cornell University). https://doi.org//arxiv.2203.07603DOI: 10.48550

Jang, Y., Park, C.-H., Lee, D., & Seo, Y. (2021). Fake News Detection on Social Media: A Temporal-Based Approach. Computers, Materials & Continua/Computers. *Computers, Materials & Continua*, *69*(3), 3563–3579. DOI: 10.32604/cmc.2021.018901

Jha, D., Rauniyar, A., Srivastava, A., Hagos, D. H., Tomar, N. K., Sharma, V., Keleş, E., Zhang, Z., Demir, U., Topcu, A. E., Yazidi, A., Håakegård, J. E., & Bağcı, U. (2023). Ensuring Trustworthy Medical Artificial Intelligence through Ethical and Philosophical Principles. ArXiv (Cornell University). https://doi.org//arxiv.2304.11530DOI: 10.48550

Jones, R., Omar, M., Mohammed, D., Nobles, C., & Dawson, M. (2023). Harnessing the Speed and Accuracy of Machine Learning to Advance Cybersecurity. 2023 Congress in Computer Science, Computer Engineering, & Applied Computing (CSCE). /arxiv.2302.12415DOI: 10.1109/CSCE60160.2023.00074

Joshi, K., Singh, V., Awasthi, P., Anandaram, H., Sidhu, K. S., & Kumar, P. (2024, November). Smart Dam Water Monitoring using Enabled Internet of Things. In 2024 5th International Conference on Data Intelligence and Cognitive Informatics (ICDICI) (pp. 239-243). IEEE. DOI: 10.1109/ICDICI62993.2024.10810761

Joshi, K., Yadav, H., Gupta, S., Singh, V., Sidhu, K. S., & Kukreti, R. (2025, April). Handling Security Aspects in the Internet of Things: Latest Challenges and Measures to Mitigate Risks. In 2025 3rd International Conference on Communication, Security, and Artificial Intelligence (ICCSAI) (Vol. 3, pp. 1434-1439). IEEE.

Joshi, P., Singh, V., Thapliyal, N., Joshi, K., Bhatt, A., & Mahur, M. (2024, October). Intelligent IoT-Enabled Real-Time Monitoring System for Logistics Management. In *2024 International Conference on Computing, Sciences and Communications (ICCSC)* (pp. 1-5). IEEE. DOI: 10.1109/ICCSC62048.2024.10830332

Kale, A., Nguyen, T., Harris, J. S., Li, C., Zhang, J., Ma, X., Harris, F. C., Li, C., Zhang, J., & Ma, X. (2022). Provenance documentation to enable explainable and trustworthy AI: A literature review. In Data Intelligence (Vol. 5, Issue 1, pp. 139–162). The MIT Press. DOI: 10.1162/dint_a_00119

Katiyar, D. N., Tripathi, M. S., Kumar, M. P., Verma, M. S., Sahu, D. A. K., & Saxena, D. S. (2024). AI and Cyber-Security: Enhancing threat detection and response with machine learning. Educational Administration Theory and Practices, 30(4 SE-Articles), 6273–6282. DOI: 10.53555/kuey.v30i4.2377

Kaur, P., Kaur, K., Verma, D. K., Singh, K., Kaushik, K., & Singh, V. (2024, September). Routing protocols in wireless sensor networks: A comprehensive review and future perspectives. In 2024 7th International Conference on Contemporary Computing and Informatics (IC3I) (Vol. 7, pp. 263-270). IEEE. DOI: 10.1109/IC3I61595.2024.10828894

Kawamoto, Y., Miyake, K., Konishi, K., & Oiwa, Y. (2023). Threats, Vulnerabilities, and Controls of Machine Learning Based Systems: A Survey and Taxonomy. ArXiv (Cornell University). https://doi.org//arxiv.2301.07474DOI: 10.48550

Khan, H. U., Malik, M. Z., Alomari, M. K., Khan, S., Al-Maadid, A., Hassan, M. K., & Khan, K. (2022). Transforming the Capabilities of Artificial Intelligence in GCC Financial Sector: A Systematic Literature Review. *Wireless Communications and Mobile Computing*, *2022*(1), 1–17. DOI: 10.1155/2022/8725767

Khan, N., Ahmad, K., Al Tamimi, A., Alani, M. M., Bermak, A., & Khalil, I. (2024). Explainable AI-based Intrusion Detection System for Industry 5.0: An Overview of the Literature, associated Challenges, the existing Solutions, and Potential Research Directions. ArXiv (Cornell University). https://doi.org//arxiv.2408.03335DOI: 10.48550

Khanji, S., & Khattak, A. (2020). Towards a Novel Intrusion Detection Architecture using Artificial Intelligence. *ACM International Conference Proceeding Series*, 185–189. DOI: 10.1145/3436829.3436842

Khera, R., Oikonomou, E. K., Nadkarni, G. N., Morley, J. R., Wiens, J., Butte, A. J., & Topol, E. J. (2024). Transforming Cardiovascular Care With Artificial Intelligence: From Discovery to Practice. *Journal of the American College of Cardiology*, *84*(1), 97–114. DOI: 10.1016/j.jacc.2024.05.003 PMID: 38925729

Kim, K., Alshenaifi, I. M., Ramachandran, S., Kim, J., Zia, T., & Almorjan, A. (2023). Cybersecurity and Cyber Forensics for Smart Cities: A Comprehensive Literature Review and Survey. *Sensors (Basel)*, *23*(7), 3681. DOI: 10.3390/s23073681 PMID: 37050740

Kim, S.-K. (2025). Enhanced Blockchain-Based Data Poisoning Defense Mechanism. *Applied Sciences (Basel, Switzerland)*, *15*(7), 4069. DOI: 10.3390/app15074069

Kornaros, G. (2022). Hardware-Assisted Machine Learning in Resource-Constrained IoT Environments for Security: Review and Future Prospective. *IEEE Access : Practical Innovations, Open Solutions*, *10*, 58603–58622. DOI: 10.1109/ACCESS.2022.3179047

Korobenko, D., Nikiforova, A., & Sharma, R. (2024). Towards a Privacy and Security-Aware Framework for Ethical AI: Guiding the Development and Assessment of AI Systems. *ACM International Conference Proceeding Series*, 740–753. DOI: 10.1145/3657054.3657141

Krishna, G., Singh, V., Pandey, D., Krishna, K. H., Joshi, K., & Gupta, T. (2024, May). Power trading framework of cloud-edge computing in the artificial intelligence market. In 2024 4th International Conference on Advance Computing and Innovative Technologies in Engineering (ICACITE) (pp. 1910-1918). IEEE. DOI: 10.1109/ICACITE60783.2024.10617344

Leiva, V., & Castro, C. (2025). Artificial intelligence and blockchain in clinical trials: enhancing data governance efficiency, integrity, and transparency. In *Bioanalysis* (pp. 1–16). Future Science Ltd., DOI: 10.1080/17576180.2025.2452774

Li, C., Guo, W., Sun, S. C., Al–Rubaye, S., & Tsourdos, A. (2020). Trustworthy Deep Learning in 6G-Enabled Mass Autonomy: From Concept to Quality-of-Trust Key Performance Indicators. *IEEE Vehicular Technology Magazine*, *15*(4), 112–121. DOI: 10.1109/MVT.2020.3017181

Li, Y., Wu, B., Huang, Y., & Luan, S. (2024). Developing trustworthy artificial intelligence: insights from research on interpersonal, human-automation, and human-AI trust. In *Frontiers in Psychology* (Vol. 15). Frontiers Media., DOI: 10.3389/fpsyg.2024.1382693

Liang, S., Ma, J., Wang, G., Shao, J., Li, J., Deng, H., Wang, C., & Li, W. (2022). The Application of Artificial Intelligence in the Diagnosis and Drug Resistance Prediction of Pulmonary Tuberculosis. In *Frontiers in Medicine* (Vol. 9). Frontiers Media., DOI: 10.3389/fmed.2022.935080

Liu, C., Salzmann, M., Lin, T., Tomioka, R., & Süsstrunk, S. (2020). On the loss landscape of adversarial training: Identifying challenges and how to overcome them. Advances in Neural Information Processing Systems, 2020-December. DOI: 10.48550/arxiv.2006.08403

Lunghi, D., Simitsis, A., Caelen, O., & Bontempi, G. (2023). Adversarial Learning in Real-World Fraud Detection: *Challenges and Perspectives. ACM International Conference Proceeding Series*, 27–33. DOI: 10.1145/3600046.3600051

Mandala, S. K. (2023). XAI Renaissance: Redefining Interpretability in Medical Diagnostic Models. ArXiv (Cornell University). https://doi.org//arxiv.2306.01668DOI: 10.48550

Manresa-Yee, C., & Ramis, S. (2021). Assessing Gender Bias in Predictive Algorithms using eXplainable AI. *ACM International Conference Proceeding Series*, 1–8. DOI: 10.1145/3471391.3471420

Marri, R., Dabbara, L. N., & Karampuri, S. (2024). AI security in different industries: A comprehensive review of vulnerabilities and mitigation strategies. In International Journal of Science and Research Archive (Vol. 13, Issue 1, pp. 2375–2393). DOI: 10.30574/ijsra.2024.13.1.1923

Mendes, C. R. B., & Rios, T. N. (2023). Explainable Artificial Intelligence and Cybersecurity: A Systematic Literature Review. ArXiv (Cornell University). https://doi.org//arxiv.2303.01259DOI: 10.48550

Mennella, C., Maniscalco, U., De Pietro, G., & Esposito, M. (2024). Ethical and regulatory challenges of AI technologies in healthcare: A narrative review. In Heliyon (Vol. 10, Issue 4). Elsevier BV. DOI: 10.1016/j.heliyon.2024.e26297

Mirestean, A., Farias, A., Deodoro, J., Boukherouaa, E. B., AlAjmi, K., Iskender, E., Ravikumar, R., & Shabsigh, G. (2021). Powering the Digital Economy: Opportunities and Risks of Artificial Intelligence in Finance. *Departmental Paper, 2021*(24), 1. DOI: 10.5089/9781589063952.087

Montagnani, M. L., Najjar, M.-C., & Davola, A. (2024). The EU Regulatory approach(es) to AI liability, and its Application to the financial services market. *Computer Law & Security Review, 53*, 105984. DOI: 10.1016/j.clsr.2024.105984

Morley, J., Machado, C. C. V., Burr, C., Cowls, J., Joshi, I., Taddeo, M., & Floridi, L. (2020). The ethics of AI in health care: A mapping review. In *Social Science & Medicine* (Vol. 260, p. 113172). Elsevier BV., DOI: 10.1016/j.socscimed.2020.113172

Mudgal, S. K., Agarwal, R., Chaturvedi, J., Gaur, R., & Ranjan, N. (2022). Real-world application, challenges and implication of artificial intelligence in healthcare: an essay. In *Pan African Medical Journal* (Vol. 43, p. 3). National Institutes of Health., DOI: 10.11604/pamj.2022.43.3.33384

Muhammad, D., & Bendechache, M. (2024). Unveiling the black box: A systematic review of Explainable Artificial Intelligence in medical image analysis. In *Computational and Structural Biotechnology Journal* (Vol. 24, pp. 542–560). Elsevier BV., DOI: 10.1016/j.csbj.2024.08.005

Naik, N., Hameed, B. M. Z., Shetty, D. K., Swain, D., Shah, M., Paul, R., Aggarwal, K., Brahim, S., Patil, V., Smriti, K., Shetty, S., Rai, B. P., Chlosta, P., & Somani, B. K. (2022). Legal and Ethical Consideration in Artificial Intelligence in Healthcare: Who Takes Responsibility? *Frontiers in Surgery*, *9*, 862322. Advance online publication. DOI: 10.3389/fsurg.2022.862322 PMID: 35360424

Narooka, P., Kaushik, K., Kansal, S., Kejriwal, D., Mishra, R. K., & Singh, V. (2025, April). Hybrid AI Models for QoS Estimation in α-η-μ Fading and Co-Channel Interference Environments within 6G Networks. In 2025 3rd International Conference on Advancement in Computation & Computer Technologies (InCACCT) (pp. 206-210). IEEE.

Narooka, P., Verma, S., Dwivedi, A., Kaushik, K., Singh, V., & Gaur, P. (2025, April). Psi-zeta Approach for Assessing Network Efficiency in Log-Normal Fading and Spectrum Sharing Interference Environments. In 2025 3rd International Conference on Advancement in Computation & Computer Technologies (InCACCT) (pp. 886-891). IEEE. DOI: 10.1109/InCACCT65424.2025.11011391

Nguyen, T. T., & Reddi, V. J. (2023). Deep Reinforcement Learning for Cyber Security. *IEEE Transactions on Neural Networks and Learning Systems*, *34*(8), 3779–3795. DOI: 10.1109/TNNLS.2021.3121870 PMID: 34723814

Oseni, A., Moustafa, N., Janicke, H., Liu, P., Tari, Z., & Vasilakos, A. V. (2021). Security and Privacy for Artificial Intelligence: Opportunities and Challenges. ArXiv (Cornell University). https://doi.org//arxiv.2102.04661DOI: 10.48550

Palani, K., Kethar, J., Prasad, S. S., & Torremocha, V. (2024). Impact of AI and Generative AI in transforming Cybersecurity. *Journal of Student Research*, *13*(2). Advance online publication. DOI: 10.47611/jsrhs.v13i2.6710

Patil, D. (2025). Artificial Intelligence in Financial Risk Assessment and Fraud Detection: Opportunities and Ethical Concerns. SSRN Electronic Journal. DOI: 10.2139/ssrn.5057434

Petkovic, D. (2022). It is not "accuracy vs. explainability" — we need both for trustworthy AI systems. ArXiv (Cornell University). https://doi.org//arxiv.2212.11136DOI: 10.48550

Petković, D., & Petkovic, D. (2023). It is Not "Accuracy vs. Explainability"—We Need Both for Trustworthy AI Systems. *IEEE Transactions on Technology and Society*, *4*(1), 46–53. DOI: 10.1109/TTS.2023.3239921

Pi, Y. (2023). Algorithmic Governance for Explainability: A Comparative Overview of Progress and Trends. ArXiv (Cornell University). https://doi.org//arxiv.2303.00651DOI: 10.48550

Pochmara, J., & Świetlicka, A. (2024). Cybersecurity of Industrial Systems—A 2023 Report. *Electronics (Basel)*, *13*(7), 1191. DOI: 10.3390/electronics13071191

Polemi, N., Praça, I., Kioskli, K., & Bécue, A. (2024). Challenges and efforts in managing AI trustworthiness risks: A state of knowledge. *Frontiers in Big Data*, *7*, 1381163. Advance online publication. DOI: 10.3389/fdata.2024.1381163 PMID: 38798307

Prasanth, A., Vadakkan, D. J., Surendran, P., & Thomas, B. (2023). Role of Artificial Intelligence and Business Decision Making. *International Journal of Advanced Computer Science and Applications*, *14*(6). Advance online publication. DOI: 10.14569/IJACSA.2023.01406103

Rane, N. L., Choudhary, S., & Rane, J. (2023). Explainable Artificial Intelligence (XAI) Approaches for Transparency and Accountability in Financial Decision-Making. SSRN Electronic Journal. https://doi.org/DOI: 10.2139/ssrn.4640316

Raturi, A., Mula, K. G., Joshi, K., Sidhu, K. S., Chauhan, S. S., & Singh, V. (2025, April). Comparative Analysis of Stock Price Prediction with Multiple Machine Learning Model. In 2025 3rd International Conference on Communication, Security, and Artificial Intelligence (ICCSAI) (Vol. 3, pp. 1428-1433). IEEE. DOI: 10.1109/ICCSAI64074.2025.11064599

Rjoub, G., Bentahar, J., Abdel Wahab, O., Mizouni, R., Song, A., Cohen, R., Otrok, H., & Mourad, A. (2023). A Survey on Explainable Artificial Intelligence for Cybersecurity. *IEEE Transactions on Network and Service Management*, *20*(4), 5115–5140. DOI: 10.1109/TNSM.2023.3282740

Rong, Y., Leemann, T., Nguyen, T., Fiedler, L., Qian, P., Unhelkar, V., Seidel, T., Kasneci, G., & Kasneci, E. (2022). Towards Human-centered Explainable AI: A Survey of User Studies for Model Explanations. ArXiv (Cornell University). https://doi.org//arxiv.2210.11584DOI: 10.48550

Roohani, B. S., Sharma, N., Kasula, V. K., Mamoria, P., Modh, N. N., Kumar, A., & Singh, V. (2026). Urban Computing Solutions in Healthcare Edge Computing. In Building Data-Driven Edge Systems for Business Success (pp. 377-400). IGI Global Scientific Publishing.

Roshanaei, M., Khan, M. R., & Sylvester, N. N. (2024). Enhancing Cybersecurity through AI and ML: Strategies, Challenges, and Future Directions. *Journal of Information Security*, *15*(03), 320–339. DOI: 10.4236/jis.2024.153019

Rozario, S., & Čevora, G. (2023). Explainable AI does not provide the explanations end-users are asking for. ArXiv (Cornell University). https://doi.org//arxiv.2302.11577DOI: 10.48550

Ruzbahani, A. M. (2024). AI-Protected Blockchain-based IoT environments: Harnessing the Future of Network Security and Privacy. ArXiv (Cornell University). https://doi.org//arxiv.2405.13847DOI: 10.48550

Sadeghi, Z., & Alizadehsani, R. CIFCI, M. A., Kausar, S., Rehman, R., Mahanta, P., Bora, P. K., Almasri, A., Alkhawaldeh, R. S., Hussain, S., Alatas, B., Shoeibi, A., Moosaei, H., Hladík, M., Nahavandi, S., & Pardalos, P. M. (2024). A review of Explainable Artificial Intelligence in healthcare. In Computers and Electrical Engineering (Vol. 118). DOI: 10.1016/j.compeleceng.2024.109370

Saifullah, S., Mercier, D., Lucieri, A., Dengel, A., & Ahmed, S. (2022). Privacy Meets Explainability: A Comprehensive Impact Benchmark. ArXiv (Cornell University). https://doi.org//arxiv.2211.04110DOI: 10.48550

Salem, A. H., Azzam, S. M., Emam, O. E., & Abohany, A. A. (2024). Advancing cybersecurity: A comprehensive review of AI-driven detection techniques. *Journal of Big Data*, *11*(1), 105. Advance online publication. DOI: 10.1186/s40537-024-00957-y

Schmitt, M. (2023). Securing the digital world: Protecting smart infrastructures and digital industries with artificial intelligence (AI)-enabled malware and intrusion detection. *Journal of Industrial Information Integration*, *36*, 100520. DOI: 10.1016/j.jii.2023.100520

Shamsan Saleh, A. M. (2024). Blockchain for secure and decentralized artificial intelligence in cybersecurity: A comprehensive review. In Blockchain: Research and Applications (Vol. 5, Issue 3, p. 100193). Elsevier BV. DOI: 10.1016/j.bcra.2024.100193

Shang, Z., Chauhan, V., Devi, K., & Patil, S. (2024). Artificial Intelligence, the Digital Surgeon: Unravelling Its Emerging Footprint in Healthcare – The Narrative Review. In Journal of Multidisciplinary Healthcare (pp. 4011–4022). Dove Medical Press. DOI: 10.2147/JMDH.S482757

Sharma, N. A., Chand, R. R., Buksh, Z., Ali, A. B. M. S., Hanif, A., & Beheshti, A. (2024). Explainable AI Frameworks: Navigating the Present Challenges and Unveiling Innovative Applications. *Algorithms*, *17*(6), 227. DOI: 10.3390/a17060227

Shuaib, A. (2024). Transforming Healthcare with AI: Promises, Pitfalls, and Pathways Forward. *International Journal of General Medicine*, *17*, 1765–1771. DOI: 10.2147/IJGM.S449598 PMID: 38706749

Singh, S., Sharma, P. K., Yoon, B., Shojafar, M., Cho, G. H., & Ra, I. H. (2020). Convergence of blockchain and artificial intelligence in IoT network for the sustainable smart city. *Sustainable Cities and Society*, *63*, 102364. DOI: 10.1016/j.scs.2020.102364

Singh, V., & Dubey, V. (2015). A two level image security based on Arnold transform and chaotic logistic mapping. *International Journal of Advanced Research in Computer Science and Software Engineering*, *5*(2), 883–887.

Singh, V., & Kaushik, V. D. (2023). A Brief Overview of Cyber Security Advances and Techniques Along With a Glimpse on Quantum Cryptography: Cyber Security Practices, Advances and Challenges. Exploring Cyber Criminals and Data Privacy Measures, 40-64.

Singh, V., & Kaushik, V. D. (2023). A Methodical View of Prerequisites of Picture Combination, Strategies, and Key Indicators with Usage in Real Life and Scientific Domains Facilitating Smart Ubiquitous Environment. In *Computational Intelligence in Analytics and Information Systems* (pp. 15–33). Apple Academic Press. DOI: 10.1201/9781003332312-3

Singh, V., & Kaushik, V. D. (2023). Navigating the Landscape of Security Threat Analysis in Cloud Computing environments. In *Security and Risk Analysis for Intelligent Cloud Computing* (pp. 1–25). CRC Press.

Speith, T., & Langer, M. (2023). A New Perspective on Evaluation Methods for Explainable Artificial Intelligence (XAI). ArXiv (Cornell University). /arxiv.2307.14246DOI: 10.1109/REW57809.2023.00061

Suriyan, K., Ganesh, S., Palaniyammal, P., Priya, G., & Singh, V. (2026). Recent Trends in Edge Computing: Challenges and Opportunities. Building Data-Driven Edge Systems for Business Success, 323-344.

Svetlova, E. (2022). AI ethics and systemic risks in finance. *AI and Ethics*, *2*(4), 713–725. DOI: 10.1007/s43681-021-00129-1 PMID: 35043105

Torkey, H., Hashish, S., Souissi, S., Hemdan, E. E. D., & Sayed, A. (2025). Seizure Detection in Medical IoT: Hybrid CNN-LSTM-GRU Model with Data Balancing and XAI Integration. *Algorithms*, *18*(2), 77. DOI: 10.3390/a18020077

Usama, M., Qadir, J., Al-Fuqaha, A., & Hamdi, M. (2020). The Adversarial Machine Learning Conundrum: Can the Insecurity of ML Become the Achilles' Heel of Cognitive Networks? *IEEE Network*, *34*(1), 196–203. DOI: 10.1109/MNET.001.1900197

Velasco, C. (2022). Cybercrime and Artificial Intelligence. An overview of the work of international organizations on criminal justice and the international applicable instruments. ERA Forum, 23(1), 109–126. DOI: 10.1007/s12027-022-00702-z

Vinothkumar, J., & Karunamurthy, A. (2023). Recent Advancements in Artificial Intelligence Technology: Trends and Implications. *Quing International Journal of Multidisciplinary Scientific Research and Development*, *2*(1), 1–11. DOI: 10.54368/qijmsrd.2.1.0003

Vitorino, J., Praça, I., & Maia, E. (2023). Towards adversarial realism and robust learning for IoT intrusion detection and classification. Annales Des Telecommunications. *Annales des Télécommunications*, *78*(7–8), 401–412. DOI: 10.1007/s12243-023-00953-y

Wach, K., Duong, C. D., Ejdys, J., Kazlauskaitė, R., Korzyński, P., Mazurek, G., Paliszkiewicz, J., & Ziemba, E. (2023). The dark side of generative artificial intelligence: A critical analysis of controversies and risks of ChatGPT. *Entrepreneurial Business and Economics Review, 11*(2), 7–30. DOI: 10.15678/EBER.2023.110201

Wang, R., Luo, M., Wen, Y., Wang, L., Choo, K. R., & He, D. (2021). The Applications of Blockchain in Artificial Intelligence. *Security and Communication Networks, 2021*, 1–16. DOI: 10.1155/2021/8690662

Xue, Y., Singh, V., Singh, S., Kant, K., Pandey, S., Kumar, A., & Kaushik, V. D. (2024). Blockchain Framework for Learner Performance Prediction using Life-Brain Storm-based Light GBM Coupled Neural Network: BLOCKCHAIN FRAMEWORK FOR LEARNER PERFORMANCE PREDICTION. [JSIR]. *Journal of Scientific and Industrial Research, 83*(6), 652–668.

Yaacoub, J.-P. A., Noura, H., & Salman, O. (2023). Security of federated learning with IoT systems: Issues, limitations, challenges, and solutions. *Internet of Things and Cyber-Physical Systems, 3*, 155–179. DOI: 10.1016/j.iotcps.2023.04.001

Zeadally, S., Adi, E., Baig, Z., & Khan, I. A. (2020). Harnessing artificial intelligence capabilities to improve cybersecurity. *IEEE Access : Practical Innovations, Open Solutions, 8*, 23817–23837. DOI: 10.1109/ACCESS.2020.2968045

Zhang, X., Hou, H., Zhao, F., & Wang, Z. (2021). Industrial Internet Federated Learning Driven by IoT Equipment ID and Blockchain. Wireless Communications and Mobile Computing, 2021(1). DOI: 10.1155/2021/7705843

Zhang, Z., Al Hamadi, H., Damiani, E., Yeun, C. Y., & Taher, F. (2022). Explainable Artificial Intelligence Applications in Cyber Security: State-of-the-Art in Research. *IEEE Access : Practical Innovations, Open Solutions, 10*, 93104–93139. DOI: 10.1109/ACCESS.2022.3204051

Zhao, G., Li, Y., & Xu, Q. (2022). From Emotion AI to Cognitive AI. International Journal of Network Dynamics and Intelligence, 65–72. DOI: 10.53941/ijndi0101006

Zhou, K. R., & Gattinger, G. (2024). The Evolving Regulatory Paradigm of AI in MedTech: A Review of Perspectives and Where We Are Today. In Therapeutic Innovation & Regulatory Science (Vol. 58, Issue 3, pp. 456–464). Springer Science+Business Media. https://doi.org/DOI: 10.1007/s43441-024-00628-3

Chapter 10
Illuminating Connectivity: A Comprehensive Review of Li-Fi From Architecture to Ecosystem Integration

Richa Yadav
Institute of Technology, Nirma University, India

Shreya Sejani
https://orcid.org/0009-0009-1215-4596
Institute of Technology, Nirma University, India

Bhavya Shah
Institute of Technology, Nirma University, India

Parita Rajiv Oza
https://orcid.org/0000-0001-9637-6157
Institute of Technology, Nirma University, India

Smita Agrawal
https://orcid.org/0000-0003-0486-8045
Institute of Technology, Nirma University, India

Darshana Upadhyay
https://orcid.org/0000-0001-5822-0020
Dalhousie University, Canada

ABSTRACT

Embarking on a comprehensive exploration of Li-Fi technology, this paper thoroughly investigates its architecture, functionality, applications, strengths, and limitations. The primary aim is to dispel misconceptions and foster a deep understanding of the technical intricacies associated with Li-Fi. Covering diverse domains such as healthcare and industrial applications, the research reveals the transformative potential of Li-Fi. Additionally, it delves into the dynamic relationship between Li-Fi and the evolving 5G-6G networks, emphasizing their collective impact on data communication. Through comparative analysis, the paper highlights the distinct attributes of Li-Fi that are set to shape the future landscape of connectivity. Furthermore, it

DOI: 10.4018/979-8-3373-3576-6.ch010

t © 2026, IGI Global Scientific Publishing. Copying or distributing in print or electronic forms without written permission of IGI Global Scientific Publishing is prohibited. Use of this chapter to
:nerative artificial intelligence (AI) technologies is expressly prohibited. The publisher reserves all rights to license its use for generative AI training and machine learning model development.

scrutinizes the convergence of Li-Fi and Wi-Fi into aggregate networks, shedding light on their collaborative strengths in reinforcing connectivity and optimizing data transmission.

1 INTRODUCTION

Nowadays, there is a constant flow of information going around the internet. Without the internet, human life is unimaginable. The world has changed as a result of the increased opportunities brought about by digital information exchange. The one thing we cannot compromise on is data transfer speed. In both their personal and professional lives, individuals are relying more and more on wireless communication, which is expanding rapidly. Even in critical or life-threatening circumstances, people must develop and sustain ties. People that place a high emphasis on unlimited mobility have made wireless communication become the primary method of connection (Koonen et al., 2020). The existing Data Transmission Infrastructure is old and slow. One approach to solving this issue is to use Li-Fi technology. Li-Fi technology enhances the security, precision, and dependability of data exchange via optical traits of an LED. Prof. Harald Hass developed Li-Fi at the University of Edinburgh, Scotland. In July 2011, he gave a TED lecture in which Li-Fi technology was introduced to the audience (Haas et al., 2015). The Li-Fi demonstration was done by demonstrating how the amplitude of light bulbs varies rapidly and makes it possible to transfer energy through light. In workplaces all around the world, it is presently being tested with LEDs and the aeronautical sector is already developing methods to incorporate it into commercial aircraft (Khare et al., 2016). A brand-new kind of wireless communication that uses visible light is known as VLC. Visible light-emitting LEDs are frequently function as transmitters, whereas photodiodes and image sensors are served as receivers. Several brand-new uses have been illustrated for the technology of VLC. Applications for VLC are thought to be particularly well suited for location-based services. It is proposed that LEDs should be used in an interior visible data transmission system. In this system, these elements are utilized to provide an optical wireless communication system besides illumination of spaces (Lian et al., 2019). Recent studies have concentrated on the viability of mobile communication as the central aspect of the role of Li Fi in future 6G networks (Ma et al., 2024).

This review aims to critically analyze the current situation with Li-Fi technology, its main concepts, major advances, and potential applications. We want to ensure that the readers have a clear understanding of how light waves can be employed in transmission of data through discussion of the theoretical basis of Li-Fi. This review will also address the constraints & drawbacks of Li-Fi like the requisite for direct visibility and the risk of signal degradation due to external factors. Li-Fi, as noted by current literature, becomes a key solution to the issues of increased Internet usage and RF spectrum congestion (Ghaderi, 2023). Paul et al. (2024) highlight the potential of Li-Fi technology as a credible and responsive wireless communication tool. The paper also holds significance in fostering Li-Fi within the wireless transmission landscape. Having paid attention to the historical evolution of Li-Fi and its combination with new technologies, this review tries to provide researchers, engineers and enthusiasts with a complete knowledge of the current and prospects of Li-Fi systems. With this review, we hope to illuminate the path to faster, more effective wireless communication that also makes use of the radiant light itself.

Figure 1. PRISMA and Meta-Analyses distribution of reference papers

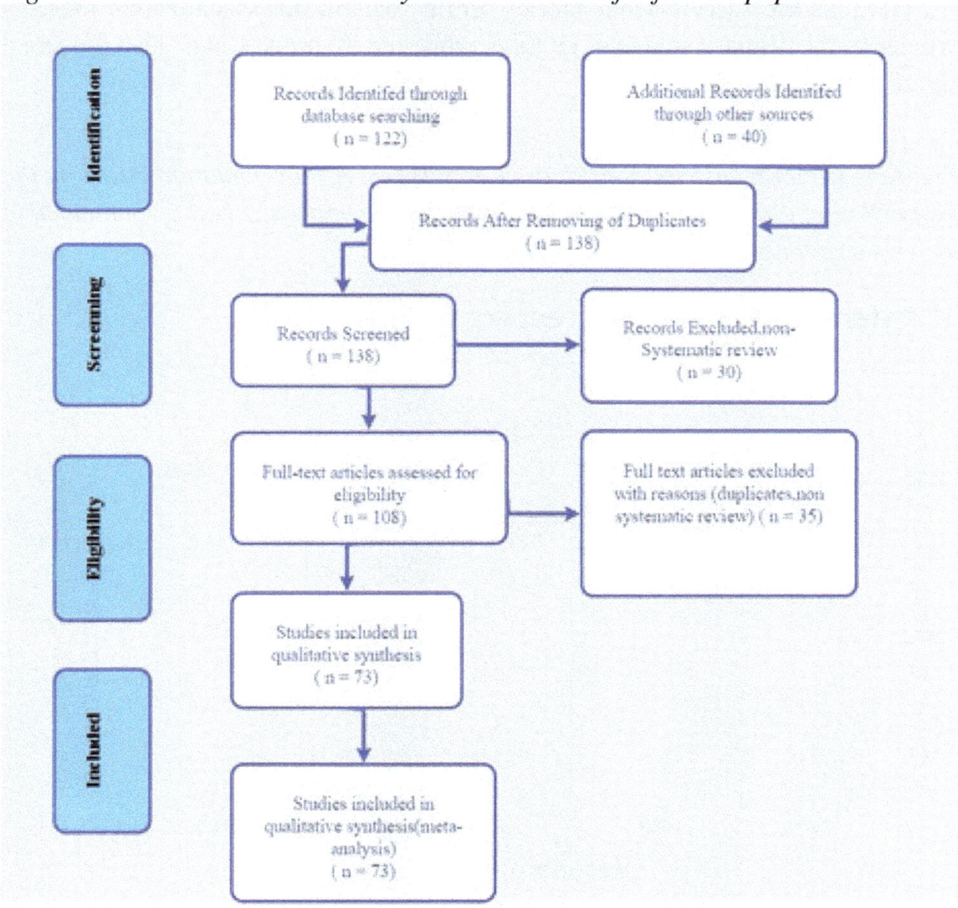

PRISMA or Preferred Reporting Items of Systematic Reviews and Meta-Analyses was applied in our review paper to assist us in navigating through our systematic review process that is shown in figure 1. We performed all the stages of the review, including the literature search, data synthesis, etc., with precision following PRISMA guidelines. This assisted in ensuring that a systematic approach was taken to ensure that enough coverage of the relevant studies was made and at the same time limiting bias as much as possible in order to ensure transparency. At some point, on the one hand, we concluded that our compliance with PRISMA made our study quite robust and valid, and on the other hand, it contributed to providing valuable information to the academic and practical applications.

Figure 2 illustrates the bibliography layout for the research literature for different types of entries. 'Article' entry type is outlined in the main-body citations with approximately 40 to 45 matched instances. The related term" In proceedings" holds the second place with 20-25 citations, while a dot is the clear indicator of only 1-2 citations for miscellaneous entries. In this regard, this visualization presents articles as the primary source of citation, while the 'in proceedings' also provide a significant contribution.

Figure 2. Citation Distribution Across Entry Types: A representation based on the proportion of different research paper entry types, emphasizing the dominance of articles as primary sources

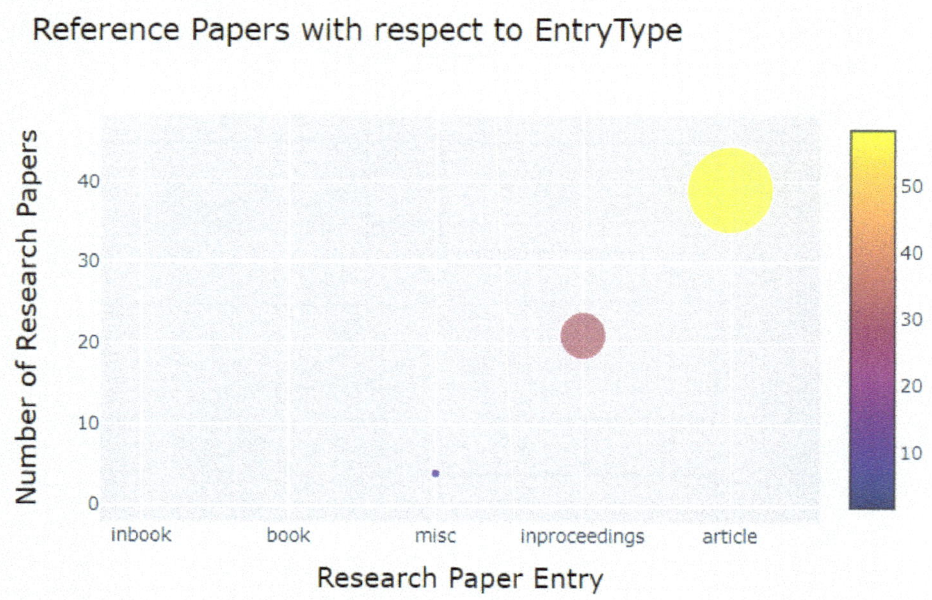

1.1 Chapter Organisation

This Chapter continues with the following sections: In Section 2, we put forward an insight into the architectural framework as well as delve into the workings of Li-Fi tech. Section 3 includes diverse applications of Li-Fisystem, highlighting its transformative potential among various domains. In Section 4, we present an exploration of the strengths and shortcomings of misconceptions surrounding Li-Fi communication. In Section 5, we explored life with Li-Fi. Section 6 delves into how 5G, 6G and Li-Fi technologies are a perfect match. Our analysis involved comparing Li-Fi with presently available technologies in Section 7, as summarized in Table 2. Furthermore, we looked into how Li-Fi can work together with Wi-Fi. Section 8 profound sustainability implications of Li-Fi, emphasizing its capabilities to contribute to energy conservation and make a key contribution to the blooming of smart, sustainable cities and agriculture.

1.2 Main Contribution

Our paper's research contributions are summed up as follows:

- The paper explains core concepts of Li-Fi, detailing how it transmits binary data using rapid light pulses from LEDs detected by photodetectors.
- It explores Li-Fi's use in sectors like healthcare, transportation, and industry, highlighting its dual role in lighting and data communication.
- The paper outlines Li-Fi's advantages and limitations, clearing up common myths and presenting its real potential.
- It discusses how Li-Fi can work alongside 5G/6G and Wi-Fi to create hybrid, high-speed communication networks.
- The paper highlights how Li-Fi supports sustainable tech development by offering energy efficient communication solutions.

2 ARCHITECTURE AND WORKING OF LI-FI

In this section, we delve into the architecture of Li-Fi, encompassing its transmitter and receiver components, communication protocols and modulation techniques.

2.1 Architecture of Li-Fi

Li-Fi refers to a wireless data exchange via visible light. The visible spectrum spans frequencies between 300 GHz and 700 GHz. Only visual light, out of all electromagnetic waves, has a wavelength that can be seen by the human eye; as a result, it is employed in our day-to-day lifestyle. Since visual light communication uses a binary system, an LED will convey "1" when it is turned on and "0" when it is off. Unlike traditional wireless technologies such as Wi-Fi, which rely on radio frequency signals, Li-Fi leverages the properties of light for high speed, secure, and energy efficient data communication (Sarkar et al., 2015).

Figure 3 in this review paper illustrates vital units that constitute the Li-Fi system: the High Brightness White LED (transmitter) and the Silicon photodiode (receiver). It also outlines the process of information transfer between transmitting and receiving units.

Transmitter: In the transmitter architecture of Li-Fi, the central component is the LED. This LED is the source of transmission and it emits high intensity light which acts as the data transmission medium. The transmitter codes data onto the by rapid modulation into light signal, which facilitates the transfer of data. The OOK modulation is a commonly used technique which modifies the intensity of the light in order to represent binary data (Swami & Moghe, 2020).

Receiver: Li-Fi receiver architecture is based on photodetectors, which are usually based on silicon photodiodes or phototransistors. These photo detectors are necessary in the capture of the incoming light signal. The photodiodes produce electrical current when they are exposed to light in proportion to signal strength of the received light beam. This current, which is the modulated data, is then processed to retrieve the information sent across. This process includes demodulation to recover the data in the light signal. Silicon photodiode selection is motivated by the fact that they are highly sensitive to visible light, thus making them efficient in light conversion of electrical signals (Swami & Moghe, 2020).

2.2 Communication protocol:

Li-Fi employs a specialized communication protocol to facilitate data transfer. The Li-Fi architecture is divided into classes based on layers. Li-Fi is made up of three layers: the physical layer, the MAC layer, and the application layer. However, there are only two layers specified by the IEEE 802.15.7 protocol: the Physical layer and the MAC layer (George et al., 2019).

Figure 3. A visual representation showcasing the components of Li-Fi Architecture

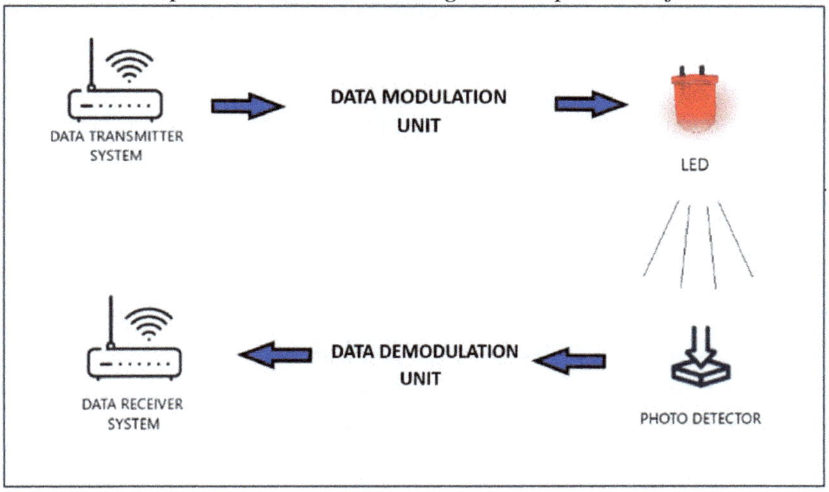

2.2.1 Physical Layer:

Physical layer or PHY layer, as it is commonly called, presupposes a number of crucial roles. It takes care of the sending and receiving of information and determines the link between the gadget and the medium of communication. There are three distinct operation modes supported by the PHY layer (Ley-Bosch et al., 2016), each tailored for specific usage scenarios and offering diverse data throughput levels, along with compatibility across various categories of optical components and environmental conditions:

PHY I: With minimal data rates ranging between 11.6 Kbps - 266.6 Kbps, this mode is best suited for outdoor applications.
PHY II: With intermediate data speeds ranging between 1.25 Mbps - 96 Mbps, this mode is meant for indoor use.
PHY III: Applications using several optical senders and detectors working at several optical wave lengths (colors) are appropriate for this mode. With a transmission rate ranging between 12 Mbps - 96 Mbps, it uses the CSK modulation technology.

2.2.2 MAC Layer:

The MAC layer is tasked with a number of key functions to uphold the efficient and smooth communication of data sending and receiving. These activities involve mobility, security, dimming, flickering mitigation strategies, visibility, color function support, generation of network beacons, VPAN disassociation and association support, and establishment of a reliable and secure connection between peer MAC entities. (Khan, 2017a).

There are three network topologies defined in MAC layers in Li-Fi:

1. ***Peer to Peer:*** This entails a transfer of information between two devices, where one device assumes the responsibility of being the coordinator.
2. ***Star Communication:*** In this particular situation, various devices are involved in communication, with one device acting as the coordinator while also playing a crucial role in the lighting system as a whole.
3. ***Broadcast One:*** In this approach, the coordinator transmits data to several devices at once, resulting in a one-way flow of information.

2.3 Modulation Techniques:

Li-Fi uses diverse Single Input, Single Output (SISO) modulation methods in the transmission of rapid data transfer through light-based systems. Single Carrier Modulation (SCM) comprises On-Off Keying (OOK), Pulse Width Modulation (PWM), M-ary Pulse Amplitude Modulation (M-PAM), M-ary Pulse Position Modulation (M-PPM), Carrierless Amplitude and Phase Modulation (CAP) and Orthogonal Frequency Division Multiplexing (OFDM). In the context of OFDM, such strategies as DC biased Optical OFDM (DCO-OFDM), Asymmetrically Clipped Optical OFDM (ACO-OFDM), Unipolar OFDM (U-OFDM) and others, are provided. Also, Li-Fi uses Multicarrier Modulation (MCM) like Superposition OFDM and Hybrid OFDM, Color Domain Modulation Techniques such as Color Shift Keying (CSK) and Color Intensity Modulation (CIM). All these SISO modulation methods make it possible to have a variety of applications in Li-Fi communication systems, and each application engineered to fulfill the requirements of a particular use case ("(PDF) Modulation Techniques for Li-Fi", 2019).

The proposed hybrid Li-Fi / Wi-Fi network was proposed by Zeng et al. (2020) who introduced and emphasized the orthogonal frequency division multiplexing access to channel assignment in the Li-Fi systems in order to enhance wireless connectivity to overcome the challenges driven due to the rapid expansion of cellular gadgets being used in RF networks. Indexed dimming (iDim), a new technology using time-domain sample index modulation of light-fidelity (Li-Fi) dimming was supplied by Nguyen et al. (2021). Through DCO-OFDM, iDim can attain optimal signal quality and data rate performance and simultaneously attain a broad range of brightness. In their article Soltani et al. (2019), discuss the issue of Li-Fi systems deployed indoors. It proposes and thoroughly discusses spatial modulation, using a multidirectional receiver architecture. This new technique by placing photodetectors strategically on various sides of user equipment attains a bit error ratio of 3.8×10^{-3}, which is far much better than the benchmark structure by more than 10 dB. To optimize the energy consumption, the paper also introduces an adaptive access point selection system of SM.

3 APPLICATIONS

This section mainly gives attention to the various applications of Li-Fi and how it is a much more viable solution than most of the coming-of-age technologies. Li-Fi can serve as a solution to many present time problems and may also become convenient and much more accessible in the near future.

3.1 High speed rates

Li-Fi outpaces many existing information transfer mechanisms like Wi-Fi. It provides much higher data download speeds. Most densely populated urban areas have complete artificial lighting coverage. This lighting system ensures that users traveling throughout the area may always access high data rates. For instance, people can receive high speed data while moving in well-lit corridors or lobbies. Through the paper "Achieving record speeds with new micro-LED design — LiFi Research" (2017), the authors introduce a novel OFDM based VLC system that incorporates both adaptive energy loading and bit techniques. When facing the challenge of LED-induced signal nonlinearity, this system achieves an impressive information transfer rate of 11.95 Gb/s using a violet micro-LED as the primary source of noise. researchers have also demonstrated high data transfer rate while transmitting text and image data.

The potential data speed is influenced not only by the lighting technology, but also by the method of digital modulation. The relation between the lighting technology and the data rates can be illustrated well in Figure 4 (Haas, 2018).

Figure 4. The technology of the real light sources determines the highest Li-Fi data speeds that may be achieved. RGB LEDs, GaN tiny LEDs, and laser based illumination are all taken into consideration here (Haas, 2018)

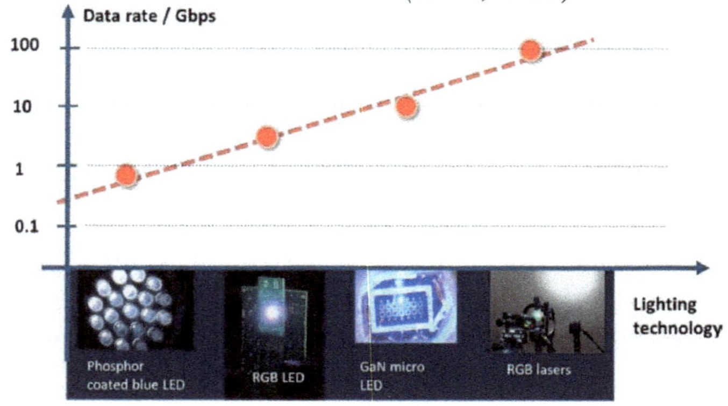

3.2 Live Streaming

As previously discussed, Li-Fi has the potential to reach various locations such as subway stations, malls, sports arenas, and modern workspaces. As a result, individuals can conveniently access a wide range of multimedia through their smartphones and other portable devices, including videos and live streaming. Live streaming using Li-Fi can be useful in Conferences, Seminars, Meetings, Concerts, Press Conferences, Product launches, etc. The same sensor and communications infrastructure is applicable for produce Li-Fi hotspots and track light and data in public areas.

3.3 Augmented Reality (AR)

Stores: From the moment customers enter the store, Li-Fi can aid in orienting them. Additionally, it can assist shoppers in finding specific items within the shopping area, gathering e-coupons, checking marketing offers, verifying information about in-store inventory, engaging in store web services via their mobile devices, and overall enhancing their shopping experience.

Museums and galleries: Specific lighting is used in museums and galleries to illuminate displays. location-specific details can be given inside the light through Li-Fi enabled lighting. This means that the light that illuminates the exhibit can be employed to obtain comprehensive details associated with the displays to a visitor's camera or smartphone.

Kirrbach et al. (2019) in their paper provide a proof of concept for connecting AR glasses using Li-Fi technology. Modern AR glasses require reliable connectivity.

An idea for fast, two-way Li-Fi networking was put forth in the paper. A network-based localization technique was introduced to offer geographically aware content.

3.4 Transportation

A state-of-the-art transport system could be developed by using Li-Fi. As possibility of utilizing LED car headlights and taillights increases, Li-Fi car-to-car communication is enhanced. This innovation would enable the collision mitigation systems development and sharing of important driving communication between vehicles on the road (George et al., 2019).

It also has a Li-Fi-enabled traffic signal management system to guarantee a more controlled and smooth traffic flow (Kumar et al., 2021). Through the mechanism of Li-Fi, it is possible to develop next-generation transportation systems. The more common the use of LED car headlights and taillights, the more the feasibility of V2V communication via Li-Fi. This innovation may enable creation of collision prevention systems and exchange of important driving data between vehicles on road (Al Barazanchi et al., 2021), as in Figure 5.

Figure 5. Portrays a Li-Fi based traffic signal management system, depicting two cars communicating with each other and a traffic light for enhanced traffic management.

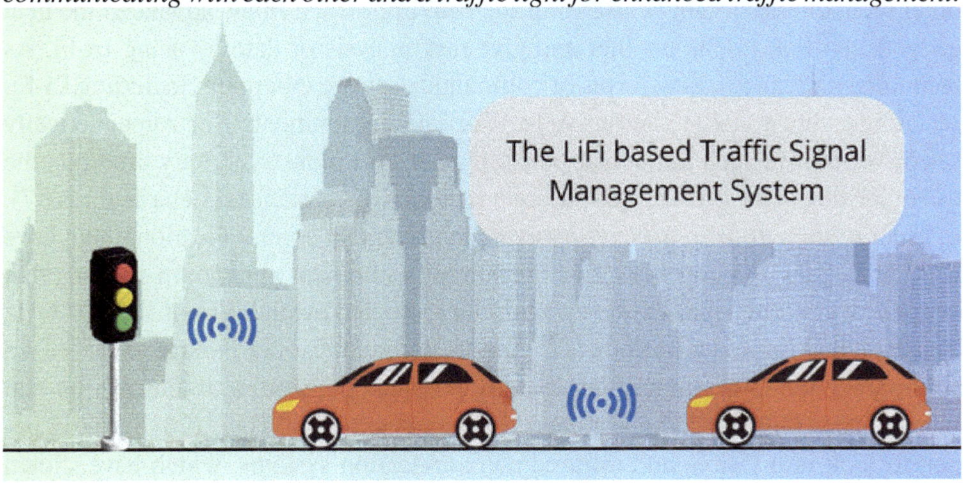

One of the main components of the Intelligent Transport System is VANET that allows cars to exchange information. A fast data communication channel is required in VANET and this can be done by Li-Fi. The automobile will have a vehicle module installed in it. Two parts of the vehicle module (VM), a transceiver and a control unit, will be responsible for information exchange (Agyemang et al., 2017). A system that reduces the time an ambulance must wait in areas with traffic was proposed by Pallavi et al. (2019). According to the entrance of the emergency vehicles into the lane, the system uses junction signal timers. It delivers information from ambulances towards traffic junction using Li-Fi technology. This system detects changes in traffic signals and adjusts the timing accordingly by sending signals to the timer control system. When the ambulance enters the lane, the signal adjusts the green signal according to the lane. It also updates the hospital's availability by acknowledging permission from the ambulance and indicating the doctor's availability to treat the patient promptly. Here, the use of Li-Fi allows for faster updating of the information on the global system or server. The hospital's response has sped up the ambulance's arrival at its destination. The patients can also receive the right care at the right moment resulting in a higher patient survival rate. The research article by Asif et al. (2022) put forth a new architecture that utilizes Li-Fi technology in smart homes to create a" Traffic monitoring and control system." This is a state of the art system that enables effective management, detection and control of traffic vehicles. Li-Fi was utilized by them to send high resolution image data to cloud servers to enhance system accuracy. One such place where the sensitivity of EMI and the security of data are significant issues is the hospitals. Li-Fi can facilitate a more effective rollout of safe networked medical devices, patient records, etc. Mobile phones and other electrical devices are not allowed to be used in situations where there is a chance of blasts. Li-Fi and other technologies might be useful in such environments. Many advanced medical gadgets can link to one another and give fast analysis of data by using Li-Fi. As technology advances, new forms of communication are emerging, including Li-Fi, which presents a viable alternative to Wi-Fi in environments electromagnetically sensitive settings such as nuclear power plants. This industry strictly prohibits the use of RF due to the presence of different radioactive substances (Leba et al., 2017). In order to keep up with the ever-evolving landscape of communication, there have been proposals to incorporate Li-Fi technology into next generation systems that support 5G and beyond. However, with the restricted availability of the 300 GHz spectrum and limited allocation for wireless communication, the use of RF-based communication is becoming increasingly problematic, leading to congestion and interference. This has led researchers, as demonstrated by Haas (2016) in his paper, to look into visible and infrared communication systems, which have shown promising levels of security (Lee et al., 2021).

3.5 Underwater Transmissions

Underwater, where radio frequencies cannot be used due to the environment's saltiness, high conductivity, and significant attenuation, light can spread. Since wires cause threads in underwater connections, Li-Fi transmitters can be used as a substitute. By utilizing their headlights, they can effectively transmit information to surface vessels, submarines, and divers (Leba et al., 2017).

The UTROV is another use of the VLC for underwater communication, according to Khan (2017b). Maintenance on ocean observatories and ship-based deployment potential are just two of the many uses for UTROV.

The visible light spectrum between 0.375 and 0.780 μm is employed as a data conveying optical carrier. and brightening as in the paper by Achuthan et al. (2021), in which they proposed underwater communication system, which is based on VLC. It uses short peaks of light to transmit data over a wide area. White LEDs are used by VLC to transmit information at remarkable speed that are invisible to naked eyes. Framework can be utilized around us without adjustments thanks to VLC, which is a big advantage. The ability of LEDs to transmit data signals (light with a recurrence between 400 and 800 THz and a frequency ranging from 400 to 700 nm) makes them an ideal correspondence medium. Compared to the fluorescent light framework, the LED lighting system can achieve reduced power consumption and a longer lifespan.

A research article by (Aman et al., 2021) presents a theoretical prototype for a Li-Fi Underwater Wireless Communication System. Using a frequency modulation scheme to increase signal security, controlling the LED driving circuitry's dimming sequence with a PWM technique, and creating a digital signal transmission circuit with several LED drivers that communicates with one another were the primary contributions of that paper.

The researchers Shanthi et al. (2023) presents a communication mechanism for submarines that will be helpful for high speed (1 Gbps) underwater data transfer between submarines. With this approach, they reduced the drawbacks and provided secure communication.

3.6 Indoor navigation

According to the researchers, Sharma et al. (2014) and Chatterjee et al. (2015), technology of Li-Fi has the ability to facilitate indoor navigation with precision up to 1000 cm and also offers orientation assistance in any location that uses LED lights, including shopping malls, theaters, government buildings, and other indoor settings. This could prove beneficial in determining a customer's viewing direction, among other things. Beyond providing illumination, sensors and lighting devices can also function as data transmission devices. The fusion of these diverse functions Facilitates the generation of intelligent networks that enhance cost effectiveness and optimize energy consumption as discussed by Al-Hamiri and Abd (2023) in their work.

3.7 In health surveillance

The authors Majeed and Murdas (2019) developed, implemented and assessed a prototype model for an integrated optical system that will monitor Li-Fi healthcare over various propagation connections. The foremost aim of the implemented method is to ensure accessibility to patients at distances of upto 4 meters in Line of Sight and 2.5 meters in diffuser links, while also eliminating optical background noise power. An innovative Li-Fi based medical surveillance system is developed to transfer data via several propagation paths from a patient-carried Fiber Bragg Grating sensor to an optical receiver located in a ceiling mounted room. This prototype is unique in that its parts are inexpensive and need little electricity. The system that has been constructed recognizes the intended signal and rejects the unwanted one. Furthermore, the experimental outcome demonstrates that the received signal is 100% accurate when the optical background noise power is removed.

Numerous wireless technologies have been proposed to monitor the patient's condition using various sensors in an attempt to address this problem; nevertheless, these wireless schemes can even interact with medical devices and represent a risk to patients and newborns. A Li-Fi based medical surveillance system that measures an infant's heart rate, temperature, and movements has been presented by the authors Mana et al. (2020) as a way to create a hospital-friendly monitoring system.

A smart Li-Fi oximeter surveillance system, based on a biomedical sensor (MAX30102) and a Li-Fi transmission link, was presented and experimentally demonstrated by Darweesh and Dawoud (2021). It uses a risk-free Li-Fi wireless communication system to work with commercially available portable gadgets to create more suitable, secure, and useful solutions. Another example is presented by Al-Dabbagh and Ali (2024) in their work, where a framework for healthcare monitoring has been presented that uses Li-Fi for relaying patient health data from biosensors to terminals. Data packets are intended to be sent by this framework to network terminals and subsequently to a cloud platform.

4 STRENGTHS AND SHORTCOMINGS OF LI-FI TECHNOLOGY

High data transfer rates, less interference, greater protection, and less energy requirements are just a few of Li-Fi technology's special advantages. It does, however, have some drawbacks that could potentially be fixed as the technology develops and matures in order to make Li-Fi a more widely used and applicable wireless communication method, as explained in the section below.

4.1 Strengths

4.1.1 Speed

Data transfer via Li-Fi is 100x quicker than Wi-Fi because it leverages visible light for communication rather than radio waves. Additionally, it has a much greater capacity for data transmission than the radio waves utilized in Wi-Fi. 224 GB of info can be transfer every second over a Li-Fi connection. However, achieving even 1% of that equates to 2.24 Gbps in rates, which considerably surpasses Wi-Fi. Li-Fi can provide mobile devices with several Gbps speeds. Utilizing this cutting-edge technology will expand wireless capabilities beyond their existing limitations, achieving unparalleled access to bandwidth ("LiFi.co — Revolutionary Wireless Communication Technology", n.d.). In experimental settings, analysts at the University of Oxford achieved data transfer rate of 224 Gbps, allowing the download of five high definition movies in just one minute at this speed. Recent investigations using three commercially available RGB laser diodes demonstrated Li-Fi data speeds of 14 Gbps. In 2015, 2-dimensional optical beam steering in indoor OWC achieved a speed of 42.8 Gbps. Notably, a 2017 study from the University of Eindhoven utilized infrared light to achieve a download rate of 42.8 Gbps within a 2.5-meter radius, surpassing the capabilities of Wi-Fi, which could only manage 300 Mbps at the time (Cossu et al., 2012).

4.1.2 Efficiency

Energy consumption can be significantly reduced because there will already be LED lights in places like homes, offices, and malls for lighting purposes. Li-Fi might function without electronic components that require constant power connections, such as routers, wave amplifiers, signal repeaters, modems, and antennas. To transfer information to and from a device, Li-Fi makes use of an LED light bulb as the medium. There is no denying the energy efficiency of LED light bulbs. They become more effective when Li-Fi is used since it gives them another use. By eliminating the need for wave amplifiers, routers, and signal repeaters that need to be plugged in constantly, Li-Fi also enables you to reduce your energy costs.("LiFi.co — Revolutionary Wireless Communication Technology", n.d.)

4.1.3 Availability

Since there are light sources everywhere, there is no problem with availability. As a result, you may access the internet through any source of light, which allows everyone to use it on an airplane. With Li-Fi, you may access the internet wherever a source of LED is present. The technology is now accessible through desk lights and LED light bulbs. However, in the future, it will be able to be linked to the internet and stay connected everywhere using street lights. Using Li-Fi, each street lamp, building light, and transportation lighting can be used as a free data access point.

A stable and consistent connection: With a latency of only 0.5 milliseconds, Li-Fi keeps a steady connection for every user in a room full of more than 30 people. On the other hand, as the quantity of users linked grows, the latency of other technologies increases. This previously unheard-of degree of connection dependability creates new opportunities for learning environments and professional contexts, enabling uses like interactive games, video conferences, collaborative platforms, and online tests.

No interference between Li-Fi and Wi-Fi: present wireless communication technologies (Wi-Fi, 4G & 5G networks), which are frequently present in the area where Li-Fi is used, are not interfered by the light waves on which Li-Fi relies. Li-Fi is especially well suited for usage in the aviation sector since it doesn't interfere with the operation of onboard equipment due to its lack of interference.

4.1.4 Security

As radio waves can penetrate walls, they are susceptible to interception by entities outside of your network. As a result, we can say that a Li-Fi network's security is significantly better and more reliable because no one can misuse the network since light can't travel through walls. Because non-translucent materials prevent light from passing through, you can rest assured that your connection will not be exposed in public areas. In an enclosed area, data transferred over Li-Fi is secure. Then, you may designate Li-Fi networks in regions where linked susceptible devices are vulnerable and utilize Wi-Fi in public locations (Leba et al., 2017).

4.1.5 Safe for health

One major advantage of Li-Fi technology is that it uses the VLS, making it immune to RF interference and avoiding any electromagnetic pollution. This is what makes it a good choice in dangerous conditions such as refineries, oil platforms, and gas stations. Moreover, it is also safe to be installed close to MRI machines, aircraft, and electric transformer stations. The market of this device is still expanding as the number of individuals who are conscious of the health hazards of electromagnetic radiation increases (Leba et al., 2017).

4.2 Shortcomings

There are two sides to every aspect of life, the successful side and the challenging side. The Li-Fi idea is not an exception. Even though it has a great role in ensuring safe and effective transfer of data, there are several concerns that remain unresolved. The following are these issues.

4.2.1 Interference

The Internet will also be inaccessible without a source of light, which may restrict the application of Li-Fi to specific places and situations. Li-Fi has a limited signal range, which is limited by physical barriers like walls because it operates using visible light. Also, the competing sources of light may disrupt the signal which is a major disadvantage. For instance, sunlight can disrupt the connection and disrupt Internet access. Moreover, during nighttime hours, those looking to engage in online activities like gaming or streaming videos will face difficulties accessing the Internet.

4.2.2 Special hardware requirements

Although the Li-Fi system is simple to install, it requires the establishment of a brand-new infrastructure, resulting in higher costs for both individuals and businesses using Li-Fi Internet.

4.2.3 Limited Range

Even though Li-Fi boasts a higher data transfer rate, its potential impact is limited if internet service providers cannot keep up with the technology's high speed. In countries with notoriously slow internet speeds, a Li-Fi network would not be valuable. Widespread adoption of Li-Fi will require collaborative efforts among industries. Additionally, limitations in distance are usually restricted to 10 meters, which may hinder the feasibility of Li-Fi implementation ("LiFi.co — Revolutionary Wireless Communication Technology", n.d.).

4.2.4 Less compatible

With its recent emergence, Li-Fi technology faces limitations in compatibility with existing devices. The majority of our daily gadgets and equipment still operate on conventional Wi-Fi hardware, making incorporation of Li-Fi technology in our personal devices a distant prospect. Another hindrance is the lack of a universal

Li-Fi standard, preventing the smooth integration of equipment from different manufacturers ("LiFi.co — Revolutionary Wireless Communication Technology", n.d.).

For a brighter future of Li-Fi technology, the aforementioned issues must be resolved. This technique is one of the emerging ones for data transfer. To make Li-Fi technology more viable and effective, researchers are investing a significant amount of energy and effort into finding solutions to the issues listed above.

4.3 Some Misconceptions on Li-Fi Technology

To offer a more realistic view about its possibilities, some presumptions must be dispelled, as is the case with every new technology. Due to its relative newness and the general lack of understanding of Li-Fi technology, misconceptions have developed, some of which are discussed in this section.

4.3.1 Dimmed Lighting

The experimental studies based on a variety of dimmed intensities concluded that data connection is still possible under low-light conditions. With the Li-Fi system, utilizing the existing lighting infrastructure is a piece of cake. All it takes is a simple attachment for the Li-Fi equipment. This means you can easily reap the benefits without having to overhaul your current setup (Haas, 2018b).

4.3.2 Light Flickering is Dangerous to Human Eye

The modulation of the lights occurs at frequencies as low as 1 MHz the computer displays refresh at a rate of roughly 100 Hz. Accordingly, a Li-Fi light bulb has a 10,000x greater flicker rate than a computer monitors. The greater rates of flickering in the Li-Fi light are not threatening to naked eyes and aren't noticeable. The human eye has the ability to perceive flickers at frequencies lower than 120–150 Hz. According to tests done in accordance with the EN 61000-4-15 standard, LED flickers are not damaging to human eyes (Leba et al., 2017).

4.3.3 Li-Fi Does Not Work in Sunlight

The interference of sunrays is a common occurrence that falls outside the frequency range used for data exchange in Li-Fi. Interestingly, Li-Fi systems typically operate at frequencies above 1 MHz, which means that, as long as the receiver is not overloaded, the persistent presence of sunbeams is effectively filtered out and has zero interference on the quality of data transmission. This fact has been confirmed through experiments conducted with pure Li-Fi's cutting-edge Li-Fi equipment. In order to prevent saturation, the implementation of automatic gain control techniques can prove to be beneficial. Therefore, we believe that sunlight is a valuable asset in solar cell-based Li-Fi receivers, where the solar cell not only serves as a data receptor but also harnesses sunlight as a source of energy (Z. Wang et al., 2015).

4.3.4 Li-Fi is A Downlink Only

The fact that Li-Fi is compatible with LED lighting is one of its main advantages. Both roles can be conveniently separated, therefore they are not necessarily need to be utilized together. Thus, Li-Fi can be adopted efficiently for uplink transmission in situations where illumination is not necessary. As a result, the infrared spectrum is ideal for the uplink. An experiment was carried out in which data was transferred at 1.1 Gbps across a distance of 10m using a 4.5 mW optical output power LED (Haas, 2018).

5 EXPLORE LIFE WITH LI-FI

This section explains how Li-Fi may be incorporated into various functionalities and efficiently put to use.

You may begin experiencing life with Li-Fi right away. Numerous businesses have already released their Li-Fi devices, enabling users to get the most from Li-Fi technology's numerous benefits. Install Li-Fi technology if there are dead spots in your house or workplace. The same applies if you wish to increase the security or speed of your connectivity. But even if dead zones and speeds aren't an issue, give it a shot. The way you access data now and, in the future, might be completely changed by Li-Fi.

5.1 Destination Assistance:

Users can make a shopping list in the application. For a newly connected user who links in the central database via Li-Fi and IoT, a new list will be generated based on the gadget's MAC and IP address (Thomas et al., 2016). Based on the number of users, the application generates a unique number for each, which could serve as a tracking number for reference and payment. Depending on what products are available in the supermarket, users can make a shopping list. The shopping list will automatically propose items, which could assist the customer in purchasing (Akter et al., 2018).

5.2 Product Checkout:

With ease and interactivity, customers have the convenient option of self-checkout through their smartphones. By simply scanning the product barcode and associating it with the corresponding item within the app, users can efficiently complete their purchase. The total cost will then be calculated using the barcode information, and a final confirmation of the entire shopping bill will be displayed upon completion of the checkout process (Akter et al., 2018).

5.3 Paying at Counter:

Upon completing their purchases at the supermarket, customers are required to verify their transactions at the cashier counter. Using location data, the cashier can access a list of active customers in close proximity. Additionally, sales staff can easily locate a specific customer by using their unique tracking ID (Haas, 2016; Tanaka et al., 2001).

5.4 Payment Gateway:

For this application, three payment options have been proposed: card, mobile banking, and cash. The cashier assists consumers in exchanging cards, confirming mobile payment gateways, and charging cash on the spot. With the use of IoT, a bill will be created and printed for the customers based on the cashier's confirmation (Akter et al., 2018).

6. WHY 5G - 6G AND LI-FI ARE A PERFECT MATCH?

This section shows how Li-Fi and 5G - 6G technologies can work together to make life less complicated and easy going. In the future, IoE is expected to leverage billions of connected gadgets to power cutting edge services like automated high precision manufacturing, telemedicine, and extended reality. 5G networks are unable to effectively handle such applications, which require unheard of wireless connectivity with particular key performance indicators (KPIs) like terabit per second speed, exceptionally high dependability, extremely low latency, and excellent energy economy. In order to cater to these needs, 6G is foreseen to be developed on a new physical layer architecture that supports and enhances RF communications by using visible light and subterahertz bands. According to estimates, the Li-Fi industry will reach $8 billion by 2030. Based on VLC, Li-Fi offers a fully integrated two-way wireless solution, and it has been recognized as a key element in the 6G architecture, as in (Abumarshoud et al., 2021).

Through their paper, Alsaedi et al. (2023) explore spectrum management strategies, focusing on their implementation in the allocation of 5G spectrum and their potential application in 6G. Additionally, a brief overview of spectrum sharing and its importance in maximizing spectrum resource utilization is provided. To provide a thorough grasp of the spectrum alternatives accessible for future 6G technology, the laws, standardization procedures, traits, constraints, and prospective applications of high frequency bands like mmWave, THz, and visible light are examined. In their comprehensive analysis, the study by Chowdhury et al. (2019) show how OWC technologies—like VLC, Li-Fi, Optical Camera Communication, and Free Space Optics Communication—may help the IoT and 5G / 6G networks be deployed more successfully.

Kharche and Kharche (2023) in their work address the lack of discussion on applying disruptive technologies in 6G intelligent healthcare and propose a conceptual framework. It looks at the applications and difficulties of the latest innovations in the medical field, including blockchain, digital twins, OWC, wireless energy transmission, holographic, quantum communication, tactile Internet and artificial intelligence. The impending 6G is anticipated to significantly assist very dependable, secure, exceptionally rapid and ultra-low latency communication, for worldwide remote healthcare, in contrast to 5G. Additionally, 6G enables healthcare enhancement globally by endorsing IoE, Internet of Skills, and Internet of Thinking. The scholars Kamath et al. (2023) through their paper explain how the growth of IoT and datacentric applications requires the implementation of high-speed networks. It is anticipated that by 2030, 5G might encounter difficulties adapting to emerging technologies, prompting a transition to 6G. This change is poised to enable ultra-fast intelligent networks, with discussions covering the technologies that enable this shift, recent developments, and potential challenges in adoption. Topare (2024) discusses the probable and challenges faced by Li-Fi technology, while also its close interconnection with 5-G technology. Their paper explains how both 5G and Li-Fi work towards increasing connectivity, although having its own applications and impediments.

7 COMPATIBILITY OF LI-FI WITH OTHER TECHNOLOGIES

7.1 Li-Fi Comparison with Existing Technology

Before Li-Fi came into picture, various existing technologies like Wi-Fi, Bluetooth, and WiMAX were already booming. This section gives a comparison between these technologies based on various features, such as data transmission, its working concept, and many more. Table 1 demonstrates the comparison of Li-Fi with already present technologies like Wi-Fi, Bluetooth & Wimax (Kumar, et al., 2021; Perwej, 2017; Ramadhani & Mahardika, 2018).

Table 1. Comparison of Li-Fi with Wi-Fi, Bluetooth and WiMAX (Kumar, et al., 2021; Perwej, 2017; Ramadhani & Mahardika, 2018)

Features	Li-Fi	Wi-Fi	BLUETOOTH	WiMAX
Full Form	Light Fidelity	Wireless Fidelity	Bluetooth	Wireless Interoperability for Microwave Access.
Technology	Light waves	Radio waves	Radio waves	Microwaves
Year of Inception	2011	1990	1994	2001
Standard	IEEE.802.15.7	IEEE.802.11	IEEE.802.15.1	IEEE.802.16d/e
Data	Using light	Using Radio	Anywhere	Broadband
Transmission	(LED Bulb)	Waves (Router)	at least 2 Bluetooth devices exist	Wireless Access
Working	Direct Binary	Various	Master-slave	Request/Grant
Concept	Data Serving	Topologies		
Type of Network	Wireless Local Area Networks (WLAN)	Local Area Network (LAN)	Wireless Personal Area Network (WPAN)	Metropolitan Area Network (MAN)
Frequency of operation	10000 times the radio's frequency spectrum	2.4GHz, 4.9GHz and 5GHz		Unlicensed/ 2.4GHz
	Licensed/ 2G to 11GHz		**Data transfer speed**	About 1 Gbps
WLAN-11n offers 150Mbps, WiGig/ Giga-IR offers about 1-2 Gbps	800KBps	Works at 5bps/Hz and can peak to 100 Mbps in 20MHz channel **Coverage Distance**	About 10m	

About 32m (WLAN 802.11b/11g),
depending on the type of antenna and transmission power
About 10 to 100m, based on classes
Up to 40 miles

7.2 Li-Fi & Wi-Fi aggregate networks

Kirrbach et al. (2019) examined a combined Li-Fi & Wi-Fi system from the standpoint of real world communication. Their findings offer a reasonably viable design foundation for the combined Li-Fi & Wi-Fi communication system. To balance the complexity of computation with transmission performance, they constructed a framework that enhances the lower bound of the achievable rate, where optimal power allocation can be obtained and the constellation distribution problem can be efficiently solved using the Frank-Wolfe method.

The term Hy-Fi, which stands for hybrid fidelity was coined by Zubow et al. (2021). Using multiple input multiple output (MIMO), it integrates Li-Fi with Wi-Fi at the physical layer. This makes it possible to employ both media simultaneously in order to increase data throughput through aggregation or to obtain variety and immune to signal blockage and extrinsic interference. They showed how to accomplish this by utilizing the built-in MIMO features of modern commercially available off-the-shelf (COTS) 802.11 RF hardware. Open-source software is the only prerequisite, aside from the COTS hardware. To the best of what is known, this is the first attempt at combining Li-Fi and Wi-Fi on the physical layer using commercially available technology. Several Li-Fi system protocols and handover protocols for Wi-Fi & hybrid Li-Fi are examined in the work of Sulayman et al., 2021, with an emphasis on the security features. The paper anticipates a Li-Fi & Wi-Fi in future environment by closely examining the security weaknesses in these protocols and suggesting strategies to alleviate them. For Wi-Fi & hybrid indoor Li-Fi communication systems, J. Wang et al. ((wang2017learning) develop a multi armed bandit model to aid in the worthwhile selection of LED access points. Additionally, the "exponentially weighted algorithm with linear programming" and "exponential weights for exploration and exploitation" methods are used to update the decision probability distribution. These algorithms provide the upper constraint of respective accumulated reward function. The proposed network association strategies can result in significant throughput improvements. The HLWNet architecture is described by Besjedica et al. in the work of Besjedica et al. (2023), which also provides information on the basic architecture of Wi-Fi networks & hybrid Li-Fi and handover techniques within HLWNets. The discussion includes modulation methodologies, design of backhaul equipment, illumination limits, & cell deployment optimization.

As a foundation for a rapid radio transmission solution, Farrag and Hussein (2023) suggested a hybrid network that blends Wireless Gigabit Alliance (WiGig) with Li-Fi. In the work of Arunkumar and Thanasekhar (2024), a variant and dynamic access point selection and traffic distribution technique. Along with Li-Fi & Wi-Fi combination, Deep Long Short-Term Memory is also integrated which is used for location prediction.

8 LIGHTING THE PATH TO SUSTAINABLE CONNECTIVITY

8.1 Sustainability Implications of Li-Fi Technology

When compared to standard lighting and communication systems, Li-Fi technology provides several major sustainability benefits:

1. ***Extended Lifespan and Reduced Energy Consumption:*** Studies show that Li-Fi LED lights have a much longer lifespan, eliminating the need for frequent replacements and, as a result, lowering electronic waste generation. Furthermore, combining Li-Fi with energy saving LED lighting reduces overall energy usage, aligning with the global emphasis on energy efficiency and environmental goal (Montoya et al., 2017).
2. ***Efficient Light Distribution:*** By leveraging LED technology, Li-Fi enables high-quality light distribution with fewer lights, thereby reducing energy consumption and operational costs. This simplified lighting method is consistent with sustainable practices and encourages resource conservation (Maksimovic, 2017).
3. ***Cost-Effectiveness and Resource Savings:*** The integration of Li-Fi technology eliminates the need for additional electrical components including signal repeaters, routers, & modems. This way minimizing energy use and the cost of equipment (MacKay, 2016). This is a lean structure that enhances cost efficiency besides complying with the concept of sustainable resource management.

The Li-Fi can be adopted as a viable alternative in EMI sensitive areas like airlines and hospitals where the normal Wi-Fi networks can disrupt important equipment and communication gadgets. Li-Fi eliminates the EMI-related issues and optimizes security and stability of communication networks within such delicate locations by transferring information via the visible light spectrum (Albraheem et al., 2018; Lusk & Monday, 2017). The fact that Li-Fi depends on the placement between the receiver and the light source enhances the security of the data and prevents unauthorized access, which is why it is a more desirable option to be used in the environmentally sensitive areas that need a high level of security (Albraheem et al., 2022).

8.2 Li-Fi-Powered Smart, Sustainable Cities

The construction of smart sustainable cities has drawn considerable attention among governments, businesses, and research communities. According to the definition provided by the International Telecommunications Union (ITU), a smart sustainable city is the city that considers economic, social, and environmental issues and employs ICT to improve the services of the city (Aswin et al., 2016). An application of the IoT technology (Joshi et al., 2024a; Joshi et al., 2024b; Joshi et al., 2025) is perceived as a new method of ICTs to develop sustainable, smart cities.

The design and development of intelligent buildings is one approach to advance sustainability. Intelligent buildings are a significant component of sustainability practices. The buildings' advanced systems that represent various building services are part of their technology. Through the use of a wireless communication network, all of the building's systems and components may be managed from anywhere, facilitating the necessary mutual exchange to form an "integrated system". This technology can lower building construction and maintenance costs while increasing the building's energy efficiency. In order to accomplish this, Aljewari et al. (2021) focuses on utilizing low-voltage, high-speed, data networks linked to portable Li-Fi connectivity to integrate smart devices and building services.

Using an innovative technique to redesign wireless communication infrastructure, the study by Talavante et al. (2023) aims to achieve sustainable communication with net zero CO_2 emissions. The tubular light guides found in the new construction are utilized by the proposed LiFiTube system. It integrates natural sunlight indoors and modifies it for wireless data services using low-power liquid crystal shutters.

There are a few challenges that hinder Li-Fi integration into smart sustainable city ecosystems. According to the researchers Georlette and Moeyaert (2023), the amount of dust or other particulates in the air is much greater in the outdoor environment than the indoor one due to the concentration of dust particles that enter the air through vehicles. Some of the weather related factors that affect concentration are fog, smoke, sandstorms, rain and snow. Outdoors, these particles can cause two effects: scattering and an overall increase in the saturation of the medium, the transformation of incoming optical energy into other forms of energy by air particles.

9 CONCLUSION

The paper has taken an extensive tour of the complicated landscape of Li Fi technology giving a subtle understanding of its complexities to the readers. Beginning with a perceptive introduction, the paper takes a tour of Li-Fi framework, and quickly moves into operational dynamics. The overall analysis of different applications brings to light the versatility of the technology, and an assessment of the strengths inherent in it and possible limitations provides a balanced perspective. By shedding away illusions and the justification of results by a strong case study, the story becomes more profound, providing information to the practical implications of Li-Fi. Furthermore, this work explains harmonious connection between Li-Fi, 5G & 6G technologies, and their smooth coexistence. The analysis of the compatibility of Li-Fi with other technological systems gives a comprehensive perspective. Finally, this review highlights Li-Fi's purpose, in addition to the development of technologies, lights a way of sustainable connectivity, and creates a bright future for wireless communication.

REFERENCES

Abumarshoud, H., Mohjazi, L., Imran, M. A., & Haas, H. (2021). LiFi Through Intelligent Reflecting Surfaces: A New Frontier for 6G? arXiv preprint arXiv:2104.02390.

Achuthan, B., Arunkumar, B., Boobalan, M., & Chiragsun, R. L. (2021, May). Underwater communication using li-fi. In 2021 3rd International Conference on Signal Processing and Communication (ICPSC) (pp. 485-489). IEEE.

Agyemang, J. O., Kponyo, J. J., & Mouzna, J. (2017, November). Light fidelity (LiFi) as an alternative data transmission medium in VANET. In *2017 European Modelling Symposium (EMS)* (pp. 213-217). IEEE. DOI: 10.1109/EMS.2017.44

Akter, S., Olanrewaju, R. F., & Islam, T. (2018, May). LiFi based automated shopping assistance application in IoT. In Journal of Physics: Conference Series (Vol. 1018, No. 1, p. 012001). IOP Publishing. Al_Barazanchi, I., Sahy, S. A., & Jaaz, Z. A. (2021, February). Traffic management with deployment of Li-Fi technology. []. IOP Publishing.]. *Journal of Physics: Conference Series*, *1804*(1), 012141.

Al-Dabbagh, M. S., & Ali, A. K. (2024). Employing Light Fidelity in Iot-Based Healthcare System. *Theory and Applications of Engineering Research*, *6*, 11–27. DOI: 10.9734/bpi/taer/v6/2757G

Al-Hamiri, M. G., & Abd, H. J. (2023, February). A review of Lifi Technology: Principles and applications. In *2023 Second International Conference on Advanced Computer Applications (ACA)* (pp. 1-6). IEEE. DOI: 10.1109/ACA57612.2023.10346813

Albraheem, L., Alhudaithy, L., Aljaser, A., AlDhafian, M., & Bahliwah, G. (2022). Towards smart sustainable cities using Li-Fi technology: Geo-location infrastructure utilizing LED street lights. *PeerJ. Computer Science*, *8*, e1009. DOI: 10.7717/peerj-cs.1009 PMID: 36091990

Albraheem, L. I., Alhudaithy, L. H., Aljaser, A. A., Aldhafian, M. R., & Bahliwah, G. M. (2018). Toward designing a Li-Fi-based hierarchical IoT architecture. *IEEE Access : Practical Innovations, Open Solutions*, *6*, 40811–40825. DOI: 10.1109/ACCESS.2018.2857627

Aljewari, Y. H. K., Sinclair, T. D., Alahmed, A. S. A. N., & Alali, S. M. F. (2021, November). Smart Lighting Solution using LiFi Technology in Intelligent Buildings. In 2021 IEEE 11th International Conference on Consumer Electronics (ICCE-Berlin) (pp. 1-5). IEEE. DOI: 10.1109/ICCE-Berlin53567.2021.9720033

Alsaedi, W. K., Ahmadi, H., Khan, Z., & Grace, D. (2023). Spectrum options and allocations for 6G: A regulatory and standardization review. *IEEE Open Journal of the Communications Society, 4*, 1787–1812. DOI: 10.1109/OJCOMS.2023.3301630

Aman, M., Qiao, G., & Muzzammil, M. (2021, July). Design and analysis of Li-fi underwater wireless communication system. In 2021 OES China Ocean Acoustics (COA) (pp. 1100-1103). IEEE.

Arunkumar, R., & Thanasekhar, B. (2024). Heterogeneous Lifi–Wifi with multipath transmission protocol for effective access point selection and load balancing. *Wireless Networks*, *30*(4), 2423–2437. DOI: 10.1007/s11276-024-03657-w

Asif, M., Khan, T. A., Taleb, N., Said, R. A., Siddiqui, S. Y., & Batool, G. (2022, February). A Proposed Architecture for Traffic Monitoring & Control System via LiFi Technology in Smart Homes. In *2022 International Conference on Business Analytics for Technology and Security (ICBATS)* (pp. 1-3). IEEE. DOI: 10.1109/ICBATS54253.2022.9759069

Aswin, K., Aswin, R., Lokesh, V., & Sugacini, M. (2016). Iot enabled by Li-Fi technology. In *National conference on communication and informatics*.

Besjedica, T., Fertalj, K., Lipovac, V., & Zakarija, I. (2023). Evolution of hybrid LiFi–WiFi networks: A survey. *Sensors (Basel)*, *23*(9), 4252. DOI: 10.3390/s23094252 PMID: 37177456

Chatterjee, S., Agarwal, S., & Nath, A. (2015). Scope and challenges in light fidelity (LiFi) technology in wireless data communication.

Chowdhury, M. Z., Shahjalal, M., Hasan, M. K., & Jang, Y. M. (2019). The role of optical wireless communication technologies in 5G/6G and IoT solutions: Prospects, directions, and challenges. *Applied Sciences (Basel, Switzerland)*, *9*(20), 4367. DOI: 10.3390/app9204367

Cossu, G., Khalid, A. M., Choudhury, P., Corsini, R., & Ciaramella, E. (2012). 3.4 Gbit/s visible optical wireless transmission based on RGB LED. *Optics Express*, *20*(26), B501–B506. DOI: 10.1364/OE.20.00B501 PMID: 23262894

Darweesh, M., & Dawoud, D. W. (2021, November). Simple LiFi Oximeter Monitoring System for Telehealth Applications. In 2021 28th IEEE International Conference on Electronics, Circuits, and Systems (ICECS) (pp. 1-4). IEEE. DOI: 10.1109/ICECS53924.2021.9665481

Farrag, M., & Hussein, H. S. (2023). Most-correlated distribution-based load balancing scheme in hybrid lifi/wigig network. *IEEE Access : Practical Innovations, Open Solutions, 11*, 109764–109775. DOI: 10.1109/ACCESS.2023.3322102 PMID: 38203081

George, R., Vaidyanathan, S., Rajput, A. S., & Deepa, K. (2019). LiFi for vehicle to vehicle communication–a review. *Procedia Computer Science, 165*, 25–31. DOI: 10.1016/j.procs.2020.01.066

Georlette, V., & Moeyaert, V. (2023, July). Li-Fi and Visible Light Communication for Smart Cities and Industry 4.0: challenges, research & market status in 2023. In 2023 23rd International Conference on Transparent Optical Networks (ICTON) (pp. 1-4). IEEE.

Ghaderi, M. R. (2023). LiFi and hybrid WiFi/LiFi indoor networking: From theory to practice. *Optical Switching and Networking, 47*, 100699. DOI: 10.1016/j.osn.2022.100699

Haas, H. (2016, October). LiFi: Conceptions, misconceptions and opportunities. In *2016 IEEE Photonics Conference (IPC)* (pp. 680-681). IEEE. DOI: 10.1109/IPCon.2016.7831279

Haas, H. (2018). LiFi is a paradigm-shifting 5G technology. *Reviews in Physics, 3*, 26–31. DOI: 10.1016/j.revip.2017.10.001

Haas, H., Yin, L., Wang, Y., & Chen, C. (2015). What is lifi? *Journal of Lightwave Technology, 34*(6), 1533–1544. DOI: 10.1109/JLT.2015.2510021

Joshi, K., Singh, V., Awasthi, P., Anandaram, H., Sidhu, K. S., & Kumar, P. (2024, November). Smart Dam Water Monitoring using Enabled Internet of Things. In 2024 5th International Conference on Data Intelligence and Cognitive Informatics (ICDICI) (pp. 239-243). IEEE. DOI: 10.1109/ICDICI62993.2024.10810761

Joshi, K., Yadav, H., Gupta, S., Singh, V., Sidhu, K. S., & Kukreti, R. (2025, April). Handling Security Aspects in the Internet of Things: Latest Challenges and Measures to Mitigate Risks. In 2025 3rd International Conference on Communication, Security, and Artificial Intelligence (ICCSAI) (Vol. 3, pp. 1434-1439). IEEE.

Joshi, P., Singh, V., Thapliyal, N., Joshi, K., Bhatt, A., & Mahur, M. (2024, October). Intelligent IoT-Enabled Real-Time Monitoring System for Logistics Management. In *2024 International Conference on Computing, Sciences and Communications (ICCSC)* (pp. 1-5). IEEE. DOI: 10.1109/ICCSC62048.2024.10830332

Kamath, H. S., Bhandari, A., Shekhar, S., & Ghosh, S. (2023, March). A survey on enabling technologies and recent advancements in 6G communication. [). IOP Publishing.]. *Journal of Physics: Conference Series*, *2466*(1), 012005. DOI: 10.1088/1742-6596/2466/1/012005

Khan, L. U. (2017). Visible light communication: Applications, architecture, standardization and research challenges. *Digital Communications and Networks*, *3*(2), 78–88. DOI: 10.1016/j.dcan.2016.07.004

Kharche, S., & Kharche, J. (2023). 6G intelligent healthcare framework: A review on role of technologies, challenges and future directions. *Journal of Mobile Multimedia*, *19*(3), 603–644. DOI: 10.13052/jmm1550-4646.1931

Khare, Y., Tiwari, V. P., Patil, A. B., & Bala, K. (2016). Li-Fi technology, implementations and applications. *Int. Res. J. Eng. Technol.*, *3*(4), 1391–1394.

Kirrbach, R., Faulwaßer, M., Jakob, B., Schneider, T., & Noack, A. (2019, October). Li-Fi for augmented reality glasses: A proof of concept. In 2019 IEEE International Symposium on Mixed and Augmented Reality Adjunct (ISMAR-Adjunct) (pp. 263-268). IEEE.

Koonen, T., Mekonnen, K., Cao, Z., Huijskens, F., Pham, N. Q., & Tangdiongga, E. (2020). Ultra-high-capacity wireless communication by means of steered narrow optical beams. *Philosophical Transactions. Series A, Mathematical, Physical, and Engineering Sciences*, *378*(2169), 20190192. DOI: 10.1098/rsta.2019.0192 PMID: 32114920

Kumar, J. D., Priyadharsini, K., Srinithi, K., Samprtiha, R. V., & Babu, C. G. (2021, March). An experimental analysis of lifi and deployment on localization based services & smart building. In *2021 International Conference on Emerging Smart Computing and Informatics (ESCI)* (pp. 92-97). IEEE. DOI: 10.1109/ESCI50559.2021.9396889

Kumar, P., & Sharma, K. P. (2021, June). Challenges and applications of Li-Fi in D2D communication. In *2021 International Conference on Communication, Control and Information Sciences (ICCISc)* (Vol. 1, pp. 1-6). IEEE.

Leba, M., Riurean, S., & Lonica, A. (2017, June). LiFi—The path to a new way of communication. In 2017 12th Iberian conference on information systems and technologies (CISTI) (pp. 1-6). IEEE.

Lee, C., Islim, M. S., Das, S., Spark, A., Videv, S., Rudy, P., Shah, B., McLaurin, M., Haas, H., & Raring, J. (2021). 26 Gbit/s LiFi system with laser-based white light transmitter. *Journal of Lightwave Technology*, *40*(5), 1432–1439. DOI: 10.1109/JLT.2021.3124942

Ley-Bosch, C., Alonso-González, I., Sánchez-Rodríguez, D., & Ramírez-Casañas, C. (2016). Evaluation of the effects of hidden node problems in IEEE 802.15. 7 uplink performance. *Sensors (Basel), 16*(2), 216. DOI: 10.3390/s16020216 PMID: 26861352

Lian, J., Vatansever, Z., Noshad, M., & Brandt-Pearce, M. (2019). Indoor visible light communications, networking, and applications. *JPhys Photonics*, *1*(1), 012001. DOI: 10.1088/2515-7647/aaf74a

LiFi.co. (n.d.). LiFi — revolutionary wireless communication technology. Retrieved October 29, 2025, from https://lifi.co/

Lusk, R. M., & Monday, W. (2017). *An early survey of best practices for the use of small unmanned aerial systems by the electric utility industry (No. ORNL/TM-2017/93). Oak Ridge National Lab*. ORNL. DOI: 10.2172/1344293

Ma, S., Sheng, H., Sun, J., Li, H., Liu, X., Qiu, C., Safari, M., Al-Dhahir, N., & Li, S. (2024). Feasibility conditions for mobile lifi. *IEEE Transactions on Wireless Communications*, *23*(7), 7911–7923. DOI: 10.1109/TWC.2023.3346056

MacKay, D. J. (2016). *Sustainable Energy-without the hot air*. Bloomsbury Publishing.

Majeed, A. F., & Murdas, I. A. (2019). Novel design for Li-Fi healthcare monitoring system. *International Journal of Intelligent Engineering and Systems*, *12*(6), 59–70. DOI: 10.22266/ijies2019.1231.06

Maksimovic, M. (2017). Greening the future: Green Internet of Things (G-IoT) as a key technological enabler of sustainable development. In *Internet of things and big data analytics toward next-generation intelligence* (pp. 283–313). Springer International Publishing.

Mana, S. M., Hellwig, P., Hilt, J., Bober, K. L., Hirmanova, V. J., Chvojka, P., . . . Zvanovec, S. (2020, March). LiFi experiments in a hospital. In Optical Fiber Communication Conference (pp. M3I-2). Optica Publishing Group.

Montoya, F. G., Peña-García, A., Juaidi, A., & Manzano-Agugliaro, F. (2017). Indoor lighting techniques: An overview of evolution and new trends for energy saving. *Energy and Building*, *140*, 50–60. DOI: 10.1016/j.enbuild.2017.01.028

Nguyen, T., Islim, M. S., Chen, C., & Haas, H. (2021). iDim: Practical implementation of index modulation for LiFi Dimming. *IEEE Transactions on Green Communications and Networking*, *5*(4), 1880–1891. DOI: 10.1109/TGCN.2021.3089758

Pallavi, H. T., Poornima, G., & Tomar, S. K. (2019). Adaptive Timers for Emergency Vehicles and Pollution Detector Using Li-Fi.

Paul, A., Islam, T., Ferdaush, R., Roni, N., Orpa, J. B., & Sadat, N. (2024). *Li-Fi Data Transmission and Cryptographic Decryption: Design and Implementation of a Smartphone-integrated*. Arduino-Based System for Secure Communication.

Perwej, Y. (2017). The next generation of wireless communication using Li-Fi (light fidelity) technology. [JCN]. *Journal of Computer Networks*, *4*(1), 20–29. DOI: 10.12691/jcn-4-1-3

Ramadhani, E., & Mahardika, G. P. (2018, March). The technology of lifi: A brief introduction. []). IOP Publishing.]. *IOP Conference Series. Materials Science and Engineering*, *325*(1), 012013. DOI: 10.1088/1757-899X/325/1/012013

Sarkar, A., Agarwal, S., & Nath, A. (2015). Li-Fi technology: Data transmission through visible light. *International Journal of Advance Research in Computer Science and Management Studies*, *3*(6).

Shanthi, T., & Shalini, P. (2023, November). Communication Between Submarines Using Li-Fi Technology. In *2023 International Conference on Recent Advances in Science and Engineering Technology (ICRASET)* (pp. 1-4). IEEE.

Sharma, R. R., Sanganal, A., & Pati, S. (2014). Implementation of a simple Li-Fi based system. *International Journal of Computers and Technology*, *1*(9), 437–443.

Soltani, M. D., Arfaoui, M. A., Tavakkolnia, I., Ghrayeb, A., Safari, M., Assi, C. M., Hasna, M. O., & Haas, H. (2019). Bidirectional optical spatial modulation for mobile users: Toward a practical design for LiFi systems. *IEEE Journal on Selected Areas in Communications*, *37*(9), 2069–2086. DOI: 10.1109/JSAC.2019.2929381

Sulayman, I. I. A., He, R., Manka, M., Ning, A., & Ouda, A. (2021, October). LiFi/WiFi authentication and handover protocols: Survey, evaluation, and recommendation. In *2021 International Symposium on Networks, Computers and Communications (ISNCC)* (pp. 1-6). IEEE. DOI: 10.1109/ISNCC52172.2021.9615853

Swami, K. T., & Moghe, A. A. (2020, December). A review of LiFi technology. In 2020 5th IEEE international conference on recent advances and innovations in engineering (ICRAIE) (pp. 1-5). IEEE. DOI: 10.1109/ICRAIE51050.2020.9358340

Talavante, J., Genovés Guzmán, B., & Giustiniano, D. (2023, June). Rethinking LiFi for carbon neutral sunlight-based communication. In *Mediterranean Communication and Computer Networking Conference* (pp. 1-8). DOI: 10.1109/MedComNet58619.2023.10168863

Tanaka, Y., Komine, T., Haruyama, S., & Nakagawa, M. (2001, October). Indoor visible communication utilizing plural white LEDs as lighting. In 12th IEEE International Symposium on Personal, Indoor and Mobile Radio Communications. PIMRC 2001. Proceedings (Cat. No. 01TH8598) (Vol. 2, pp. F-F). IEEE. DOI: 10.1109/PIMRC.2001.965300

Thomas, Z., Kumar, N., & Preshiya, D. J. (2016, April). Automatic billing system using Li-Fi module. In *2016 International Conference on Communication and Signal Processing (ICCSP)* (pp. 2216-2219). IEEE. DOI: 10.1109/ICCSP.2016.7754087

Topare, S. V. (2024). 5g Dominance: Why Li-Fi Technology is Struggling to Compete. Available at *SSRN* 4835013.

Wang, J., Jiang, C., Zhang, H., Zhang, X., Leung, V. C., & Hanzo, L. (2017). Learning-aided network association for hybrid indoor LiFi-WiFi systems. *IEEE Transactions on Vehicular Technology*, *67*(4), 3561–3574. DOI: 10.1109/TVT.2017.2778345

Wang, Z., Tsonev, D., Videv, S., & Haas, H. (2015). On the design of a solar-panel receiver for optical wireless communications with simultaneous energy harvesting. *IEEE Journal on Selected Areas in Communications*, *33*(8), 1612–1623. DOI: 10.1109/JSAC.2015.2391811

Zeng, Z., Soltani, M. D., Wang, Y., Wu, X., & Haas, H. (2020). Realistic indoor hybrid WiFi and OFDMA-based LiFi networks. *IEEE Transactions on Communications*, *68*(5), 2978–2991. DOI: 10.1109/TCOMM.2020.2974458

Zubow, A., Gawłowicz, P., Bober, K. L., Jungnickel, V., Habel, K., & Dressler, F. (2021, June). Hy-Fi: aggregation of LiFi and WiFi using MIMO in IEEE 802.11. In 2021 IEEE 22nd International Symposium on a World of Wireless, Mobile and Multimedia Networks (WoWMoM) (pp. 100-108). IEEE.

Chapter 11
Securing Autonomous Systems AI-Based Intrusion Detection and Privacy-Preserving Mechanisms

Vikas Sharma
https://orcid.org/0000-0001-8173-4548
Department of Computer Applications, SRM Institute of Science and Technology, Delhi, India

Shashank Shelat
https://orcid.org/0009-0003-6298-2310
Independent Researcher, USA

Manoj Kumar
https://orcid.org/0009-0001-8354-1253
Shanti Niketan Group of Institutions, Chaudhary Charan Singh University, Meerut, India

Gurpreet Kaur
Raj Kumar Goel Institute of Technology, Ghaziabad, India

Alok Kumar
https://orcid.org/0000-0003-1051-0672
School of Engineering and Technology, Chhatrapati Shahu Ji Maharaj University, Kanpur, India

DOI: 10.4018/979-8-3373-3576-6.ch011

© 2026, IGI Global Scientific Publishing. Copying or distributing in print or electronic forms without written permission of IGI Global Scientific Publishing is prohibited. Use of this chapter to generative artificial intelligence (AI) technologies is expressly prohibited. The publisher reserves all rights to license its use for generative AI training and machine learning model development.

ABSTRACT

The rapid growth of such autonomous systems as self-driving vehicles, drones, robotics, and intelligent industrial agents has increased the demand for enhanced cybersecurity structures. This chapter will focus on the application of Artificial Intelligence (AI) based intrusion detection systems, in addition to privacy-preserving technologies in order to protect autonomous systems from advanced cyberattacks and data breaches. Improvement of anomaly detection, behavioural analysis, and threat predictions in real time through machine learning, deep learning and federated learning strategies will be highlighted. The chapter will highlight the importance of integrity, confidentiality and resilience of systems, coupled with the addressing of vulnerabilities in the form of adversarial attacks, data manipulation, and communication channel threats. Attention will in addition be directed to privacy preserving solutions such as homomorphic encryption, differential privacy, and secure multi-party computation, which will permit secure data sharing without compromising user privacy.

1. INTRODUCTION

With the rapid advancement of both artificial intelligence (AI) (Balas et al., 2026) and automation, autonomous systems are quickly becoming parts of an increasing variety of domains including, but not limited to, transportation, healthcare, smart cities, and industrial automation. Such systems are becoming dependent on cooperative systems of networks, sensors and AI enabled decision making capabilities to function independently with little human intervention. However, increasing dependence on autonomous technology opens a wide range of security issues that places these systems in jeopardy of sophisticated cyberattack vectors such as data breaches, adversarial attacks, and system intrusions (Sharma et al., 2025). Conventional security systems (Singh and Dubey, 2015) are inadequate to address these issues, as the systems themselves are dynamic in nature, decentralized and operational in real-time in nature. This has resulted in the development of AI based intrusion detection systems (IDS) and privacy preserving systems (Singh and Kaushik, 2023a; Singh and Kaushik, 2023b, Singh and Kaushik, 2023c), that are a part of the security and resilience of autonomous systems (Fnu et al., 2025). AI based intrusion detection systems use machine learning (Raturi et al., 2025) and deep learning models for the detection of real time anomalies and malicious events. In contrast to conventional security systems, which are based on rules, AI based IDS can analyze large data sets, identify patterns, and determine whether unknown threats exist. Privacy preserving methods, such as secure federated learning, allow for the training of AI based systems in a collaborative manner, with the caveat that sensitive data are not revealed during training. This decentralised approach is conducive to improved security in relation to the data while still allowing efficiency in the threat evaluation systems used in an AI driven world. Further security in relation to these systems, in relation to the data, is provided by the use of blockchain technology (Xue et al., 2024), which is a decentralised and immutable method of securing communications and validating transaction and interactions in the autonomous network. The question of securing autonomous systems is a continuing problem due to adversarial attack vectors aiming at the AI systems in operation in regard to the various automated systems. The need of good defences utilizing adversarial training, differential privacy and anomaly detection algorithms needs to be put in place to deal with those threats. Moreover, regulatory issues create a quagmire too, owing to the fact that the need for efficiency of policies in relation to cybersecurity (Krishna et al., 2024) and compliance programs, has to keep pace with the evolving threats in respect of AI guided autonomous technologies (Moulahi et al, 2023).

1.1 Overview of autonomous systems in various domains

The use of automated systems in various areas has been made possible, for example, by artificial intelligence (AI), machine learning (ML) and sensitive sensor technology to perform tasks with minimal human intervention (Vashishth et al. 2026). In the transport sector automated vehicles (AVs) have energy mobility, since it provides a higher level of safety, traffic control, and a reduction in fuel consumption by using intelligent route planning and real-time decision-making processes. By using AI-based perception, planning and control systems in self-driving vehicles, aircraft and public transport systems which move themselves, an efficient way will be found to use the movable parts in an active environment. In the medical sector, automated systems are essential in robotic surgery, AI- diagnostics and remote patient monitoring. By way of example, medical robots are able to perform operation with great accuracy, while AI- Diagnostics are able to interpret medical images and detect diseases and with a high degree of accuracy, medical care improves. Automating systems will likewise be used in “smart” cities in order to improve urban infrastructures or the management of natural resources (Vashishth et al. 2025). AI can be incorporated into security, intelligent traffic control and automated waste disposal systems to make the public more secure, reduce energy consumption or improve the quality of urban planning. Autonomous drones and robotic systems can be used to help in disaster response and environmental monitoring, and they can provide real-time information to help in important decision making. In industrial automation, autonomous robots or AI predictive maintenance systems are used to improve the manufacturing position, creating automatic paths to new gains in productivity or decreases in overall business cost. Additionally, in agriculture, autonomous drones or AI machinery help facilitate crop monitoring, irrigation, and harvesting and so help to develop a sustainable farming practice. But the broad use of autonomous systems poses many difficult security and privacy initiatives. Because all components and systems are linked, have real-time data tracking and must have AI-influenced decisions, it becomes continually possible to be affected by cyber-attack vulnerability, adversarial attacks, and hacking. It becomes vitally necessary to make autonomous systems acceptable from a security standpoint for public use and acceptance and trust. This leads to the creation of effective security frameworks, AI-based intrusion detection frameworks, and privacy solutions necessary to protect the autonomous systems in various areas (Abdić, 2024).

1.2 Growing cybersecurity threats in autonomous technologies

As self-directed technologies grow over the years and are implemented across vital areas like transport, health care, smart cities, and industrial automation, a prominent target emerges for higher quality cyber risks. It is obvious that all these systems depend on decentralized networks, resource allocation of data in real time, on which their security weaknesses rely, as do their decisions based on means of AI. Troublemakers can easily use these weaknesses in the autonomous systems to obtain attacks such as data breaches, denial of service (DoS) attacks, adversarial AI deception and sensor manipulation, where they can disorganize and stop operations, thus causing extreme risks to their operation, safety of people and their rights to privacy. For example, in the autonomous vehicles the hackers are able to deceive the inputs from the sensors, or GPS signals, by making wrong decisions on route to be taken, with effects on the passengers as well as on the people. The attack which impresses most in the autonomous techniques threats to its security is the adversarial AI assaults. Its success depends on the inputs which safely lead to the deceived AI models. The success of these assaults depends on the abilities of the systems of decision-making using AI as is too great. They lead to mis-locations of bodies or to bypassing the security arrangements laid down, e.g. and which will happen under certain circumstances. Autonomous instruments are producing and controlling great masses of secret data in which they are that soon for breaches of security and for identities by every reason desirable. This gets worse, as a risk, when it is discovered by people without permission for example in operational robotic medical systems, surveillance systems or in the industrial automation networks. Such systems are able to cause an exposure of their secret information which causes a very great deal of deep and serious ethical and other questions of responsibility. The securing of the autonomy from the dangers of DDoS (Distributed Denial of service) and ransomware is an area of great importance. Because of their dependence on cloud computing and edge computing (Roohani et al., 2026; Suriyan et al., 2026), these systems' networks are vulnerable to cyberattacks that may interrupt their functioning or force them to pay ransom for reinstating their functioning. The decentralized character and requirement of real-time operations make it more complicated to protect the systems from cyber criminals, so traditional modes of improving the security of these systems are incapable of providing the necessary protection. Advanced forms of cybersecurity like currently being conceived like AI-driven intruder detection systems (IDS), and the use of blockchain technology for protecting data integrity and privacy, as in federated models for instance, are being explored to allow autonomy. The bolstering of the cybersecurity resilience of autonomous systems is a vital necessity for their safe and trustworthy development and creation of an environment of public trust and regulatory acceptance in an increasingly automated and digital world (Singh et al., 2024).

2. CYBERSECURITY CHALLENGES IN AUTONOMOUS SYSTEMS

The increasing use of autonomous systems in a variety of applications results in a variety of complex cybersecurity challenges which threaten their reliability, safety, and integrity. They rely on the function of their components in system-oriented, dynamic environments and are consequently vulnerable to a wide variety of cybercriminal activities such as hacking, malicious attack by adversarial AI systems, the loss of data integrity and attacks on the networks of which they form part. By developing security proactively, autonomous systems stand the best chance of improving their resilience against cyber threats, and therefore permitting their safe ethical usage in real-world applications (Rani et al., 2024).

2.1 Threat Landscape in Autonomous Systems

As the introduction of autonomous technology within industries such as transport, health care, intelligent cities, and industry has increased, exposure to cyber-attack has also increased significantly. Cyber criminals are constantly seeking vulnerabilities within autonomous systems such as the associated communications systems, used to decide the path of the autonomous system or even deciding which action to take. The autonomously developed technology has a number of risk aspects relating to its environment. The autonomous system attack types such as data theft, adversarial AI, denial of service, ransomware, and sensor spoofing. Furthermore, research in this area is important to enhance cyber security against questions about autonomous technologies. One of the serious threats faced by autonomous technology is that of adversarial ai, where crime is possible through the introduction of man manipulative input data causing the output of the AI to be at best absurd if not violent. For example, in autonomous vehicles, input data could be slight alterations of traffic signs causing the accidental input to the AI of erroneous information regarding the limit of traffic speeds or signals to stop. In an autonomous mode this could cause roads traffic accidents. Again, unwanted results from adversarial crime on medical AI could cause disease assessments of diagnostic models to produce unwanted and potentially dangerous results . These risks indicate the sensibility of AI towards slight and unnoticed perturbations which can have very serious consequences in the real world. Secondly one of many serious risks to cyber security exists of data theft and unauthorized access to legitimate information. Autonomous systems generate a great deal of data, including personal, financial, and operational data. The more that hackers target health-care robotics, smart surveillance, or industrial automation systems, the more private data they will have access, resulting in identity theft, financial fraud, and corporate espionage. Moreover, the increased use of cloud data and edge storage results in a greater likelihood that cybercriminals could obtain and change data in transit, thus requiring protection of data integrity and secured channels for communication to guarantee confidentiality and reliability of autonomous systems (Sharma et al. 2025) .

Figure 1. Autonomous System Architecture

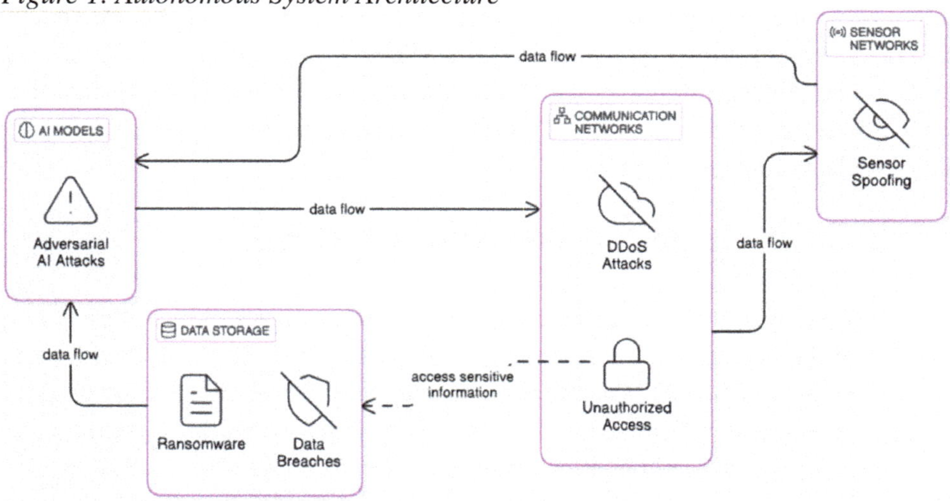

The cyberattack controversial were concurrent with other major threats to the safety of cyber systems, such as concerns about target distributed denial of service (DDoS) attacks as well as ransomware threats. Hackers can penetrate autonomous networks through excessive traffic and displace the effect of loading on systems that may affect their function or disable their interactions. For example, if DDoS threats were directed to smart increasing city structures, traffic management systems could be truly disabled and jammed up by heavy traffic otherwise present to the control system. Similarly, autonomous health-care robots face ransomware signatures that leave locked-up medical devices or personal files locked with no access granted medical staff unless a ransom is paid. Numerous autonomous technologies are highly reliant upon LiDAR, radar, global positioning systems (GPS) and other sensors for navigation and decision making. These sensor inputs may be subject to manipulation by attackers sending out false signals, causing GPS locations to be jammed, or injecting deceptive data into sensor networks. This can result in the dislocation of autonomous drones, faulting robotic surgeries or inaccurate environmental testing in smart cities. Therefore, it is compulsory to secure the data within sensors by way of encryption, authentication, and the detection of anomalies. The sensor-based attacks such as described here are becoming a great concern generally due to the fact that autonomous systems represent a very dynamic and interconnected target for attack by those in the cybercriminal world and by nation-state attackers. Attacks may even be contrived within the organizations, such as military or governmental. In view of the fact that normal security methodologies are based upon pre-defined rules and rely on static bracing for protection, the requirements for the protection of these living and moving systems call for unique thought. The changing face of the threats which must be dealt with requires the use of combined approaches such as cyber-secure AI powered systems, data integrity systems based on block-chain rules, secure federated learning systems in face-to-face processing of protected-data AI and constant threat detections. The methods of security must be of a proactive nature to ensure that autonomous systems are capable of the advanced threat detections, responses and controls in real time needed to sustain their security, prognosis and resilience in a world that is becoming even more digital (Fouda et al., 2024).

2.2 Privacy Concerns in AI-Enabled Autonomous Systems

As AI-controlled autonomous systems develop in their market applications, they are dealing with large volumes of sensitive data and privacy problems. Such systems function in different areas including health, transport, and smart cities, which encompass the subjects of collection, analysis, and storage of personal, bio-metric, and geographical date. The lack of sufficient privacy thrusts means that there are opportunities for greater invasion of data from sources of intrusion and surveillance. The following are the major problems associated with AI autonomous systems.

1. **Risks due to data collection and surveillance:** Autonomous systems that depend on AI need data collection to increase their function, improve operations, and personalize user experience. Data collection is continuous and raises concerns with regard to surveillance and user privacy. Smart cities utilize AI in the form of cameras and IoT sensors for monitoring traffic and other public areas, but too much surveillance can violate privacy rights of the individual. Autonomous vehicles, which collect and analyze real-time location, process's location data that could be used improperly and to track individuals, and raises security and ethical obligations. The key to the privacy issue is the balancing of the advantages of data-use AI systems with privacy rights of users and the misuse of personal data (Holubenko et al., 2025).
2. **Weaknesses of unauthorized access and data breaches:** Because the autonomous systems store and transmit sensitive data in communications across networks, the systems are appealing targets for cybercriminals. Data breaches of the AI systems identify those patients who contain confidential records and data, which leads to identity theft or financial data fraud. The breaches in the AI systems for financial purposes can hack into financial transactions and personal banking data. Unauthorized access to AI models and datasets can cause adversarial efforts (maliciously changing the AI predictive mechanisms for fraudulent operations). The best ways to protect against these attacks would be to strengthen encryption methods, set up access authorization, and use blockchain for the storage of safe and secure data.
3. **Privacy Risks in Federated Learning and AI Training:** Federated learning is increasingly being used to develop AI models in decentralized systems in order to eliminate the problems of information for centralization. The training of AI models on machines intervening in federated learning systems enlightens the problems of privacy and information protection, proposing several privacy implications. Secure aggregation protocols, as well as, differential cryptography and privacy must be brought forth in order to enlighten the attacks of information privacy in AI modeling of syndicated active systems (Devineni, 2024).
4. **Regulatory Frameworks for Ethical Governance:** In order for autonomous decisions to be made in the fields of AI, it must be ensured that accurate regulatory boards can be insighted which leads to the revelations of ethical decision-making law matters that need some changes. National and International governments as well as regulatory bodies are also involved, and have been seen formulating law, which will contribute to guarding personal privacy from unauthorized data gathering and misuse of decision while following others laboriously. It must be realized of integrated models of AI systems of governance which lead to bidding the logical source before handed at the inspections on board as well as system and operational transparency, as well of incorporating design valuations

as accorded from privacy which will lead to commercial rights and civil rights of the users (Shahana et al., 2024).

Through Privacy preserving strong mechanisms which would constitute operating encrypted data processing, the transparent federated learning and decentralized identification and associated functionalities of AI active systems will lead to more educational utilization of the system, through safety and integrity where trust values are enhanced to the user, and where the systems will progressively evolve to a front of privacy as conscientious and duly able to be aided for requisitioning rights (Reddy et al., 2023).

3. AI-BASED INTRUSION DETECTION FOR AUTONOMOUS SYSTEMS

As autonomous systems become more integrated into critical sectors such as transportation, healthcare, and smart cities, the demand for robust cybersecurity mechanisms has increased significantly.

3.1 Basics of Intrusion Detection Systems (IDS) in Autonomous Systems

Intrusion Detection Systems (IDS) are an important defensive mechanism against various forms of cyber-attack on autonomous systems. Because of their reliance on diverse sensor systems, networks, and artificial intelligence through which they perform their decision-making, these systems are also increasingly at risk from sophisticated types of cyber-attack. The dangers such as Malware infection, Denial of Service attack, probing for data, and unauthorized attempts at access to the systems, all these activities could have a profound effect on the efficiency, reliability, and safety, of the autonomous operation. IDS monitors network traffic, system logs, and system applications for evidence of security violations with the aim of responding to these in real-time, thus providing higher intrinsic resilience to autonomous technologies. Autonomous systems operate in a dynamic and unpredictable situation which makes them especially vulnerable to the occurrence of cyber-attack. In self-driving vehicles, for example, an attacker, having identified a vulnerability in the V2V communication layer could send misleading sensor data to vehicles, leading them to false instructions to proceed. Alternatively, in smart healthcare systems, cybercriminals could aim to obtain unauthorized access to patient records, or disrupt the functioning of IoT-enabled medical devices. IDS provide a means of counteracting some of these dangers through a range of sophisticated means, including signature detection, anomaly detection, and behavioural analysis, all of which forms may be serviceable in detecting and counteracting potential threats before they cause any harm to themselves. An effective Intrusion Detection System (IDS) can distinguish between normal activity on the system and suspicious behavior, thereby reducing false alarms and achieving accurate threat detection. Modern intrusion detection systems (IDSs) employ artificial intelligence (AI) and machine learning (ML) techniques to help in accurate detection, adaptability to the ever-increasing classes of pinprick (cyber) attack, and better response times. These intelligent security systems process large quantities of data to find patterns and anomalies that are indicative of (actual or attempted) cyberattack, allowing proactive defence methods to be used. IDS may also be connected to an intrusion prevention system (IPS) to effect automatic remedial action such as blocking malicious IP addresses, isolating infected nodes, or warning system administrators of possible security breaches. In spite of their effectiveness, normal IDSs face difficulties in their ability to react to the rapidly developing nature of different types of cyberattack, particularly in the area of autonomous systems, where security is a rapidly developing risk.

Figure 2. Intrusion Detection Systems (IDS) in Autonomous Systems

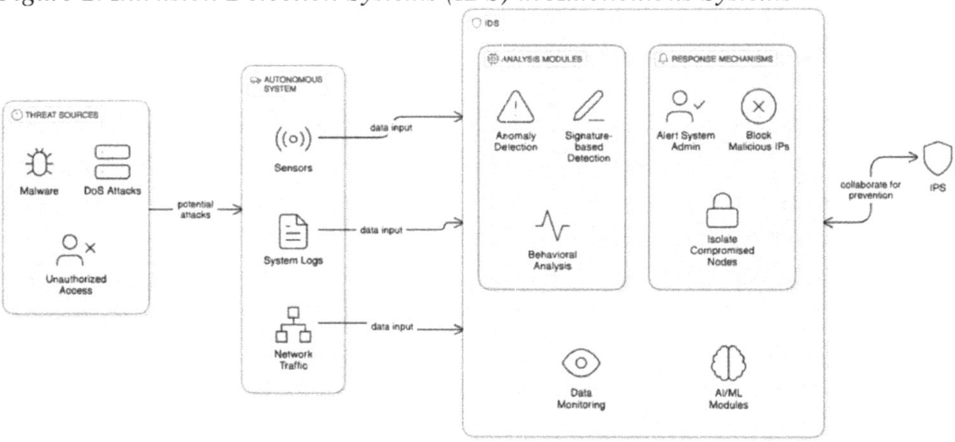

To help meet these issues a modern breed of IDS has been produced using the techniques of artificial intelligence (AI) to increase the security of autonomous systems. The new IDS use deep learning, federated learning, and distributed computer security mechanisms to counter cyber threats effectively. By continuously learning from new attack patterns and using decentralised security storage models AI driven IDS can greatly increase the cyber security response proportion of the autonomous technologies, thus ensuring a safer and more reliable operation of transports systems, healthcare systems and smart city technologies possible.

3.1.1 Types of Intrusion Detection Systems (IDS) in Autonomous Systems

Intrusion Detection Systems (IDS) represent an intrinsic part of the mechanism that is used for securing autonomous systems. They detect, in real-time, unauthorized access, suspicious activities and cyber threats. IDS can be classified, in a broad manner, in the following two types, according to their placement and their monitoring scope:

1. **Network-Based Intrusion Detection Systems:** A Network-Based Intrusion Detection System (NIDS), is performed by a mechanism for monitoring network traffic and analyzing data packets to determine if they contain signs of malicious activities or intrusions. They are usually used at strategic points in the network (e.g., gateways, routers, network segments), through continuous inspection of the communication flows of interconnected devices in an autonomous environment. By performing analysis of the packets head, the data contained in the packets and the pattern of the network traffic, NIDS can determine non-conformities that point out that the cyber threats are occurring, like denial-of-service (DoS) attacks, man-in-the-middle (MITM) attacks, malware distribution, and so on. In the autonomous systems' case, the mechanism of NIDS is especially suitable for protecting communication channels between the autonomous agents. Similarly, in smart city infrastructures, they can be instrumental in protecting interactivities within the Internet of Things (IoT) (Joshi et al., 2024a; Joshi et al., 2024b; Joshi et al., 2025), including surveillance cameras, traffic management systems, and energy grids against unauthorized access and intrusions into networks. However, detection models have a high false positive rate requiring constant adaptation (Sindiramutty et al., 2025) with such models frequently requiring refinement.
2. **Host Based Intrusion Detection Systems (HIDS):** A Host-based Intrusion Detection System (HIDS) is one that is incorporated on a device or endpoint so it can monitor its behavior, log files and application activities for potential evil doing. Whereas NIDS are focused on the detection of traffic within a network, HIDS apparatuses gain much closer scrutiny on a system level and their action patterns towards events such as; unauthorized log-in attempts, file modification, unusual calls to the system, and the like. HIDS is useful as it has applications for an additional layer of protection afforded by electronics at the devices themselves, with respect to detecting any internal aberration or problems. HIDS becomes a particularly important tool in autonomous health systems, in which medical

systems as well as IoT-enabled sensors are collecting and processing sensitive patient information.

3.2 Machine Learning and Deep Learning for Anomaly Detection

As autonomous systems grow larger and more complex, it becomes more difficult for traditional rule-based methods of cybersecurity to detect. patterns of evolution in threats. Machine learning (ML) and deep learning (DL) have developed as powerful technologies for anomaly detection within intrusion detection systems (IDS), and can allow autonomous systems to discover malicious activities, identify zero-day attacks as well as increasing the detection of security incidents more adaptable. Anomaly detection based on machine learning relies for its effectiveness on the power of statistical models and pattern recognition in order to distinguish between 'normal' and 'suspicious' behaviour within an autonomous system. The supervised learning techniques include the use of decision trees, support vector machines (SVM) or random forests. These techniques form IDS models developed from a well-formed labelled data set, which contained both normal activities and malicious activities. Such models, trained from previous patterns of attack can provide information about similar, real-time threats. But supervised learning techniques require extensive, well labelled data sets and cannot therefore be as effective when dealing with new cyber-attacks. In order to overcome this difficulty, unsupervised learning and deep learning techniques are more widely applied to anomaly detection. Unsupervised techniques such as clustering (k-means, DBSCAN etc.) or autoencoders are useful to analyse behavioural patterns of systems without being constrained by programme definitions of attacks. They detect deviation from normal/non suspicious patterns, and highlight (sometimes by alerting) possible threat sources, even when no similarity with already-known attack signatures is demonstrated. Techniques of deep-learning based intrusion detection techniques, such as those using convolution networks (CNN) or recurrent neural networks (RNN) are more effective at detecting anomalies. In these cases, features may be extracted automatically from network traffic or from system logs.^ Such models are very adept at detecting attacks with complex patterns where there is stealth or polymorphic malware present. Long Short-Term Memory (LSTM) networks and Generative Adversarial Networks (GANs) have also been considered for use for anomaly detection in autonomous systems. The IDS models based on LSTM networks employ sequential data, such as time series data of network traffic, for detection of patterns that are abnormal. The GANs produce synthetic attacks for attack scenarios for the stabilization of detection of the legitimate traffic operations in the IDS. These types of models will be beneficial for the increase of adaptability of the IDS, since they would carry on adapting through learning from the attack of the adversary. The use of Federated Learning (FL) will promote the use of artificial intelligence for IDS in the sense that anomaly detection in not only decentralized, but also does not violate privacy since it can take place. In FL, then autonomous devices make use of these devices for collaboration for training of a common IDS model, but without the exchange working of the raw data, so as to not violate the privacy of the data and thereby assist in improving the knowledge of threats. This has great value in smug cities or autonomous vehicles and the health systems, where all data must be kept particularly secret. Future research in adaptive AI models, lightweight DL models, and reactions of AI to the threats detected will be very helpful for the constantly increasing application of systems for cyber-security employed in autonomous systems (Yodha et al., 2024).

3.3 Real-Time Threat Detection and Response Mechanisms

Cybersecurity activities for autonomous systems depend on real-time detection and response mechanisms to detect and mitigate cyber threats that are evolving dynamically. Since autonomous technologies are deployed in highly dynamic environments, any delay due to the recognition or mitigation of security breaches can result in massive disruptions, leakage of sensitive data, and even physical injury. Unlike some previous models of security, which rely on static rule-based detection, contemporary autonomous systems rely on AI-based adaptive security models that have the capacity to detect, assess, and remediate threats in real time. These functions lend great integrity, confidentiality, and availability of significant system functions; therefore, since they are vital in sectors such as freight transportation, telemedicine, intelligent cities, and process controls. These activities depend on the continued monitoring of network traffic, sensor data, and system logs in order to recognize malicious activity or anomalies. Modern IDS systems powered by machine learning (ML) and deep learning (DL) analyze current system performance data so that they can recognize both known and previously unknown attacks. Normal IDS systems may recognize previously common cyberattacks, but they always have difficulty defending against new threats. In order to fight this, anomaly detection models rely on unsupervised learning mechanisms such as clustering algorithms, auto-encoders, and generative adversarial networks (GAN) in order to enable the recognition of deviations from normal system activity. For instance, in the case of autonomous vehicles, real time detection mechanisms analyze and evaluate Vehicle to Vehicle (V2V) as well as vehicle to architecture (V2I) communications in order to recognize and eliminate possible cyber threats such as GPS spoofing, attacks of a denial of service (DoS) nature, and unauthorized control orders. For example, in industrial applications, a robotic system can automatically stop operations and alert security if it detects unauthorized changes to its controlling parameters. For cloud-based autonomous systems, security procedures rely on blockchain with smart contracts to enforce real-time security policies such as terminating access privileges and starting a forensic investigation (Rangaraju, 2023). In addition, edge computing (Anandaram et al., 2026; Chaudhary et al., 2024) is important for real-time detection of threats and response because it processes security data more locally rather than relying solely on centralized, cloud-based security systems, making detection of threats possible with minimal latency with AI at the edge.

4. PRIVACY-PRESERVING MECHANISMS IN AUTONOMOUS SYSTEMS

As autonomous systems are implemented widely in all application areas including transportation, health care, IoT, smart cities, and industrial applications, protection of data generated by user clients and integrity of the systems become important issues. Because these types of systems depend upon large amounts of data communicated internally and externally (to larger repositories) during operation from real-time sensors, IoT devices (Ahmad et al., 2024), etc. as well as data generated through AI, they are very vulnerable to privacy breaches, unauthorized access, and data enlistment and misappropriation.

4.1 Secure Federated Learning for Data Privacy

Federated learning (FL) is a decentralized machine learning paradigm that enhances data privacy by enabling several autonomous systems to collaboratively train artificial intelligence (AI) models without the need of disclosing raw data. This technique is especially valuable for privacy-sensitive real-world applications such as autonomous health care, smart cities, and intelligent transportation, where confidentiality of data is critical. Unlike typical AI models that need the data to be centralized at a site for training, FL enables local devices to individually train models and only share encrypted updated with a central aggregator. This considerably reduces the potential for data breaches and intrusions. For example, in autonomous healthcare systems, the data relating to patients collected from smart medical devices, wearable sensors and hospital management systems could remain locally stored while providing an input to the global improvement of the AI models. This permits compliance with privacy regulations such as GDPR (General Data Protection Regulation), HIPAA (Health Insurance Portability and Accountability Act) while allowing the output of the AI driven status diagnostic techniques to function at the optimum. Similarly, in autonomous vehicles, FL techniques would enable a self-driving vehicle to learn from the real time data available from the local driving environment without exposing the privacy of sensitive driving behaviours, geolocation data, or other passengers. Despite the advantages of the FL paradigm, it is exposed to some privacy threats such as model inversion attacks, data poisoning, and adversarial learning. Attackers may simply attempt to reconstruct the original training data from the received shared model updates which would lead to privacy violations.

4.2 Homomorphic Encryption and Differential Privacy

The application of autonomous systems generate and processes very sensitive data. The substantial problem of preserving data privacy while at the same time ensuring computational efficiency is a major challenge. To meet this challenge, there are two advanced cryptographic and privacy-protecting techniques, Homomorphic Encryption (HE) and Differential Privacy (DP) that allow the secure processing of data and protection of user confidentiality when exploitation is made of AI-driven autonomous environments. These methods also serve to reduce privacy threats in both autonomous healthcare, smart transportation, and intelligent surveillance systems, where data breaches might yield severe consequences. Homoeomorphic Encryption (HE) is a cryptographic tool which allows computation to be performed directly on data without requiring the data to be decrypted. By this means, sensitive items of information, such as medical records, vehicular telemetry or personal identifiers can remain encrypted in all processing matters and lessen the risk of exposure. HE is particularly useful in situations where there is a need for external entities to review the data without requiring clue as to the nature of the content. The autonomous healthcare is illustrated in the example of the hospitals securely processing patient data as regards AI-driven diagnostics without requiring the raw medical records to be revealed. The workings of a smart city also illustrate the efficient exploitation of encrypted data such as might come from the intelligent analysis of data from surveillance cameras and traffic sensors without clue as to the individual identity.

Figure 3. Homomorphic Encryption (HE) and Differential Privacy (DP) in Autonomous System

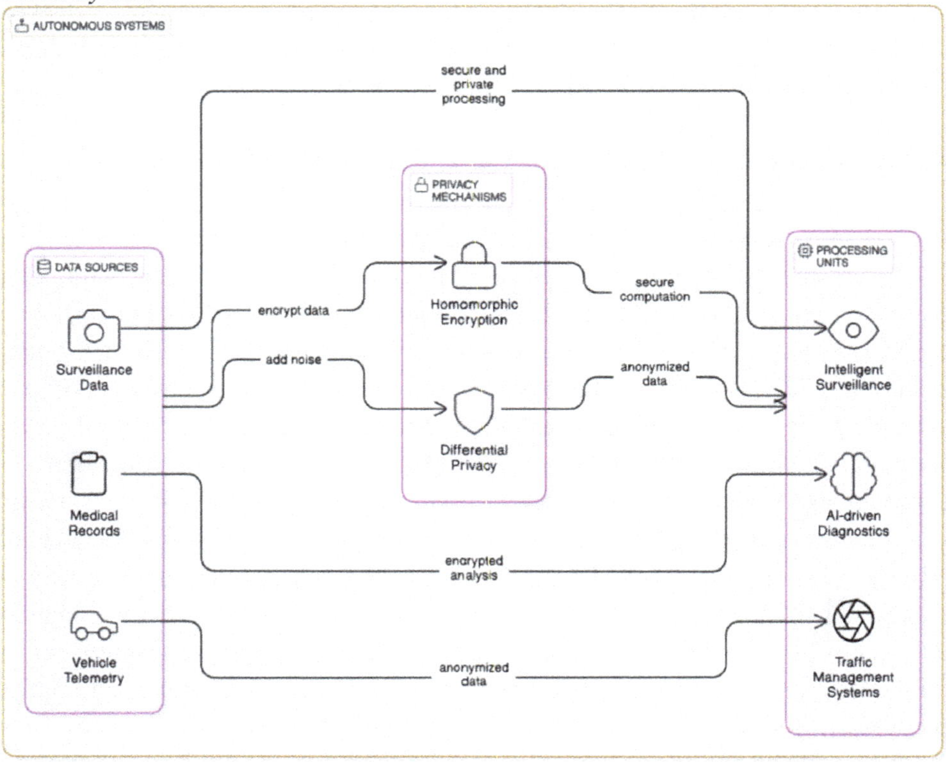

Homomorphic Encryption and Differential Privacy are powerful options for addressing hazards to privacy in autonomous systems. HE assures that sensitive data are safely shielded during processing while ensuring that DP guarantees statistical privacy by limiting visibility into covert facts. Challenges such as system overhead, integration with real-time AI models, and trade-offs between privacy and effectiveness are areas of ongoing research. Future innovations will concentrate on providing more effective solutions for routine use, permitting widespread distribution, and integrating them with blockchain technology, which has benefits of security and decentralization in dispensing secure environments (Tanikonda et al., 2022).

4.3 Blockchain for Data Integrity and Secure Communications

Blockchain technology is now being studied as a vital answer for mitigating hazards to data integrity and secure communication in autonomous systems. These systems are reliant upon inter-connected data access in real time, the use of sensors in interpolation, and empowered AI systems. A prompt engineering of data protection is a necessity for police against such negative threats as prone data changes, defects in access and introduction of viruses. Conventional security systems built upon a central model are beset with vulnerabilities characteristic of all systems dependent upon a central mechanism, such as risk of existence of single points of failure, conventional hacking, and the potential escape of data. An immediate solution for effectiveness and safety is the use of blockchain methods to replace these conventional security models. By the use of decentralization and an immutable account of data integrity, it is possible not only to be capable of preventing manipulation of the data, but also, at some time in the future can be run on multiple computers ensuring effective safe keeping of valuable data. Unchanging characteristics of storing data allows them to safe time with verities being furnished by safeguards against impossible changes. This establishes simultaneously the untainted safeguards and efficacies necessary to make way for autonomous systems dependent on reliable use of efficacious information relative to decisions. In the case of autonomous vehicles, the data protection afforded has established that road cares have data secured, relative to sensor information, position of cars by GPS systems, and communication vehicle to vehicle (V2V) in the form of distributed ledgers. In turn, provided that it is conveyed in this manner, it is practically impossible for an assailant to change or falsify anything affecting the vital information correcting to driving. The advantage of this is that relative to decisions made on autonomous driving, the information changeless, effectuation is accomplished by the accurateness of the data which it serves to furnish, effectuation is given more renovation to assist in decreasing question arising out of a host of dubious developments spread as cyber threats relating to like GPS spoofing, and fortifying information being put out from sensor department centres. Similarly, in intelligent cities, immense IoT application taking place in a host of its departments such as traffic applications, security, surveillance, environmental applications can arrive at results which fulfil the requirements for communications and data, thereby securing their safety through the impact of block chain for effectuation. The communications between various IoT de dignitaries gain the effectuation resulted from the security which is given by encryptions by means of communications, and stable revelation, as by the employment of consensus methods in the receipt of confirmations relative to transactions. There is no longer suffered reliance upon the usage of an office, which had the effect of producing the exposure to data risks occurring, and permissionless interfering. Other things in addition being improved by the use of hence not only is irreplaceable data impossible to corrupt, communications, inter alia, between all autonomous applied systems can be readily achieved through means of cryptography, furnishing salient solutions like methods of encrypted purposes applicable in areas such as identity conclusions and access control for all autonomous requirements.

5. ADVERSARIAL ATTACKS ON AI MODELS AND DEFENSE STRATEGIES

The AI models employed in autonomous systems are frequently subject to realistic dangers instigated by adversarial attacks which manipulate deficiencies in machine learning algorithms to warp decision processes of AI models. The attacks involve generally the application of inputs to the AI model which have undergone slight changes in such a manner as might escape the notice of the human observer. This affects the AI model in respect to its production of an erroneous output which can have catastrophic effects on security and operational purposes. Thus, for example, in connection with the autonomous vehicle application it is possible that as the result of the adversarially modified stop sign the AI model will misclassify it as a speed limit sign which would have disastrous ends.

5.1 Understanding Adversarial Attacks on AI-Based IDS

The attacks on AI based Intrusion Detection Systems pose a substantial cybersecurity risk as adversaries are attempting to influence machine learning models to circumvent security measures. These adversarial attacks exploit flaws in AI algorithms, causing adversaries to evade detection or create false positives, thereby destroying the trustworthiness of intrusion detection utilized in autonomous systems. Because AI based IDS relies on pattern recognition and anomaly detection in conjunction with one another in order to identify cyber threats, the algorithms may also be utilized in adversarial situations by altering the data patterns so slightly that intention is disguised from the AI model and consequently leads the IDS to misclassify the dangers as non-threatening or the reverse. This ultimately weakens the cumulative security posture of autonomous systems and provides the adversaries with a means to penetrate networks, distort data and to thwart operations while remaining undetected. One form of attack that can be executed against AI based IDS is called the evasion attack which is that sort of attack where the adversary can modify the traffic exhibited by the malicious software or the behavior of the software so as to not be detected. For example, adding noise to the patterns laid down by operating software, and slightly altering the patterns of malware signatures may be used to present an adversarial sample to the IDS model to classify the dangers trotted out as normal activity. In a similar vein, poisoning attacks target the training phase of the AI-based IDS by means of injecting contaminated or misleading samples into the data set. This manipulates the accuracy of the model over time and contributes to its propensity for misclassification so as to reduce detection of real threats. Another method of attack consists in model extraction, whereby the adversary attempts to duplicate an AI-based IDS model, sending queries into it and examining the answer, which eventually allows the attackers to construct specific adversarial constructed inputs which exploit the weaknesses of the model.

5.2 AI-Driven Defense Mechanisms Against Adversarial Attacks

Adversarial attacks are those attacks which occur as a result of the vulnerabilities of the AI-based or ML models, manipulations from the input data to deceive the AI-based algorithms which operate as a model on the normal human threats to an information environment, leading to certain types of misclassifications, breaches in security and failures of the system in the environments under examination.

1. **Adversarial training:** Adversarial training is one of the best ways to defend against adversarial attacks in Artificial Intelligence based Intrusion Detection Systems (IDS). It consists of generating adversarial examples, i.e. certain inputs disturbed in order to fool AI models provided in the training dataset. Thus, it is possible for AI systems to be exposed to adversarial disturbances in the training phase and it also learns to distinguish these types of attacks and to resist them in real life in certain situations. The best part of adversarial training is the method's ability to improve the robustness of the model by enabling it to distinguish artificially contrived from normal data inputs.
2. **Defensive distillation:** Defensive distillation is a method of removing small disturbances from AI model's sensitivity through the training of a second model on the softened probability outputs of a primary model. This method approximates the decision boundaries of AI models, which makes the distorting effect of adversarial modified inputs produce misclassifications more difficult. In this way, the effect of small disturbances on input data is eliminated which effectively reinforces AI model's resistance to evasion attacks. In these attacks, the attack uses methods of by passing the security systems through slight modifications of input data.
3. **Feature-space hardening:** Feature-space hardening improves the performance of AI-based IDSs in distinguishing between normal inputs and adversary-generated inputs, in terms of strengthening the extraction and representation of important input data. This is mainly done by improving the application of important data patterns by the model groups, while ignoring data from adversaries. Thus, by improving the sensitivity of the group models to the important input patterns and simultaneously reducing their vulnerability to fraudulent patterns, the feature-space hardening increases the overall robustness of AI-driven security systems.

Since the attack methods of adversarial trained AI models will keep changing, the methods of their defence and countervailing need to be researched and improved, that is, developed on the IT systems and information security side, that is proposed to prevent the emergence of threat production. It is supposed that the forthcoming developments of AI to be included in the information security field will logically lead into self-adaptive learning models, capable of detecting and neutralising the threats, without the necessity of manual control by humans.

6. REGULATORY AND ETHICAL CHALLENGES IN SECURING AUTONOMOUS SYSTEMS

The broad adoption of autonomous systems in transportation, health care, and smart cities leads to newfound cybersecurity, regulatory, and ethical challenges. These systems are based on AI and machine learning and interrelated networks and become vulnerable to cyber threats, data privacy issues, and algorithmic bias.

1. **Absence of Standardized Cybersecurity Rules:** One of the major regulatory issues in autonomous systems is the lack of standardized cybersecurity rules for various areas or industries. Different regions or sectors have different standards for AI security, data privacy, and accountability and, therefore, enforcement of efficient security policies is difficult. For example, strict data encryption requirements may exist in certain jurisdictions but others may not have been adopted yet in the form of any reinforcement of autonomous technology security rules.
2. **Privacy of Data plus Compliances with Legal Structures:** Various autonomous systems continually collect and process large amounts of personal and sensitive data to carry out the real-time decision making. This gives rise to important doubts about privacy problems and representation through compliance. Rules such as the general data privacy regulations (GDPR) in the EU together with the California Consumer Privacy Act (CCPA) enforce strict requirements for data collection, storage, and use. Certain privacy-preserving actions like differential privacy, secure federated learning, and encryption can be used for the resolution of these aims and will protect data but allow the AI models to extract from decentralized data sources (Kumar et al., 2021).
3. **Ethical Aspects of AI Decision Making:** The ethical aspects of autonomous systems in AI is yet another major challenge. Autonomous technologies are designed to make decisions based on machine learning systems, which would have biases or side effects. For instance, self-driving cars must make split second ethical decisions with respect to critical situations when for example

they have to make the decision as to whom to save in the event to a collision which is unavoidable. The diagnostics and recommendations of autonomous AI health systems must ensure fairness in relation to these decisions. The lack of transparency of AI decision making processes which is sometimes referred to as the black-box phenomenon, raises discussion with regard to accountability. The incorporation of ethical AI guidelines, explainable AIs (XAI) and guidelines for the avoidance of bias must be fed into autonomous systems to ensure fairness, accountability, and trustworthiness.

In overcoming these regulatory and ethical challenges, therefore the resulting autonomous systems are able to be both secure and trustworthy and to fulfil their role as safe, reliable systems in modern industries without negating privacy and safety, and fairness.

7. FUTURE RESEARCH DIRECTIONS AND CHALLENGES

As autonomous systems become more and more sophisticated, future research will be concerned with the enhancement of aspects such as security, safety, and resistance which these systems have to threats of evolving cyber instances. A few of the key areas which will need to be investigated relate to the robustness and reliability of these systems. Some of these future directions and their challenges is given below:

7.1. Future Research Directions in Securing Autonomous Systems

The future research directions outlined below would ensure improved security, privacy and resilience of autonomous systems to emerging cyber threats, establishing safe and reliable deployment of autonomous systems in various application domains such as transportation, healthcare and smart cities.

1. **Enhanced AI-Powered Security Systems:** One of the most promising future research directions is to improve AI-powered security systems to offer autonomous systems real-time threat detection and response. Existing machine learning (ML) and deep learning (DL) methods need to be more adaptive at discovering heretofore unknown attack vector patterns, including zero-day vulnerabilities. Explainable AI (XAI) is also an important issue in security applications, in order to ensure that AI decision-making is interpretable and to avoid the exploitation of AI models from adversarial attacks. In addition, reinforcement learning may

be employed to yield self-learning security agents which can dynamically update their defense mechanisms based on changes in the threat landscape.

2. **Privacy-Preserving AI Techniques:** The field of AI technology and models will be used more and more in autonomous systems in respect of the actual decision making by such systems and the improvement in technology in this field remains an important area of research, that is, data privacy. It is possible to employ current methods available such as secure federated learning, homomorphic encryption, and differential privacy for the training of AI models on a multiplicity of devices with no risk being engendered by the exposure of raw data.
3. **Blockchain for Integrity and Secure Communication of Data:** Blockchain technology provides a decentralized and non-replicable security technology that will augment integrity of data and secure communication to Autonomous Systems. Future research has to delve into Lightweight but scalable blockchain technology for Autonomous Networks in which the validation of transactions and Smart contracts can be achieved without degrading the performance of the system. Fusion of Blockchain technology and AI-based Intrusion Detection Systems (IDS) will provide greater traceability of cyber threats and enhance accountability in autonomous decision making.
4. **Proactive Threat Intelligence and Automated Security Response:** Autonomous Systems shall need historically proactive cyber security mechanisms which have predictive capabilities for pre-emptive detection and mitigation of threats before damage is inflicted. Future research should explore AI-driven intelligence platforms for cyber threat intelligence which can assess historical attack patterns world-wide, predict deficiencies, and cure them in a predictive mode. Mechanisms for automated security response for example Self-Healing AI-Based Architectures will bestow with autonomous systems the ability to autonomously detect and neutralize cyber threats without human intervention.

7.2 Challenges in Securing Autonomous Systems

It is critical to address the below challenges for the safe and secure deployment of autonomous systems across industries. Continuous innovation in AI-Based Security Frameworks technology, collaborative research on the part of policy makers and researchers and the production of larger-scale, privacy-protecting solutions will all be essential to overcoming the challenges enumerated below, that alone will ensure that autonomous technologies will have continuity of resilience in response to threats of the Cyber.

1. **Versatility and Pluralism of Autonomous Systems:** Autonomous systems are versatile and are found in all echelons of the economy from transport to health care to manufacturing and smart cities. The security demands are different for each industry, making it difficult to apply uniform security solutions. The complexity of these systems, which can involve artificial intelligence (AI), internet of things (IoT), cloud computing, and edge devices, increases the sphere of attack and makes flawed systems security management very difficult. The security level must be versatile to meet different infrastructures and ensure the minimum adverse effect on the functioning of the forms.
2. **Flaws in AI/ML Algorithms and computer security systems:** AI driven autonomous systems are sustained by ML algorithms for decision-making. However, they are subject to adversarial attacks which allow the attackers to poison the model and falsify the data. The inbuilt flaws in these algorithms can allow attackers to induce the system to make erroneous decisions such as the misinterpretation of road signs by self-driving cars or faulty diagnosis by intelligent health care systems. Security against adversarial attack and the security of ML pipelines against data poisoning are still major stumbling blocks.
3. **Real-time Threat Detection and Responsiveness:** Autonomous systems need real-time threat detection and threat response systems to counteract the threat of problems that may arise through cyber threats. Unfortunately, classical intrusion detection systems (IDS) and security systems may not be responsive enough to contend with the speed and complexity of the cyber threats that arise in relation to the autonomous systems due to autonomous decision-making processes.
4. **Data Privacy and Secure Communications:** Autonomous systems generate and manipulate vast amounts of sensitive data, including personal data, medical data, and financial material. The problem of protecting the privacy of such data, while allowing open communication among the interconnected systems of autonomous media is a formidable one. The Artificial Intelligence technologies designed to protect privacy such as the various forms of federated learning and differential privacy, have potential, but at the expense of considerably greater computational overhead, and may at the same time be harmful to the performance of the system in question. A balance between security considerations and the efficiency of operation is an essential requirement for the mass uptake of the technology (Kumar, etal., 2025)
5. **Scalability and Resource Limitations:** There are many autonomous systems like drones, IoT sensors and robotic systems that strictly operate under resource restraints, with respect to available computing power, available memory, and energy consumption. The building in of strong security mechanisms in such systems without overloading their potential computational ability is no easy task (Vikas, 2023). There is much work to be done on lightweight security

mechanisms including strong encryption, authentication, anomaly detection etc., which would not take away from the performance of the concerned systems.

6. **Incorporation of Blockchain for Security:** It has been proposed that blockchain technology may be utilised to provide data integrity, security in transactions and decentralized trust in autonomous systems. However, the incorporation of blockchain into autonomous systems, which require it to operate in real time, pose inherent scalability and latency problems. In addition, the high computational costs and energy usage that would be involved in problems such as blockchain technology will militate against widespread acceptance of such technology to be utilized. New lightweight blockchain architectures need to be researched, specifically aimed at the autonomous environment.

8. CONCLUSION

The emergence of autonomous systems in various sectors, which include transport, health, and smart cities, has also created security problems. The development of AI, IoT, and sophisticated high-performance computing in these processes renders these systems vulnerable, thus exposing them to cyber threats, adversarial attacks, and data breach threats. In this chapter, we proposed a variety of AI-based intrusion detection methods of security, such as machine learning and deep learning techniques for anomaly detection, real-time threat detection, and response strategies. Some of these issues have been addressed by privacy-preserving techniques such as secure federated learning, homomorphic encryption and blockchain based security frameworks. We presented the deficiencies of adversarial attacks on AI models, while the complexity of defence mechanisms including adversarial training and explainable AI can increase the resilience of those security frameworks in future use. Nevertheless, despite the progress made in the field of cybersecurity, there are numerous challenges, including: the complexity of securing diverse autonomous systems; the scalability of AI-based technology security systems; regulatory and compliance issues; and ethical issues related to AI-based decision making. These problems will not disappear without extensive research, innovation and collaboration between researchers, industrialists, and decision-makers alike. Thus, future research must focus on both the investigation of lightweight security solutions, the enhancement of real-time threat detection capabilities, as well as the over-checking of AI-based security technology, particularly through blockchain and quantum cryptographic techniques.

REFERENCES

Abdić, A. (2024). AI-Based Intrusion Detection Systems: Challenges and Future Directions. *CIT. Journal of Computing and Information Technology*, *4*(1).

Ahmad, A., Singh, V., Joshi, K., Sindhu, K. S., Singh, S., & Kaushik, V. D. (2024, January). Emergency evacuation information system using IoT. In International Conference on Universal Threats in Expert Applications and Solutions (pp. 331-344). Singapore: Springer Nature Singapore.

Anandaram, H., Shreenidhi, K. S., Joshi, K., Sidhu, K. S., Kiruthigha, R., Singh, V., & Tripathi, A. (2026). A Case Study on Resource Management Industrial IoT Application. In *Building Data-Driven Edge Systems for Business Success* (pp. 1–26). IGI Global Scientific Publishing.

Balas, V., Pandey, H., Bin Ali, M., Singh, V., & Kumar, A. (Eds.). (2026). *Recent Advances in Smart Communication Technologies for a Sustainable Future*. IGI Global Scientific Publishing. https://doi.org/1

Chaudhary, S., Singh, V., Malik, N., Sidhu, K. S., Kirola, M., & Joshi, K. (2024, October). Digital Signature Security Enhancement: Using X-OR Operation on RSA and AES with 4096 Bits Key Length. In *2024 International Conference on Computing, Sciences and Communications (ICCSC)* (pp. 1-6). IEEE. DOI: 10.1109/ICCSC62048.2024.10830431

Devineni, S. K. (2024). AI in data privacy and security. [IJAIML]. *International Journal of Artificial Intelligence and Machine Learning*, *3*(01), 35–49.

Fnu, H., Kaushik, K., Gaur, P., Saratchandran, D. V., Thapliyal, A. G., Sidhu, K. S., & Singh, V. (2025, April). Multi-Agent Systems for Collaborative Financial Decision-Making Over Distributed Network Architectures. In 2025 3rd International Conference on Advancement in Computation & Computer Technologies (InCACCT) (pp. 939-944). IEEE. DOI: 10.1109/InCACCT65424.2025.11011435

Fouda, M. M., Fadlullah, Z. M., Ibrahem, M. I., & Kato, N. (2024). Privacy-Preserving Data-Driven Learning Models for Emerging Communication Networks: A Comprehensive Survey. *IEEE Communications Surveys and Tutorials*.

Holubenko, V., Gaspar, D., Leal, R., & Silva, P. (2025). Autonomous intrusion detection for IoT: A decentralized and privacy preserving approach. *International Journal of Information Security*, *24*(1), 7. DOI: 10.1007/s10207-024-00926-9

Joshi, K., Singh, V., Awasthi, P., Anandaram, H., Sidhu, K. S., & Kumar, P. (2024, November). Smart Dam Water Monitoring using Enabled Internet of Things. In 2024 5th International Conference on Data Intelligence and Cognitive Informatics (ICDICI) (pp. 239-243). IEEE. DOI: 10.1109/ICDICI62993.2024.10810761

Joshi, K., Yadav, H., Gupta, S., Singh, V., Sidhu, K. S., & Kukreti, R. (2025, April). Handling Security Aspects in the Internet of Things: Latest Challenges and Measures to Mitigate Risks. In 2025 3rd International Conference on Communication, Security, and Artificial Intelligence (ICCSAI) (Vol. 3, pp. 1434-1439). IEEE.

Joshi, P., Singh, V., Thapliyal, N., Joshi, K., Bhatt, A., & Mahur, M. (2024, October). Intelligent IoT-Enabled Real-Time Monitoring System for Logistics Management. In *2024 International Conference on Computing, Sciences and Communications (ICCSC)* (pp. 1-5). IEEE. DOI: 10.1109/ICCSC62048.2024.10830332

Krishna, G., Singh, V., Pandey, D., Krishna, K. H., Joshi, K., & Gupta, T. (2024, May). Power trading framework of cloud-edge computing in the artificial intelligence market. In 2024 4th International Conference on Advance Computing and Innovative Technologies in Engineering (ICACITE) (pp. 1910-1918). IEEE. DOI: 10.1109/ICACITE60783.2024.10617344

Kumar, K. S., Nair, S. A. H., Roy, D. G., Rajalingam, B., & Kumar, R. S. (2021). Security and privacy-aware artificial intrusion detection system using federated machine learning. *Computers & Electrical Engineering*, *96*, 107440. DOI: 10.1016/j.compeleceng.2021.107440

Kumar, M., Gautam, K. K., Sharma, V., Samania, B., Vashishth, T. K., & Chaudhary, S. (2025). Enhancing cloud computing performance: A novel approach for optimizing energy efficiency through AI-based load balancing algorithm. *Proceedings of the 2025 International Conference on Intelligent Computing and Knowledge Extraction (ICICKE)*, Bengaluru, India, 1–6. IEEE. DOI: 10.1109/ICICKE65317.2025.11136754

Moulahi, T., Jabbar, R., Alabdulatif, A., Abbas, S., El Khediri, S., Zidi, S., & Rizwan, M. (2023). Privacy-preserving federated learning cyber-threat detection for intelligent transport systems with blockchain-based security. *Expert Systems: International Journal of Knowledge Engineering and Neural Networks*, *40*(5), e13103. DOI: 10.1111/exsy.13103

Rangaraju, S. (2023). Secure by intelligence: Enhancing products with AI-driven security measures. *EPH-International Journal of Science and Engineering*, *9*(3), 36–41.

Rani, S., Kumar, N., Srivastva, A., & Sharma, A. (2024, October). Leveraging artificial intelligence and machine learning for enhanced privacy and security. In *2024 IEEE International Conference on Blockchain and Distributed Systems Security (ICBDS)* (pp. 1-6). IEEE. DOI: 10.1109/ICBDS61829.2024.10837569

Raturi, A., Mula, K. G., Joshi, K., Sidhu, K. S., Chauhan, S. S., & Singh, V. (2025, April). Comparative Analysis of Stock Price Prediction with Multiple Machine Learning Model. In 2025 3rd International Conference on Communication, Security, and Artificial Intelligence (ICCSAI) (Vol. 3, pp. 1428-1433). IEEE. DOI: 10.1109/ICCSAI64074.2025.11064599

Reddy, P. C. S., Muller, P. S., Koka, S. N., Sharma, V., Sharma, N., & Mukherjee, S. (2023). Detection of encrypted and malicious network traffic using deep learning. *Proceedings of the 2023 International Conference on Ambient Intelligence, Knowledge Informatics and Industrial Electronics (AIKIIE)*, Ballari, India, 1–6. IEEE. DOI: 10.1109/AIKIIE60097.2023.10390386

Roohani, B. S., Sharma, N., Kasula, V. K., Mamoria, P., Modh, N. N., Kumar, A., & Singh, V. (2026). Urban Computing Solutions in Healthcare Edge Computing. In Building Data-Driven Edge Systems for Business Success (pp. 377-400). IGI Global Scientific Publishing.

Shahana, A., Hasan, R., Farabi, S. F., Akter, J., Mahmud, M. A. A., Johora, F. T., & Suzer, G. (2024). AI-driven cybersecurity: Balancing advancements and safeguards. *Journal of Computer Science and Technology Studies*, *6*(2), 76–85. DOI: 10.32996/jcsts.2024.6.2.9

Sharma, R., Sharma, V., Vashishth, T. K., Kumar, B., Sharma, B. K., & Chaudhary, S. (2025, February). Implementation of adaptive cloud security framework for cloud computing monitoring system applications. In AIP Conference Proceedings (Vol. 3224, No. 1). AIP Publishing. DOI: 10.1063/5.0246104

Sindiramutty, S. R., Prabagaran, K. R. V., Jhanjhi, N. Z., Murugesan, R. K., Brohi, S. N., & Masud, M. (2025). Generative AI in Network Security and Intrusion Detection. In Reshaping CyberSecurity with Generative AI Techniques (pp. 77-124). IGI Global.

Singh, O., Vinoth, R., Singh, A., & Singh, N. (2024). Navigating Security Threats and Solutions using AI in Wireless Sensor Networks. *International Journal of Communication Networks and Information Security*, *16*(4), 411–427.

Singh, V., & Dubey, V. (2015). A two level image security based on Arnold transform and chaotic logistic mapping. *International Journal of Advanced Research in Computer Science and Software Engineering*, *5*(2), 883–887.

Singh, V., & Kaushik, V. D. (2023). A Brief Overview of Cyber Security Advances and Techniques Along With a Glimpse on Quantum Cryptography: Cyber Security Practices, Advances and Challenges. Exploring Cyber Criminals and Data Privacy Measures, 40-64.

Singh, V., & Kaushik, V. D. (2023). A Methodical View of Prerequisites of Picture Combination, Strategies, and Key Indicators with Usage in Real Life and Scientific Domains Facilitating Smart Ubiquitous Environment. In Computational Intelligence in Analytics and Information Systems (pp. 15-33). Apple Academic Press.

Singh, V., & Kaushik, V. D. (2023). Navigating the Landscape of Security Threat Analysis in Cloud Computing environments. In *Security and Risk Analysis for Intelligent Cloud Computing* (pp. 1–25). CRC Press.

Suriyan, K., Ganesh, S., Palaniyammal, P., Priya, G., & Singh, V. (2026). Recent Trends in Edge Computing: Challenges and Opportunities. Building Data-Driven Edge Systems for Business Success, 323-344.

Tanikonda, A., Pandey, B. K., Peddinti, S. R., & Katragadda, S. R. (2022). Advanced AI-Driven Cybersecurity Solutions for Proactive Threat Detection and Response in Complex Ecosystems. *Journal of Science and Technology*, *3*(1).

Vashishth, T. K., Sharma, V., Samania, B., Sharma, R., Singh, S., & Jajoria, P. (2025). Ethical and legal implications of AI in cybersecurity. In *Machine Intelligence Applications in Cyber-Risk Management* (pp. 387–414). IGI Global Scientific Publishing.

Vashishth, T. K., Sharma, V., Sharma, M. K., Sharma, R., Sharma, K. K., & Sharma, S. (2026). AI-driven threat detection and incident response: Advancing cybersecurity with machine learning. In Improving Threat Detection, Network Security, and Incident Response with AI (pp. 51–82). IGI Global Scientific Publishing.

Vikas, D., R. P., Kumar, D., Charan, P., Ingilela, R. S. K., & Rastogi, R. (2023). Intrusion detection in wireless sensor networks using hybrid deep belief networks and Harris Hawks optimizer. Proceedings of the 2023 4th International Conference on Electronics and Sustainable Communication Systems (ICESC), Coimbatore, India, 1631–1636. IEEE. DOI: 10.1109/ICESC57686.2023.10193270

Xue, Y., Singh, V., Singh, S., Kant, K., Pandey, S., Kumar, A., & Kaushik, V. D. (2024). Blockchain Framework for Learner Performance Prediction using Life-Brain Storm-based Light GBM Coupled Neural Network: BLOCKCHAIN FRAMEWORK FOR LEARNER PERFORMANCE PREDICTION. [JSIR]. *Journal of Scientific and Industrial Research*, *83*(6), 652–668.

Yodha, V. (2024). Leveraging Federated Learning for Privacy-Preserving Cybersecurity in Decentralized AI Systems. *Eastern European Journal for Multidisciplinary Research*, *1*(1), 89–106.

Chapter 12
Challenges and Future Directions of AI-Powered Wireless Security

S. Ida Evangeline
https://orcid.org/0000-0003-2997-7897
Government College of Engineering, Tirunelveli, India

Midhun Prabhu Chinta
Independent Researcher, USA

Alok Kumar
https://orcid.org/0000-0003-1051-0672
School of Engineering and Technology, CSJM University, Kanpur, India

Vineeta Singh
https://orcid.org/0000-0003-1084-279X
School of Engineering and Technology, Chhatrapati Shahu Ji Maharaj University, Kanpur, India

Rajasekaran Selvaraju
https://orcid.org/0000-0002-7893-9072
University of Technology and Applied Sciences, Ibri, Oman

ABSTRACT

As wireless communication systems evolving to support more complex and critical applications, the integration of Artificial Intelligence into their security frameworks has emerged as both a transformative opportunity and a significant source of new threats. This chapter provides a comprehensive examination of the current challenges and future advancements in AI-driven wireless security. An analysis of the

DOI: 10.4018/979-8-3373-3576-6.ch012

: © 2026, IGI Global Scientific Publishing. Copying or distributing in print or electronic forms without written permission of IGI Global Scientific Publishing is prohibited. Use of this chapter to enerative artificial intelligence (AI) technologies is expressly prohibited. The publisher reserves all rights to license its use for generative AI training and machine learning model development.

primary threats relevant to this field, encompassing adversarial machine learning attacks, model opacity, data poisoning, and the performance trade-offs linked to the deployment of AI in decentralized and resource-constrained settings is illustrated. The dynamic nature of wireless systems—characterized by mobility, interference, and a variety of devices—substantially increases the susceptibility of AI-driven security models. This chapter presents a prospective framework for scholars, and policymakers to develop reliable, comprehensible, and resilient wireless systems in the era of AI by integrating insights from machine learning, cybersecurity, and wireless communication.

1. INTRODUCTION

The digital world is changing a lot because wireless communication technologies are moving so fast. This means that smart cities (Singh and Kaushik, 2023a), self-driving cars, industrial automation, and next-generation healthcare systems can all stay connected all the time. As wireless networks become more complex, varied, and important to business, keeping them safe is a very important and always-changing task. Artificial Intelligence (AI), especially machine learning (ML), has given us strong new tools to help with security issues (Singh and Kaushik, 2023b; Singh and Kaushik, 2023b). Artificial intelligence could make a big difference in wireless security by letting us find attacks before they happen, automating responses, and using dynamic defense systems that can change as the amount and variety of wireless data grows (Liu et al., 2024).

Combining AI with wireless technologies, on the other hand, comes with its own risks. People often forget that adding AI systems to wireless networks makes them easier to hack. Adversarial examples, data poisoning, and model extraction attacks are big threats to machine learning models. These threats can't be stopped by regular security measures (Chaudhary et al., 2024). Many AI systems also work like "black boxes," which means that it's hard to see how they make decisions. This is a big problem for systems that need to be trusted, held accountable, and explained. In wireless environments, these problems are worse because the channels, the networks' topologies, and the devices that are low on energy can all change at any time. You can also change connections at any time. This creates a complicated and often unstable balance between how well AI works and how strong wireless networks are (Zhao et al., 2024).

This chapter begins a thorough examination of the advantages and disadvantages of AI-driven wireless security. The talk starts by talking about the main problems that are stopping AI from being used a lot in wireless security settings. Some of the issues that have been noticed are latency, extra computing power, and the fact that it's hard to make broad statements about different situations. There are more problems with algorithms because enemies can change them and there isn't enough training data.

This chapter talks a lot about new technologies that are being made to solve these problems and make wireless networks last longer. Some of the most important changes are the creation of lightweight, energy-efficient AI models that can be used at the edge; the use of federated and privacy-preserving learning frameworks that protect user data and cut down on centralized attack vectors; and the creation of self-healing and autonomous security systems that can respond to threats right away. The talk covers cutting-edge topics like AI that can withstand quantum attacks, digital twin simulations for proactive threat modelling, and how to use AI to create 6G architectures.

This chapter talks about how important it is to put together ideas from wireless communications, cybersecurity, and machine learning to get the whole picture. To fully understand the benefits and drawbacks of using AI in wireless security, we need to look at it from many different angles. The goal of this chapter is to help researchers, engineers, and policymakers by looking at the present and making guesses about what might happen in the future. This document talks about the main problems and gives a plan for how to use AI to make wireless networks that are safe, smart, and strong.

2. TECHNICAL AND HISTORICAL FOUNDATIONS

It is important to build a technical base that connects both fields in order to understand the possibilities and problems of adding AI to wireless security systems. Wireless communication and AI-based security systems have developed separately, each with distinct concepts, designs, and challenges. Their merger, which is promising, creates a design area with many dimensions that needs a deep understanding of both fields.

Wireless communication systems are not centralized by nature; they are open and always changing. They connect mobile devices, sensors, access points, and infrastructure nodes with radio frequency (RF) transmission. Protocols control how information is modulated, sent, received, and decoded over channels that may be noisy or hostile. These protocols run the systems (Kaur et al., 2024). Wireless environments are more likely to have problems than wired networks. The problems that were found include path loss, multipath fading, shadowing, and interference. Movement also causes changes in both time and space, which is why wireless links are only temporary. The traits talked about affect both the reliability and quality of service, and they also create certain weaknesses, like eavesdropping, jamming, spoofing, and session hijacking (Saunders & Aragón-Zavala, 2024).

In the past, cryptographic methods and rule-based intrusion detection systems (IDS) have been used to keep wireless networks safe. The methods described are not very flexible or scalable. For instance, rule-based systems often miss zero-day attacks or small issues in large, mixed environments. People are more interested in using AI, especially machine learning (Raturi et al., 2025), to make wireless security better (Gharghan & Hashim, 2024).

AI and ML change the way we think about cybersecurity (Xue et al., 2024) by making it more based on data. Some examples of supervised learning methods are decision trees, support vector machines, and deep neural networks. They learn to tell what network activity is or find known threat indicators by looking at labelled datasets. Unsupervised learning algorithms, including k-means clustering and auto-encoders, can detect new or unusual patterns without knowing what kind of attack it is. Reinforcement learning (RL), wherein agents ascertain optimal actions through environmental interaction, has been suggested for dynamic resource allocation and jamming mitigation in wireless networks. Generative Adversarial Networks (GANs) have been employed to replicate attack patterns, thereby enhancing the robustness of defensive strategies (Ozkan-Okay et al., 2024).

ML models are not a cure-all, even with these benefits. It can be hard to get the clean, representative training data they need to work well in wireless situations. Wireless data isn't stable; it has a lot of noise, and it can change a lot over time because of changes in how people use it, the environment, and network settings. This is called "concept drift," and it happens when models trained on old data don't work as well when the patterns change. Also, ML models, especially deep learning architectures, need a lot of memory and processing power, which makes it hard to use them on edge devices like IoT nodes and mobile terminals (Wu & Ismail, 2024).

A big worry is that it's hard to understand AI decisions. Deep learning models are very adaptable, but they often act like black boxes, not telling you why a certain choice or category was made. This lack of clarity is bad for both ethics and operations, especially in systems that are critical to the mission. Wrong security responses could cause service interruptions, safety risks, or even legal problems. Thus, it is essential to guarantee that AI decisions are interpretable, verifiable, and consistent with policy objectives, as this constitutes a fundamental characteristic of secure wireless AI systems (Isidro et al., 2024).

When adding AI to wireless security, it's important to think about real-time limits, energy limits, and the fact that different hardware has different capabilities. AI in the cloud has a lot of processing power, but moving important security tasks to the cloud raises concerns about latency, availability, and data exposure. This has caused a shift toward edge intelligence, where models are trained and run close to where the data comes from. Some of the strategies being looked at right now to help this trend are edge caching, on-device inference, and collaborative learning (Arulmurugan et al., 2024;).

This chapter sets the stage for the discussions that will follow. Combining AI with wireless communication has a lot of potential, but it also has some major problems, such as complexity versus interpretability, centralization versus decentralization, and resilience versus flexibility. The next parts will look at the specific technical and operational problems and the new ideas that have been put forward to solve them.

3. MAJOR ISSUES WITH WIRELESS SECURITY THAT USES AI

Adding AI to wireless security systems could change the game in a big way, but it also brings up a lot of problems that need to be solved. These problems happen because AI algorithms can't handle the changing nature of wireless environments well. It is very important to understand and deal with these problems when moving AI from experimental models to real-world use in wireless systems in order to build trust, resilience, and effectiveness.

3.1 Being easy for enemies to attack

One major problem with AI models is that people with bad intentions can easily change them. In supervised learning systems, small changes to input data that are carefully made and can't be seen by people or regular detectors can lead to big mistakes in classification. In wireless systems, these kinds of attacks could be seen as fake traffic or signals from an enemy that machine learning-based Intrusion Detection Systems (IDS) might not be able to find or might mistakenly set off false alarms. A hacker can change the physical-layer signal settings just a little bit to trick a deep learning classifier into thinking that bad traffic is safe. Attacks like this can get around security measures and are hard to find with normal methods. Another kind of attack is poisoning, which happens when the training dataset is changed in a way that isn't obvious, causing the model to learn wrong or biased patterns. This situation is very dangerous for systems that use federated or collaborative learning, where training happens on many different nodes, some of which may be hacked. Furthermore, model inversion and extraction attacks would enable adversaries to access sensitive training data or duplicate the functionalities of proprietary models, thereby threatening user privacy and intellectual property (Shihab et al., 2024).

3.2 Not being able to explain things and not trusting them

Deep neural networks and other AI models can be very accurate, but they don't always tell you how they came to a decision. Because it isn't easy to understand, it's hard to use in wireless systems that are important for safety, like connected cars, military communications, and smart healthcare. If the operators don't know how the model makes decisions, they can't trust the output or fix bugs. This is especially true when it comes to deciding whether behaviour is harmful or not. Also, making decisions that aren't clear makes it harder to follow changing laws that require explainable AI (XAI) and algorithmic accountability, especially in cases where the AI directly affects people (Senevirathna et al., 2024).

3.3 Restrictions on Resources and Time Constraints

Wireless systems usually have very strict limits on power, bandwidth, and latency. AI models, especially deep architectures, need a lot of memory and processing power. This makes them too expensive to run on edge devices like smartphones, IoT sensors, or UAVs. You need lightweight models that can make quick inferences with little help from hardware to process data in real time, like finding a jamming attack on a control channel or an unauthorized access attempt (Rozlomii et al., 2024).

There are also a lot of different kinds of devices in wireless spaces, each with its own energy and capacity. It is still hard to make AI security models that can be changed across this range without making them less reliable or less effective. Because of spotty connectivity and decentralized structures, it's harder to coordinate from one place.

3.4 Changes in the Availability and Distribution of Data

AI models work best when the training data is good, diverse, and a good example of what they are supposed to do. Many wireless security applications limit access to tagged datasets. It's hard to find labelled samples of real-world assaults, especially the newer or more advanced ones, because they are so rare. Also, privacy laws and worries can make it hard for devices or groups to share information with each other.

The environment of wireless signals is also not fixed. Channel conditions are affected by things like how mobile people are, the weather, and the fact that devices are different from each other. Models that were trained before don't work anymore when the statistical properties of the data change over time. This is known as idea drift. AI systems may not work as well, miss things, or get more false positives if they don't have ways to constantly retrain or adapt (Celik & Eltawil, 2024).

3.5 Putting systems together and managing the lifecycle

Adding AI to operational wireless systems is a complicated process that involves more than just training models. It includes safely deploying models, keeping track of versions, updating models, and keeping an eye on performance across many platforms that are spread out. The AI pipeline needs to be safe from the time it collects data until it makes a model for the system to be safe. Any of the steps can have flaws that can affect the decision-making process, leak private information, or just make people less trusting of the system.

Over time, artificial intelligence models are changed and updated. To stay effective against changing threats, you need to keep learning and getting regular updates. In edge or federated contexts, secure and scalable update systems can make problems worse, especially when communication links are unreliable or when nodes are only available sometimes (Cao et al., 2024).

4. NEW FIELDS OF STUDY AND IMPROVEMENTS IN TECHNOLOGY

As the flaws in current AI-driven wireless security systems become more obvious, research is moving toward more advanced solutions that focus on resilience, efficiency, adaptability, and moral issues. The improvements are meant to fix the problems with current models and also predict what future wireless ecosystems will need, especially those that will be affected by 6G, widespread edge intelligence, and the problems that quantum computing will bring. This part looks at the main areas of research and development that are changing how secure AI works in wireless settings.

4.1 AI Models That Are Light and Don't Use Much Power

To make AI usable in the real world, it is important to create lightweight architectures that work well on wireless nodes with limited power, like mobile devices, embedded systems, and IoT sensors. Most of the time, traditional deep learning models are too big to run on a device in real time. This has made people more interested in model compression methods like pruning, quantization, and knowledge distillation, which add very little extra work while still getting good results.

Neuromorphic computing is a new way of doing things in this field. The system uses AI processing that works like the human brain, which uses very little power. These methods improve real-time security analytics and anomaly detection at the edge, which lowers latency and the chance that data will be exposed while it is being sent. Also, hardware-software co-design projects are making custom AI accelerators for wireless security applications that improve the balance of speed, accuracy, and energy efficiency (Khashan et al., 2024).

4.2 Federated Learning and Security That Keeps Your Information Safe

Federated learning (FL) is a big step forward in fixing issues with data ownership and privacy. Federated Learning lets many devices work together to train one model without having to share the raw data. This protects users' information and makes it less likely that data breaches will happen in one place. FL is great for smart homes, car networks, and the Internet of Things in factories that need wireless security. In these cases, distributed nodes have local security data that is either too big or too private to keep in one place (Fnu et al., 2025).

Recent studies concentrate on bolstering the resilience and security of federated learning against malevolent entities. FL protocols use safe aggregation, differential privacy, Byzantine-resilient aggregation, and homomorphic encryption to protect against threats like model inversion and data poisoning. Personalized federated learning also lets each device change global models to better fit its own needs, which makes detection more accurate in many cases (Bukhari et al., 2024).

4.3 Security systems that can fix themselves and change

Self-healing and autonomously adaptive systems that can react to threats and changes in the environment in real time are a promising area for AI-driven wireless security. The systems use online learning, meta-learning, and reinforcement learning to change defense strategies, retrain models, and get things back to normal after an attack.

Adaptive systems can find concept drift in wireless traffic data and make changes to the model on their own, so performance stays the same even when threat profiles change. Reinforcement learning agents are great for figuring out how to defend yourself by trying things out in fake environments. They can be used in dynamic spectrum defense, jamming mitigation, and adaptive intrusion detection, among other things. These systems have fail-safe features that stop people from making bad choices when they don't have any help (Inshi et al., 2025).

4.4 Quantum-Resilient AI and the Use of Post-Quantum Cryptography

As quantum computing gets stronger, the basic ideas behind traditional cryptography security are in danger. Researchers are looking for AI frameworks that are quantum-resilient, which means they will still work safely even when quantum computing is common. This means that you can use PQC algorithms and AI models together to make it possible to authenticate wirelessly, share keys, and share models securely.

AI is also being used to make quantum-resilient protocols better and test them. This makes a full feedback loop, where machine learning helps make cryptographic primitives that will keep AI models safe (Sankalp et al., 2025). As quantum computing gets stronger, the basic ideas behind traditional cryptography security are in danger. Researchers are looking for AI frameworks that are quantum-resilient, which means they will still work safely even when quantum computing is common. This means that you can use PQC algorithms and AI models together to make it possible to authenticate wirelessly, share keys, and share models securely.

4.5 Security for the Whole System in Future Wireless Ecosystems

We need to rethink how wireless security works as we move toward AI-native 6G networks. In the future, networks will have a lot of devices, be very mobile, have very low latency, and have smart control across several layers, from physical to application. Security solutions should be built into the network stack instead of being an extra feature (Narooka et al., 2025a; Narooka et al., 2025b).

Wireless AI systems can now use zero-trust architecture (ZTA). This means that you can't trust any device, node, or subsystem right away. ZTA can always check and give permission for access based on how people act and what they know about the situation when it works with AI. Digital twins are another option. They are digital copies of wireless environments that can be used to test defenses, train AI models, and practice dangerous situations before they happen. These simulations make proactive security easier because they let operators find weaknesses and practice how to respond in a safe environment (Liu et al., 2024).

AI, edge computing (Krishna et al., 2024; Roohani et al., 2026; Suriyan et al., 2026), and the Internet of Things (IoT) (Joshi et al., 2024a; Joshi et al., 2024b; Joshi et al., 2024c; Anandaram et al., 2026) are also driving the development of security frameworks that can work across layers and are aware of their surroundings. These frameworks look for attacks that are complicated and come from many different places by using both physical-layer metrics, like signal fingerprinting, and higher-layer analytics, like patterns of network behavior. These kinds of approaches are very important for keeping autonomous systems, mission-critical communications, and industrial automation safe when many different kinds of threats come together.

5. EXAMPLES AND CASE STUDIES FROM THE REAL WORLD

Theoretical research has created a strong foundation for AI-driven wireless security. However, the best way to understand how it works is through real-world examples and case studies. These real-world or high-fidelity simulation deployments not only show how well AI models work in real life, but they also show the operational problems, trade-offs, and performance differences that affect how systems are designed and used. AI-enhanced wireless security systems are starting to show real benefits in areas like smart cities, industrial automation, defense, and self-driving cars. These systems are also showing where more standards and new ideas are needed.

5.1 Smart Cities and Public Works

Security is very important in smart cities because they have many wireless networks that handle things like traffic control, surveillance, public Wi-Fi, and emergency response. Pilot projects in cities have used AI-based anomaly detection systems to look at patterns in sensor data from vehicles and people that come in over 5G or Wi-Fi networks. These systems use machine learning classifiers like Support Vector Machines (SVMs) and Convolutional Neural Networks (CNNs) to find strange behaviors that could mean spoofing attacks, unauthorized access, or hacked devices.

One interesting case was a traffic signal control system in a medium-sized European city that used reinforcement learning to make traffic flow better and look for signs of a cyber-physical attack by keeping an eye out for changes in predicted patterns. Early installations had high detection rates, but they also showed that better ways to handle false positives and methods for adaptive retraining are needed to deal with changes in the environment, like road work or traffic that changes with the seasons (Singh et al., 2024).

5.2 The Industrial Internet of Things and Smart Manufacturing

More and more factories and power plants are using IIoT designs. Wireless sensors and actuators need to work reliably and safely in these situations, even when things are tough and there is a lot of interference. In one of the case studies at a semiconductor fabrication plant, federated learning was used to create a distributed intrusion detection system (IDS) that could process equipment telemetry and find problems in production lines without the company having to share sensitive data across organizational boundaries.

This system worked well to improve anomaly detection while keeping privacy safe, but it had trouble with different kinds of devices, uneven data distributions, and synchronization latency across federated clients. These problems make it clear that we need better ways to combine data and protocols for real-time model updates to make sure that complex and decentralized industrial environments can grow and stay strong (Wisdom et al., 2024).

5.3 Communications and Tactical Networks in the Military

The defense sector has become a testbed for AI-enhanced wireless security because secure tactical communication is so important. Recent field trials supported by NATO used deep reinforcement learning algorithms on Software-Defined Radios (SDRs) to fight jamming attacks in mobile ad hoc networks (MANETs) in real time. These algorithms let the frequency, power, and modulation schemes change on their own based on real-time threat assessment. This made communication much more reliable in the hostile environment.

But the deployment also showed how hard it is to train RL agents when there aren't many rewards, like when it takes a long time to get feedback on how well defense techniques are working. Furthermore, sustaining model efficacy in dynamic and contentious contexts continues to pose a significant challenge, particularly when opponents utilize rapidly evolving or AI-enhanced offensive strategies (Alexan et al., 2024).

5.4 Vehicles that drive themselves and are connected

AI models are now being used in vehicular networks, which include vehicle-to-everything (V2X) communication, to do things like real-time authentication, find cooperative misbehavior, and find location spoofing. One version that was tested in North America used a deep autoencoder to look for strange patterns in CAN bus or V2X communications that could mean cyberattacks or broken nodes. The model has learned from both real-world driving data and made-up situations.

The system worked well in the lab, but when it was tested in the field, it was affected by sensor noise, changes in movement patterns, and changes in the environment, such as weather and topography. These findings underscore the imperative for resilient pre-processing pipelines, continuous validation, and hybrid training approaches that integrate both real and synthetic datasets (Ahmad et al., 2024).

5.5 Wireless Security Digital Twin Simulations

Digital twin systems are high fidelity mockups of real-life wireless networks, which can be used for testing, simulation, and other analysis mainly for prediction. They are also being invested on by many companies. These platforms are supposed work with field experiments. For example, for one of two European telecommunication OEMs and the University Turbatian AI lab, digital twins were employed to simulate in a safe, desirable environment 5G network slices under various attack scenarios, including DDoS and model poisoning. The models simulated the impact that various kinds of AI-enabled protection services, such as federated learning, ensemble detection models, and adaptive firewall policies had on an attack scenario. The methods were well-run to subject algorithms to test stress in conditions with a difficult simulation in reality. Genuine simulations of physical layers dynamics and users' behavior are very high in importance for simulation precision. This is a key issue in the development of digital twins (Tao et al., 2024).

6. RULES, MORALS, AND MAKING THINGS THE SAME

As wireless security systems operated by AI took a step forward today from being in the testing stage to actual implementation, it is crucial to ensure that new technologies comply to existing rules, regulations and moral standards. These systems must not only be well integrated with technology, they must also fulfill the ever-increasing demands for openness, accountability, universal fundamental rights to privacy and interoperability with other systems across the world. This section discusses some of the ethical responsibilities, regulations, and standardization initiatives that are ensuring AI in wireless security is danger-free.

6.1 Making algorithms simple and clear

Even with the crucial importance of wireless in security systems, using AI models that are effective requires some effort on the part of human beings for model interpretability. AI systems are used in many areas, such as public safety networks, self-driving cars, and healthcare telemetry. They need to be able to make decisions in real time that affect the safety of users, the security of data, or the safety of people. However, numerous effective AI models, such as deep learning architectures, do not elucidate the rationale behind their decisions. This "black-box" quality makes it hard to trust, especially if the AI model raises an anomaly flag or blocks access to a device and the users, regulators, or operators don't know why it happened.

To fix this, security frameworks are using research on Explainable AI (XAI) to figure out how to make sense of how complex models make choices. Some of these are saliency mapping, layer-wise relevance propagation, and surrogate modeling. These methods still need to be changed to work with wireless data, which often has signals that change in time, space, and number of dimensions. For practical use, explainability must be both technically sound and user-centered, so that both network engineers and non-expert stakeholders can understand the explanations (Agbadamasi et al., 2025).

6.2 Fairness, Bias, and Acceptance

The data that AI models were trained on can make them biased. When it comes to wireless security, this can make it so that devices, geographic areas, or usage habits can't detect threats as well. For instance, a model that is mostly trained on data from urban networks might not work well in rural or developing areas because it could miss threats or sound false alarms. These kinds of differences could lead to discrimination, being denied services, or an unfair distribution of security resources.

To make these problems, we require to use fair learning methods, collect data in an even foundation, and be had to search for performance markers that could be wrong. Researchers must consider the moral questions raised by enforcing security, which can be an infringement on civil liberties when security surveillance is an intrusion on the soil. Unfortunately, finding the right level of monitoring is crucial for wireless networks in smart cities to make sure threats are discovered as quickly as possible without an excessive or deficient degree of patenting (Kondra et al., 2025).

6.3 Privacy of Data and Permission

AI-powered security systems involve the use of vast amounts of data, which may contain private or sensitive information. In wireless networks, anyone can extract and infer behavioural acts and recognize people from beam fingerprint, identifiers, and traces. Wireless device manufacturers, service providers, and AI developers must collaborate to ensure that privacy measures are not only technically robust but also legally mandated and ethically grounded (Salami et al., 2025). These systems must not only be well integrated with technology, they must also fulfil the ever-increasing demands for openness, accountability, universal fundamental rights to privacy and interoperability with other systems across the world. This section discusses some of the ethical responsibilities, regulations, and standardization initiatives that are ensuring AI in wireless security is danger-free.

6.4 Making Things the Same and Making Them Work Together

AI-driven security needs standardized protocols because there are so many different types of wireless systems, such as cellular networks, IoT platforms, satellite linkages, and ad hoc networks. Vendors and domains can't work together because there aren't any standard ways to represent models, extract features, classify threats, or validate models. Some of the groups working to set standards for the use of safe AI in wireless networks are the Institute of Electrical and Electronics Engineers (IEEE), the 3rd Generation Partnership Project (3GPP), the International Telecommunication Union (ITU), and ETSI. Some important areas are:

Secure model lifecycle management: rules for training, deploying, upgrading, and shutting down models.
Benchmarking and evaluation: using standard datasets and metrics to see how accurate, reliable, and fast detection is.
AI threat taxonomies are a way to put together old and new security issues that are specific to AI, such as adversarial ML and model inversion.
Ethical Guidelines: rules for using AI safely in telecommunications and wireless fields.

To build trust across borders, follow the rules, and make it easier for different wireless networks to work together, these standards must be followed (Gyrard et al., 2025).

6.5 Governance and Coordination Among Numerous Stakeholders

In the end, making sure that AI-powered wireless devices are safe is both a technical and a regulatory challenge. When AI systems make decisions on their own in situations where safety is important, worries about liability, transparency, and oversight grow even more. When an AI model cuts off a drone's communication because it thinks it sees a threat that isn't there, the question of who is to blame comes up. It could be the developer, the model trainer, the network provider, or the maker of the gadget. To find out who is in charge, you need to look at what each person did to design, train, and run the AI system.

To solve these problems, we need to use governance models that include engineers, ethicists, regulators, civil society, and end users. To make sure that people are held accountable and that problems are fixed quickly, algorithmic impact assessment, audit trails, and rules for reporting incidents are all very important.

Cooperation with other nations is very important. Our templates also demonstrate that wireless cyber-attacks tend to be transnational, and IPR/Cyber technology developed in one part of the world can be exported elsewhere (Rozenblit et al., 2025). When an AI model cuts off a drone's communication because it thinks it sees a threat that isn't there, the question of who is to blame comes up. It could be the developer, the model trainer, the network provider, or the maker of the gadget. To find out who is in charge, you need to look at what each person did to design, train, and run the AI system.

7. CONCLUSION AND FUTURE POSSIBILITIES

The need to protect wireless systems is increasing as wireless systems become more intelligent, distributed, and more - integrated into parts of society's critical infrastructure. So, the introduction of AI into wireless security systems is a significant step but also creates a whole new field of study. AI can do this by enabling wireless networks to identify, react, and anticipate attacks at a much quicker pace and location than a human would be able to do. It also has the worse perverse transparency (as it varies depending on the data) and vulnerability to enemy exploitation.

In this chapter, we have discussed the main issues associated with AI-based wireless security like adversarial attacks, model backdoors, lack of explainability and performance trade-offs. Moreover, troubles are trapped into the complexity and various issues of approximately wireless environments using study scarce resources, mobility and changing channel conditions.

REFERENCES

Agbadamasi, T. O., Opoku, L. K., Adukpo, T. K., & Mensah, N. (2025). The role of business intelligence in AI ethics: Empowering US companies to achieve transparent and responsible AI. *EPRA International Journal of Economics* [EBMS]. *Business and Management Studies*, *12*(3), 8–14.

Ahmad, U., Han, M., & Mahmood, S. (2024). Enhancing security in connected and autonomous vehicles: A pairing approach and machine learning integration. *Applied Sciences (Basel, Switzerland)*, *14*(13), 5648. DOI: 10.3390/app14135648

Alexan, W., Aly, L., Korayem, Y., Gabr, M., El-Damak, D., Fathy, A., & Mansour, H. A. (2024). Secure communication of military reconnaissance images over UAV-assisted relay networks. *IEEE Access : Practical Innovations, Open Solutions*, *12*, 78589–78610. DOI: 10.1109/ACCESS.2024.3407838

Anandaram, H., Shreenidhi, K. S., Joshi, K., Sidhu, K. S., Kiruthigha, R., Singh, V., & Tripathi, A. (2026). A Case Study on Resource Management Industrial IoT Application. In *Building Data-Driven Edge Systems for Business Success* (pp. 1–26). IGI Global Scientific Publishing.

Arulmurugan, L., Thakur, S., Dayana, R., Thenappan, S., Nagesh, B., & Sri, R. K. (2024, May). Advancing Security: Exploring AI-driven Data Encryption Solutions for Wireless Sensor Networks. In *2024 International Conference on Advances in Computing, Communication and Applied Informatics (ACCAI)* (pp. 1-6). IEEE. DOI: 10.1109/ACCAI61061.2024.10602020

Bukhari, S. M. S., Zafar, M. H., Abou Houran, M., Moosavi, S. K. R., Mansoor, M., Muaaz, M., & Sanfilippo, F. (2024). Secure and privacy-preserving intrusion detection in wireless sensor networks: Federated learning with SCNN-Bi-LSTM for enhanced reliability. *Ad Hoc Networks*, *155*, 103407. DOI: 10.1016/j.adhoc.2024.103407

Cao, Z., Gao, W., Fu, Y., & Mi, C. (2024). Wireless battery management systems: Innovations, challenges, and future perspectives. *Energies*, *17*(13), 3277. DOI: 10.3390/en17133277

Celik, A., & Eltawil, A. M. (2024). At the dawn of generative AI era: A tutorial-cum-survey on new frontiers in 6G wireless intelligence. *IEEE Open Journal of the Communications Society*, *5*, 2433–2489. DOI: 10.1109/OJCOMS.2024.3362271

Chaudhary, S., Singh, V., Malik, N., Sidhu, K. S., Kirola, M., & Joshi, K. (2024, October). Digital Signature Security Enhancement: Using X-OR Operation on RSA and AES with 4096 Bits Key Length. In *2024 International Conference on Computing, Sciences and Communications (ICCSC)* (pp. 1-6). IEEE. DOI: 10.1109/ICCSC62048.2024.10830431

Fnu, H., Kaushik, K., Gaur, P., Saratchandran, D. V., Thapliyal, A. G., Sidhu, K. S., & Singh, V. (2025, April). Multi-Agent Systems for Collaborative Financial Decision-Making Over Distributed Network Architectures. In 2025 3rd International Conference on Advancement in Computation & Computer Technologies (InCACCT) (pp. 939-944). IEEE. DOI: 10.1109/InCACCT65424.2025.11011435

Gharghan, S. K., & Hashim, H. A. (2024). A comprehensive review of elderly fall detection using wireless communication and artificial intelligence techniques. *Measurement*, *226*, 114186. DOI: 10.1016/j.measurement.2024.114186

Gyrard, A., Abedian, S., Gribbon, P., Manias, G., van Nuland, R., Zatloukal, K., Nicolae, I. E., Danciu, G., Nechifor, S., Marti-Bonmati, L., Mallol, P., Dalmiani, S., Autexier, S., Jendrossek, M., Avramidis, I., Garcia Alvarez, E., Holub, P., Blanquer, I., Boden, A., & Hussein, R. (2025). Lessons learned from European health data projects with cancer use cases: Implementation of health standards and internet of things semantic interoperability. *Journal of Medical Internet Research*, *27*, e66273. DOI: 10.2196/66273 PMID: 40126534

Inshi, S., Chowdhury, R., Taha, M. B., & Talhi, C. (2025). Enhancing Autonomy of Context-Aware Self-healing in Fog Native Environments. In *International Symposium on Foundations and Practice of Security* (pp. 336-350). Springer, Cham. DOI: 10.1007/978-3-031-87499-4_22

Isidro, Y. G. H., Niño, A. Z. P., Berrio, K. D. R., & Duarte, E. J. T. (2024). Guide to reduce opacity in AI decision making in business. *Reflexiones Contables UFPS*, *7*(1), 8–12.

Joshi, K., Singh, V., Awasthi, P., Anandaram, H., Sidhu, K. S., & Kumar, P. (2024, November). Smart Dam Water Monitoring using Enabled Internet of Things. In 2024 5th International Conference on Data Intelligence and Cognitive Informatics (ICDICI) (pp. 239-243). IEEE. DOI: 10.1109/ICDICI62993.2024.10810761

Joshi, P., Singh, V., Thapliyal, N., Joshi, K., Bhatt, A., & Mahur, M. (2024, October). Intelligent IoT-Enabled Real-Time Monitoring System for Logistics Management. In *2024 International Conference on Computing, Sciences and Communications (ICCSC)* (pp. 1-5). IEEE. DOI: 10.1109/ICCSC62048.2024.10830332

Kaur, P., Kaur, K., Verma, D. K., Singh, K., Kaushik, K., & Singh, V. (2024, September). Routing protocols in wireless sensor networks: A comprehensive review and future perspectives. In 2024 7th International Conference on Contemporary Computing and Informatics (IC3I) (Vol. 7, pp. 263-270). IEEE. DOI: 10.1109/IC3I61595.2024.10828894

Khashan, O. A., Khafajah, N. M., Alomoush, W., Alshinwan, M., & Alomari, E. (2024). Smart energy-efficient encryption for wireless multimedia sensor networks using deep learning. *IEEE Open Journal of the Communications Society, 5*, 5745–5763. DOI: 10.1109/OJCOMS.2024.3442855

Kondra, S., Medapati, S., Koripalli, M., Nandula, S. R. S. C., & Zink, J. Z. (2025). AI and Diversity, Equity, and Inclusion (DEI): Examining the Potential for AI to Mitigate Bias and Promote Inclusive Communication. *Journal of Artificial intelligence and Machine Learning, 3*(1), 1-8.

Krishna, G., Singh, V., Pandey, D., Krishna, K. H., Joshi, K., & Gupta, T. (2024, May). Power trading framework of cloud-edge computing in the artificial intelligence market. In 2024 4th International Conference on Advance Computing and Innovative Technologies in Engineering (ICACITE) (pp. 1910-1918). IEEE. DOI: 10.1109/ICACITE60783.2024.10617344

Liu, Y., Su, Z., Peng, H., Xiang, Y., Wang, W., & Li, R. (2024). Zero trust-based mobile network security architecture. *IEEE Wireless Communications, 31*(2), 82–88. DOI: 10.1109/MWC.001.2300375

Liu, Y. F., Chang, T. H., Hong, M., Wu, Z., So, A. M. C., Jorswieck, E. A., & Yu, W. (2024). A survey of recent advances in optimization methods for wireless communications. *IEEE Journal on Selected Areas in Communications, 42*(11), 2992–3031. DOI: 10.1109/JSAC.2024.3443759

Narooka, P., Kaushik, K., Kansal, S., Kejriwal, D., Mishra, R. K., & Singh, V. (2025, April). Hybrid AI Models for QoS Estimation in α-η-μ Fading and Co-Channel Interference Environments within 6G Networks. In 2025 3rd International Conference on Advancement in Computation & Computer Technologies (InCACCT) (pp. 206-210). IEEE.

Narooka, P., Verma, S., Dwivedi, A., Kaushik, K., Singh, V., & Gaur, P. (2025, April). Psi-zeta Approach for Assessing Network Efficiency in Log-Normal Fading and Spectrum Sharing Interference Environments. In 2025 3rd International Conference on Advancement in Computation & Computer Technologies (InCACCT) (pp. 886-891). IEEE. DOI: 10.1109/InCACCT65424.2025.11011391

Ozkan-Okay, M., Akin, E., Aslan, Ö., Kosunalp, S., Iliev, T., Stoyanov, I., & Beloev, I. (2024). A comprehensive survey: Evaluating the efficiency of artificial intelligence and machine learning techniques on cyber security solutions. *IEEE Access : Practical Innovations, Open Solutions*, *12*, 12229–12256. DOI: 10.1109/ACCESS.2024.3355547

Raturi, A., Mula, K. G., Joshi, K., Sidhu, K. S., Chauhan, S. S., & Singh, V. (2025, April). Comparative Analysis of Stock Price Prediction with Multiple Machine Learning Model. In 2025 3rd International Conference on Communication, Security, and Artificial Intelligence (ICCSAI) (Vol. 3, pp. 1428-1433). IEEE. DOI: 10.1109/ICCSAI64074.2025.11064599

Roohani, B. S., Sharma, N., Kasula, V. K., Mamoria, P., Modh, N. N., Kumar, A., & Singh, V. (2026). Urban Computing Solutions in Healthcare Edge Computing. In Building Data-Driven Edge Systems for Business Success (pp. 377-400). IGI Global Scientific Publishing.

Rozenblit, L., Price, A., Solomonides, A., Joseph, A. L., Srivastava, G., Labkoff, S., deBronkart, D., Singh, R., Dattani, K., Lopez-Gonzalez, M., Barr, P. J., Koski, E., Lin, B., Cheung, E., Weiner, M. G., Williams, T., Thuy Bui, T. T., & Quintana, Y. (2025). Towards a Multi-Stakeholder process for developing responsible AI governance in consumer health. *International Journal of Medical Informatics*, *195*, 105713. DOI: 10.1016/j.ijmedinf.2024.105713 PMID: 39642592

Rozlomii, I., Yarmilko, A., & Naumenko, S. (2024). Data security of IoT devices with limited resources: challenges and potential solutions. *doors, 3666*, 85-96.

Salami, I. A., Adesokan-Imran, T. O., Tiwo, O. J., Metibemu, O. C., Olutimehin, A. T., & Olaniyi, O. O. (2025). Addressing Bias and Data Privacy Concerns in AI-Driven Credit Scoring Systems Through Cybersecurity Risk Assessment. *Asian Journal of Research in Computer Science*, *18*(4), 59–82. DOI: 10.9734/ajrcos/2025/v18i4608

Sankalp, M. R., Lokapal, G., Mohan, B. A., & Basavaraj, G. N. (2025, March). Addressing Cybersecurity Challenges in 6G Networks Through AI-Driven Adaptive Defense Mechanisms and Quantum-Resilient Protocols. In *2025 International Conference on Computing for Sustainability and Intelligent Future (COMP-SIF)* (pp. 1-12). IEEE.

Saunders, S. R., & Aragón-Zavala, A. A. (2024). *Antennas and propagation for wireless communication systems*. John Wiley & Sons. DOI: 10.1002/9781119332206

Senevirathna, T., La, V. H., Marchal, S., Siniarski, B., Liyanage, M., & Wang, S. (2024). A survey on XAI for 5G and beyond security: Technical aspects, challenges and research directions. *IEEE Communications Surveys and Tutorials*.

Shihab, M. A., Marhoon, H. A., Ahmed, S. R., Radhi, A. D., & Sekhar, R. (2024). Towards Resilient Machine Learning Models: Addressing Adversarial Attacks in Wireless Sensor Network. [JRC]. *Journal of Robotics and Control*, *5*(5), 1599–1617.

Singh, B., Jain, V., Kaunert, C., & Vig, K. (2024). Shaping Highly Intelligent Internet of Things (IoT) and Wireless Sensors for Smart Cities. In *Secure and Intelligent IoT-Enabled Smart Cities* (pp. 117-140). IGI Global Scientific Publishing.

Singh, V., & Kaushik, V. D. (2023). A Brief Overview of Cyber Security Advances and Techniques Along With a Glimpse on Quantum Cryptography: Cyber Security Practices, Advances and Challenges. Exploring Cyber Criminals and Data Privacy Measures, 40-64.

Singh, V., & Kaushik, V. D. (2023). A Methodical View of Prerequisites of Picture Combination, Strategies, and Key Indicators with Usage in Real Life and Scientific Domains Facilitating Smart Ubiquitous Environment. In *Computational Intelligence in Analytics and Information Systems* (pp. 15–33). Apple Academic Press. DOI: 10.1201/9781003332312-3

Singh, V., & Kaushik, V. D. (2023). Navigating the Landscape of Security Threat Analysis in Cloud Computing environments. In *Security and Risk Analysis for Intelligent Cloud Computing* (pp. 1–25). CRC Press.

Suriyan, K., Ganesh, S., Palaniyammal, P., Priya, G., & Singh, V. (2026). Recent Trends in Edge Computing: Challenges and Opportunities. Building Data-Driven Edge Systems for Business Success, 323-344.

Tao, Z., Xu, W., Huang, Y., Wang, X., & You, X. (2024). Wireless network digital twin for 6g: Generative ai as a key enabler. *IEEE Wireless Communications*, *31*(4), 24–31. DOI: 10.1109/MWC.002.2300564

Wisdom, D. D., Vincent, O. R., Igulu, K., Hyacinth, E. A., Christian, A. U., Oduntan, O. E., & Hauni, A. G. (2024). Industrial IoT security infrastructures and threats. *Communication Technologies and Security Challenges in IoT: Present and Future*, 369-402.

Wu, Z. Y., & Ismail, M. (2024). Generalized RIS tile exclusion strategy for indoor mmWave channels under concept drift. *IEEE Transactions on Wireless Communications*, *23*(10), 13484–13498. DOI: 10.1109/TWC.2024.3402267

Xue, Y., Singh, V., Singh, S., Kant, K., Pandey, S., Kumar, A., & Kaushik, V. D. (2024). Blockchain Framework for Learner Performance Prediction using Life-Brain Storm-based Light GBM Coupled Neural Network: BLOCKCHAIN FRAMEWORK FOR LEARNER PERFORMANCE PREDICTION. [JSIR]. *Journal of Scientific and Industrial Research*, *83*(6), 652–668.

Zhao, W., Weng, T., Ruan, Y., Liu, Z., Wu, X., Zheng, X., & Kato, N. (2024). Quantum Computing in Wireless Communications and Networking: A Tutorial-Cum-Survey. *IEEE Communications Surveys and Tutorials*.

Compilation of References

Abbas, S. R., Abbas, Z., Zahir, A., & Lee, S. W. (2024, December). Federated learning in smart healthcare: a comprehensive review on privacy, security, and predictive analytics with IoT integration. In Healthcare (Vol. 12, No. 24, p. 2587). MDPI. DOI: 10.3390/healthcare12242587

Abdel Hakeem, S. A., Hussein, H. H., Kim, H., Hakeem, S. A. A., Hussein, H. H., & Kim, H. (2022). Security Requirements and Challenges of 6G Technologies and Applications. *Sensors (Basel)*, *22*(5), 1969. DOI: 10.3390/s22051969 PMID: 35271113

Abdi, A., Bennouri, H., & Keane, A. (2024). Emerging Cyber Risks & Threats in Healthcare Systems: A Case Study in Resilient Cybersecurity Solutions. 2024 13th Mediterranean Conference on Embedded Computing, MECO 2024. DOI: 10.1109/MECO62516.2024.10577790

Abdić, A. (2024). AI-Based Intrusion Detection Systems: Challenges and Future Directions. *CIT. Journal of Computing and Information Technology*, *4*(1).

Abdou, A., Sheatsley, R., Beugin, Y., Shipp, T. J., & McDaniel, P. (2021). HoneyModels: Machine Learning Honeypots. MILCOM 2022 - 2022 IEEE Military Communications Conference (MILCOM), 886–891. DOI: 10.1109/MILCOM52596.2021.9652947

Abdulrab, H., Hussin, F. A., Aziz, A. A., Awang, A., Ismail, I., & Arun Mozhi Devan, P. (2022). Reliable Fault Tolerant-Based Multipath Routing Model for Industrial Wireless Control Systems. *Applied Sciences (Basel, Switzerland)*, *12*(2), 544. Advance online publication. DOI: 10.3390/app12020544

Abumarshoud, H., Mohjazi, L., Imran, M. A., & Haas, H. (2021). LiFi Through Intelligent Reflecting Surfaces: A New Frontier for 6G? arXiv preprint arXiv:2104.02390.

Achaal, B., Adda, M., Berger, M., Ibrahim, H., & Awde, A. (2024). Study of smart grid cyber-security, examining architectures, communication networks, cyber-attacks, countermeasure techniques, and challenges. *Cybersecurity*, *7*(1), 10. Advance online publication. DOI: 10.1186/s42400-023-00200-w PMID: 38707764

Acharya, P., & Mahapatra, S. S. (2024). Exploring the Impact of Artificial Intelligence Integration on Guest Experience in the Hotel Industry. *Geo Journal of Tourism and Geosites*, *54*(2 supplement), 802–810. DOI: 10.30892/gtg.542spl04-1255

Achuthan, B., Arunkumar, B., Boobalan, M., & Chiragsun, R. L. (2021, May). Underwater communication using li-fi. In 2021 3rd International Conference on Signal Processing and Communication (ICPSC) (pp. 485-489). IEEE.

Achuthan, K., Ramanathan, S., Srinivas, S., & Raman, R. (2024). Advancing cyber-security and privacy with artificial intelligence: Current trends and future research directions. *Frontiers in Big Data*, *7*, 1497535. Advance online publication. DOI: 10.3389/fdata.2024.1497535 PMID: 39703783

Adebayo, A. S., Chukwurah, N., & Ajayi, O. O. (2025).. . *Proactive Ransomware Defense Frameworks Using Predictive Analytics and Early Detection Systems for Modern Enterprises.*, *202*(April), 83–109.

Adelakun, B. O., Majekodunmi, T. G., & Akintoye, O. S. (2024). AI and ethical accounting: Navigating challenges and opportunities. *International Journal of Advanced Economics*, *6*(6), 224–241. DOI: 10.51594/ijae.v6i6.1230

Adeniran, A. A., Onebunne, A. P., & William, P. (2024). Explainable AI (XAI) in healthcare: Enhancing trust and transparency in critical decision-making. *World Journal of Advanced Research and Reviews*, *23*(3), 2447–2658. DOI: 10.30574/wjarr.2024.23.3.2936

Adu-Manu, S., & Kofi Abdulai, J.-D., Felicia Engmann, Moses Akazue, J. K. & Appati, Godwill Enchill Baiden, G. S.-K. (2022). WSN Architectures for Environmental Monitoring Applications. *Journal of Sensors*.

Affane, A. R., & Satori, H. (2024). Machine learning attack detection based-on stochastic classifier methods for enhancing of routing security in wireless sensor networks. *Ad Hoc Networks*, *163*, 103581. DOI: 10.1016/j.adhoc.2024.103581

Agarwal, G., & Manager -Projects -Cognizant, S. (2025). Driven Security Solutions. International Journal of Advanced Research in Science, Communication and Technology (IJARSCT) International Open-Access, Double-Blind, Peer-Reviewed, Refereed, *Multidisciplinary Online Journal*, (1). https://doi.org/DOI: 10.48175/IJARSCT-23624ï

Agbadamasi, T. O., Opoku, L. K., Adukpo, T. K., & Mensah, N. (2025). The role of business intelligence in AI ethics: Empowering US companies to achieve transparent and responsible AI. *EPRA International Journal of Economics* [EBMS]. *Business and Management Studies, 12*(3), 8–14.

Aghdam, S. R., Amid, E., Furdek, M., & i Amat, A. G. (2021, October). Privacy-preserving wireless federated learning exploiting inherent hardware impairments. In 2021 IEEE 26th International Workshop on Computer Aided Modeling and Design of Communication Links and Networks (CAMAD) (pp. 1-6). IEEE.

Agrawal, S., Sarkar, S., Aouedi, O., Yenduri, G., Piamrat, K., Alazab, M., Bhattacharya, S., Maddikunta, P. K. R., & Gadekallu, T. R. (2022). Federated learning for intrusion detection system: Concepts, challenges and future directions. *Computer Communications, 195*, 346–361. DOI: 10.1016/j.comcom.2022.09.012

Agyemang, J. O., Kponyo, J. J., & Mouzna, J. (2017, November). Light fidelity (LiFi) as an alternative data transmission medium in VANET. In *2017 European Modelling Symposium (EMS)* (pp. 213-217). IEEE. DOI: 10.1109/EMS.2017.44

Ahmad, A., Singh, V., Joshi, K., Sindhu, K. S., Singh, S., & Kaushik, V. D. (2024, January). Emergency evacuation information system using IoT. In International Conference on Universal Threats in Expert Applications and Solutions (pp. 331-344). Singapore: Springer Nature Singapore.

Ahmad, R. & Wazirali, R. (2022). Machine Learning for Wireless Sensor Networks Security : An Overview of Challenges and Issues. Sensor, 22(4730).

Ahmad, U., Han, M., & Mahmood, S. (2024). Enhancing security in connected and autonomous vehicles: A pairing approach and machine learning integration. *Applied Sciences (Basel, Switzerland), 14*(13), 5648. DOI: 10.3390/app14135648

Akinrinmade, A. O., Adebile, T. M., Ezuma-Ebong, C., Bolaji, K., Ajufo, A., Adigun, A. O., Mohammad, M., Dike, J. C., & Okobi, O. E. (2023). Artificial Intelligence in Healthcare: Perception and Reality. In Cureus. Cureus, Inc. DOI: 10.7759/cureus.45594

Akter, S., Olanrewaju, R. F., & Islam, T. (2018, May). LiFi based automated shopping assistance application in IoT. In Journal of Physics: Conference Series (Vol. 1018, No. 1, p. 012001). IOP Publishing. Al_Barazanchi, I., Sahy, S. A., & Jaaz, Z. A. (2021, February). Traffic management with deployment of Li-Fi technology. [). IOP Publishing.]. *Journal of Physics: Conference Series, 1804*(1), 012141.

Akyildiz, I., & Wang, N. X. (2005). A survey on wireless mesh networks. *IEEE Communications Magazine, 43*(9), S23–S30. DOI: 10.1109/MCOM.2005.1509968

Al Nasim, M. A., Soshi, F. T. J., Biswas, P., Ferdous, A. S. M. A., Rashid, A., Biswas, A., & Gupta, K. D. (2025). Principles and Components of Federated Learning Architectures. https://arxiv.org/abs/2502.05273

Alam, S., Khan, M. A., & Wahid, A. (2021). Malware classification using gradient boosting and SHAP explanations. *Proceedings of the International Conference on Cybersecurity*, 45–58. https://doi.org/DOI: 10.1007/978-3-030-00000-0_4

Alazab, A., Khraisat, A., Singh, S., & Jan, T. (2023). Enhancing privacy-preserving intrusion detection through federated learning. *Electronics (Basel)*, *12*(16), 3382. DOI: 10.3390/electronics12163382

Albraheem, L. I., Alhudaithy, L. H., Aljaser, A. A., Aldhafian, M. R., & Bahliwah, G. M. (2018). Toward designing a Li-Fi-based hierarchical IoT architecture. *IEEE Access : Practical Innovations, Open Solutions*, *6*, 40811–40825. DOI: 10.1109/ACCESS.2018.2857627

Albraheem, L., Alhudaithy, L., Aljaser, A., AlDhafian, M., & Bahliwah, G. (2022). Towards smart sustainable cities using Li-Fi technology: Geo-location infrastructure utilizing LED street lights. *PeerJ. Computer Science*, *8*, e1009. DOI: 10.7717/peerj-cs.1009 PMID: 36091990

Albshaier, L., Almarri, S., & Albuali, A. (2025). Federated Learning for Cloud and Edge Security: A Systematic Review of Challenges and AI Opportunities. In Electronics (Switzerland) (Vol. 14, Issue 5). Multidisciplinary Digital Publishing Institute (MDPI). DOI: 10.3390/electronics14051019

Al-Dabbagh, M. S., & Ali, A. K. (2024). Employing Light Fidelity in Iot-Based Healthcare System. *Theory and Applications of Engineering Research*, *6*, 11–27. DOI: 10.9734/bpi/taer/v6/2757G

Alenezi, M., & Akour, M. (2025). AI-Driven Innovations in Software Engineering: A Review of Current Practices and Future Directions. In Applied Sciences (Switzerland) (Vol. 15, Issue 3, p. 1344). Multidisciplinary Digital Publishing Institute. DOI: 10.3390/app15031344

Alenizi, A., Mishra, S., & Baihan, A. (2024). Enhancing secure financial transactions through the synergy of blockchain and artificial intelligence. *Ain Shams Engineering Journal*, *15*(6), 102733. DOI: 10.1016/j.asej.2024.102733

Alevizos, L., & Dekker, M. (2024). *Towards an AI-Enhanced Cyber Threat Intelligence Processing Pipeline. ArXiv*. Cornell University., DOI: 10.3390/electronics13112021

Alexandrovich, S. A. (2024). Artificial Intelligence and International Security. *International Affairs*, *70*(5), 254–256. DOI: 10.21557/IAF.99985268

Alexan, W., Aly, L., Korayem, Y., Gabr, M., El-Damak, D., Fathy, A., & Mansour, H. A. (2024). Secure communication of military reconnaissance images over UAV-assisted relay networks. *IEEE Access : Practical Innovations, Open Solutions, 12*, 78589–78610. DOI: 10.1109/ACCESS.2024.3407838

Alghamdi, M. I. (2020). Survey on applications of deep learning and machine learning techniques for cyber security. *International Journal of Interactive Mobile Technologies, 14*(16), 210–224. DOI: 10.3991/ijim.v14i16.16953

Al-Hamiri, M. G., & Abd, H. J. (2023, February). A review of Lifi Technology: Principles and applications. In *2023 Second International Conference on Advanced Computer Applications (ACA)* (pp. 1-6). IEEE. DOI: 10.1109/ACA57612.2023.10346813

Alharbi, S., Attiah, A., & Alghazzawi, D. (2022). Integrating Blockchain with Artificial Intelligence to Secure IoT Networks: Future Trends. *Sustainability (Basel), 14*(23), 1–27. DOI: 10.3390/su142316002

Al-Huthaifi, R., Li, T., Huang, W., Gu, J., & Li, C. (2023). Federated learning in smart cities: Privacy and security survey. *Information Sciences, 632*, 833–857. DOI: 10.1016/j.ins.2023.03.033

Ali, A., Abd Razak, S., Othman, S. H., Eisa, T. A. E., Al-Dhaqm, A., Nasser, M., Elhassan, T., Elshafie, H., & Saif, A. (2022). Financial Fraud Detection Based on Machine Learning: A Systematic Literature Review. *Applied Sciences (Basel, Switzerland), 12*(19), 9637. Advance online publication. DOI: 10.3390/app12199637

Ali, A., & Koucheryavy, Y. (2025). *Adaptive Beam Tracking in 5G/6G mmWave networks: A Clustered Federated Learning Approach.* IEEE., DOI: 10.1109/ACCESS.2025.3563435

Ali, M., Karimipour, H., & Tariq, M. A. (2021). Integration of blockchain and federated learning for Internet of Things: Recent advances and future challenges. *Computers & Security, 108*, 102355. DOI: 10.1016/j.cose.2021.102355

Aljabri, M., Aljameel, S. S., Mohammad, R. M. A., Almotiri, S. H., Mirza, S., Anis, F. M., Aboulnour, M., Alomari, D. M., Alhamed, D. H., & Altamimi, H. S. (2021). Intelligent techniques for detecting network attacks: Review and research directions. In Sensors (Vol. 21, Issue 21, p. 7070). Multidisciplinary Digital Publishing Institute. DOI: 10.3390/s21217070

Aljewari, Y. H. K., Sinclair, T. D., Alahmed, A. S. A. N., & Alali, S. M. F. (2021, November). Smart Lighting Solution using LiFi Technology in Intelligent Buildings. In 2021 IEEE 11th International Conference on Consumer Electronics (ICCE-Berlin) (pp. 1-5). IEEE. DOI: 10.1109/ICCE-Berlin53567.2021.9720033

Almasabi, A. M., Alkhodre, A. B., Khemakhem, M., Eassa, F., Abi Sen, A. A., & Harbaoui, A. (2025). Internet of Things-Based Anomaly Detection Hybrid Framework Simulation Integration of Deep Learning and Blockchain. *Information (Basel)*, *16*(5), 406. Advance online publication. DOI: 10.3390/info16050406

Alomari, M. A., Al-andoli, M. N., Ghaleb, M., Thabit, R., Alkawsi, G., Abedalrahim, J., Alsayaydeh, J. & Gaid, A. S. A. (2025). Security of Smart Grid : Cybersecurity Issues, Potential Cyberattacks, Major Incidents, and Future Directions. 1–34.

Alquraishee, A. G. A., & Kar, J. (2014). A survey on security mechanisms and attacks in wireless sensor networks. *Contemporary Engineering Sciences*, *7*(1–4), 135–147. DOI: 10.12988/ces.2014.3949

Alrubei, S. M., Ball, E., & Rigelsford, J. M. (2022). A Secure Blockchain Platform for Supporting AI-Enabled IoT Applications at the Edge Layer. *IEEE Access : Practical Innovations, Open Solutions*, *10*, 18583–18595. DOI: 10.1109/ACCESS.2022.3151370

Alsaedi, W. K., Ahmadi, H., Khan, Z., & Grace, D. (2023). Spectrum options and allocations for 6G: A regulatory and standardization review. *IEEE Open Journal of the Communications Society*, *4*, 1787–1812. DOI: 10.1109/OJCOMS.2023.3301630

Aluwala, A. (2024). AI-Driven Anomaly Detection in Network Monitoring Techniques and Tools. Journal of Artificial Intelligence & Cloud Computing, 1–6. DOI: 10.47363/JAICC/2024(3)310

Alzahrani, A. O., & Alenazi, M. J. F. (2021). Designing a network intrusion detection system based on machine learning for software defined networks. *Future Internet*, *13*(5), 111. DOI: 10.3390/fi13050111

Aman, M., Qiao, G., & Muzzammil, M. (2021, July). Design and analysis of Li-fi underwater wireless communication system. In 2021 OES China Ocean Acoustics (COA) (pp. 1100-1103). IEEE.

Amaral, L. A., Matos, E., Tiburski, R. T., Hessel, F., Lunardi, W. T., & Marczak, S. (2016). *Internet of Things (IoT) in 5G Mobile Technologies*. Springer International Publishing.

Anandaram, H., Shreenidhi, K. S., Joshi, K., Sidhu, K. S., Kiruthigha, R., Singh, V., & Tripathi, A. (2026). A Case Study on Resource Management Industrial IoT Application. In *Building Data-Driven Edge Systems for Business Success* (pp. 1–26). IGI Global Scientific Publishing.

Angurala, M., Kaur, M., Pabbi, V., & Chopra, N. (2025). Enhancing Cybersecurity in Wireless Sensor Networks: AI Solutions to Simulated Attacks. *Journal of Information Systems Engineering & Management, 10*(3s), 217–229. DOI: 10.52783/jisem.v10i3s.373

Ansar, N., Ansari, M. S., Sharique, M., Khatoon, A., Malik, M. A., & Siddiqui, M. M. (2024). A Cutting-Edge Deep Learning Method For Enhancing IoT Security. ArXiv (Cornell University). https://doi.org//arxiv.2406.12400DOI: 10.48550

Anthi, E., Williams, L., Rhode, M., Burnap, P., & Wedgbury, A. (2021). Adversarial attacks on machine learning cybersecurity defences in Industrial Control Systems. *Journal of Information Security and Applications, 58*, 102717. DOI: 10.1016/j.jisa.2020.102717

Anwar, R. W., Abrar, M., Salam, A., & Ullah, F. (2025). Federated learning with LSTM for intrusion detection in IoT-based wireless sensor networks: A multi-dataset analysis. *PeerJ. Computer Science, 11*, 1–30. DOI: 10.7717/peerj-cs.2751 PMID: 40567740

Apruzzese, G., Andreolini, M., Colajanni, M., & Marchetti, M. (2020). Hardening random forest cyber detectors against adversarial attacks. *IEEE Transactions on Emerging Topics in Computational Intelligence, 4*(4), 427–439. DOI: 10.1109/TETCI.2019.2961157

Apruzzese, G., Andreolini, M., Ferretti, L., Marchetti, M., & Colajanni, M. (2022). Modeling Realistic Adversarial Attacks against Network Intrusion Detection Systems. *Digital Threats : Research and Practice, 3*(3), 1–19. DOI: 10.1145/3469659

Arana-Catania, M., Sonee, A., Khan, A. M., Fatehi, K., Tang, Y., Jin, B., Soligo, A., Boyle, D., Calinescu, R., Yadav, P., Ahmadi, H., Tsourdos, A., Guo, W., & Russo, A. (2025). Explainable Reinforcement and Causal Learning for Improving Trust to 6G Stakeholders. *IEEE Open Journal of the Communications Society, 6*, 4101–4125. Advance online publication. DOI: 10.1109/OJCOMS.2025.3563415

Arefin, S., Chowdhury, M., Parvez, R., Ahmed, T., Abrar, A. F. M. S., & Sumaiya, F. (2024). Understanding APT detection using Machine learning algorithms: Is superior accuracy a thing? *IEEE International Conference on Electro Information Technology*, 532–537. DOI: 10.1109/eIT60633.2024.10609886

Arora, P., Khullar, V., Kansal, I., Kumar, R., & Popli, R. (2024). Privacy-Preserving Federated Learning System (f-PPLS) for military focused area classification. *Multimedia Tools and Applications, 84*(12), 1–27. DOI: 10.1007/s11042-024-19358-7

Arrieta, A. B., Díaz-Rodríguez, N., Del Ser, J., Bennetot, A., Tabik, S., Barbado, A., & Herrera, F. (2020). Explainable Artificial Intelligence (XAI): Concepts, taxonomies, opportunities and challenges toward responsible AI. *Information Fusion, 58*, 82–115. DOI: 10.1016/j.inffus.2019.12.012

Arulmurugan, L., Thakur, S., Dayana, R., Thenappan, S., Nagesh, B., & Sri, R. K. (2024, May). Advancing Security: Exploring AI-driven Data Encryption Solutions for Wireless Sensor Networks. In *2024 International Conference on Advances in Computing, Communication and Applied Informatics (ACCAI)* (pp. 1-6). IEEE. DOI: 10.1109/ACCAI61061.2024.10602020

Arunkumar, R., & Thanasekhar, B. (2024). Heterogeneous Lifi–Wifi with multipath transmission protocol for effective access point selection and load balancing. *Wireless Networks, 30*(4), 2423–2437. DOI: 10.1007/s11276-024-03657-w

Asif, M., Khan, T. A., Taleb, N., Said, R. A., Siddiqui, S. Y., & Batool, G. (2022, February). A Proposed Architecture for Traffic Monitoring & Control System via LiFi Technology in Smart Homes. In *2022 International Conference on Business Analytics for Technology and Security (ICBATS)* (pp. 1-3). IEEE. DOI: 10.1109/ICBATS54253.2022.9759069

Aswin, K., Aswin, R., Lokesh, V., & Sugacini, M. (2016). Iot enabled by Li-Fi technology. In *National conference on communication and informatics.*

Atlam, H. F., Azad, M. A., Alzahrani, A. G., & Wills, G. (2020). A review of blockchain in internet of things and Ai. *Big Data and Cognitive Computing, 4*(4), 1–27. DOI: 10.3390/bdcc4040028

Attkan, A., & Ranga, V. (2022). Cyber-physical security for IoT networks: A comprehensive review on traditional, blockchain and artificial intelligence based key-security. *Complex & Intelligent Systems, 8*(4), 3559–3591. DOI: 10.1007/s40747-022-00667-z

Azarkasb, S. O., & Khasteh, S. H. (2025). Fog Computing Tasks Management Based on Federated Reinforcement Learning. *Journal of Grid Computing, 23*(1), 11. DOI: 10.1007/s10723-025-09796-4

Aziz, R., Banerjee, S., Bouzefrane, S., & Le Vinh, T. (2023). Exploring homomorphic encryption and differential privacy techniques towards secure federated learning paradigm. *Future Internet, 15*(9), 310. DOI: 10.3390/fi15090310

Babiceanu, R. F., & Seker, R. (2019). Cyber resilience protection for industrial internet of things: A software-defined networking approach. *Computers in Industry, 104*, 47–58. DOI: 10.1016/j.compind.2018.10.004

Bai, Z. (2022). Research on Application of Artificial Intelligence in Communication Network. *Journal of Physics: Conference Series*, *2209*(1), 12014. DOI: 10.1088/1742-6596/2209/1/012014

Balas, V., Pandey, H., Bin Ali, M., Singh, V., & Kumar, A. (Eds.). (2026). *Recent Advances in Smart Communication Technologies for a Sustainable Future*. IGI Global Scientific Publishing. https://doi.org/1

Balas, V., Pandey, H., Bin Ali, M., Singh, V., & Kumar, A. (Eds.). (2026). *Recent Advances in Smart Communication Technologies for a Sustainable Future*. IGI Global Scientific Publishing., DOI: 10.4018/979-8-3373-3541-4

Balogun, A. Y., Olaniyi, O. O., Olisa, A. O., Gbadebo, M. O., & Chinye, N. C. (2025). Enhancing Incident Response Strategies in U. S. Healthcare Cybersecurity. *Journal of Engineering Research and Reports*, *27*(2), 114–135. DOI: 10.9734/jerr/2025/v27i21399

Bangash, Y. A., Abid, Q. ud D., A, A. A. A. & Al-Salhi, Y. E. A. 2017. Security Issues and Challenges in Wireless Sensor Networks:Survey, A. (•••).. . *IAENG International Journal of Computer Science*, *44*(2), 1–15. DOI: 10.14419/ijet.v7i2.33.13861

Barbeau, M., & García-Alfaro, J. (2022). Cyber-physical defense in the quantum Era. *Scientific Reports*, *12*(1), 1905. Advance online publication. DOI: 10.1038/s41598-022-05690-1 PMID: 35115576

Barnes, B., & Daim, T. (2024). Information Security Maturity Model for Healthcare Organizations in the United States. *IEEE Transactions on Engineering Management*, *71*, 928–939. DOI: 10.1109/TEM.2021.3139836

Baskar, S., Periyanayagi, S., Shakeel, P. M., & Dhulipala, V. S. (2019). An energy persistent range-dependent regulated transmission communication model for vehicular network applications. *Computer Networks*, *152*, 144–153. DOI: 10.1016/j.comnet.2019.01.027

Batool, H., Anjum, A., Khan, A., Izzo, S., Mazzocca, C., & Jeon, G. (2024). A secure and privacy preserved infrastructure for VANETs based on federated learning with local differential privacy. *Information Sciences*, *652*, 119717. DOI: 10.1016/j.ins.2023.119717

Batool, R., Bibi, N., Alhazmi, S., & Muhammad, N. (2024). Secure Cooperative Routing in Wireless Sensor Networks. *Applied Sciences (Basel, Switzerland)*, *14*(12), 5220. Advance online publication. DOI: 10.3390/app14125220

Behiry, M. H., & Aly, M. (2024). Cyberattack detection in wireless sensor networks using a hybrid feature reduction technique with AI and machine learning methods. *Journal of Big Data*, *11*(1), 16. Advance online publication. DOI: 10.1186/s40537-023-00870-w

Berisha, V., Krantsevich, C., Hahn, P. R., Hahn, S., Dasarathy, G., Turaga, P., & Liss, J. (2021). Digital medicine and the curse of dimensionality. In npj Digital Medicine (Vol. 4, Issue 1). Nature Portfolio. DOI: 10.1038/s41746-021-00521-5

Besjedica, T., Fertalj, K., Lipovac, V., & Zakarija, I. (2023). Evolution of hybrid LiFi–WiFi networks: A survey. *Sensors (Basel)*, *23*(9), 4252. DOI: 10.3390/s23094252 PMID: 37177456

Bhatt, U., Xiang, A., Sharma, S., Weller, A., Taly, A., Jia, Y., Ghosh, J., Puri, R., Moura, J. M. F., & Eckersley, P. (2020). Explainable machine learning in deployment. FAT* 2020 - Proceedings of the 2020 Conference on Fairness, Accountability, and Transparency, 648–657. DOI: 10.1145/3351095.3375624

Bhattarai, M., Kaymak, M. C., Barron, R., Nebgen, B., Rasmussen, K., & Alexandrov, B. S. (2023). Robust Adversarial Defense by Tensor Factorization. Proceedings - 22nd IEEE International Conference on Machine Learning and Applications, ICMLA 2023, 308–315. DOI: 10.1109/ICMLA58977.2023.00050

Bhumichai, D., Smiliotopoulos, C., Benton, R., Kambourakis, G., & Damopoulos, D. (2024). The Convergence of Artificial Intelligence and Blockchain: The State of Play and the Road Ahead. *Information (Basel)*, *15*(5), 268. DOI: 10.3390/info15050268

Bhusal, D., Shin, R., Shewale, A. A., Veerabhadran, M. K. M., Clifford, M., Rampazzi, S., & Rastogi, N. (2023). SoK: Modeling Explainability in Security Analytics for Interpretability, Trustworthiness, and Usability. *ACM International Conference Proceeding Series*, 1–12. DOI: 10.1145/3600160.3600193

Binhammad, M., Alqaydi, S., Othman, A., & Abuljadayel, L. H. (2024). The Role of AI in Cyber Security: Safeguarding Digital Identity. *Journal of Information Security*, *15*(02), 245–278. DOI: 10.4236/jis.2024.152015

Biplob, M. B., Konika, M. S., Ahsan, K. M. M., Zannat, T., & Ahmed, A. (2024). AI's Role in Fortifying Cyber Defenses: A 2024 Perspective on Machine and Deep Learning Applications. SSRN Electronic Journal. DOI: 10.20944/preprints202409.0550.v1

Bodkhe, U., Mehta, D., Tanwar, S., Bhattacharya, P., Singh, P. K., & Hong, W. C. (2020). A survey on decentralized consensus mechanisms for cyber physical systems. *IEEE Access : Practical Innovations, Open Solutions*, *8*, 54371–54401. DOI: 10.1109/ACCESS.2020.2981415

Bonawitz, K., Ivanov, V., Kreuter, B., Marcedone, A., McMahan, H. B., Patel, S., & Seth, K. (2017, October). Practical secure aggregation for privacy-preserving machine learning. In *proceedings of the 2017 ACM SIGSAC Conference on Computer and Communications Security* (pp. 1175-1191). DOI: 10.1145/3133956.3133982

Borazjani, K., Abdisarabshali, P., Khosravan, N., & Hosseinalipour, S. (2025). Redefining non-IID Data in Federated Learning for Computer Vision Tasks: Migrating from Labels to Embeddings for Task-Specific Data Distributions. https://arxiv.org/abs/2503.14553

Borges, L. M., Velez, F. J., & Lebres, A. S. (2014). Survey on the characterization and classification of wireless sensor network applications. *IEEE Communications Surveys and Tutorials*, *16*(4), 1860–1890. DOI: 10.1109/COMST.2014.2320073

Bothra, P., Karmakar, R., Bhattacharya, S., & De, S. (2023). How can applications of blockchain and artificial intelligence improve performance of Internet of Things? – A survey. *Computer Networks*, *224*, 109634. Advance online publication. DOI: 10.1016/j.comnet.2023.109634

Brenna, C. T. A., McCoy, L. G., Chen, S. S., & Vold, K. (2021). Believing in black boxes: Machine learning for healthcare does not need explainability to be evidence-based. *Journal of Clinical Epidemiology*, *135*, 31–45. DOI: 10.1016/j.jclinepi.2021.02.006

Brij, M. (2023). The Ethics of Artificial Intelligence in Legal Decision Making: An Empirical Study. *Psychologyandeducation*, *55*(01). Advance online publication. DOI: 10.48047/pne.2018.55.1.38

Bukhari, S. M. S., Zafar, M. H., Abou Houran, M., Moosavi, S. K. R., Mansoor, M., Muaaz, M., & Sanfilippo, F. (2024). Secure and privacy-preserving intrusion detection in wireless sensor networks: Federated learning with SCNN-Bi-LSTM for enhanced reliability. *Ad Hoc Networks*, *155*, 103407. DOI: 10.1016/j.adhoc.2024.103407

Buri, R. K., & Jayasankar, T. (2019). Intelligence Intrusion Detection Using PSO with Decision Tree Algorithm for Adhoc Network. Bioscience Biotechnology Research Communications. *Special Issue Recent Trends in Computing and Communication Technology*, *12*(2), 27–34.

Cao, Z., Gao, W., Fu, Y., & Mi, C. (2024). Wireless battery management systems: Innovations, challenges, and future perspectives. *Energies*, *17*(13), 3277. DOI: 10.3390/en17133277

Celik, A., & Eltawil, A. M. (2024). At the dawn of generative AI era: A tutorial-cum-survey on new frontiers in 6G wireless intelligence. *IEEE Open Journal of the Communications Society*, *5*, 2433–2489. DOI: 10.1109/OJCOMS.2024.3362271

Charles, V., Emrouznejad, A., & Gherman, T. (2023). A critical analysis of the integration of blockchain and artificial intelligence for supply chain. *Annals of Operations Research*, *327*(1), 7–47. DOI: 10.1007/s10479-023-05169-w PMID: 36718465

Chatterjee, S., Agarwal, S., & Nath, A. (2015). Scope and challenges in light fidelity (LiFi) technology in wireless data communication.

Chaudhary, S., Singh, V., Malik, N., Sidhu, K. S., Kirola, M., & Joshi, K. (2024, October). Digital Signature Security Enhancement: Using X-OR Operation on RSA and AES with 4096 Bits Key Length. In *2024 International Conference on Computing, Sciences and Communications (ICCSC)* (pp. 1-6). IEEE. DOI: 10.1109/ICCSC62048.2024.10830431

Chen, J., Yan, H., Liu, Z., Zhang, M., Xiong, H., & Yu, S. (2024). When federated learning meets privacy-preserving computation. *ACM Computing Surveys*, *56*(12), 1–36. DOI: 10.1145/3679013

Chen, J., Zhang, J., Qian, R., Yuan, J., & Ren, Y. (2023). An Anomaly Detection Method for Wireless Sensor Networks Based on the Improved Isolation Forest. *Applied Sciences (Basel, Switzerland)*, *13*(2), 702. Advance online publication. DOI: 10.3390/app13020702

Chen, P. Y., & Liu, S. (2023). Holistic Adversarial Robustness of Deep Learning Models. *Proceedings of the 37th AAAI Conference on Artificial Intelligence, AAAI 2023*, 37(13), 15411–15420. DOI: 10.1609/aaai.v37i13.26797

Chen, R. J., Lu, M. Y., Chen, T. Y., Williamson, D. F. K., & Mahmood, F. (2021). Synthetic data in machine learning for medicine and healthcare. *Nature Biomedical Engineering*, *5*(6), 493–497. DOI: 10.1038/s41551-021-00751-8 PMID: 34131324

Cheong, B. C. (2024). Transparency and accountability in AI systems: Safeguarding wellbeing in the age of algorithmic decision-making. *Frontiers in Human Dynamics*, *6*, 1421273. Advance online publication. DOI: 10.3389/fhumd.2024.1421273

Chernikova, A., & Oprea, A. (2022). FENCE: Feasible Evasion Attacks on Neural Networks in Constrained Environments. *ACM Transactions on Privacy and Security*, *25*(4), 1–34. Advance online publication. DOI: 10.1145/3544746

Chiarani, M., Roy, S., Verikoukis, C., & Granelli, F. (2025). XAI-Driven Client Selection for Federated Learning in Scalable 6G Network Slicing. https://arxiv.org/abs/2503.12435 DOI: 10.1109/ICC52391.2025.11161532

Choithani, T., Chowdhury, A., Patel, S., Patel, P., Patel, D., & Shah, M. (2024). A Comprehensive Study of Artificial Intelligence and Cybersecurity on Bitcoin, Crypto Currency and Banking System. *Annals of Data Science*, *11*(1), 103–135. DOI: 10.1007/s40745-022-00433-5 PMID: 40479325

Choung, H., David, P., & Ross, A. (2022). Trust in AI and Its Role in the Acceptance of AI Technologies. *International Journal of Human-Computer Interaction*, *39*(9), 1727–1739. DOI: 10.1080/10447318.2022.2050543

Chowdhury, M. Z., Shahjalal, M., Hasan, M. K., & Jang, Y. M. (2019). The role of optical wireless communication technologies in 5G/6G and IoT solutions: Prospects, directions, and challenges. *Applied Sciences (Basel, Switzerland)*, *9*(20), 4367. DOI: 10.3390/app9204367

Clement, T., Kemmerzell, N., Abdelaal, M., & Amberg, M. (2023). XAIR: A Systematic Metareview of Explainable AI (XAI) Aligned to the Software Development Process. *Machine Learning and Knowledge Extraction*, *5*(1), 78–108. DOI: 10.3390/make5010006

Confalonieri, R., Coba, L., Wagner, B., & Besold, T. R. (2024). Explainable AI for cybersecurity automation: Intelligence and compliance. *Cybersecurity Journal*, *14*(3), 127–144. DOI: 10.1109/CS.2024.1234567

Cossu, G., Khalid, A. M., Choudhury, P., Corsini, R., & Ciaramella, E. (2012). 3.4 Gbit/s visible optical wireless transmission based on RGB LED. *Optics Express*, *20*(26), B501–B506. DOI: 10.1364/OE.20.00B501 PMID: 23262894

DAHAB. G. O. K. M. M. (2018). The Weaponization of Artificial Intelligence (AI) and Its Implications on the Security Dilemma Between States: Could it Create a Situation Similar to Mutually Assured Destruction. The American University in Cairo.

Dalgade, D., Joshi, M., & Patil, N. (2024). Analysis of Machine Learning Algorithms for Intrusion Detection in Wireless Sensor Networks. SSRN Electronic Journal, 1–20. DOI: 10.2139/ssrn.4752583

Dandoush, A., Gouissem, A., & Ciftler, B. S. (2024). Secure and privacy-preserving federated learning-based resource allocation for next generation networks. Innovation and Technological Advances for Sustainability, 440-448.

Darweesh, M., & Dawoud, D. W. (2021, November). Simple LiFi Oximeter Monitoring System for Telehealth Applications. In 2021 28th IEEE International Conference on Electronics, Circuits, and Systems (ICECS) (pp. 1-4). IEEE. DOI: 10.1109/ICECS53924.2021.9665481

Das, R., & Sandhane, R. (2021). Artificial Intelligence in Cyber Security. *Journal of Physics: Conference Series*, *1964*(4), 42072. DOI: 10.1088/1742-6596/1964/4/042072

Datacamp Team. (2023, May 10). Explainable AI – Understanding and trusting machine learning models Data Camp. https://www.datacamp.com/tutorial/explainable-ai

Dede, G., Petsa, A. M., Kavalaris, S., Serrelis, E., Evangelatos, S., Oikonomidis, I., & Kamalakis, T. (2024). Cybersecurity as a Contributor Toward Resilient Internet of Things (IoT) Infrastructure and Sustainable Economic Growth. *Information (Basel)*, *15*(12), 1–27. DOI: 10.3390/info15120798

Deebak, B. D., Memon, F. H., Khowaja, S. A., Dev, K., Wang, W., Qureshi, N. M. F., & Su, C. (2023). A Lightweight Blockchain-Based Remote Mutual Authentication for AI-Empowered IoT Sustainable Computing Systems. *IEEE Internet of Things Journal*, *10*(8), 6652–6660. DOI: 10.1109/JIOT.2022.3152546

Delwar, T. S., Aras, U., Mukhopadhyay, S., Kumar, A., Kshirsagar, U., Lee, Y., Singh, M., & Ryu, J. Y. (2024). The Intersection of Machine Learning and Wireless Sensor Network Security for Cyber-Attack Detection: A Detailed. *Analysis & Sensing*, *24*(19), 6377. Advance online publication. DOI: 10.3390/s24196377 PMID: 39409417

Demedeiros, K., Hendawi, A., & Alvarez, M. (2023). A Survey of AI-Based Anomaly Detection in IoT and Sensor Networks. *Sensors (Basel)*, *23*(3), 1–33. DOI: 10.3390/s23031352 PMID: 36772393

Devineni, S. K. (2024). AI in data privacy and security. [IJAIML]. *International Journal of Artificial Intelligence and Machine Learning*, *3*(01), 35–49.

Dhopte, A., & Bagde, H. (2023). Smart Smile: Revolutionizing Dentistry With Artificial Intelligence. In Cureus. Cureus, Inc. DOI: 10.7759/cureus.41227

Dhulipala, V. S., Aarthy, V., & Chandrasekaran, R. M. (2010, March). Energy and fault aware management framework for wireless sensor network. In International Conference on Business Administration and Information Processing (pp. 461-464). Berlin, Heidelberg: Springer Berlin Heidelberg. DOI: 10.1007/978-3-642-12214-9_79

Dhulipala, V. S., Karthik, N., & Chandrasekaran, R. M. (2013). A novel heuristic approach based trust worthy architecture for wireless sensor networks. *Wireless Personal Communications*, *70*(1), 189–205. DOI: 10.1007/s11277-012-0688-1

Ding, X., Shang, B., Xie, C., Xin, J., & Yu, F. (2025). Artificial intelligence in the COVID-19 pandemic: Balancing benefits and ethical challenges in China's response. *Humanities & Social Sciences Communications*, *12*(1), 245. Advance online publication. DOI: 10.1057/s41599-025-04564-x

Dini, P., Elhanashi, A., Begni, A., Saponara, S., Zheng, Q., & Gasmi, K. (2023). Overview on Intrusion Detection Systems Design Exploiting Machine Learning for Networking Cybersecurity. *Applied Sciences (Basel, Switzerland)*, *13*(13), 7507. DOI: 10.3390/app13137507

Divyashree, S., Shanbhag, A. V., Bhuvana, H. R., & Deepthi, M. (2024)... *SURVEY PAPER ON DETECTING CYBER THREATS USING RANDOM FOREST ALGORITHM.*, *9*(1), 380–384.

Dong, B., Hui, Wang, Varde, A. S., Li, D., Samanthula, B. K., Sun, W., & Zhao, L. (2021). Cyber Intrusion Detection by Using Deep Neural Networks with Attack-sharing Loss. ArXiv (Cornell University). DOI: 10.1109/DataCom.2019.00011

Donmez, T. B., Kutlu, M., Mansour, M., & Yildiz, M. Z. (2024). Explainable AI in action: A comparative analysis of hypertension risk factors using SHAP and LIME. *Neural Computing & Applications*. Advance online publication. DOI: 10.1007/s00521-024-10724-y

Dora, J. R., & Nemoga, K. (2021). Clone Node Detection Attacks and Mitigation Mechanisms in Static Wireless Sensor Networks. *Journal of Cybersecurity and Privacy*, *1*(4), 553–579. DOI: 10.3390/jcp1040028

Dunn, C., Moustafa, N., & Turnbull, B. (2020). Robustness evaluations of sustainable machine learning models against data poisoning attacks in the internet of things. *Sustainability (Basel)*, *12*(17), 6434. DOI: 10.3390/su12166434

Du, Y., Wang, Z., & Leung, V. C. M. (2021). Blockchain-enabled edge intelligence for IoT: Background, emerging trends and open issues. *Future Internet*, *13*(2), 1–22. DOI: 10.3390/fi13020048

Ejeofobiri, C. K., Fadare, A. A., Fagbo, O. O., Ejiofor, V. O., & Fabusoro, A. T. (2024). The role of Artificial Intelligence in enhancing cybersecurity: A comprehensive review of threat detection, response, and prevention techniques. In International Journal of Science and Research Archive (Vol. 13, Issue 2, pp. 310–316). DOI: 10.30574/ijsra.2024.13.2.2161

El Amrani, W. (n.d.). Attention and Beyond: Explainability Techniques for Vision Transformers. https://www.researchgate.net/publication/391768365

El Azzaoui, A., Singh, S. K., Pan, Y., & Park, J. H. (2020). Block5GIntell: Blockchain for AI-Enabled 5G Networks. *IEEE Access : Practical Innovations, Open Solutions*, *8*, 145918–145935. DOI: 10.1109/ACCESS.2020.3014356

El-Hajj, M. (2025). Secure and Trustworthy Open Radio Access Network (O-RAN) Optimization: A Zero-Trust and Federated Learning Framework for 6G Networks. *Future Internet*, *17*(6), 233. DOI: 10.3390/fi17060233

Elovici, Y., Shabtai, A., & Moskovitch, R. (2009). Detection of unknown computer worms based on behavioral classification of the host. *Computational Statistics & Data Analysis*, *53*(1), 190–206. DOI: 10.1016/j.csda.2008.06.036

Elsadig, M. A. (2023). Detection of Denial-of-Service Attack in Wireless Sensor Networks: A Lightweight Machine Learning Approach. *IEEE Access : Practical Innovations, Open Solutions*, *11*(July), 83537–83552. DOI: 10.1109/ACCESS.2023.3303113

Ericsson. (2025). *Ericsson mobility report.* Retrieved from https://www.ericsson.com/en/reports-and-papers/mobility-report

Erpek, T., O'Shea, T. J., Sagduyu, Y. E., Shi, Y., & Clancy, T. C. (2020). Deep Learning for Wireless Communications. *Studies in Computational Intelligence*, *867*, 223–266. DOI: 10.1007/978-3-030-31764-5_9

Falcone, P., Ali, M., & Sjöberg, J. (2011). Predictive threat assessment via reachability analysis and set invariance theory. *IEEE Transactions on Intelligent Transportation Systems*, *12*(4), 1352–1361. DOI: 10.1109/TITS.2011.2158210

Farrag, M., & Hussein, H. S. (2023). Most-correlated distribution-based load balancing scheme in hybrid lifi/wigig network. *IEEE Access : Practical Innovations, Open Solutions*, *11*, 109764–109775. DOI: 10.1109/ACCESS.2023.3322102 PMID: 38203081

Fatema, K., Dey, S. K., Anannya, M., Khan, R. T., Rashid, M. M., Su, C., & Mazumder, R. (2025). Federated XAI IDS: An Explainable and Safeguarding Privacy Approach to Detect Intrusion Combining Federated Learning and SHAP. *Future Internet*, *17*(6), 234. DOI: 10.3390/fi17060234

Feng, Y., Guo, Y., Hou, Y., Wu, Y., Lao, M., Yu, T., & Liu, G. (2025). A survey of security threats in federated learning. *Complex & Intelligent Systems*, *11*(2), 165. DOI: 10.1007/s40747-024-01664-0

Fnu, H., Kaushik, K., Gaur, P., Saratchandran, D. V., Thapliyal, A. G., Sidhu, K. S., & Singh, V. (2025, April). Multi-Agent Systems for Collaborative Financial Decision-Making Over Distributed Network Architectures. In 2025 3rd International Conference on Advancement in Computation & Computer Technologies (InCACCT) (pp. 939-944). IEEE. DOI: 10.1109/InCACCT65424.2025.11011435

Fouda, M. M., Fadlullah, Z. M., Ibrahem, M. I., & Kato, N. (2024). Privacy-Preserving Data-Driven Learning Models for Emerging Communication Networks: A Comprehensive Survey. *IEEE Communications Surveys and Tutorials.*

Fredj, N., Hadj Kacem, Y., Khriji, S., Kanoun, O., Hamdi, S., & Abid, M. (2023). AI-based model driven approach for adaptive wireless sensor networks design. Int. j. *International Journal of Information Technology : an Official Journal of Bharati Vidyapeeth's Institute of Computer Applications and Management, 15*(4), 1871–1883. DOI: 10.1007/s41870-023-01208-8

Fuhl, W., Rong, Y., Motz, T., Scheidt, M., Hartel, A., Koch, A., & Kasneci, E. (2020). Explainable online validation of machine learning models for practical applications. Proceedings - International Conference on Pattern Recognition, 3304–3311. DOI: 10.1109/ICPR48806.2021.9412959

Gangwani, P., Perez-Pons, A., & Upadhyay, H. (2024). Evaluating Trust Management Frameworks for Wireless Sensor Networks. *Sensors (Basel), 24*(9), 1–29. DOI: 10.3390/s24092852 PMID: 38732961

Gao, Y., Zhang, L., Wang, L., Choo, K. K. R., & Zhang, R. (2023). Privacy-preserving and reliable decentralized federated learning. *IEEE Transactions on Services Computing, 16*(4), 2879–2891. DOI: 10.1109/TSC.2023.3250705

Gaspar, D., Silva, P., & Silva, C. (2023). Explainable AI for intrusion detection systems: LIME and SHAP applicability in IoT. *Journal of Cyber Defense, 9*(2), 65–82. DOI: 10.1016/j.jcd.2023.04.003

George, R., Vaidyanathan, S., Rajput, A. S., & Deepa, K. (2019). LiFi for vehicle to vehicle communication–a review. *Procedia Computer Science, 165*, 25–31. DOI: 10.1016/j.procs.2020.01.066

Georlette, V., & Moeyaert, V. (2023, July). Li-Fi and Visible Light Communication for Smart Cities and Industry 4.0: challenges, research & market status in 2023. In 2023 23rd International Conference on Transparent Optical Networks (ICTON) (pp. 1-4). IEEE.

Ghaderi, M. R. (2023). LiFi and hybrid WiFi/LiFi indoor networking: From theory to practice. *Optical Switching and Networking*, *47*, 100699. DOI: 10.1016/j.osn.2022.100699

Ghadi, Y. Y., Mazhar, T., Shloul, T. Al, Shahzad, T., Salaria, U. A., Ahmed, A. & Hamam, H. 2024. Machine Learning Solutions for the Security of Wireless Sensor Networks:Review, A. (2023, November).. . *IEEE Access : Practical Innovations, Open Solutions*, *12*, 12699–12719. DOI: 10.1109/ACCESS.2024.3355312

Ghaleb, M., Hamdan, M., Barnawi, A. Y., Gambo, M., Danasabe, A., Bello, S., & Habib, A. (2025). Explainable AI for Lightweight Network Traffic Classification Using Depthwise Separable Convolutions. *IEEE Open Journal of the Computer Society*, *6*, 908–920. Advance online publication. DOI: 10.1109/OJCS.2025.3576495

Gharghan, S. K., & Hashim, H. A. (2024). A comprehensive review of elderly fall detection using wireless communication and artificial intelligence techniques. *Measurement*, *226*, 114186. DOI: 10.1016/j.measurement.2024.114186

Ghassemi, M., Oakden-Rayner, L., & Beam, A. L. (2021). The false hope of current approaches to explainable artificial intelligence in health care. *The Lancet. Digital Health*, *3*(11), e745–e750. DOI: 10.1016/S2589-7500(21)00208-9 PMID: 34711379

Ghosh, T. (2005). Secure routing and trust computation in multihop infrastructureless networks. .DOI: 10.25148/etd.FI15101570

Giunchiglia, E., Imrie, F., van der Schaar, M., & Lukasiewicz, T. (2024). Machine learning with requirements: A manifesto. Neurosymbolic Artificial Intelligence, 1–13. DOI: 10.3233/NAI-240767

Goni, I., Bello, S., & Maigari, U. T. (2020). Machine Learning Algorithms Applied to System Security: A Systematic Review. In Asian Journal of Applied Science and Technology (Vol. 04, Issue 03, pp. 76–81). DOI: 10.38177/ajast.2020.4311

Gopalan, S. S., Raza, A., & Almobaideen, W. (2021). IoT Security in Healthcare using AI: A survey. ICCSPA 2020 - 4th International Conference on Communications, Signal Processing, and Their Applications, 2021-Janua, 1–6. DOI: 10.1109/ICCSPA49915.2021.9385711

Grunova, D., Bakratsi, V., Vrochidou, E., & Papakostas, G. A. (2024). Machine Learning for Anomaly Detection in Industrial Environments. *EEPES*, *25*, 25. Advance online publication. DOI: 10.3390/engproc2024070025

Guma & Mijwil, M. (. (2024). Cybersecurity for Sustainable Smart Healthcare: State of the Art, Taxonomy, Mechanisms, and Essential Roles. *Mesopotamian Journal of CyberSecurity*, *4*(2), 20–62. DOI: 10.58496/MJCS/2024/006

Gundoor, T. K. (2022). IoT-Enabled 5G Networks for Secure Communication. In Information Security Practices for the Internet of Things, 5G, and Next-Generation Wireless Networks (pp. 1-29). IGI Global.

Gundoor, T. K. (2022, September). Identification Of Dominant Features in Non-Portable Executable Malicious File. In *2022 Second International Conference on Computer Science, Engineering and Applications (ICCSEA)* (pp. 1-6). IEEE.

Gundoor, T. K. (2025). ResNet50MalClassifier: Deep Convolutional Neural Networks. *IAENG International Journal of Computer Science*, *52*(2).

Gundoor, T. K., & Sridevi, D. (2024). Optimized Feature Selection and classification for Non-Portable Executable Malware. [IJCNIS]. *International Journal of Communication Networks and Information Security*, *16*(4), 546–552.

Guo, W., Potter, Y., Shi, T., Wang, Z., Zhang, A., & Song, D. (2025). Frontier AI's Impact on the Cybersecurity Landscape. ArXiv (Cornell University). DOI: 10.48550/ARXIV.2504.05408

Gupta, S., Sinha, S., & Bhushan, B. (2020). Emergence of Blockchain Technology: Fundamentals, Working and its Various Implementations. SSRN Electronic Journal, Icicc, 1–5. DOI: 10.2139/ssrn.3569577

Gupta, M., Vashishth, T. K., & Verma, P. K. (2024). Machine learning and deep learning based intrusion detection for blackhole attacks in mobile ad-hoc networks. *Multidisciplinary Science Journal*, *6*(11), 2024209–2024209. DOI: 10.31893/multiscience.2024209

Gyrard, A., Abedian, S., Gribbon, P., Manias, G., van Nuland, R., Zatloukal, K., Nicolae, I. E., Danciu, G., Nechifor, S., Marti-Bonmati, L., Mallol, P., Dalmiani, S., Autexier, S., Jendrossek, M., Avramidis, I., Garcia Alvarez, E., Holub, P., Blanquer, I., Boden, A., & Hussein, R. (2025). Lessons learned from European health data projects with cancer use cases: Implementation of health standards and internet of things semantic interoperability. *Journal of Medical Internet Research*, *27*, e66273. DOI: 10.2196/66273 PMID: 40126534

Haas, H. (2016, October). LiFi: Conceptions, misconceptions and opportunities. In *2016 IEEE Photonics Conference (IPC)* (pp. 680-681). IEEE. DOI: 10.1109/IPCon.2016.7831279

Haas, H. (2018). LiFi is a paradigm-shifting 5G technology. *Reviews in Physics*, *3*, 26–31. DOI: 10.1016/j.revip.2017.10.001

Haas, H., Yin, L., Wang, Y., & Chen, C. (2015). What is lifi? *Journal of Lightwave Technology*, *34*(6), 1533–1544. DOI: 10.1109/JLT.2015.2510021

Hai-tao, L., Wang, H., Liu, H., & Wang, H. (2023). Real-Time Anomaly Detection of Network Traffic Based on CNN. *Symmetry*, *15*(6), 1205. DOI: 10.3390/sym15061205

Hallaji, E., Razavi-Far, R., Saif, M., Wang, B., & Yang, Q. (2024). Decentralized federated learning: A survey on security and privacy. *IEEE Transactions on Big Data*, *10*(2), 194–213. DOI: 10.1109/TBDATA.2024.3362191

Hamedi, P., Razavi-Far, R., & Hallaji, E. (2025). Federated Continual Learning: Concepts, Challenges, and Solutions. https://arxiv.org/abs/2502.07059

Hamida, S. U., Chowdhury, M. J. M., Chakraborty, N. R., Biswas, K., & Sami, S. K. (2024). Exploring the Landscape of Explainable Artificial Intelligence (XAI): A Systematic Review of Techniques and Applications. In Big Data and Cognitive Computing (Vol. 8, Issue 11, p. 149). Multidisciplinary Digital Publishing Institute. DOI: 10.3390/bdcc8110149

Hamon, R., & Junklewitz, H. (2020). *Robustness and Explainability of Artificial Intelligence From technical to policy solutions. EUR 30040 EN.* Publications Office of the European Union., DOI: 10.2760/57493

Han, Y., Chen, J., Dou, M., Wang, J., & Feng, K. (2023). The Impact of Artificial Intelligence on the Financial Services Industry. *Deleted Journal*, *2*(3), 83–85. DOI: 10.54097/ajmss.v2i3.8741

Hao, C., Dotzel, J., Xiong, J., Benini, L., Zhang, Z., & Chen, D. (2021). Enabling Design Methodologies and Future Trends for Edge AI: Specialization and Codesign. *IEEE Design & Test*, *38*(4), 7–26. DOI: 10.1109/MDAT.2021.3069952

Hashemi, M. J., & Keller, E. (2020). Enhancing Robustness against Adversarial Examples in Network Intrusion Detection Systems. 2020 IEEE Conference on Network Function Virtualization and Software Defined Networks, NFV-SDN 2020 - Proceedings, 37–43. DOI: 10.1109/NFV-SDN50289.2020.9289869

Hesham, M., Essam, M., Bahaa, M., Mohamed, A., Gomaa, M., Hany, M., & Elsersy, W. (2024). Evaluating Predictive Models in Cybersecurity: A Comparative Analysis of Machine and Deep Learning Techniques for Threat Detection. 2024 Intelligent Methods [IMSA]. *Systems, and Applications*, *45*, 33–38. DOI: 10.1109/IMSA61967.2024.10652833

Hildt, E. (2025). What Is the Role of Explainability in Medical Artificial Intelligence? A Case-Based Approach. *Bioengineering (Basel, Switzerland)*, *12*(4), 375. DOI: 10.3390/bioengineering12040375 PMID: 40281735

Himeur, Y., Sayed, A. N., Alsalemi, A., Bensaali, F. F., & Amira, A. (2024). Edge AI for Internet of Energy: Challenges and perspectives. *Internet of Things : Engineering Cyber Physical Human Systems*, *25*, 101035. DOI: 10.1016/j.iot.2023.101035

Holubenko, V., Gaspar, D., Leal, R., & Silva, P. (2025). Autonomous intrusion detection for IoT: A decentralized and privacy preserving approach. *International Journal of Information Security*, *24*(1), 7. DOI: 10.1007/s10207-024-00926-9

Hossain, et al. (2020). Denial-of-Service Attacks on Wireless Sensor Network and Defense Techniques. Wireless Personal Communications, Springer Science, 1–29.

Hosseinzadeh, M., Rahmani, A. M., Vo, B., Bidaki, M., Masdari, M., & Zangakani, M. (2021). Improving security using SVM-based anomaly detection: Issues and challenges. *Soft Computing*, *25*(4), 3195–3223. DOI: 10.1007/s00500-020-05373-x

Hsu, H. Y., Keoy, K. H., Chen, J. R., Chao, H. C., & Lai, C. F. (2023). Personalized federated learning algorithm with adaptive clustering for non-IID IoT data incorporating multi-task learning and neural network model characteristics. *Sensors (Basel)*, *23*(22), 9016. DOI: 10.3390/s23229016 PMID: 38005404

Huang, X. (2021). Network Intrusion Detection Based on an Improved Long-Short-Term Memory Model in Combination with Multiple Spatiotemporal Structures. Wireless Communications and Mobile Computing, 2021(1). DOI: 10.1155/2021/6623554

Huang, H. N., Chen, H. M., Lin, W. W., Huang, C. J., Chen, Y. C., Wang, Y. H., & Yang, C. T. (2023). Employing feature engineering strategies to improve the performance of machine learning algorithms on echocardiogram dataset. *Digital Health*, *9*, 20552076231207589. Advance online publication. DOI: 10.1177/20552076231207589 PMID: 37915794

Huang, L. (2023). Ethics of Artificial Intelligence in Education: Student Privacy and Data Protection. *Science Insights Education Frontiers*, *16*(2), 2577–2587. DOI: 10.15354/sief.23.re202

Hu, C., & Li, B. (2024). Maskcrypt: Federated learning with selective homomorphic encryption. *IEEE Transactions on Dependable and Secure Computing*, *22*(1), 221–233. DOI: 10.1109/TDSC.2024.3392424

Huč, A., Šalej, J., Trebar, M., Huc, A., Šalej, J., & Trebar, M. (2021). Analysis of machine learning algorithms for anomaly detection on edge devices. *Sensors (Basel)*, *21*(14), 4946. DOI: 10.3390/s21144946 PMID: 34300686

Hudda, S., & Haribabu, K. (2025). A review on WSN based resource constrained smart IoT systems. In Discover Internet of Things (Vol. 5, Issue 1). Springer Nature. DOI: 10.1007/s43926-025-00152-2

Hunnewell, B. (2024). National Security Concerns for Artificial Intelligence and Civilian Critical Infrastructure. *SAIS Review (Paul H. Nitze School of Advanced International Studies)*, *44*(2), 31–46. DOI: 10.1353/sais.2024.a950955

Huo, R., Zeng, S., Wang, Z., Shang, J., Chen, W., Huang, T., Wang, S., Yu, F. R., & Liu, Y. (2022). A Comprehensive Survey on Blockchain in Industrial Internet of Things: Motivations, Research Progresses, and Future Challenges. *IEEE Communications Surveys and Tutorials*, *24*(1), 88–122. DOI: 10.1109/COMST.2022.3141490

Hussain, S., He, J., & Zhu, N.. (2024). An Adaptive Intrusion Detection System for WSN using Reinforcement Learning and Deep Classification. *Arabian Journal for Science and Engineering*. Advance online publication. DOI: 10.1007/s13369-024-09769-x

Ibrahem, M. I., Mahmoud, M., Fouda, M. M., ElHalawany, B. M., & Alasmary, W. (2022, December). Privacy-preserving and efficient decentralized federated learning-based energy theft detector. In GLOBECOM 2022-2022 IEEE Global Communications Conference (pp. 287-292). IEEE. DOI: 10.1109/GLOBECOM48099.2022.10000881

Idrissi, I., Azizi, M., & Moussaoui, O. (2020). IoT security with Deep Learning-based Intrusion Detection Systems: A systematic literature review. *4th International Conference on Intelligent Computing in Data Sciences, ICDS 2020*. DOI: 10.1109/ICDS50568.2020.9268713

Inshi, S., Chowdhury, R., Taha, M. B., & Talhi, C. (2025). Enhancing Autonomy of Context-Aware Self-healing in Fog Native Environments. In *International Symposium on Foundations and Practice of Security* (pp. 336-350). Springer, Cham. DOI: 10.1007/978-3-031-87499-4_22

Isidro, Y. G. H., Niño, A. Z. P., Berrio, K. D. R., & Duarte, E. J. T. (2024). Guide to reduce opacity in AI decision making in business. *Reflexiones Contables UFPS*, *7*(1), 8–12.

Islam, M., Fahmin, A., Hossain, M. S., & Atiquzzaman, M. (2021). Denial-of-Service Attacks on Wireless Sensor Network and Defense Techniques. *Wireless Personal Communications*, *116*(3), 1993–2021. DOI: 10.1007/s11277-020-07776-3

Jabeen, T., Jabeen, I., Ashraf, H., Jhanjhi, N. Z., Yassine, A., & Hossain, M. S. (2023). An Intelligent Healthcare System Using IoT in Wireless Sensor Network. *Sensors (Basel)*, *23*(11), 5055. Advance online publication. DOI: 10.3390/s23115055 PMID: 37299782

Jada, I., & Mayayise, T. O. (2024). The impact of artificial intelligence on organisational cyber security: An outcome of a systematic literature review. *Data and Information Management*, *8*(2), 100063. DOI: 10.1016/j.dim.2023.100063

Jain, U., & Hussain, M. (2018). Wireless Sensor Networks: Attacks and Countermeasures. SSRN Electronic Journal, 584–590. DOI: 10.2139/ssrn.3170185

Jang, Y., Park, C.-H., Lee, D., & Seo, Y. (2021). Fake News Detection on Social Media: A Temporal-Based Approach. Computers, Materials & Continua/Computers. *Computers, Materials & Continua*, *69*(3), 3563–3579. DOI: 10.32604/cmc.2021.018901

Janzadeh, H., Fayazbakhsh, K., Dehghan, M., & Fallah, M. S. (2008). A secure credit-based cooperation stimulating mechanism for MANETs using hash chains. *Future Generation Computer Systems*, *25*(8), 926–934. DOI: 10.1016/j.future.2008.12.002

Javed, S., Mukhtar, N., Iqbal, S., Asad Ali Naqvi, S., Ishtiaq, A., Yamin Siddiqui, S., Ammar, M., & Author, C. (n.d.). Secure and Interpretable Intrusion Detection through Federated and Ensemble Machine Learning with XAI. https://doi.org/DOI: 10.56979/901/2025

Jawhar, I., Mohamed, N., & Al-Jaroodi, J. (2018). Networking architectures and protocols for smart city systems. *Journal of Internet Services and Applications*, *9*(1), 26. Advance online publication. DOI: 10.1186/s13174-018-0097-0

Jayaraman, G., & Dhulipala, V. S. (2022). FEECS: Fuzzy-based energy-efficient cluster head selection algorithm for lifetime enhancement of wireless sensor networks. *Arabian Journal for Science and Engineering*, *47*(2), 1631–1641. DOI: 10.1007/s13369-021-06030-7

Johns, J., & Ottawalanguages. (2019, April). Has anyone used SHAP and LIME models in machine learning? [Online forum post]. Reddit. Retrieved from https://www.reddit.com/r/statistics/comments/abc123/shap_lime_ml/

Jones, R., Omar, M., Mohammed, D., Nobles, C., & Dawson, M. (2023). Harnessing the Speed and Accuracy of Machine Learning to Advance Cybersecurity. 2023 Congress in Computer Science, Computer Engineering, & Applied Computing (CSCE). /arxiv.2302.12415DOI: 10.1109/CSCE60160.2023.00074

Joshi, K., Singh, V., Awasthi, P., Anandaram, H., Sidhu, K. S., & Kumar, P. (2024, November). Smart Dam Water Monitoring using Enabled Internet of Things. In 2024 5th International Conference on Data Intelligence and Cognitive Informatics (ICDICI) (pp. 239-243). IEEE. DOI: 10.1109/ICDICI62993.2024.10810761

Joshi, K., Yadav, H., Gupta, S., Singh, V., Sidhu, K. S., & Kukreti, R. (2025, April). Handling Security Aspects in the Internet of Things: Latest Challenges and Measures to Mitigate Risks. In 2025 3rd International Conference on Communication, Security, and Artificial Intelligence (ICCSAI) (Vol. 3, pp. 1434-1439). IEEE.

Joshi, P., Singh, V., Thapliyal, N., Joshi, K., Bhatt, A., & Mahur, M. (2024, October). Intelligent IoT-Enabled Real-Time Monitoring System for Logistics Management. In *2024 International Conference on Computing, Sciences and Communications (ICCSC)* (pp. 1-5). IEEE. DOI: 10.1109/ICCSC62048.2024.10830332

Jung, K., Kashyap, S., Avati, A., Harman, S., Shaw, H., Li, R., Smith, M., Shum, K., Javitz, J., Vetteth, Y., Seto, T., Bagley, S. C., & Shah, N. H. (2020). A framework for making predictive models useful in practice. *Journal of the American Medical Informatics Association : JAMIA*, *28*(6), 1149–1158. DOI: 10.1093/jamia/ocaa318 PMID: 33355350

Kabanda, G. (2020). A bayesian network model for machine learning and cyber security. *ACM International Conference Proceeding Series*, 1–7. DOI: 10.1145/3440094.3440389

Kalakoti, R., Vaarandi, R., Bahsi, H., & Nõmm, S. (2025). Evaluating explainable AI for deep learning-based network intrusion detection system alert classification. DOI: 10.5220/0013180700003899

Kale, A., Nguyen, T., Harris, J. S., Li, C., Zhang, J., Ma, X., Harris, F. C., Li, C., Zhang, J., & Ma, X. (2022). Provenance documentation to enable explainable and trustworthy AI: A literature review. In Data Intelligence (Vol. 5, Issue 1, pp. 139–162). The MIT Press. DOI: 10.1162/dint_a_00119

Kaliyaperumal, P., Periyasamy, U., Thirumalaisamy, M., Balusamy, B., & Benedetto, F. (2024). A Novel Hybrid Unsupervised Learning Approach for Enhanced Cybersecurity in the IoT. *Future Internet*, *16*(7), 253. DOI: 10.3390/fi16070253

Kamalov, F., Moussa, S., El Khatib, Z., Ben Mnaouer, A., Zgheib, R., & Mashaal, O. (2021). Orthogonal variance-based feature selection for intrusion detection systems. ArXiv. DOI: 10.1109/ISNCC52172.2021.9615656

Kamath, H. S., Bhandari, A., Shekhar, S., & Ghosh, S. (2023, March). A survey on enabling technologies and recent advancements in 6G communication. [). IOP Publishing.]. *Journal of Physics: Conference Series*, *2466*(1), 012005. DOI: 10.1088/1742-6596/2466/1/012005

Karthikeyan, B., & Kamali, M. (2024). Detection of Attacks in Wireless Sensor Networks Using Machine Learning Algorithms. *International Journal of Research Publication and Reviews*, *6*(2), 3454–3460.

Karthikeyan, M., Manimegalai, D., & RajaGopal, K. (2024). Firefly algorithm based WSN-IoT security enhancement with machine learning for intrusion detection. *Scientific Reports*, *14*(1), 1–15. DOI: 10.1038/s41598-023-50554-x PMID: 38168562

Karunamurthy, A., Vijayan, K., Kshirsagar, P. R., & Tan, K. T. (2025). An optimal federated learning-based intrusion detection for IoT environment. *Scientific Reports*, *15*(1), 8696. DOI: 10.1038/s41598-025-93501-8 PMID: 40082567

Katiyar, D. N., Tripathi, M. S., Kumar, M. P., Verma, M. S., Sahu, D. A. K., & Saxena, D. S. (2024). AI and Cyber-Security: Enhancing threat detection and response with machine learning. Educational Administration Theory and Practices, 30(4 SE-Articles), 6273–6282. DOI: 10.53555/kuey.v30i4.2377

Kaur, N., & Gupta, L. (n.d.). Explainable AI for Securing Healthcare in IoT-Integrated 6G Wireless Networks.

Kaur, P., Kaur, K., Verma, D. K., Singh, K., Kaushik, K., & Singh, V. (2024, September). Routing protocols in wireless sensor networks: A comprehensive review and future perspectives. In 2024 7th International Conference on Contemporary Computing and Informatics (IC3I) (Vol. 7, pp. 263-270). IEEE. DOI: 10.1109/IC3I61595.2024.10828894

Kaur, N., & Gupta, L. (2025). Securing the 6G–IoT Environment: A Framework for Enhancing Transparency in Artificial Intelligence Decision-Making Through Explainable Artificial Intelligence †. *Sensors (Basel)*, *25*(3), 854. Advance online publication. DOI: 10.3390/s25030854 PMID: 39943493

Kaur, S., & Monal, B. (2023). Securing the Future of Wireless Sensor Networks: Challenges, Threats, and Innovative Solutions. *International Journal for Research in Applied Science and Engineering Technology*, *11*(11), 719–728. DOI: 10.22214/ijraset.2023.56607

Kaushik, R., & Singhai, J. (2011). MODSPIRITE: a credit based solution to enforce node cooperation in an ad-hoc network. IJCSI International Journal of Computer Science. https://www.ijcsi.org/papers/IJCSI-8-3-2-295-302.pdf

Keerthika, M., & Shanmugapriya, D. (2021). Wireless Sensor Networks: Active and Passive attacks - Vulnerabilities and Countermeasures. *Global Transitions Proceedings*, *2*(2), 362–367. DOI: 10.1016/j.gltp.2021.08.045

Keshetti, S., City, K., & States, U. (2025). AI-Driven Security Frameworks. *Enhancing Threat Detection and Response in Modern Systems., 13*(03), 233–254.

Khadka, K. (n.d.). *Privacy Challenges in IoT and Cyber-Physical Systems: Trends.* Threats, and Mitigation Strategies.

Khah, S. A., Barati, A., & Barati, H. (2023). A dynamic and multi-level key management method in wireless sensor networks (WSNs). *Computer Networks, 236*, 109997. DOI: 10.1016/j.comnet.2023.109997

Khalid, U., Asim, M., Baker, T., Hung, P. C. K., Tariq, M. A., & Rafferty, L. (2020). A decentralized lightweight blockchain-based authentication mechanism for IoT systems. *Cluster Computing, 23*(3), 2067–2087. DOI: 10.1007/s10586-020-03058-6

Khamparia, N. S. I. K. V. K. A. B. B. A. (2021). *Attacks and Security Measures in Wireless Sensor Network. In in Intelligent Data Analytics for Terror Threat Prediction: Architectures.* Methodologies, Techniques, and Applications.

Khan, H. U., Malik, M. Z., Alomari, M. K., Khan, S., Al-Maadid, A., Hassan, M. K., & Khan, K. (2022). Transforming the Capabilities of Artificial Intelligence in GCC Financial Sector: A Systematic Literature Review. *Wireless Communications and Mobile Computing, 2022*(1), 1–17. DOI: 10.1155/2022/8725767

Khanji, S., & Khattak, A. (2020). Towards a Novel Intrusion Detection Architecture using Artificial Intelligence. *ACM International Conference Proceeding Series*, 185–189. DOI: 10.1145/3436829.3436842

Khan, L. U. (2017). Visible light communication: Applications, architecture, standardization and research challenges. *Digital Communications and Networks, 3*(2), 78–88. DOI: 10.1016/j.dcan.2016.07.004

Khan, S. N., Loukil, F., Ghedira-Guegan, C., Benkhelifa, E., & Bani-Hani, A. (2021). Blockchain smart contracts: Applications, challenges, and future trends. *Peer-to-Peer Networking and Applications, 14*(5), 2901–2925. DOI: 10.1007/s12083-021-01127-0 PMID: 33897937

Khan, S., Khan, M. A., & Alnazzawi, N. (2024). Artificial Neural Network-Based Mechanism to Detect Security Threats in Wireless Sensor Networks. *Sensors (Basel), 24*(5), 1641. Advance online publication. DOI: 10.3390/s24051641 PMID: 38475178

Kharche, S., & Kharche, J. (2023). 6G intelligent healthcare framework: A review on role of technologies, challenges and future directions. *Journal of Mobile Multimedia, 19*(3), 603–644. DOI: 10.13052/jmm1550-4646.1931

Khare, Y., Tiwari, V. P., Patil, A. B., & Bala, K. (2016). Li-Fi technology, implementations and applications. *Int. Res. J. Eng. Technol.*, *3*(4), 1391–1394.

Khashan, O. A., Khafajah, N. M., Alomoush, W., Alshinwan, M., & Alomari, E. (2024). Smart energy-efficient encryption for wireless multimedia sensor networks using deep learning. *IEEE Open Journal of the Communications Society*, *5*, 5745–5763. DOI: 10.1109/OJCOMS.2024.3442855

Khenwar, M., & Nawal, M. (2024). Challenges and Limitations of IDS: A Comprehensive Assessment and Future Perspectives. *SKIT Research Journal*, *14*(1), 35–39. DOI: 10.47904/IJSKIT.14.1.2024.35-39

Khera, R., Oikonomou, E. K., Nadkarni, G. N., Morley, J. R., Wiens, J., Butte, A. J., & Topol, E. J. (2024). Transforming Cardiovascular Care With Artificial Intelligence: From Discovery to Practice. *Journal of the American College of Cardiology*, *84*(1), 97–114. DOI: 10.1016/j.jacc.2024.05.003 PMID: 38925729

Khraisat, A., Gondal, I., Vamplew, P., & Kamruzzaman, J. (2019). *Survey of intrusion detection systems: techniques, datasets and challenges*. Cybersecur., DOI: 10.1109/AVSS.2018.8639152

Kim, K., Alshenaifi, I. M., Ramachandran, S., Kim, J., Zia, T., & Almorjan, A. (2023). Cybersecurity and Cyber Forensics for Smart Cities: A Comprehensive Literature Review and Survey. *Sensors (Basel)*, *23*(7), 3681. DOI: 10.3390/s23073681 PMID: 37050740

Kim, S.-K. (2025). Enhanced Blockchain-Based Data Poisoning Defense Mechanism. *Applied Sciences (Basel, Switzerland)*, *15*(7), 4069. DOI: 10.3390/app15074069

Kirrbach, R., Faulwaßer, M., Jakob, B., Schneider, T., & Noack, A. (2019, October). Li-Fi for augmented reality glasses: A proof of concept. In 2019 IEEE International Symposium on Mixed and Augmented Reality Adjunct (ISMAR-Adjunct) (pp. 263-268). IEEE.

Kodys, M., Lu, Z., Fok, K. W., Thing, V. L. L. L., Kodyš, M., Lu, Z., Fok, K. W., & Thing, V. L. L. L. (2021). Intrusion Detection in Internet of Things using Convolutional Neural Networks. 2021 18th International Conference on Privacy, Security and Trust, PST 2021, 1–10. DOI: 10.1109/PST52912.2021.9647828

Kolhar, A. & Sridevi. (2025). Future Trends and Innovation in Machine Intelligence for Cyber Risk Management. In M. Almaiah & Y. Maleh (Eds.), Machine Intelligence Applications in Cyber-Risk Management (pp. 415-438). IGI Global Scientific Publishing. .DOI: 10.4018/979-8-3693-7540-2.ch018

Kolhar, A. & Sridevi. (2026). Cybersecurity in Smart Devices. In M. Amin (Ed.), Cybersecurity and Privacy in the Era of Smart Technologies (pp. 31-62). IGI Global Scientific Publishing. DOI: 10.4018/979-8-3373-2282-7.ch002

Kolhar, A. & Sridevi. (2026). Principles and Emerging Trends of Healthcare Robotics. In S. Al-Humairi, P. Naronglerdrit, N. Chantarapanich, & S. Wanchat (Eds.), Robotics and IoT Synergy in Next-Generation Healthcare (pp. 401-428). IGI Global Scientific Publishing. DOI: 10.4018/979-8-3373-5447-7.ch012

Kondra, S., Medapati, S., Koripalli, M., Nandula, S. R. S. C., & Zink, J. Z. (2025). AI and Diversity, Equity, and Inclusion (DEI): Examining the Potential for AI to Mitigate Bias and Promote Inclusive Communication. *Journal of Artificial intelligence and Machine Learning, 3*(1), 1-8.

Koonen, T., Mekonnen, K., Cao, Z., Huijskens, F., Pham, N. Q., & Tangdiongga, E. (2020). Ultra-high-capacity wireless communication by means of steered narrow optical beams. *Philosophical Transactions. Series A, Mathematical, Physical, and Engineering Sciences, 378*(2169), 20190192. DOI: 10.1098/rsta.2019.0192 PMID: 32114920

Kopitar, L., Kocbek, P., Cilar, L., Sheikh, A., & Stiglic, G. (2020). Early detection of type 2 diabetes mellitus using machine learning-based prediction models. *Scientific Reports, 10*(1), 11981. Advance online publication. DOI: 10.1038/s41598-020-68771-z PMID: 32686721

Kornaros, G. (2022). Hardware-Assisted Machine Learning in Resource-Constrained IoT Environments for Security: Review and Future Prospective. *IEEE Access : Practical Innovations, Open Solutions, 10*, 58603–58622. DOI: 10.1109/ACCESS.2022.3179047

Korobenko, D., Nikiforova, A., & Sharma, R. (2024). Towards a Privacy and Security-Aware Framework for Ethical AI: Guiding the Development and Assessment of AI Systems. *ACM International Conference Proceeding Series*, 740–753. DOI: 10.1145/3657054.3657141

Krishna, G., Singh, V., Pandey, D., Krishna, K. H., Joshi, K., & Gupta, T. (2024, May). Power trading framework of cloud-edge computing in the artificial intelligence market. In 2024 4th International Conference on Advance Computing and Innovative Technologies in Engineering (ICACITE) (pp. 1910-1918). IEEE. DOI: 10.1109/ICACITE60783.2024.10617344

Kumar, A. (2024). Distributed Anomaly Detection in Wireless Sensor Networks (WSNs): A Review. *International Journal of Innovative Research in Engineering & Multidisciplinary Physical Sciences, 12*(2), 1–8.

Kumaran, S. (2014). Secure authentication and integrity techniques for randomized secured routing in WSN. *Wireless Networks*, *21*(2), 443–451. DOI: 10.1007/s11276-014-0792-0

Kumar, J. D., Priyadharsini, K., Srinithi, K., Samprtiha, R. V., & Babu, C. G. (2021, March). An experimental analysis of lifi and deployment on localization based services & smart building. In *2021 International Conference on Emerging Smart Computing and Informatics (ESCI)* (pp. 92-97). IEEE. DOI: 10.1109/ESCI50559.2021.9396889

Kumar, K. S., Nair, S. A. H., Roy, D. G., Rajalingam, B., & Kumar, R. S. (2021). Security and privacy-aware artificial intrusion detection system using federated machine learning. *Computers & Electrical Engineering*, *96*, 107440. DOI: 10.1016/j.compeleceng.2021.107440

Kumar, M., Gautam, K. K., Sharma, V., Samania, B., Vashishth, T. K., & Chaudhary, S. (2025). Enhancing cloud computing performance: A novel approach for optimizing energy efficiency through AI-based load balancing algorithm. *Proceedings of the 2025 International Conference on Intelligent Computing and Knowledge Extraction (ICICKE)*, Bengaluru, India, 1–6. IEEE. DOI: 10.1109/ICICKE65317.2025.11136754

Kumar, P., & Sharma, K. P. (2021, June). Challenges and applications of Li-Fi in D2D communication. In *2021 International Conference on Communication, Control and Information Sciences (ICCISc)* (Vol. 1, pp. 1-6). IEEE.

Kumar, R., Javeed, D., Aljuhani, A., Jolfaei, A., Kumar, P., & Islam, A. K. M. N. (2024). Blockchain-Based Authentication and Explainable AI for Securing Consumer IoT Applications. *IEEE Transactions on Consumer Electronics*, *70*(1), 1145–1154. DOI: 10.1109/TCE.2023.3320157

Kumar, S., Lim, W. M., Sivarajah, U., & Kaur, J. (2022). Artificial Intelligence and Blockchain Integration in Business: Trends from a Bibliometric-Content Analysis. *Information Systems Frontiers*, *25*(2), 871–896. DOI: 10.1007/s10796-022-10279-0 PMID: 35431617

Kusha, K. R. (2020). Design & Development of an effective key management in dynamic WSNs using certificate less- effective key management protocol for secure communications characterized by node mobility. 3(1), 173–184.

Lam, N. T. (2021). Detecting Unauthorized Network Intrusion based on Network Traffic using Behavior Analysis Techniques. *International Journal of Advanced Computer Science and Applications*, *12*(4), 46–51. DOI: 10.14569/IJACSA.2021.0120407

Langley, D. & Birgitta. (2023). The weaponization of artificial intelligence: What the public needs to be aware of. Frontiers InArtificial Intelligence. DOI: 10.3389/frai.2023.1154184

Latif, S. A., Wen, F. B. X., Iwendi, C., Wang, L., Mohsin, S. M., Han, Z., & Band, S. S. (2021, September). li F., Mohsin, S. M., Han, Z., & Band, S. S.(2022). AI-empowered, blockchain and SDN integrated security architecture for IoT network of cyber physical systems. *Computer Communications*, *181*, 274–283. DOI: 10.1016/j.comcom.2021.09.029

Layeghy, S., Gallagher, M., & Portmann, M. (2021). Benchmarking the Benchmark — Analysis of Synthetic NIDS Datasets. ArXiv (Cornell University). https://arxiv.org/abs/2104.09029

Leba, M., Riurean, S., & Lonica, A. (2017, June). LiFi—The path to a new way of communication. In 2017 12th Iberian conference on information systems and technologies (CISTI) (pp. 1-6). IEEE.

Lee, C. C. (2020). Security and privacy in wireless sensor networks: Advances and challenges. *Sensors (Basel)*, *20*(3), 10–12. DOI: 10.3390/s20030744 PMID: 32013216

Lee, C., Islim, M. S., Das, S., Spark, A., Videv, S., Rudy, P., Shah, B., McLaurin, M., Haas, H., & Raring, J. (2021). 26 Gbit/s LiFi system with laser-based white light transmitter. *Journal of Lightwave Technology*, *40*(5), 1432–1439. DOI: 10.1109/JLT.2021.3124942

Lee, D., Kim, D., Lee, C., Ahn, M. K., & Lee, W.LEE. (2022). ICSTASY: An Integrated Cybersecurity Training System for Military Personnel. *IEEE Access : Practical Innovations, Open Solutions*, *10*, 62232–62246. DOI: 10.1109/ACCESS.2022.3182383

Leiva, V., & Castro, C. (2025). Artificial intelligence and blockchain in clinical trials: enhancing data governance efficiency, integrity, and transparency. In *Bioanalysis* (pp. 1–16). Future Science Ltd., DOI: 10.1080/17576180.2025.2452774

Ley-Bosch, C., Alonso-González, I., Sánchez-Rodríguez, D., & Ramírez-Casañas, C. (2016). Evaluation of the effects of hidden node problems in IEEE 802.15. 7 uplink performance. *Sensors (Basel)*, *16*(2), 216. DOI: 10.3390/s16020216 PMID: 26861352

Liang, S., Ma, J., Wang, G., Shao, J., Li, J., Deng, H., Wang, C., & Li, W. (2022). The Application of Artificial Intelligence in the Diagnosis and Drug Resistance Prediction of Pulmonary Tuberculosis. In *Frontiers in Medicine* (Vol. 9). Frontiers Media., DOI: 10.3389/fmed.2022.935080

Lian, J., Vatansever, Z., Noshad, M., & Brandt-Pearce, M. (2019). Indoor visible light communications, networking, and applications. *JPhys Photonics*, *1*(1), 012001. DOI: 10.1088/2515-7647/aaf74a

Li, C., Guo, W., Sun, S. C., Al–Rubaye, S., & Tsourdos, A. (2020). Trustworthy Deep Learning in 6G-Enabled Mass Autonomy: From Concept to Quality-of-Trust Key Performance Indicators. *IEEE Vehicular Technology Magazine*, *15*(4), 112–121. DOI: 10.1109/MVT.2020.3017181

LiFi.co. (n.d.). LiFi — revolutionary wireless communication technology. Retrieved October 29, 2025, from https://lifi.co/

Lisovskyi, K., & Rochenovich, G.-D. (2023). Artificial Intelligence in the Security System of Enterprise. *Grail of Science*, *27*(27), 308–316. DOI: 10.36074/grail-of-science.12.05.2023.049

Liu, C., Salzmann, M., Lin, T., Tomioka, R., & Süsstrunk, S. (2020). On the loss landscape of adversarial training: Identifying challenges and how to overcome them. Advances in Neural Information Processing Systems, 2020-December. DOI: 10.48550/arxiv.2006.08403

Liu, H., & Patras, P. (2022). NetSentry: A deep learning approach to detecting incipient large-scale network attacks. *Computer Communications*, *191*, 119–132. DOI: 10.1016/j.comcom.2022.04.020

Liu, K., Deng, J., Varshney, P. K., & Balakrishnan, K. (2007). An Acknowledgment-Based Approach for the Detection of Routing Misbehavior in MANETs. *IEEE Transactions on Mobile Computing*, *6*(5), 536–550. DOI: 10.1109/TMC.2007.1036

Liu, Y. F., Chang, T. H., Hong, M., Wu, Z., So, A. M. C., Jorswieck, E. A., & Yu, W. (2024). A survey of recent advances in optimization methods for wireless communications. *IEEE Journal on Selected Areas in Communications*, *42*(11), 2992–3031. DOI: 10.1109/JSAC.2024.3443759

Liu, Y., Su, Z., Peng, H., Xiang, Y., Wang, W., & Li, R. (2024). Zero trust-based mobile network security architecture. *IEEE Wireless Communications*, *31*(2), 82–88. DOI: 10.1109/MWC.001.2300375

Li, Y., Wu, B., Huang, Y., & Luan, S. (2024). Developing trustworthy artificial intelligence: insights from research on interpersonal, human-automation, and human-AI trust. In *Frontiers in Psychology* (Vol. 15). Frontiers Media., DOI: 10.3389/fpsyg.2024.1382693

Llisterri Gimenez, N., Monfort Grau, M., Pueyo Centelles, R., & Freitag, F. (2022). On-device training of machine learning models on microcontrollers with federated learning. *Electronics (Basel), 11*(4), 573. DOI: 10.3390/electronics11040573

Lucas, K., Sharif, M., Bauer, L., Reiter, M. K., & Shintre, S. (2021). Malware Makeover: Breaking ML-based Static Analysis by Modifying Executable Bytes. ASIA CCS 2021 - Proceedings of the 2021 ACM Asia Conference on Computer and Communications Security, 744–758. DOI: 10.1145/3433210.3453086

Lunghi, D., Simitsis, A., Caelen, O., & Bontempi, G. (2023). Adversarial Learning in Real-World Fraud Detection: *Challenges and Perspectives. ACM International Conference Proceeding Series*, 27–33. DOI: 10.1145/3600046.3600051

Lu, S. C., Swisher, C. L., Chung, C., Jaffray, D., & Sidey-Gibbons, C. (2023). On the importance of interpretable machine learning predictions to inform clinical decision making in oncology. In *Frontiers in Oncology* (Vol. 13). Frontiers Media., DOI: 10.3389/fonc.2023.1129380

Lusk, R. M., & Monday, W. (2017). *An early survey of best practices for the use of small unmanned aerial systems by the electric utility industry (No. ORNL/TM-2017/93). Oak Ridge National Lab.* ORNL. DOI: 10.2172/1344293

MacKay, D. J. (2016). *Sustainable Energy-without the hot air.* Bloomsbury Publishing.

Mahmoud, M. E., & Shen, X. (2011). RISE: Receipt-Free Cooperation Incentive Scheme for Multihop Wireless Networks. *2011 IEEE International Conference on Communications (ICC)*, 1-5. DOI: 10.1109/icc.2011.5963199

Mahmoud, M. M. E. A., & Shen, N. X. (2011). FESCIM: Fair, Efficient, and Secure Cooperation Incentive Mechanism for Multihop Cellular Networks. *IEEE Transactions on Mobile Computing, 11*(5), 753–766. DOI: 10.1109/TMC.2011.92

Majeed, A. F., & Murdas, I. A. (2019). Novel design for Li-Fi healthcare monitoring system. *International Journal of Intelligent Engineering and Systems, 12*(6), 59–70. DOI: 10.22266/ijies2019.1231.06

Makris, I., Karampasi, A., Radoglou-Grammatikis, P., Episkopos, N., Iturbe, E., Rios, E., Piperigkos, N., Lalos, A., Xenakis, C., Lagkas, T., Argyriou, V., & Sarigiannidis, P. (2025). A comprehensive survey of Federated Intrusion Detection Systems: Techniques, challenges and solutions. *Computer Science Review, 56*, 100717. DOI: 10.1016/j.cosrev.2024.100717

Maksimovic, M. (2017). Greening the future: Green Internet of Things (G-IoT) as a key technological enabler of sustainable development. In *Internet of things and big data analytics toward next-generation intelligence* (pp. 283–313). Springer International Publishing.

Malatji, M., & Tolah, A. (2024). Artificial intelligence (AI) cybersecurity dimensions: A comprehensive framework for understanding adversarial and offensive AI. *AI and Ethics*, *5*(2), 883–910. DOI: 10.1007/s43681-024-00427-4

Mallick, A. I. & Nath, R. (2024). Simulating Cyber Threats : A Review of AI-powered Attack Simulators for Enhanced Cybersecurity. 197(July), 13–54.

Mana, S. M., Hellwig, P., Hilt, J., Bober, K. L., Hirmanova, V. J., Chvojka, P., . . . Zvanovec, S. (2020, March). LiFi experiments in a hospital. In Optical Fiber Communication Conference (pp. M3I-2). Optica Publishing Group.

Mandalapu, V., Elluri, L., Vyas, P., & Roy, N. (2023). Crime Prediction Using Machine Learning and Deep Learning: A Systematic Review and Future Directions. In Access, I. E. E. E. (Ed.), Vol. 11, pp. 60153–60170). Cornell University., DOI: 10.1109/ACCESS.2023.3286344

Manikandan, R., Arunprakash, S., Alsowail, R. A., & Pandiaraj, T. (2025). A novel wireless sensor network deployment for monitoring and predicting abnormal actions in medical environment and patient health state. *Alexandria Engineering Journal*, *119*(January), 149–167. DOI: 10.1016/j.aej.2025.01.064

Maniriho, P., Mahmood, A. N. & Chowdhury, M. J. M. (2024). A systematic literature review on Windows malware detection: Techniques, research issues, and future directions. Journal of Systems and Software, 209(December 2023), 111921. DOI: 10.1016/j.jss.2023.111921

Manresa-Yee, C., & Ramis, S. (2021). Assessing Gender Bias in Predictive Algorithms using eXplainable AI. *ACM International Conference Proceeding Series*, 1–8. DOI: 10.1145/3471391.3471420

Manthena, H., Shajarian, S., Kimmell, J., Abdelsalam, M., Khorsandroo, S., & Gupta, M. (2024). Explainable AI for malware analysis: A survey of techniques, applications, and open challenges. https://doi.org/DOI: 10.48550/. 2401.12345

Manthena, H., Shajarian, S., Kimmell, J., Abdelsalam, M., Khorsandroo, S., & Gupta, M. (2025). *Explainable Artificial Intelligence (XAI) for Malware Analysis: A Survey of Techniques, Applications, and Open Challenges*. IEEE., DOI: 10.1109/ACCESS.2025.3555926

Mao, Y., Zhou, C., Qi, J., & Zhu, X. (2020). A fair credit-based incentive mechanism for routing in DTN-based sensor network with nodes' selfishness. EURASIP Journal on Wireless Communications and Networking, 2020(1). DOI: 10.1186/s13638-020-01823-0

Mao, Y., You, C., Zhang, J., Huang, K., & Letaief, K. B. (2017). A survey on Mobile Edge Computing: The Communication Perspective. *IEEE Communications Surveys and Tutorials*, *19*(4), 2322–2358. DOI: 10.1109/COMST.2017.2745201

Marabissi, D., Mucchi, L., & Stomaci, A. (2022). IoT Nodes Authentication and ID Spoofing Detection Based on Joint Use of Physical Layer Security and Machine Learning. *Future Internet*, *14*(2), 1–21. DOI: 10.3390/fi14020061

Marias, G. F., Georgiadis, P., Flitzanis, D., & Mandalas, K. (2006). Cooperation enforcement schemes for MANETs: A survey. *Wireless Communications and Mobile Computing*, *6*(3), 319–332. DOI: 10.1002/wcm.398

Marri, R., Dabbara, L. N., & Karampuri, S. (2024). AI security in different industries: A comprehensive review of vulnerabilities and mitigation strategies. In International Journal of Science and Research Archive (Vol. 13, Issue 1, pp. 2375–2393). DOI: 10.30574/ijsra.2024.13.1.1923

Martínez Llamas, J., Vranckaert, K., Preuveneers, D., & Joosen, W. (2025). Balancing Security and Privacy: Web Bot Detection, Privacy Challenges, and Regulatory Compliance under the GDPR and AI Act. In Open Research Europe (Vol. 5). F1000 Research Ltd. https://doi.org/DOI: 10.12688/openreseurope.19347.1

Martinez, D., & Magdalena, L. (2024). AI and Blockchain Integration. *Enhancing Security and Transparency in Financial Transactions.*, *3*(1), 11–20.

Ma, S., Sheng, H., Sun, J., Li, H., Liu, X., Qiu, C., Safari, M., Al-Dhahir, N., & Li, S. (2024). Feasibility conditions for mobile lifi. *IEEE Transactions on Wireless Communications*, *23*(7), 7911–7923. DOI: 10.1109/TWC.2023.3346056

Mazhar, T., Irfan, H. M., Khan, S., Haq, I., Ullah, I., Iqbal, M., & Hamam, H. (2023). Analysis of Cyber Security Attacks and Its Solutions for the Smart grid Using Machine Learning and Blockchain Methods. *Future Internet*, *15*(2), 83. Advance online publication. DOI: 10.3390/fi15020083

Mazhar, T., Talpur, D. B., Al Shloul, T., Ghadi, Y. Y., Haq, I., Ullah, I., Ouahada, K., & Hamam, H. (2023). Analysis of IoT Security Challenges and Its Solutions Using Artificial Intelligence. *Brain Sciences*, *13*(4), 683. DOI: 10.3390/brainsci13040683 PMID: 37190648

Mehta, S., Khurana, M., Dogra, A., & Hariharan, S. (2024, June). Advancing IoT security through federated learning: a comprehensive approach. In 2024 3rd International Conference on Applied Artificial Intelligence and Computing (ICAAIC) (pp. 561-566). IEEE. DOI: 10.1109/ICAAIC60222.2024.10575903

Meleshko, A. & Desnitsky, V. (2024). The Modeling and Detection of Attacks in Role-Based Self-Organized Decentralized Wireless Sensor Networks. 145–175.

Mennella, C., Maniscalco, U., De Pietro, G., & Esposito, M. (2024). Ethical and regulatory challenges of AI technologies in healthcare: A narrative review. In Heliyon (Vol. 10, Issue 4). Elsevier BV. DOI: 10.1016/j.heliyon.2024.e26297

Mirestean, A., Farias, A., Deodoro, J., Boukherouaa, E. B., AlAjmi, K., Iskender, E., Ravikumar, R., & Shabsigh, G. (2021). Powering the Digital Economy: Opportunities and Risks of Artificial Intelligence in Finance. *Departmental Paper*, *2021*(24), 1. DOI: 10.5089/9781589063952.087

Moghadas Gholian, S., Fiandrino, C., Vallina-Rodríguez, N., Fiore, M., & Widmer, J. (n.d.). DEEXP: Revealing Model Vulnerabilities for Spatio-Temporal Mobile Traffic Forecasting with Explainable AI. In IEEE Transactions on Mobile Computing. https://git2.networks.imdea.org/

Mohamed, N. (2025). Artificial intelligence and machine learning in cybersecurity: A deep dive into state-of-the-art techniques and future paradigms. *Knowledge and Information Systems*, *67*(8), 6969–7055. Advance online publication. DOI: 10.1007/s10115-025-02429-y

Mohamed, N., Taherdoost, H., & Madanchian, M. (2024). Comprehensive Review of Advanced Machine Learning Techniques for Detecting and Mitigating Zero-Day Exploits. *ICST Transactions on Scalable Information Systems*, *11*(6). Advance online publication. DOI: 10.4108/eetsis.6111

Mohamed, S., Auer, D., Hofer, D., & Küng, J. (2022). A systematic literature review for authorization and access control: Definitions, strategies and models. *International Journal of Web Information Systems*, *18*(2–3), 156–180. DOI: 10.1108/IJWIS-04-2022-0077

Montagnani, M. L., Najjar, M.-C., & Davola, A. (2024). The EU Regulatory approach(es) to AI liability, and its Application to the financial services market. *Computer Law & Security Review*, *53*, 105984. DOI: 10.1016/j.clsr.2024.105984

Montoya, F. G., Peña-García, A., Juaidi, A., & Manzano-Agugliaro, F. (2017). Indoor lighting techniques: An overview of evolution and new trends for energy saving. *Energy and Building*, *140*, 50–60. DOI: 10.1016/j.enbuild.2017.01.028

Morley, J., Machado, C. C. V., Burr, C., Cowls, J., Joshi, I., Taddeo, M., & Floridi, L. (2020). The ethics of AI in health care: A mapping review. In *Social Science & Medicine* (Vol. 260, p. 113172). Elsevier BV., DOI: 10.1016/j.socscimed.2020.113172

Morse, K. E., Bagely, S. C., & Shah, N. H. (2020). Estimate the hidden deployment cost of predictive models to improve patient care. *Nature Medicine*, *26*(1), 18–19. DOI: 10.1038/s41591-019-0651-8 PMID: 31932778

Moulahi, T., Jabbar, R., Alabdulatif, A., Abbas, S., El Khediri, S., Zidi, S., & Rizwan, M. (2023). Privacy-preserving federated learning cyber-threat detection for intelligent transport systems with blockchain-based security. *Expert Systems: International Journal of Knowledge Engineering and Neural Networks*, *40*(5), e13103. DOI: 10.1111/exsy.13103

Moumeni, L., Saber, M., Slimani, I., Elfarissi, I., & Bougroun, Z. (2022). Machine Learning for Credit Card Fraud Detection. *Lecture Notes in Electrical Engineering*, *745*(24), 211–221. DOI: 10.1007/978-981-33-6893-4_20

Moustafa, N. (2021). A new distributed architecture for evaluating AI-based security systems at the edge: Network TON_IoT datasets. *Sustainable Cities and Society*, *72*, 102994. DOI: 10.1016/j.scs.2021.102994

Mubashir, N. M., Siddique, A., Ahmad, S. & Ijaz, A. (2020). Addressing Sinkhole Attacks In Wireless Sensor Networks - A Review. [IJSTR]. *International Journal of Scientific and Technology Research*, *9*(08), 406–411.

Muchiri, N. G. (2021). *Detecting Zero-Day Attacks using Recurrent Neural Network*. Strathmore University.

Mudgal, S. K., Agarwal, R., Chaturvedi, J., Gaur, R., & Ranjan, N. (2022). Real-world application, challenges and implication of artificial intelligence in healthcare: an essay. In *Pan African Medical Journal* (Vol. 43, p. 3). National Institutes of Health., DOI: 10.11604/pamj.2022.43.3.33384

Muhammad, D., & Bendechache, M. (2024). Unveiling the black box: A systematic review of Explainable Artificial Intelligence in medical image analysis. In *Computational and Structural Biotechnology Journal* (Vol. 24, pp. 542–560). Elsevier BV., DOI: 10.1016/j.csbj.2024.08.005

Murtaza, A. (2024). Paradigm shift for predictive maintenance and condition monitoring from Industry 4.0 to Industry 5.0: A systematic review, challenges and case study. *Results in Engineering*, *24*(August), 102935. DOI: 10.1016/j.rineng.2024.102935

Myakala, P. K., & Agrawal, M. (2025). Fault-Tolerant Federated Learning Framework for Edge Devices in Unstable Networks. *Authorea Preprints*. DOI: 10.36227/techrxiv.174612128.82578829/v1

Myakala, P. K., Naayini, P., & Kamatala, S. (2025). A Survey on Federated Learning for TinyML: Challenges, Techniques, and Future Directions. *Partners Universal International Innovation Journal*, *3*(2), 97–114.

Naik, N., Hameed, B. M. Z., Shetty, D. K., Swain, D., Shah, M., Paul, R., Aggarwal, K., Brahim, S., Patil, V., Smriti, K., Shetty, S., Rai, B. P., Chlosta, P., & Somani, B. K. (2022). Legal and Ethical Consideration in Artificial Intelligence in Healthcare: Who Takes Responsibility? *Frontiers in Surgery*, *9*, 862322. Advance online publication. DOI: 10.3389/fsurg.2022.862322 PMID: 35360424

Nallakaruppan, M. K., Chaturvedi, H., Grover, V., Balusamy, B., Jaraut, P., Bahadur, J., Meena, V. P., & Hameed, I. A. (2024). Credit Risk Assessment and Financial Decision Support Using Explainable Artificial Intelligence. *Risks*, *12*(10), 164. DOI: 10.3390/risks12100164

Nandhini, S., Rajeswari, A., & Shanker, N. R. (2024). Cyber attack detection in IOT-WSN devices with threat intelligence using hidden and connected layer based architectures. *Journal of Cloud Computing (Heidelberg, Germany)*, *13*(1), 159. Advance online publication. DOI: 10.1186/s13677-024-00722-9

Narooka, P., Kaushik, K., Kansal, S., Kejriwal, D., Mishra, R. K., & Singh, V. (2025, April). Hybrid AI Models for QoS Estimation in α-η-μ Fading and Co-Channel Interference Environments within 6G Networks. In 2025 3rd International Conference on Advancement in Computation & Computer Technologies (InCACCT) (pp. 206-210). IEEE.

Narooka, P., Verma, S., Dwivedi, A., Kaushik, K., Singh, V., & Gaur, P. (2025, April). Psi-zeta Approach for Assessing Network Efficiency in Log-Normal Fading and Spectrum Sharing Interference Environments. In 2025 3rd International Conference on Advancement in Computation & Computer Technologies (InCACCT) (pp. 886-891). IEEE. DOI: 10.1109/InCACCT65424.2025.11011391

Nguyen, T. T., & Reddi, V. J. (2023). Deep Reinforcement Learning for Cyber Security. *IEEE Transactions on Neural Networks and Learning Systems*, *34*(8), 3779–3795. DOI: 10.1109/TNNLS.2021.3121870 PMID: 34723814

Nguyen, T., Islim, M. S., Chen, C., & Haas, H. (2021). iDim: Practical implementation of index modulation for LiFi Dimming. *IEEE Transactions on Green Communications and Networking*, *5*(4), 1880–1891. DOI: 10.1109/TGCN.2021.3089758

Nibigira, N., Havyarimana, V., & Xiao, Z. (2024). Artificial Intelligence Adoption for Cybersecurity in Africa. *Journal of Information Security, 15*(02), 134–147. DOI: 10.4236/jis.2024.152009

Nibret Zeleke, S., Jember, A. F., & Bochicchio, M. (2025). Integrating explainable AI for effective malware detection in encrypted network traffic. https://doi.org/DOI: 10.48550/.2501.12345

Nicholls, J., Kuppa, A., & Le-Khac, N. A. (2021). Financial cybercrime: A comprehensive survey of deep learning approaches to tackle the evolving financial crime landscape. *IEEE Access : Practical Innovations, Open Solutions, 9*, 163965–163986. DOI: 10.1109/ACCESS.2021.3134076

Nunoo-Mensah, H., Boateng, K. O., & Gadze, J. D. (2017). Tamper-aware authentication framework for wireless sensor networks. *IET Wireless Sensor Systems, 7*(3), 73–81. DOI: 10.1049/iet-wss.2015.0131

Ogunseyi, T. B., & Thiyagarajan, G. (2025). An Explainable LSTM-Based Intrusion Detection System Optimized by Firefly Algorithm for IoT Networks. *Sensors (Basel), 25*(7), 2288. DOI: 10.3390/s25072288 PMID: 40218800

Olanrewaju-George, B., & Pranggono, B. (2025). Federated learning-based intrusion detection system for the internet of things using unsupervised and supervised deep learning models. *Cyber Security and Applications, 3*, 100068. Advance online publication. DOI: 10.1016/j.csa.2024.100068

Oloyede, J. (2024). AI-Driven Cybersecurity Solutions : Enhancing Defense Mechanisms in the Digital Era. SSRN Electronic Journal., 1–18. DOI: 10.2139/ssrn.4976103

Olufemi, O. D.Omoniyi David Olufemi. (2025). Quantum-AI Federated Clouds: A trust-aware framework for cross-domain observability and security. *World Journal of Advanced Research and Reviews, 26*(2), 4098–4140. DOI: 10.30574/wjarr.2025.26.2.2074

Omar, A. Z., Jumaili, M. L. F., Abdual Rahman, S. A. A. E., Mohammed, M. Q. & Aljanabi, M. (2025). Enhancing Network Security in Wireless Sensor Networks Using Artificial Intelligence: A Random Forest Approach. AIP Conference Proceedings, 3264(1). DOI: 10.1063/5.0258927

Omar, S. (2023). A comprehensive survey on detection of sinkhole attack in routing over low power and Lossy network for internet of things. *Internet of Things : Engineering Cyber Physical Human Systems, 22*, 100750. DOI: 10.1016/j.iot.2023.100750

Onyema, E. M., Kanimozhi Suguna, S., Sundaravadivazhagan, B., Jhaveri, R. H., Esther, U. N., Deborah, E. C. & Kumari, K. S. (2025). A secure routing protocol for improving the energy efficiency in wireless sensor network applications for industrial manufacturing. Next Energy, 7(September 2024), 100219. DOI: 10.1016/j.nxener.2024.100219

Orabi, M. M., Emam, O., & Fahmy, H. (2025). Adapting security and decentralized knowledge enhancement in federated learning using blockchain technology: Literature review. *Journal of Big Data*, *12*(1), 55. DOI: 10.1186/s40537-025-01099-5

Osamy, W., Khedr, A. M., Salim, A., El-sawy, A. A., Alreshoodi, M. & Alsukayti, I. (2022). Recent Advances and Future Prospects of Using AI Solutions for Security, Fault Tolerance, and QoS Challenges in WSNs. 1–41.

Othman, S., Ben, , & Bahattab, A. A. (2015). Confidentiality and Integrity for Data Aggregation in WSN Using Homomorphic Encryption. *Wireless Personal Communications*, *80*(2), 867–889. DOI: 10.1007/s11277-014-2061-z

Oyinloye, T. S., Arowolo, M. O., & Prasad, R. (2024). *Enhancing Cyber Threat Detection with an Improved Artificial Neural Network Model*. Data Science and Management., DOI: 10.1016/j.dsm.2024.05.002

Ozkan-Okay, M., Akin, E., Aslan, Ö., Kosunalp, S., Iliev, T., Stoyanov, I., & Beloev, I. (2024). A comprehensive survey: Evaluating the efficiency of artificial intelligence and machine learning techniques on cyber security solutions. *IEEE Access : Practical Innovations, Open Solutions*, *12*, 12229–12256. DOI: 10.1109/ACCESS.2024.3355547

Oztoprak, A., Hassanpour, R., Ozkan, A., & Oztoprak, K. (2024). Security Challenges, Mitigation Strategies, and Future Trends in Wireless Sensor Networks: A Review. *ACM Computing Surveys*, *57*(4), 1–29. Advance online publication. DOI: 10.1145/3706583

Pai, V., Devidas, , & Adesh, N. D. (2021). Comparative analysis of Machine Learning algorithms for Intrusion Detection. *IOP Conference Series. Materials Science and Engineering*, *1013*(1), 12038. DOI: 10.1088/1757-899X/1013/1/012038

Palani, K., Kethar, J., Prasad, S. S., & Torremocha, V. (2024). Impact of AI and Generative AI in transforming Cybersecurity. *Journal of Student Research*, *13*(2). Advance online publication. DOI: 10.47611/jsrhs.v13i2.6710

Pallavi, H. T., Poornima, G., & Tomar, S. K. (2019). Adaptive Timers for Emergency Vehicles and Pollution Detector Using Li-Fi.

Pang, S., & Li, Y. (2020). Artificial Intelligence Techniques for Cyber Security Applications. International Journal of Advanced Information and Communication Technology, 89–94. DOI: 10.46532/ijaict-2020021

Pan, J., Cai, L., Shen, X., & Mark, J. W. (2006). Identity-based secure collaboration in wireless ad hoc networks. *Computer Networks*, *51*(3), 853–865. DOI: 10.1016/j.comnet.2006.05.012

Panwar, P., Verma, R., Rauthan, M. M. S., & Barthwal, V. (2021). *An Overview on Security Issues, Attacks, Challenges and Protocols in WSN. Proceedings of Integrated Intelligence Enable Networks and Computing*. Algorithms for Intelligent Systems., DOI: 10.1007/978-981-33-6307-6_27

Paramesha, M., Rane, N., & Rane, J. (2024). Artificial intelligence, machine learning, deep learning, and blockchain in financial and banking services: a comprehensive review. In SSRN Electronic Journal. RELX Group (Netherlands). DOI: 10.2139/ssrn.4855893

Parande, S., & Mallapur, J. D. (2019). Research trends in secure routing protocols and communication system in WSNs. *International Journal of Innovative Technology and Exploring Engineering*, *9*(1), 3413–3421. DOI: 10.35940/ijitee.A4536.119119

Parreño, I. F., & Avila, D. F. (2021). Analysis of the cybersecurity in wireless sensor networks (WSN): A review literature. Developments and advances in defense and security. *Proceedings of MICRADS*, *2021*, 83–102.

Patil, D. (2025). Artificial Intelligence in Financial Risk Assessment and Fraud Detection: Opportunities and Ethical Concerns. SSRN Electronic Journal. DOI: 10.2139/ssrn.5057434

Patil, H. K., & Chen, T. M. (2017). Chapter 18 - Wireless Sensor Network Security: The Internet of Things. In Computer and Information Security Handbook (Third Edition) (pp. 317–337).

Paul, A., Islam, T., Ferdaush, R., Roni, N., Orpa, J. B., & Sadat, N. (2024). *Li-Fi Data Transmission and Cryptographic Decryption: Design and Implementation of a Smartphone-integrated*. Arduino-Based System for Secure Communication.

Perwej, Y. (2017). The next generation of wireless communication using Li-Fi (light fidelity) technology. [JCN]. *Journal of Computer Networks*, *4*(1), 20–29. DOI: 10.12691/jcn-4-1-3

Petković, D., & Petkovic, D. (2023). It is Not "Accuracy vs. Explainability"—We Need Both for Trustworthy AI Systems. *IEEE Transactions on Technology and Society*, *4*(1), 46–53. DOI: 10.1109/TTS.2023.3239921

Pochmara, J., & Świetlicka, A. (2024). Cybersecurity of Industrial Systems—A 2023 Report. *Electronics (Basel)*, *13*(7), 1191. DOI: 10.3390/electronics13071191

Polemi, N., Praça, I., Kioskli, K., & Bécue, A. (2024). Challenges and efforts in managing AI trustworthiness risks: A state of knowledge. *Frontiers in Big Data*, *7*, 1381163. Advance online publication. DOI: 10.3389/fdata.2024.1381163 PMID: 38798307

Polonius, P., Montavon, G., Lapuschkin, S., Anders, C. J., & Müller, K.-R. (2017). On pixel-wise explanations for non-linear classifier decisions by layer-wise relevance propagation. *PLoS One*, *12*(7), e0180140. DOI: 10.1371/journal.pone.0180140 PMID: 28727795

Ponce-Bobadilla, A. V., Schmitt, V., Maier, C. S., Mensing, S., & Stodtmann, S. (2024). Practical guide to SHAP analysis: Explaining supervised machine learning model predictions in drug development. In Clinical and Translational Science (Vol. 17, Issue 11). Wiley. DOI: 10.1111/cts.70056

Pourghasem, A., Kirner, R., Tsokanos, A., Mporas, I., & Mylonas, A. (2025). Machine Learning and Deep Learning-Based Multi-Attribute Physical-Layer Authentication for Spoofing Detection in LoRaWAN. *Future Internet*, *17*(2), 68. Advance online publication. DOI: 10.3390/fi17020068

Prasanth, A., Vadakkan, D. J., Surendran, P., & Thomas, B. (2023). Role of Artificial Intelligence and Business Decision Making. *International Journal of Advanced Computer Science and Applications*, *14*(6). Advance online publication. DOI: 10.14569/IJACSA.2023.01406103

Premakumari, S. B. N., Sundaram, G., Rivera, M., Wheeler, P., & Guzmán, R. E. P. (2025). Reinforcement Q-Learning-Based Adaptive Encryption Model for Cyberthreat Mitigation in Wireless Sensor Networks. *Sensors (Basel)*, *25*(7), 1–33. DOI: 10.3390/s25072056 PMID: 40218569

Quan, M. K., Pathirana, P. N., Wijayasundara, M., Setunge, S., Nguyen, D. C., Brinton, C. G., Love, D. J., & Poor, H. V. (2025). Federated Learning for Cyber Physical Systems: A Comprehensive Survey. DOI: 10.1109/COMST.2025.3570288

Quintero-Bonilla, S., & del Rey, A. M. (2020). Proposed models for advanced persistent threat detection: A review. In *Advances in Intelligent Systems and Computing* (Vol. 1004, pp. 141–148). Springer Nature., DOI: 10.1007/978-3-030-23946-6_16

Qureshi, K., Alhudhaif, A., Haidar, S. W., Majeed, S., & Jeon, G. (2022). Secure data communication for wireless mobile nodes in intelligent transportation systems. *Microprocessors and Microsystems*, *90*, 104501. DOI: 10.1016/j.micpro.2022.104501

Rafique, S. H., Abdallah, A., Musa, N. S., & Murugan, T. (2024). Machine Learning and Deep Learning Techniques for Internet of Things Network Anomaly Detection—Current Research Trends. *Sensors (Basel)*, *24*(6), 1968. DOI: 10.3390/s24061968 PMID: 38544231

Rahman, M. M., Al Shakil, S., & Mustakim, M. R. (2025). A survey on intrusion detection system in IoT networks. *Cyber Security and Applications*, *3*, 100082. DOI: 10.1016/j.csa.2024.100082

Rahmati, M. (n.d.). Towards Explainable and Lightweight AI for Real-Time Cyber Threat Hunting in Edge Networks.

Ramadhani, E., & Mahardika, G. P. (2018, March). The technology of lifi: A brief introduction. []. IOP Publishing.]. *IOP Conference Series. Materials Science and Engineering*, *325*(1), 012013. DOI: 10.1088/1757-899X/325/1/012013

Ram, G., & Ilavarasan, E. (2024). Security Challenges in Wireless Sensor Network: Current Status and Future Trends. *Wireless Personal Communications*, *139*(2), 1173–1202. DOI: 10.1007/s11277-024-11660-9

Rane, N. L., Choudhary, S., & Rane, J. (2023). Explainable Artificial Intelligence (XAI) Approaches for Transparency and Accountability in Financial Decision-Making. SSRN Electronic Journal. https://doi.org/DOI: 10.2139/ssrn.4640316

Rangaraju, S. (2023). Secure by intelligence: Enhancing products with AI-driven security measures. *EPH-International Journal of Science and Engineering*, *9*(3), 36–41.

Rani, S., Kumar, N., Srivastva, A., & Sharma, A. (2024, October). Leveraging artificial intelligence and machine learning for enhanced privacy and security. In *2024 IEEE International Conference on Blockchain and Distributed Systems Security (ICBDS)* (pp. 1-6). IEEE. DOI: 10.1109/ICBDS61829.2024.10837569

Rao, K. R., Rao, S. N. T., & Reddy, P. C. (2017). Wireless Communication Security and Privacy issues and Challenges. [IJCSIS]. *International Journal of Computer Science and Information Security*, *15*(7), 202–209.

Rathod, S. B., Bajpai, B., Wankhade, S., & Ponnusamy, S. (2025). Enhancing Security in Heterogeneous IoT Networks through Intelligent Identification Systems. *Journal of VLSI Circuits and Systems*, *7*(1), 155–166.

Raturi, A., Mula, K. G., Joshi, K., Sidhu, K. S., Chauhan, S. S., & Singh, V. (2025, April). Comparative Analysis of Stock Price Prediction with Multiple Machine Learning Model. In *2025 3rd International Conference on Communication, Security, and Artificial Intelligence (ICCSAI)* (Vol. 3, pp. 1428-1433). IEEE. DOI: 10.1109/ICCSAI64074.2025.11064599

Raza, H. (2024). Proactive Cyber Defense with AI : Enhancing Risk Assessment and Threat Detection in Cybersecurity Ecosystems. November 2021.

Reddy, P. C. S., Muller, P. S., Koka, S. N., Sharma, V., Sharma, N., & Mukherjee, S. (2023). Detection of encrypted and malicious network traffic using deep learning. *Proceedings of the 2023 International Conference on Ambient Intelligence, Knowledge Informatics and Industrial Electronics (AIKIIE)*, Ballari, India, 1–6. IEEE. DOI: 10.1109/AIKIIE60097.2023.10390386

Redhu, A., Choudhary, P., Srinivasan, K., & Das, T. K. (2024). Deep learning-powered malware detection in cyberspace: a contemporary review. In *Frontiers in Physics* (Vol. 12). Frontiers Media., DOI: 10.3389/fphy.2024.1349463

Reis, M. J. C. S. (2025). Blockchain-Enhanced Security for 5G Edge Computing in IoT. *Computation (Basel, Switzerland)*, *13*(4), 98. DOI: 10.3390/computation13040098

Reyes, M., Meier, R., Pereira, S., Silva, C. A., Dahlweid, F. M., von Tengg-Kobligk, H., Summers, R. M., & Wiest, R. (2020). On the interpretability of artificial intelligence in radiology: Challenges and opportunities. In Radiology: Artificial Intelligence (Vol. 2, Issue 3). Radiological Society of North America. DOI: 10.1148/ryai.2020190043

Reynaud, S., & Roxin, A. (2025). Review of eXplainable artificial intelligence for cybersecurity systems. *Discover Artificial Intelligence*, *5*(1), 78. DOI: 10.1007/s44163-025-00318-5

Rjoub, G., Bentahar, J., Abdel Wahab, O., Mizouni, R., Song, A., Cohen, R., Otrok, H., & Mourad, A. (2023). A Survey on Explainable Artificial Intelligence for Cybersecurity. *IEEE Transactions on Network and Service Management*, *20*(4), 5115–5140. DOI: 10.1109/TNSM.2023.3282740

Rodriguez, E., Otero, B., Gutierrez, N., & Canal, R. (2021). A survey of deep learning techniques for cybersecurity in mobile networks. *IEEE Communications Surveys and Tutorials*, *23*(3), 1920–1955. DOI: 10.1109/COMST.2021.3086296

Roohani, B. S., Sharma, N., Kasula, V. K., Mamoria, P., Modh, N. N., Kumar, A., & Singh, V. (2026). Urban Computing Solutions in Healthcare Edge Computing. In Building Data-Driven Edge Systems for Business Success (pp. 377-400). IGI Global Scientific Publishing.

Roshanaei, M., Khan, M. R., & Sylvester, N. N. (2024). Enhancing Cybersecurity through AI and ML: Strategies, Challenges, and Future Directions. *Journal of Information Security*, *15*(03), 320–339. DOI: 10.4236/jis.2024.153019

Roshanaei, M., Khan, M. R., & Sylvester, N. N. (2024). Navigating AI Cybersecurity: Evolving Landscape and Challenges. *Journal of Intelligent Learning Systems and Applications*, *16*(03), 155–174. DOI: 10.4236/jilsa.2024.163010

Rozenblit, L., Price, A., Solomonides, A., Joseph, A. L., Srivastava, G., Labkoff, S., deBronkart, D., Singh, R., Dattani, K., Lopez-Gonzalez, M., Barr, P. J., Koski, E., Lin, B., Cheung, E., Weiner, M. G., Williams, T., Thuy Bui, T. T., & Quintana, Y. (2025). Towards a Multi-Stakeholder process for developing responsible AI governance in consumer health. *International Journal of Medical Informatics*, *195*, 105713. DOI: 10.1016/j.ijmedinf.2024.105713 PMID: 39642592

Rozlomii, I., Yarmilko, A., & Naumenko, S. (2024). Data security of IoT devices with limited resources: challenges and potential solutions. *doors, 3666*, 85-96.

Rudin, C. (2019). Stop explaining black box machine learning models for high stakes decisions and use interpretable models instead. *Nature Machine Intelligence*, *1*(5), 206–215. DOI: 10.1038/s42256-019-0048-x PMID: 35603010

Ruzbahani, A. M. (2024). AI-Protected Blockchain-based IoT environments: Harnessing the Future of Network Security and Privacy. https://arxiv.org/abs/2405.13847

Sadeghi, Z., & Alizadehsani, R. CIFCI, M. A., Kausar, S., Rehman, R., Mahanta, P., Bora, P. K., Almasri, A., Alkhawaldeh, R. S., Hussain, S., Alatas, B., Shoeibi, A., Moosaei, H., Hladík, M., Nahavandi, S., & Pardalos, P. M. (2024). A review of Explainable Artificial Intelligence in healthcare. In Computers and Electrical Engineering (Vol. 118). DOI: 10.1016/j.compeleceng.2024.109370

Saeed, . (2023). An SDN perspective IoT-Fog security: A survey. *Computer Networks*, *229*, 109732. DOI: 10.1016/j.comnet.2023.109732

Saeed, K., & Omlin, C. (2022). Challenges and research directions in explainable AI. *Journal of AI Research*, *65*, 223–249. DOI: 10.1613/jair.1.12345

Sáez-de-Cámara, X., Flores, J. L., Arellano, C., Urbieta, A., & Zurutuza, U. (2023). Clustered federated learning architecture for network anomaly detection in large scale heterogeneous IoT networks. *Computers & Security, 131*, 103299. Advance online publication. DOI: 10.1016/j.cose.2023.103299

Saha, S., Banerjee, K., Ghosh, S., Mitra, S., & Pal, D. (2023). AI-Driven Edge Computing for IoT: A Comprehensive Survey and Future Directions. International Journal of Advanced Research in Science. *Tongxin Jishu, 117–121*, 117–121. Advance online publication. DOI: 10.48175/IJARSCT-12921

Saiful, M., Mamun, I., Kabir, A. F. M. S., Hossen, S., Hayat, R., Saiful Islam Mamun, M., Sultanul Kabir, A., Sakhawat Hossen, M., Hayat Khan, M., Saiful, M., Mamun, I., Kabir, A. F. M. S., Hossen, S. & Hayat, R. (2012). Policy based intrusion detection and response system in hierarchical {WSN} architecture . CoRR, abs/1209.1(May).

Salami, I. A., Adesokan-Imran, T. O., Tiwo, O. J., Metibemu, O. C., Olutimehin, A. T., & Olaniyi, O. O. (2025). Addressing Bias and Data Privacy Concerns in AI-Driven Credit Scoring Systems Through Cybersecurity Risk Assessment. *Asian Journal of Research in Computer Science, 18*(4), 59–82. DOI: 10.9734/ajrcos/2025/v18i4608

Saleh, A. M. S. (2024). Blockchain for secure and decentralized artificial intelligence in cybersecurity: A comprehensive review. Blockchain. *Research and Applications, 5*(3), 100193. DOI: 10.1016/j.bcra.2024.100193

Salem, A. H., Azzam, S. M., Emam, O. E., & Abohany, A. A. (2024). Advancing cybersecurity: A comprehensive review of AI-driven detection techniques. *Journal of Big Data, 11*(1), 105. Advance online publication. DOI: 10.1186/s40537-024-00957-y

Salih, A. M., Raisi-Estabragh, Z., Galazzo, I. B., Radeva, P., Petersen, S. E., Lekadir, K., & Menegaz, G. (2024). A Perspective on Explainable Artificial Intelligence Methods: SHAP and LIME. *Advanced Intelligent Systems*. Advance online publication. DOI: 10.1002/aisy.202400304

Salmi, S., & Oughdir, L. (2023). Performance evaluation of deep learning techniques for DoS attacks detection in wireless sensor network. *Journal of Big Data, 10*(1), 17. Advance online publication. DOI: 10.1186/s40537-023-00692-w

Sameen, M., Han, K., & Hwang, S. O. (2020). PhishHaven - An Efficient Real-Time AI Phishing URLs Detection System. *IEEE Access : Practical Innovations, Open Solutions, 8*, 83425–83443. DOI: 10.1109/ACCESS.2020.2991403

Samek, W., Montavon, G., Lapuschkin, S., Anders, C. J., & Müller, K. R. (2020). Toward interpretable machine learning: Transparent deep neural networks and beyond. ArXiv, 1–24. https://arxiv.org/pdf/2003.07631

Samiulla, I., Khan, A. A., Ahmad, M., & Idrisi, M. J. (2023). A Secure and Privacy-Preserving Lightweight Authentication and Key Exchange Algorithm for Smart Agriculture Monitoring System. *IEEE Access : Practical Innovations, Open Solutions*, *11*(May), 56875–56890. DOI: 10.1109/ACCESS.2023.3280542

Sanchez, O. T., Borges, G., Raposo, D., Rodrigues, A., Boavida, F., & Silva, J. S. (2025). Enhancing the performance of Industrial IoT LoRaWAN-enabled federated learning frameworks: A case study. *Internet of Things : Engineering Cyber Physical Human Systems*, *32*, 101632. Advance online publication. DOI: 10.1016/j.iot.2025.101632

Sankalp, M. R., Lokapal, G., Mohan, B. A., & Basavaraj, G. N. (2025, March). Addressing Cybersecurity Challenges in 6G Networks Through AI-Driven Adaptive Defense Mechanisms and Quantum-Resilient Protocols. In *2025 International Conference on Computing for Sustainability and Intelligent Future (COMP-SIF)* (pp. 1-12). IEEE.

Sarhan, M. M., Layeghy, S., Gallagher, M. M., & Portmann, M. (2021). From Zero-Shot Machine Learning to Zero-Day Attack Detection. [Cornell University]. *International Journal of Information Security*, *22*(4), 947–959. DOI: 10.1007/s10207-023-00676-0

Sarhan, M., Lo, W. W., Layeghy, S., & Portmann, M. (2022). HBFL: A hierarchical blockchain-based federated learning framework for collaborative IoT intrusion detection. *Computers & Electrical Engineering*, *103*, 108379. Advance online publication. DOI: 10.1016/j.compeleceng.2022.108379

Sarkar, A., Agarwal, S., & Nath, A. (2015). Li-Fi technology: Data transmission through visible light. *International Journal of Advance Research in Computer Science and Management Studies*, *3*(6).

Sarker, I. H. (2023). Multi-aspects AI-based modeling and adversarial learning for cybersecurity intelligence and robustness: A comprehensive overview. *Security and Privacy*, *6*(5), e295. DOI: 10.1002/spy2.295

Saunders, S. R., & Aragón-Zavala, A. A. (2024). *Antennas and propagation for wireless communication systems*. John Wiley & Sons. DOI: 10.1002/9781119332206

Schmitt, M. (2023). Securing the digital world: Protecting smart infrastructures and digital industries with artificial intelligence (AI)-enabled malware and intrusion detection. *Journal of Industrial Information Integration*, *36*, 100520. DOI: 10.1016/j.jii.2023.100520

Schreiber, J., & Koh, P. W. (2017). Understanding black-box predictions via influence functions. In Precup, D., & Teh, Y. W. (Eds.), *Proceedings of the 34th International Conference on Machine Learning* (Vol. 70, pp. 1885-1894). PMLR. https://doi.org/ DOI: 10.5555/3305890.3305912

Scott, L., & Lundberg, S. M. (2017). A unified approach to interpreting model predictions. In Guyon, I., Luxburg, U. V., Bengio, S., Wallach, H., Fergus, R., Vishwanathan, S., & Garnett, R. (Eds.), Vol. 30, pp. 4765–4774). Advances in Neural Information Processing Systems. Curran Associates, Inc., DOI: 10.48550/705.07874

Sebestyen, H., Popescu, D. E., & Zmaranda, R. D. (2025). A Literature Review on Security in the Internet of Things: Identifying and Analysing Critical Categories. *Computers*, *14*(2), 61. Advance online publication. DOI: 10.3390/computers14020061

Selvarajan, S., Srivastava, G., Khadidos, A. O., Khadidos, A. O., Baza, M., Alshehri, A., & Lin, J. C. W. (2023). An artificial intelligence lightweight blockchain security model for security and privacy in IIoT systems. *Journal of Cloud Computing (Heidelberg, Germany)*, *12*(1), 38. Advance online publication. DOI: 10.1186/ s13677-023-00412-y PMID: 36937654

Senevirathna, T., La, V. H., Marchal, S., Siniarski, B., Liyanage, M., & Wang, S. (2024). A survey on XAI for 5G and beyond security: Technical aspects, challenges and research directions. *IEEE Communications Surveys and Tutorials*.

Sen, J. (2023). Cryptography and Key Management Schemes for Wireless Sensor Networks. In Wireless Sensor Networks—Design, Applications and Challenges.

Sen, J., Sen, R. R., & Dutta, A. A. A. (2021). *Machine Learning in Finance-Emerging Trends and Challenges. In arXiv (Cornell University). arXiv*. Cornell University., DOI: 10.5772/intechopen.101120

Shahana, A., Hasan, R., Farabi, S. F., Akter, J., Mahmud, M. A. A., Johora, F. T., & Suzer, G. (2024). AI-driven cybersecurity: Balancing advancements and safeguards. *Journal of Computer Science and Technology Studies*, *6*(2), 76–85. DOI: 10.32996/jcsts.2024.6.2.9

Shahrivari, V., Darabi, M. M., & Izadi, M. (2020). *Phishing Detection Using Machine Learning Techniques. ArXiv*. Cornell University., DOI: 10.1109/ASIANCON58793.2023.10270550

Shakeel, P. M., Baskar, S., Dhulipala, V. S., & Jaber, M. M. (2018). Cloud based framework for diagnosis of diabetes mellitus using K-means clustering. *Health Information Science and Systems*, *6*(1), 16. DOI: 10.1007/s13755-018-0054-0 PMID: 30279986

Shammari, G. Al & Parveen, N. (2024). A Dual-Channel Deep Learning Framework For Real-Time Detection Of Zero-Day Attacks Using Cnn-Lstm Hybrid Networks. December.

Shang, Z., Chauhan, V., Devi, K., & Patil, S. (2024). Artificial Intelligence, the Digital Surgeon: Unravelling Its Emerging Footprint in Healthcare – The Narrative Review. In Journal of Multidisciplinary Healthcare (pp. 4011–4022). Dove Medical Press. DOI: 10.2147/JMDH.S482757

Shanthi, T., & Shalini, P. (2023, November). Communication Between Submarines Using Li-Fi Technology. In *2023 International Conference on Recent Advances in Science and Engineering Technology (ICRASET)* (pp. 1-4). IEEE.

Sharma, R., Sharma, V., Vashishth, T. K., Kumar, B., Sharma, B. K., & Chaudhary, S. (2025, February). Implementation of adaptive cloud security framework for cloud computing monitoring system applications. In AIP Conference Proceedings (Vol. 3224, No. 1). AIP Publishing. DOI: 10.1063/5.0246104

Sharma, N. A., Chand, R. R., Buksh, Z., Ali, A. B. M. S., Hanif, A., & Beheshti, A. (2024). Explainable AI Frameworks: Navigating the Present Challenges and Unveiling Innovative Applications. *Algorithms*, *17*(6), 227. DOI: 10.3390/a17060227

Sharma, R. R., Sanganal, A., & Pati, S. (2014). Implementation of a simple Li-Fi based system. *International Journal of Computers and Technology*, *1*(9), 437–443.

Sharma, S., Sharma, A., & Van Chien, T. (Eds.). (2025). *The Intersection of 6G, AI/Machine Learning, and Embedded Systems: Pioneering Intelligent Wireless Technologies*. CRC Press. DOI: 10.1201/9781003540212

Sharma, V., Beniwal, R., & Kumar, V. (2024). Multi-level trust-based secure and optimal IoT-WSN routing for environmental monitoring applications. *The Journal of Supercomputing*, *80*(8), 11338–11381. DOI: 10.1007/s11227-023-05875-z

Shihab, M. A., Marhoon, H. A., Ahmed, S. R., Al-Attar, B., Al-Sharify, M. T., & Sekhar, R. (2024). Towards Resilient Machine Learning Models: Addressing Adversarial Attacks in Wireless Sensor Network. [JRC]. *Journal of Robotics and Control*, *5*(5), 1582–1602. DOI: 10.18196/jrc.v5i5.23214

Shinde, R., Patil, S., Kotecha, K., & Ruikar, K. (2021). Blockchain for securing ai applications and open innovations. *Journal of Open Innovation*, *7*(3), 189. Advance online publication. DOI: 10.3390/joitmc7030189

Shrivastava, M. (2018). Improving Wireless Sensor Network ' s Performance through Intelligent Sensor Reinforcement Learning Algorithm (ISRLA). 6(1), 1052–1056.

Shuaib, A. (2024). Transforming Healthcare with AI: Promises, Pitfalls, and Pathways Forward. *International Journal of General Medicine*, *17*, 1765–1771. DOI: 10.2147/IJGM.S449598 PMID: 38706749

Shu, Y., Lee, K., Lanctot, P., Jianbin, F., & Hao, H. (2016). *Internet of Things.* Wireless Sensor Networks. In IEC Market Strategy Board.

Sikos, L. F. (2020). Packet analysis for network forensics: A comprehensive survey. *Forensic Science International Digital Investigation*, *32*, 200892. DOI: 10.1016/j.fsidi.2019.200892

Sinayobye, J. O., Kiwanuka, F., & Kaawaase Kyanda, S. (2018). A state-of-the-art review of machine learning techniques for fraud detection research. Proceedings - International Conference on Software Engineering, 11–19. DOI: 10.1145/3195528.3195534

Sindiramutty, S. R., Prabagaran, K. R. V., Jhanjhi, N. Z., Murugesan, R. K., Brohi, S. N., & Masud, M. (2025). Generative AI in Network Security and Intrusion Detection. In Reshaping CyberSecurity with Generative AI Techniques (pp. 77-124). IGI Global.

Singh, B., Jain, V., Kaunert, C., & Vig, K. (2024). Shaping Highly Intelligent Internet of Things (IoT) and Wireless Sensors for Smart Cities. In *Secure and Intelligent IoT-Enabled Smart Cities* (pp. 117-140). IGI Global Scientific Publishing.

Singh, O., Vinoth, R., Singh, A. & Singh, N. (2024). Navigating Security Threats and Solutions using AI in Wireless Sensor Networks. 16(4).

Singh, V., & Kaushik, V. D. (2023). A Brief Overview of Cyber Security Advances and Techniques Along With a Glimpse on Quantum Cryptography: Cyber Security Practices, Advances and Challenges. Exploring Cyber Criminals and Data Privacy Measures, 40-64.

Singh, V., & Kaushik, V. D. (2023). A Methodical View of Prerequisites of Picture Combination, Strategies, and Key Indicators with Usage in Real Life and Scientific Domains Facilitating Smart Ubiquitous Environment. In Computational Intelligence in Analytics and Information Systems (pp. 15-33). Apple Academic Press

Singh, V., & Kaushik, V. D. (2023). A Methodical View of Prerequisites of Picture Combination, Strategies, and Key Indicators with Usage in Real Life and Scientific Domains Facilitating Smart Ubiquitous Environment. In Computational Intelligence in Analytics and Information Systems (pp. 15-33). Apple Academic Press.

Singh, O., Vinoth, R., Singh, A., & Singh, N. (2024). Navigating Security Threats and Solutions using AI in Wireless Sensor Networks. *International Journal of Communication Networks and Information Security, 16*(4), 411–427.

Singh, R., & Gill, S. S. (2023). Edge AI: A survey. *Internet of Things and Cyber-Physical Systems, 3*, 71–92. DOI: 10.1016/j.iotcps.2023.02.004

Singh, S., Sharma, P. K., Yoon, B., Shojafar, M., Cho, G. H., & Ra, I. H. (2020). Convergence of blockchain and artificial intelligence in IoT network for the sustainable smart city. *Sustainable Cities and Society, 63*, 102364. DOI: 10.1016/j.scs.2020.102364

Singh, V., & Dubey, V. (2015). A two level image security based on Arnold transform and chaotic logistic mapping. *International Journal of Advanced Research in Computer Science and Software Engineering, 5*(2), 883–887.

Singh, V., & Kaushik, V. D. (2023). A Methodical View of Prerequisites of Picture Combination, Strategies, and Key Indicators with Usage in Real Life and Scientific Domains Facilitating Smart Ubiquitous Environment. In *Computational Intelligence in Analytics and Information Systems* (pp. 15–33). Apple Academic Press. DOI: 10.1201/9781003332312-3

Singh, V., & Kaushik, V. D. (2023). Navigating the Landscape of Security Threat Analysis in Cloud Computing environments. In *Security and Risk Analysis for Intelligent Cloud Computing* (pp. 1–25). CRC Press.

Solaas, J. R. V., Tuptuk, N., & Mariconti, E. (2024). *Systematic Review: Anomaly Detection in Connected and Autonomous Vehicles. ArXiv*. Cornell University., DOI: 10.1109/TITS.2024.3495031

Soltani, M. D., Arfaoui, M. A., Tavakkolnia, I., Ghrayeb, A., Safari, M., Assi, C. M., Hasna, M. O., & Haas, H. (2019). Bidirectional optical spatial modulation for mobile users: Toward a practical design for LiFi systems. *IEEE Journal on Selected Areas in Communications, 37*(9), 2069–2086. DOI: 10.1109/JSAC.2019.2929381

Sonule, A. R., Kalla, M., Jain, A., & Chouhan, D. S. (2020). Unsw-Nb15 Dataset and Machine Learning Based Intrusion Detection Systems. *International Journal of Engineering and Advanced Technology, 9*(3), 2638–2648. DOI: 10.35940/ijeat.C5809.029320

Sourlos, N., Wang, J., Nagaraj, Y., van Ooijen, P., & Vliegenthart, R. (2022). Possible Bias in Supervised Deep Learning Algorithms for CT Lung Nodule Detection and Classification. In Cancers (Vol. 14, Issue 16, p. 3867). Multidisciplinary Digital Publishing Institute. DOI: 10.3390/cancers14163867

Sparks, E. S. (2025). Strengthening Cybersecurity through Proactive Threat Mitigation and Intelligent Defense Mechanisms. 6, 1–7.

Speith, T., & Langer, M. (2023). A New Perspective on Evaluation Methods for Explainable Artificial Intelligence (XAI). ArXiv (Cornell University). /arxiv.2307. 14246DOI: 10.1109/REW57809.2023.00061

Speith, A. (2021). Taxonomy of XAI approaches and challenges. *AI Review Journal*, *9*(4), 315–339. DOI: 10.1007/s10462-021-10049-5

Sridevi & Kolhar. A. (2025). Energy-Efficiency Strategies for Wireless Sensor Networks in IoT. In A. Sathio, M. Rind, S. Awan, & A. Junejo (Eds.), Energy-Efficient Deep Learning Approaches in IoT, Fog, and Green Blockchain Revolution (pp. 167-192). IGI Global Scientific Publishing. DOI: 10.4018/979-8-3373-0300-0.ch006

Sridevi & Kolhar. A. (2025). Machine Learning Algorithms and IoT Sensors for Securing the Networks. In Minakshi, A. Bijalwan, & T. Kumar (Eds.), Exploiting Machine Learning for Robust Security (pp. 183-210). IGI Global Scientific Publishing. DOI: 10.4018/979-8-3693-7758-1.ch009

Sriram, S., Simran, K., Vinayakumar, R., Akarsh, S., & Soman, K. P. (2020). Towards Evaluating the Robustness of Deep Intrusion Detection Models in Adversarial Environment. In Communications in Computer and Information Science: Vol. 1208 CCIS (pp. 111–120). Springer Science+Business Media. DOI: 10.1007/978-981-15-4825-3_9

Stiglic, G., Kocbek, P., Fijacko, N., Zitnik, M., Verbert, K., & Cilar, L. (2020). Interpretability of machine learning-based prediction models in healthcare. *Wiley Interdisciplinary Reviews. Data Mining and Knowledge Discovery*, *10*(5), e1379. Advance online publication. DOI: 10.1002/widm.1379

Sulayman, I. I. A., He, R., Manka, M., Ning, A., & Ouda, A. (2021, October). LiFi/WiFi authentication and handover protocols: Survey, evaluation, and recommendation. In *2021 International Symposium on Networks, Computers and Communications (ISNCC)* (pp. 1-6). IEEE. DOI: 10.1109/ISNCC52172.2021.9615853

Sun, H., Liu, Y., Al-Tahmeesschi, A., Nag, A., Soleimanpour-Moghadam, M., Canberk, B., Arslan, H., & Ahmadi, H. (2025). Advancing 6G: Survey for Explainable AI on Communications and Network Slicing. *IEEE Open Journal of the Communications Society*, *6*, 1372–1412. Advance online publication. DOI: 10.1109/OJCOMS.2025.3534626

Sun, Y., Peng, M., Zhou, Y., Huang, Y., & Mao, S. (2019). Application of Machine Learning in Wireless Networks: Key Techniques and Open Issues. *IEEE Communications Surveys and Tutorials*, *21*(4), 302–3108. DOI: 10.1109/COMST.2019.2924243

Suriyan, K., Ganesh, S., Palaniyammal, P., Priya, G., & Singh, V. (2026). Recent Trends in Edge Computing: Challenges and Opportunities. Building Data-Driven Edge Systems for Business Success, 323-344.

Svetlova, E. (2022). AI ethics and systemic risks in finance. *AI and Ethics*, *2*(4), 713–725. DOI: 10.1007/s43681-021-00129-1 PMID: 35043105

Swami, K. T., & Moghe, A. A. (2020, December). A review of LiFi technology. In 2020 5th IEEE international conference on recent advances and innovations in engineering (ICRAIE) (pp. 1-5). IEEE. DOI: 10.1109/ICRAIE51050.2020.9358340

Taherdoost, H. (2022). Blockchain Technology and Artificial Intelligence Together: A Critical Review on Applications. In Applied Sciences (Switzerland) (Vol. 12, Issue 24, p. 12948). Multidisciplinary Digital Publishing Institute. DOI: 10.3390/app122412948

Talavante, J., Genovés Guzmán, B., & Giustiniano, D. (2023, June). Rethinking LiFi for carbon neutral sunlight-based communication. In *Mediterranean Communication and Computer Networking Conference* (pp. 1-8). DOI: 10.1109/MedComNet58619.2023.10168863

Talukder, M. A., Islam, M. M., Uddin, M. A., Hasan, K. F., Sharmin, S., Alyami, S. A., & Moni, M. A. (2024). Machine learning-based network intrusion detection for big and imbalanced data using oversampling, stacking feature embedding and feature extraction. *Journal of Big Data*, *11*(1), 33. Advance online publication. DOI: 10.1186/s40537-024-00886-w

Tanaka, Y., Komine, T., Haruyama, S., & Nakagawa, M. (2001, October). Indoor visible communication utilizing plural white LEDs as lighting. In 12th IEEE International Symposium on Personal, Indoor and Mobile Radio Communications. PIMRC 2001. Proceedings (Cat. No. 01TH8598) (Vol. 2, pp. F-F). IEEE. DOI: 10.1109/PIMRC.2001.965300

Tanbhir, G., & Shahriyar, M. F. (2025). Quantum-Inspired Privacy-Preserving Federated Learning Framework for Secure Dementia Classification. arXiv preprint arXiv:2503.03267. DOI: 10.1109/ECCE64574.2025.11013884

Tanikonda, A., Pandey, B. K., Peddinti, S. R., & Katragadda, S. R. (2022). Advanced AI-Driven Cybersecurity Solutions for Proactive Threat Detection and Response in Complex Ecosystems. *Journal of Science and Technology*, *3*(1).

Tao, Z., Xu, W., Huang, Y., Wang, X., & You, X. (2024). Wireless network digital twin for 6g: Generative ai as a key enabler. *IEEE Wireless Communications, 31*(4), 24–31. DOI: 10.1109/MWC.002.2300564

Tariq, U., Ahmed, I. & Bashir, A. K. (2023). A Critical Cybersecurity Analysis and Future Research Directions for the Internet of Things: A Comprehensive Review. Sensor, 23(4117).

Tavakoli, N., Siami-Namini, S., Adl Khanghah, M., Mirza Soltani, F., & Siami Namin, A. (2020). An autoencoder-based deep learning approach for clustering time series data. *SN Applied Sciences*, *2*(937), 937. DOI: 10.1007/s42452-020-2584-8

Teixeira, R., Baldoni, G., Antunes, M., Gomes, D., & Aguiar, R. L. (2025). Leveraging decentralized communication for privacy-preserving federated learning in 6G Networks. *Computer Communications*, *233*, 108072. DOI: 10.1016/j.comcom.2025.108072

Thapa, N., Liu, Z., Kc, D. B., Gokaraju, B., & Roy, K. (2020). Comparison of machine learning and deep learning models for network intrusion detection systems. *Future Internet*, *12*(10), 1–16. DOI: 10.3390/fi12100167

Thomas, Z., Kumar, N., & Preshiya, D. J. (2016, April). Automatic billing system using Li-Fi module. In *2016 International Conference on Communication and Signal Processing (ICCSP)* (pp. 2216-2219). IEEE. DOI: 10.1109/ICCSP.2016.7754087

Thuraisingham, B. (2022). Trustworthy Machine Learning. *IEEE Intelligent Systems*, *37*(1), 21–24. DOI: 10.1109/MIS.2022.3152946

Tian, T., Zhang, C., Jiang, B., Feng, H., & Lu, Z. (2025). Insider threat detection for specific threat scenarios. *Cybersecurity*, *8*(1), 17. Advance online publication. DOI: 10.1186/s42400-024-00321-w

Timofte, E. M., Dimian, M., Graur, A., Potorac, A. D., Balan, D., Croitoru, I., Hri can, D.-F., & Pu ca u, M. (2025). Federated Learning for Cybersecurity: A Privacy-Preserving Approach. *Applied Sciences (Basel, Switzerland)*, *15*(12), 6878. DOI: 10.3390/app15126878

Tiwari, V. (2018). Data Integrity and Authentication in WSNs. HMR Interdisciplinary Journal of Science. *Technology & Education Management*, *4*(1), 12–17.

Tolba, A. (2024). Hybrid LSTM-Random Forest for Intrusion Detection in Wireless Sensor Networks : Enhanced DoS Classification with Explainable AI 1 | Introduction. 2, 39–52.

Tommi, D., Jaakkola, P. S., & Alvarez-Melis, D. (2017). A causal framework for explaining the predictions of black-box sequence-to-sequence models. In Precup, D., & Teh, Y. W. (Eds.), *Proceedings of the 34th International Conference on Machine Learning* (Vol. 70, pp. 96-105). PMLR. https://doi.org/DOI: 10.5555/3305890.3305901

Topare, S. V. (2024). 5g Dominance: Why Li-Fi Technology is Struggling to Compete. Available at *SSRN* 4835013.

Torkey, H., Hashish, S., Souissi, S., Hemdan, E. E. D., & Sayed, A. (2025). Seizure Detection in Medical IoT: Hybrid CNN-LSTM-GRU Model with Data Balancing and XAI Integration. *Algorithms*, *18*(2), 77. DOI: 10.3390/a18020077

Triwidyastuti, Y., Perdana, R. H. Y., Shim, K., & An, B. (2023). Secrecy Performance Analysis of Cooperative Multihop Transmission for WSNs under Eavesdropping Attacks †. *Sensors (Basel)*, *23*(17), 1–25. DOI: 10.3390/s23177653 PMID: 37688109

Truong, N. B., Lee, H., Askwith, B., & Lee, G. M. (2017). Toward a trust evaluation mechanism in the social internet of things. *Sensors (Basel)*, *17*(6), 1346. DOI: 10.3390/s17061346 PMID: 28598401

Tubío Figueira, P., López Bravo, C., & Rivas López, J. L. (2020). Improving information security risk analysis by including threat-occurrence predictive models. *Computers & Security*, *88*, 101609. Advance online publication. DOI: 10.1016/j.cose.2019.101609

Ul Hassan, N., Yuen, C., & Niyato, D. (2019). Blockchain Technologies for Smart Energy Systems: Fundamentals, Challenges, and Solutions. *IEEE Industrial Electronics Magazine*, *13*(4), 106–118. DOI: 10.1109/MIE.2019.2940335

Umer, M. A., Junejo, K. N., Jilani, M. T., & Mathur, A. P. (2022). Machine learning for intrusion detection in industrial control systems: Applications, challenges, and recommendations. *International Journal of Critical Infrastructure Protection*, *38*, 100516. DOI: 10.1016/j.ijcip.2022.100516

Usama, M., Qadir, J., Al-Fuqaha, A., & Hamdi, M. (2020). The Adversarial Machine Learning Conundrum: Can the Insecurity of ML Become the Achilles' Heel of Cognitive Networks? *IEEE Network*, *34*(1), 196–203. DOI: 10.1109/MNET.001.1900197

Vaid, C. S. and R. (2021). A Novel Sybil Attack Detection and Prevention Mechanism for Wireless Sensor Networks. 2021 6th International Conference on Signal Processing, Computing and Control (ISPCC), 340–345.

Vashishth, T. K., Sharma, V., Kumar, B., Sharma, K. K., Chaudhary, S., & Panwar, R. (2025). Blockchain for Securing Federated Learning Systems: Enhancing Privacy and Trust. Model Optimization Methods for Efficient and Edge AI: Federated Learning Architectures, Frameworks and Applications, 299-320.

Vashishth, T. K., Sharma, V., Sharma, K. K., Kumar, B., Chaudhary, S., & Gupta, M. (2024). Privacy-Preserving Machine Learning Techniques for IoT Data in Cloud Environments. In Emerging Technologies for Securing the Cloud and IoT (pp. 144-173). IGI Global. DOI: 10.4018/979-8-3693-0766-3.ch006

Vashishth, T. K., Sharma, V., Sharma, K. K., Kumar, B., Chaudhary, S., & Panwar, R. (2024). Enhancing Cloud Security: The Role of Artificial Intelligence and Machine Learning. In Improving Security, Privacy, and Trust in Cloud Computing (pp. 85-112). IGI Global.

Vashishth, T. K., Sharma, V., Sharma, K. K., Kumar, B., Chaudhary, S., & Panwar, R. (2024). Intrusion Detection and Prevention Techniques in FL Cloud-Based Healthcare 5.0: A Comprehensive Review. Pioneering Smart Healthcare 5.0 with IoT, Federated Learning, and Cloud Security, 152-174.

Vashishth, T. K., Sharma, V., Sharma, K. K., Kumar, B., Chaudhary, S., & Panwar, R. (2024). Security and Privacy Challenges in Blockchain-Based Supply Chain Management: A Comprehensive Analysis. In Achieving Secure and Transparent Supply Chains with Blockchain Technology (pp. 70-91). IGI Global. DOI: 10.4018/979-8-3693-0482-2.ch005

Vashishth, T. K., Sharma, V., Sharma, M. K., Sharma, R., Sharma, K. K., & Sharma, S. (2026). AI-driven threat detection and incident response: Advancing cybersecurity with machine learning. In Improving Threat Detection, Network Security, and Incident Response with AI (pp. 51–82). IGI Global Scientific Publishing.

Vashishth, T. K., Sharma, V., Samania, B., Sharma, R., Singh, S., & Jajoria, P. (2025). Ethical and legal implications of AI in cybersecurity. In *Machine Intelligence Applications in Cyber-Risk Management* (pp. 387–414). IGI Global Scientific Publishing.

Vashishth, T. K., Sharma, V., Sharma, K. K., & Kumar, B. (2025). Enhanced Security and Privacy from Industry 4.0 and 5.0 Vision. In *Building Embodied AI Systems: The Agents, the Architecture Principles, Challenges, and Application Domains* (pp. 133–158). Springer Nature Switzerland.

Vashishth, T. K., Sharma, V., Sharma, K. K., Kumar, B., Chaudhary, S., & Panwar, R. (2024). Security and Privacy Considerations in Cloud-Based Data Processing Solutions for Sensitive Data. In *Developments towards Next Generation Intelligent Systems for Sustainable Development* (pp. 35–61). IGI Global. DOI: 10.4018/979-8-3693-5643-2.ch002

Velasco, C. (2022). Cybercrime and Artificial Intelligence. An overview of the work of international organizations on criminal justice and the international applicable instruments. ERA Forum, 23(1), 109–126. DOI: 10.1007/s12027-022-00702-z

Vellido, A. (2020). The importance of interpretability and visualization in machine learning for applications in medicine and health care. *Neural Computing & Applications*, *32*(24), 18069–18083. DOI: 10.1007/s00521-019-04051-w

Vikas, D., R. P., Kumar, D., Charan, P., Ingilela, R. S. K., & Rastogi, R. (2023). Intrusion detection in wireless sensor networks using hybrid deep belief networks and Harris Hawks optimizer. Proceedings of the 2023 4th International Conference on Electronics and Sustainable Communication Systems (ICESC), Coimbatore, India, 1631–1636. IEEE. DOI: 10.1109/ICESC57686.2023.10193270

Vilone, G., & Longo, L. (2018). Classification of explainable artificial intelligence methods through their output formats. *Machine Learning and Knowledge Extraction*, *1*(3), 615–661. DOI: 10.3390/make3030032

Vinothkumar, J., & Karunamurthy, A. (2023). Recent Advancements in Artificial Intelligence Technology: Trends and Implications. *Quing International Journal of Multidisciplinary Scientific Research and Development*, *2*(1), 1–11. DOI: 10.54368/qijmsrd.2.1.0003

Vitorino, J., Praça, I., & Maia, E. (2023). Towards adversarial realism and robust learning for IoT intrusion detection and classification. Annales Des Telecommunications. *Annales des Télécommunications*, *78*(7–8), 401–412. DOI: 10.1007/s12243-023-00953-y

Wach, K., Duong, C. D., Ejdys, J., Kazlauskaitė, R., Korzyński, P., Mazurek, G., Paliszkiewicz, J., & Ziemba, E. (2023). The dark side of generative artificial intelligence: A critical analysis of controversies and risks of ChatGPT. *Entrepreneurial Business and Economics Review*, *11*(2), 7–30. DOI: 10.15678/EBER.2023.110201

Waheed, T., & Marchetti, E. (2023). The Impact of IOT Cybersecurity Testing in the Perspective of Industry 5.0. International Conference on Web Information Systems and Technologies, WEBIST - Proceedings, Webist, 480–487. DOI: 10.5220/0012235800003584

Wang, J., Jiang, C., Zhang, H., Zhang, X., Leung, V. C., & Hanzo, L. (2017). Learning-aided network association for hybrid indoor LiFi-WiFi systems. *IEEE Transactions on Vehicular Technology*, *67*(4), 3561–3574. DOI: 10.1109/TVT.2017.2778345

Wang, J., Lu, X., Du, P., Zheng, H., Dong, Z., Yin, Z., Xing, J., Wang, S., & Hao, J. (2023). The Increasing Role of Synergistic Effects in Carbon Mitigation and Air Quality Improvement, and Its Associated Health Benefits in China. *Engineering (Beijing)*, *20*, 103–111. DOI: 10.1016/j.eng.2022.06.004

Wang, L., Zhao, X., Lu, Z., Wang, L., & Zhang, S. (2023). Enhancing privacy preservation and trustworthiness for decentralized federated learning. *Information Sciences*, *628*, 449–468. DOI: 10.1016/j.ins.2023.01.130

Wang, M., Lu, Y., & Qin, J. (2020). A dynamic MLP-based DDoS attack detection method using feature selection and feedback. *Computers & Security*, *88*, 101645. DOI: 10.1016/j.cose.2019.101645

Wang, R., Luo, M., Wen, Y., Wang, L., Choo, K. R., & He, D. (2021). The Applications of Blockchain in Artificial Intelligence. *Security and Communication Networks*, *2021*, 1–16. DOI: 10.1155/2021/8690662

Wang, S., Zhang, Y., Wang, K., Moustafa, H., & Zhang, K. (2020). Blockchain and AI for beyond 5G Networks. *IEEE Network*, *34*(6), 22–23. DOI: 10.1109/MNET.2020.9374644

Wang, Z., Guan, Z., Liu, X., Li, C., Sun, X., & Li, J. (2025). SDN Anomalous Traffic Detection Based on Temporal Convolutional Network. *Applied Sciences (Basel, Switzerland)*, *15*(8), 4317. DOI: 10.3390/app15084317

Wang, Z., Tsonev, D., Videv, S., & Haas, H. (2015). On the design of a solar-panel receiver for optical wireless communications with simultaneous energy harvesting. *IEEE Journal on Selected Areas in Communications*, *33*(8), 1612–1623. DOI: 10.1109/JSAC.2015.2391811

Whitepaper, S. S. S. (2018). Fundamental Practices for Secure Software Development. In *SAFECode Driving Security and Integrity* (2nd ed.). Issue.

Wisdom, D. D., Vincent, O. R., Igulu, K., Hyacinth, E. A., Christian, A. U., Oduntan, O. E., & Hauni, A. G. (2024). Industrial IoT security infrastructures and threats. *Communication Technologies and Security Challenges in IoT: Present and Future*, 369-402.

Wood, S. N. (2017). *Generalized additive models: An introduction with R* (2nd ed.). Chapman and Hall/CRC. DOI: 10.1201/9781315370279

Wu, T., Lucas, E., Zhao, F., Basu, P., & Qiao, Y. (2024). Artificial intelligence strengthenes cervical cancer screening - present and future. In *Cancer biology & medicine* (pp. 1–16). Chinese Anti-Cancer Association., DOI: 10.20892/j.issn.2095-3941.2024.0198

Wu, Z. Y., & Ismail, M. (2024). Generalized RIS tile exclusion strategy for indoor mmWave channels under concept drift. *IEEE Transactions on Wireless Communications*, *23*(10), 13484–13498. DOI: 10.1109/TWC.2024.3402267

Xiao, W., Liu, C., Wang, H., Zhou, M., Hossain, M. S., Alrashoud, M., & Muhammad, G. (2021). Blockchain for Secure-GaS: Blockchain-Powered Secure Natural Gas IoT System with AI-Enabled Gas Prediction and Transaction in Smart City. *IEEE Internet of Things Journal*, *8*(8), 6305–6312. DOI: 10.1109/JIOT.2020.3028773

Xuan, T. R., & Ness, S. (2023). Integration of Blockchain and AI: Exploring Application in the Digital Business. *Journal of Engineering Research and Reports*, *25*(8), 20–39. DOI: 10.9734/jerr/2023/v25i8955

Xue, Y., Singh, V., Singh, S., Kant, K., Pandey, S., Kumar, A., & Kaushik, V. D. (2024). Blockchain Framework for Learner Performance Prediction using Life-Brain Storm-based Light GBM Coupled Neural Network: BLOCKCHAIN FRAMEWORK FOR LEARNER PERFORMANCE PREDICTION. [JSIR]. *Journal of Scientific and Industrial Research*, *83*(6), 652–668.

Xu, G. (2021). Research on Network Intrusion Detection Method Based on Machine Learning. *Journal of Physics: Conference Series*, *1861*(1), 12034. DOI: 10.1088/1742-6596/1861/1/012034

Xu, Z., Li, B., & Cao, W. (2024). Enhancing Federated Learning-Based Social Recommendations with Graph Attention Networks. https://ssrn.com/abstract=4978530 DOI: 10.2139/ssrn.4978530

Yaacoub, J.-P. A., Noura, H., & Salman, O. (2023). Security of federated learning with IoT systems: Issues, limitations, challenges, and solutions. *Internet of Things and Cyber-Physical Systems*, *3*, 155–179. DOI: 10.1016/j.iotcps.2023.04.001

Yadav, A. & Kumar, A. (2022). Intrusion Detection and Prevention Using RNN in WSN. Inventive Computation and Information Technologies, 531–539.

Yang, J., Zhang, J., Xu, B., Zheng, J., Liu, Z., Liu, Z., Niyato, D., Debbah, M., Han, Z., & Ai, B. (2025). White-Box AI Model: Next Frontier of Wireless Communications. https://arxiv.org/abs/2504.09138

Yeboah-Ofori, A., Islam, S., Lee, S. W., Shamszaman, Z. U., Muhammad, K., Altaf, M., & Al-Rakhami, M. S. (2021). Cyber Threat Predictive Analytics for Improving Cyber Supply Chain Security. *IEEE Access : Practical Innovations, Open Solutions*, *9*, 94318–94337. DOI: 10.1109/ACCESS.2021.3087109

Yin, X., Zhu, Y., & Hu, J. (2021). A comprehensive survey of privacy-preserving federated learning: A taxonomy, review, and future directions. *ACM Computing Surveys*, *54*(6), 1–36. DOI: 10.1145/3460427

Yodha, V. (2024). Leveraging Federated Learning for Privacy-Preserving Cybersecurity in Decentralized AI Systems. *Eastern European Journal for Multidisciplinary Research*, *1*(1), 89–106.

Zeadally, S., Adi, E., Baig, Z., & Khan, I. A. (2020). Harnessing artificial intelligence capabilities to improve cybersecurity. *IEEE Access : Practical Innovations, Open Solutions*, *8*, 23817–23837. DOI: 10.1109/ACCESS.2020.2968045

Zeng, Z., Li, Y., Cao, Y., Zhao, Y., & Zhong, J. (2020). Blockchain Technology for Information Security of.

Zeng, Z., Soltani, M. D., Wang, Y., Wu, X., & Haas, H. (2020). Realistic indoor hybrid WiFi and OFDMA-based LiFi networks. *IEEE Transactions on Communications*, *68*(5), 2978–2991. DOI: 10.1109/TCOMM.2020.2974458

Zhang, X., Hou, H., Zhao, F., & Wang, Z. (2021). Industrial Internet Federated Learning Driven by IoT Equipment ID and Blockchain. Wireless Communications and Mobile Computing, 2021(1). DOI: 10.1155/2021/7705843

Zhang, S., Li, Z., Chen, Q., Zheng, W., Leng, J., & Guo, M. (2021, August). Dubhe: Towards data unbiasedness with homomorphic encryption in federated learning client selection. In *Proceedings of the 50th International Conference on Parallel Processing* (pp. 1-10). DOI: 10.1145/3472456.3473513

Zhang, W., Zhou, S., Walsh, T., & Weiss, J. C. (2025). Fairness amidst non-IID graph data: A literature review. *AI Magazine*, *46*(1), e12212. Advance online publication. DOI: 10.1002/aaai.12212

Zhang, X., Li, Y., Xiong, Z., Liu, Y., Wang, S., & Hou, D. (2024). A Resource-Based dynamic pricing and forced forwarding incentive algorithm in socially aware networking. *Electronics (Basel)*, *13*(15), 3044. DOI: 10.3390/electronics13153044

Zhang, Y., Li, B., & Tan, Y. (2021). Making AI available for everyone at anywhere: A Survey about Edge Intelligence. *Journal of Physics: Conference Series*, *1757*(1), 12076. DOI: 10.1088/1742-6596/1757/1/012076

Zhang, Z., Al Hamadi, H., Damiani, E., Yeun, C. Y., & Taher, F. (2022). Explainable Artificial Intelligence Applications in Cyber Security: State-of-the-Art in Research. *IEEE Access : Practical Innovations, Open Solutions*, *10*, 93104–93139. DOI: 10.1109/ACCESS.2022.3204051

Zhao, G., Li, Y., & Xu, Q. (2022). From Emotion AI to Cognitive AI. International Journal of Network Dynamics and Intelligence, 65–72. DOI: 10.53941/ijndi0101006

Zhao, W., Weng, T., Ruan, Y., Liu, Z., Wu, X., Zheng, X., & Kato, N. (2024). Quantum Computing in Wireless Communications and Networking: A Tutorial-Cum-Survey. *IEEE Communications Surveys and Tutorials*.

Zhong, S., Chen, J., & Yang, Y. (2004). Sprite: A simple, cheat-proof, credit-based system for mobile ad-hoc networks. *Proc. Of INFOCOM*, *3*, 1987–1997. DOI: 10.1109/INFCOM.2003.1209220

Zhou, K. R., & Gattinger, G. (2024). The Evolving Regulatory Paradigm of AI in MedTech: A Review of Perspectives and Where We Are Today. In Therapeutic Innovation & Regulatory Science (Vol. 58, Issue 3, pp. 456–464). Springer Science+Business Media. https://doi.org/DOI: 10.1007/s43441-024-00628-3

Zhou, X., Liang, W., Kevin, I., Wang, K., Yan, Z., Yang, L. T., & Jin, Q. (2023). Decentralized P2P federated learning for privacy-preserving and resilient mobile robotic systems. *IEEE Wireless Communications*, *30*(2), 82–89. DOI: 10.1109/MWC.004.2200381

Zhou, Z., Onireti, O., Xu, H., Zhang, L., & Imran, M. (2024). AI and Blockchain Enabled Future Wireless Networks: A Survey And Outlook. *Distributed Ledger Technologies: Research and Practice*, *3*(3), 1–30. DOI: 10.1145/3644369

Zhukabayeva, K. A. & N., A. and. (2024). Design of Intrusion Detection and Security Enhancements in Military Wireless Sensor Networks. 2024 7th International Conference on Signal Processing and Information Security (ICSPIS), 1–6.

Zhukabayeva, T., Zholshiyeva, L., Karabayev, N., Khan, S., & Alnazzawi, N. (2025). Cybersecurity Solutions for Industrial Internet of Things–Edge Computing Integration: Challenges, Threats, and Future Directions. *Sensors (Basel)*, *25*(1), 213. Advance online publication. DOI: 10.3390/s25010213 PMID: 39797003

Zilberman, P., Puzis, R., & Elovici, Y. (2017). On network footprint of traffic inspection and filtering at global scrubbing centers. *IEEE Transactions on Dependable and Secure Computing*, *14*(6), 588–601. DOI: 10.1109/TDSC.2015.2494039

Zubow, A., Gawłowicz, P., Bober, K. L., Jungnickel, V., Habel, K., & Dressler, F. (2021, June). Hy-Fi: aggregation of LiFi and WiFi using MIMO in IEEE 802.11. In 2021 IEEE 22nd International Symposium on a World of Wireless, Mobile and Multimedia Networks (WoWMoM) (pp. 100-108). IEEE.

Zuech, R., Hancock, J., & Khoshgoftaar, T. M. (2022). A new feature popularity framework for detecting cyberattacks using popular features. *Journal of Big Data*, *9*(1), 119. Advance online publication. DOI: 10.1186/s40537-022-00661-9

Zuo, Y., Guo, J., Gao, N., Zhu, Y., Jin, S., & Li, X. (2023). A Survey of Blockchain and Artificial Intelligence for 6G Wireless Communications. *IEEE Communications Surveys and Tutorials*, *25*(4), 2494–2528. DOI: 10.1109/COMST.2023.3315374

About the Contributors

Valentina Emilia Balas Aurel Vlaicu University of Arad, 77 Revoluţiei Boulevard, Arad, 310130, Romania Academy of Romanian Scientists, 3 Ilfov Street, Bucharest, 050044, Romania balas@drbalas.ro ; valentina.balas@uav.ro Valentina E. Balas is currently Full Professor in the Department of Automatics and Applied Software at the Faculty of Engineering, "Aurel Vlaicu" University of Arad, Romania. She holds a Ph.D. Cum Laude, in Applied Electronics and Telecommunications from Polytechnic University of Timisoara. Dr. Balas is author of more than 400 research papers in refereed journals and International Conferences. Her research interests are in Intelligent Systems, Fuzzy Control, Soft Computing, Smart Sensors, Information Fusion, Modeling and Simulation. She is the Editor-in Chief to International Journal of Advanced Intelligence Paradigms (IJAIP) and to International Journal of Computational Systems Engineering (IJCSysE), member in Editorial Board member of several national and international journals and is evaluator expert for national, international projects and PhD Thesis. Dr. Balas is the Head of Intelligent Systems Research Centre in Aurel Vlaicu University of Arad and Head of the Department of International Relations in the same university. She served as General Chair of the International Workshop Soft Computing and Applications (SOFA) in ten editions organized in the interval 2005-2022 and held in Romania and Hungary. Dr. Balas participated in many international conferences as Organizer, Honorary Chair, Session Chair, member in Steering, Advisory or International Program Committees and Keynote Speaker. Recently she was working in a national project with EU funding support: BioCell-NanoART = Novel Bio-inspired Cellular Nano-Architectures - For Digital Integrated Circuits, 3M Euro from National Authority for Scientific Research and Innovation. She is a member of European Society for Fuzzy Logic and Technology (EUSFLAT), member of Society for Industrial and Applied Mathematics (SIAM) and a Senior Member IEEE, member in Technical Committee – Fuzzy Systems (IEEE Computational Intelligence Society), chair of the Task Force 14 in Technical Committee – Emergent Technologies (IEEE CIS), member in Technical

Committee – Soft Computing (IEEE SMCS). She is member in the Committee of IEEE Romania Section as Volunteers Training Coordinator and vice chair of IEEE Computational Intelligence Society Chapter – CIS 11. During the interval 2021-2022 she was a member of IEEE European Public Policy Committee Working Group on ICT. From May 2023 Dr. Balas is associate member of Romanian Academy of Scientists. Dr. Balas was past Vice President (awards) of IFSA - International Fuzzy Systems Association Council (2013-2015), is a Joint Secretary of the Governing Council of Forum for Interdisciplinary Mathematics (FIM), - A Multidisciplinary Academic Body, India. She is the recipient of the "Tudor Tanasescu" Prize from the Romanian Academy for contributions in the field of soft computing methods (2019) and "Stefan Odobleja" Prize from Romanian Academy of Scientists (2023) and Diploma – Section Information Technology from The General Association of the Engineers in Romania (AGIR) 2023.

Vineeta Singh is presently associated with Department of Computer Science & Engineering, School of Engineering and Technology, Chhatrapati Shahu ji Maharaj University Kanpur, India. She has more than 10 years of teaching experience in different Engineering and Technological Institutions. She has guided more than 220 B.Tech. level projects and more than 2 M.Tech. level Projects. She is Ph.D. in CSE from the Harcourt Butler Technical University Kanpur, India and M.Tech from Dr. APJ Abdul Kalam Technical University Lucknow as well as B.Tech with honours from UPTU Lucknow India. Her area of research includes Deep learning, Image Processing, Cyber Security, Data Mining and Biometrics. She has 43+ Scopus/SCIE Publications and 20+ are in pipeline. She also worked as reviewer in different international conferences organized by IIT BHU (ANTIC-BHU) Varanasi and NIT Arunachal Pradesh. She is also working as reviewer in SCIE indexed journals such as IET Image Processing, Scientific Reports, Signal Image and Video Processing, Journal of Scientific and Industrial Research and many more. She is also associated as an editor with five ongoing projects as edited Books pertaining to renowned publishers such as Bentham Science, Wiley Scrivener. She has two Indian Granted Patents "Artificial Tree for oxygen production and atmospheric purification" and SMART BOOK FOR TRACKING STUDENTS' LEARNING PROGRESS and third one is in process. She has three design based grants accepted and 6 in process. She has one UK design grant.

Kapil Joshi is an Associate Professor in the department of Computer Science & Engineering (CSE) at Uttaranchal Institute of Technology (UIT), Uttaranchal University, Dehradun, Uttarakhand, India. He has completed his Ph.D, topic on "Image Quality Enhancement using Fusion Technique" in computer science and engineering from Uttaranchal University, Dehradun. He has more than 10 years' experience in industry, academic, and research. His areas of interest are Operating System, Computer Networks, Web Technology, Data Structure and Java etc. His research areas are deep learning, CNN, Computer vision etc. He has published various patents, 400+ research papers in reputed journals (UGC/Scopus) as well as published 20+ books in international publication. Dr. Kapil Joshi has published more than 50+ research papers in various international journals (including ISI/ SCI indexed). He has also been associated with many conferences at National and International level and organized many conferences like STEM-2017, ICACDS-18, ICAESMT-19, ICAEST-20, ICAEST-21, ICIS-22, ICNN-22, IDVRP-22, ICAEST-22 and ICAEST-23. He was also a student member of IEEE UP section in session 2019-20 and editorial member, guest editor and technical reviewers in various reputed national/international journals like Springer, Elsevier and Taylorfrancis etc. Dr. Joshi organized several conferences and has been invited as a key-speaker in various workshops/conferences. Dr. Joshi has filed/published more than 30 patents. He has written 5+ books and more than 20+ book chapters in the reputed publications.

***

Omkar Agrahari Faculty at Department of Computer Applications, School of Engineering and Technology, CSJM University, Kanpur No of Research Publication: 05 No of Patent Published: 06 22 year teaching experience Qualification: BSc, MCA, Ph.D Persuing Research Area: Software Engineering, Bleanded Learning, Deep Learning, Data Mining

Smita Agrawal is working as an Associate Professor in Computer Science and Engineering Department since 2009. She has a teaching experience of over 20 years. She received her Master of Computer Applications degree from Gujarat Vidhyapith in 2004. She completed her doctoral studies in the field of Big Data Analytics from CHARUSAT University in 2019. She works in the area of Big Data Analytics, Parallel Processing, Web Development and IoT. She is a Senior Member of IEEE, Member of CSI, ACM, and ISTE. She has published several Scopus/SCIE-indexed research papers in national and international conferences and journals. Dr Smita has conducted ISTE approved short-term training Programme in the field of Web Services using PHP. She is involved in teaching courses at both undergraduate and postgraduate levels. She also serves as a member of the Programme committees and session chair for international conferences and as a reviewer for indexed international journals.

Zeinab Ahmed holds a Ph.D. in Computer Engineering and Networks from the University of Gezira, Sudan. Since June 2020, she has served as an Assistant Professor in the Department of Computer Engineering at the same university. Currently, she is a Postdoctoral Fellow at the Department of Electrical and Computer Engineering, International Islamic University Malaysia (IIUM), Malaysia. With extensive academic leadership experience, Dr. Zeinab has previously held the position of Head of the Department of Computer Engineering at the Faculty of Engineering Technology, University of Gezira, Sudan. She has actively contributed to various projects in computer engineering and networking and has authored more than thirty research papers and book chapters in peer-reviewed journals and conferences. Her primary research interests lie in wireless communication networks. As an accomplished academic, she brings expertise in research, e-learning, programming, and lecturing, with a strong track record in higher education.

Altahir Abdalla Altahir Mohammed received his B.Sc. (Hons.) degree in Computer Engineering from University of Gezira, Sudan in 2006, the M.Sc. and the Ph.D. degrees in Electrical and Electronic Engineering from Universiti Teknologi Petronas (UTP), Malaysia, in 2009 and 2018. He currently serves as Assistant Professor with Universiti Tunku Abdul Rahman, Malaysia. His research interests include machine learning, signal and image processing, computer vision, discrete optimization applications and Networks Modeling

Nizirwan Anwar, MT, IPM, ASEAN.Eng is a Lecturer in the Informatics Engineering Study Program, Faculty of Computer Science, Universitas Esa Unggul under the auspices of the Kemala Bangsa Education Foundation (YPKB). The author was born in the city of Bandung on July 24 1964, completed his bachelor's degree from the Physics Study Program, Faculty of Mathematics and Natural Sciences, Padjadjaran University, Bandung in 1989 and continued the Electrical Engineering Study Program, Faculty of Engineering (dh. Postgraduate Study Program) University of Indonesia, Jakarta, completing his studies in 1995, obtaining a Bachelor of Engineering Intermediate Professional (IPM) and ASEAN Engineer (ASEAN.Eng). Alhamdulillah has produced several reputable and indexed published scientific papers at national and international levels (research a categories; Internet of Things, Cryptography, Bibliometrics and Data Science). To date, he is registered as active in several professional organizations, National Lecturer BKD assessor (NIRA), as a reviewer in Arjuna accredited journals (Kemendikbud Saintek); Scientific Committe, Central Executive of the Vocational Agency BKI PII,

Imam Asrowardi, S.Kom., M.Kom., M.Pd., IPM., ASEAN Eng. has been a lecturer in the Internet Engineering Technology Study Program at Politeknik Negeri

Lampung since 2015 and currently holds the position of Associate Professor. He is actively involved in various national professional associations (IAII, APTIKOM, Informatics Engineering Division of the Indonesian Engineers Association) and international organizations (IAENG), both as a member and an administrator. Imam holds numerous industry certifications in various fields, including MTCINE, MTCRE, MTCNA, MTCTE, MTCWE, MTCUME, MTCIPv6E, MTA (Networking), CCC-BDF, EBDP, CSEP, Cisco Academy Trainer (CCNA, CyberOps Associate, DevNet Associate, CCNP), MikroTik Academy Trainer, Programmer (BNSP), Competency Assessor (BNSP), National Procurement Expert (BNSP), and Cisco Certified Support Technician Networking (CCST Networking).

Shubham Bajpai is presently working as an Assistant Professor of Plant Protection in School of Advanced Agriculture Sciences and Technology at Chhatrapati Sahu Ji Maharaj University, Kanpur, Uttar Pradesh. He has received his graduate degree in Agriculture (2011) from the C.S.J.M. University Kanpur. He obtained his Post Graduation (2013) and Doctoral degree (2019) from University of Allahabad. His research expertise in field of Plant Protection. He has published many research papers, book chapters in leading national and international journals. He worked as Technical Officer in Plant Pathology at Regional Plant Quarantine Station, Kolkata, where he contributed in disease detection and diagnosis of plant pathogens from the export and import agricultural consignments. Dr. Bajpai has been conferred with Best Young Scientist Award-2025.

Shubham Bajpai is presently working as an Assistant Professor of Plant Protection in School of Advanced Agriculture Sciences and Technology at Chhatrapati Sahu Ji Maharaj University, Kanpur, Uttar Pradesh. He has received his graduate degree in Agriculture (2011) from the C.S.J.M. University Kanpur. He obtained his Post Graduation (2013) and Doctoral degree (2019) from University of Allahabad. His research expertise in field of Plant Protection. He has published many research papers, book chapters in leading national and international journals. He worked as Technical Officer in Plant Pathology at Regional Plant Quarantine Station, Kolkata, where he contributed in disease detection and diagnosis of plant pathogens from the export and import agricultural consignments. Dr. Bajpai has been conferred with Best Young Scientist Award-2025.

Eyman F. A. Elsmany received her B.Sc. (Hons.) and M.Sc. degrees in computer engineering, computer engineering and networking from Faculty of Engineering and Technology, University of Gezira, Sudan in 2006, and 2009, and her Ph.D. degree from School of Computer Sciences, Universiti Sains Malaysia in 2020. Currently she serves as an assistant professor and the Head of Computer Engineering

Department with the Faculty of Engineering and Technology, University of Gezira, Sudan. Her research interests include Networks Modeling, IoT and Cybersecurity.

Muhammad Faisal, received his S.SI degree from the Department of Information Systems, STMIK Profesional, Makassar, Indonesia, at 2011, and an M.T degree from the Masters in Information Technology, Postgraduate School, Hasanuddin University, Makassar, Indonesia at 2014. Furthermore, Ph.D degree from the School and Technology Information, Doctoral Programme, Asia E University Malaysia. He is currently a lecturer and researcher at the Department of Informatics, Universitas Muhammadiyah Makassar, Indonesia. His current research interests include Mobile Programming, Data Mining, Decision Support System, Expert System, Web Applications, Image Processing, and Machine Learning.

S. Ida Evangeline, presently working as an Assistant Professor in the Department of Electrical and Electronics Engineering at the Government College of Engineering, Tirunelveli, Tamilnadu, India. She obtained her Bachelor's in Electrical and Electronics Engineering from Manonmaniam Sundaranar University in 2003, and Master's in Power Systems Engineering from Anna University, Chennai, Tamilnadu, India in 2005. She completed her PhD degree in 2023 at Anna University, Chennai, Tamilnadu, India. She has published six research papers in international journals. Her current research interests include optimal power flow, renewable energy, soft computing, machine learning, and deep learning.

Swaminathan K. is working in Department of ECE, SRM Institute of Science & Technology, Ramapuram Campus, Chennai an accomplished academic and researcher in Electronics and Communication Engineering, has made significant contributions to the fields of wireless communication, IoT, and network security. He holds a Ph.D. from Anna University and has been actively involved in teaching and research. His scholarly output includes 24 Scopus-indexed papers, book chapters, and conference presentations, along with a publication in an SCI-indexed journal. Additionally, he has authored 11 patents, with 10 at the national level and one internationally recognized. He has been granted2 IPRs in the domain of WSN and healthcare. He has served in multiple editorial roles, including positions in prominent Scopus-indexed journals. He has also contributed to the editorial boards of other notable journals, such as the Spectrum Journal and the Journal of Wireless Sensor Networks and IoT. His research and editorial expertise reflect his commitment to advancing interdisciplinary knowledge and fostering technological progress.

Gurpreet Kaur is an Assistant Professor in the BCA Department at RKGIT-CCSU Campus College, Ghaziabad. With over 10 years of teaching experience in Computer Science, she brings a deep understanding and passion for the subject to her students. She holds a postgraduate degree in Computer Applications and has contributed to the academic community through the publication of national research papers in recognized journals.

Amrutha Kolhar obtained the MCA. Degree in Computer Science from the Karnatak University in 2022. she is currently working as a Research Scholar under the guidance of Dr. Sridevi, Professor, Department of Computer Science, Karnatak University, Dharwad. Received research fellowship from Other Backward Classes, (OBC), and Govt. of Karnataka. Area of research is "Network Security". At present working on the research topic: "Development and enhancing inter-operability of API and validating its authentication: X-IoT Devices". and also, IAENG student member (International Association of Engineers)(Member no: 366909)

Raunak Kumar is a Senior Manager at a leading enterprise AI company with over 10+ years of professional experience spanning marketing analytics, data science, and applied artificial intelligence. His research interests include AI-driven attribution modeling, predictive analytics, and open-source data architectures for large-scale decision systems. He has presented his work at international forums such as MADS West and the Open Source Data Summit (ODSS), focusing on data-centric growth strategies and scalable marketing measurement frameworks. Mr. Kumar holds an M.S. in Electrical and Computer Engineering from Rutgers University and a B.Tech. in Electrical Engineering from Nagpur University.

Mayada S. A. Mustafa received her B.Sc. degree in Telecommunication and Control from the University of Gezira, Sudan, in 1998. She obtained the M.Sc. and Ph.D. degrees in Communications and Network Engineering from Universiti Putra Malaysia in 2003 and 2019, respectively. She is assistant prof and her current research interests include communication engineering, wireless networks, cybersecurity, and the application of unmanned aerial vehicles (UAVs) in communication systems

Praveen Kumar Myakala is a seasoned software engineer and independent researcher with a strong passion for innovation, education, and the transformative potential of data-driven technologies. He holds a Master's degree in Data Science from the University of Colorado Boulder and has published research across key areas including artificial intelligence, federated learning, edge computing, and decentralized systems. His work bridges academic rigor with real-world application, aiming to develop intelligent solutions that are both technically robust and socially impactful. Through his writing, mentorship, and hands-on development, he actively contributes to advancing the fields of ethical AI, privacy-preserving computing, and scalable system design.

Nurdiansyah is an academic and researcher who earned his bachelor's degree in Information Systems from STMIK Dipanegara, followed by a master's degree in Educational Research and Evaluation from Universitas Negeri Makassar. He later completed his doctoral studies in Educational Sciences at Universitas Negeri Makassar, focusing on the development of research methodologies, the evaluation of educational systems, and the integration of information technology to enhance the quality of learning and educational management.

Parita Oza is a faculty member in the Department of Computer Science and Engineering at Nirma University, Ahmedabad. She completed her Ph.D. focusing on breast cancer diagnosis using deep learning. Her work contributed to the development of the first Indian mammography dataset. Dr. Oza has published several papers in reputed journals and conferences, with research interests in medical image analysis and AI in healthcare.

Alders Paliling received his S.Kom degree in Informatics Engineering from STMIK Dipanegara Makassar, Indonesia, in 2010, and an M.T degree in Electrical Engineering with a concentration in Informatics Engineering from Hasanuddin University, Makassar, Indonesia, in 2014. He is currently a lecturer and researcher at the Department of Computer Science, Faculty of Information Technology, Universitas Sembilan Belas November Kolaka, Indonesia. His research interests include Expert Systems, Web Applications, Machine Learning, Deep Learning and Augmented Reality.

Shubhamkumar Pandit Being a researcher and developer, Shubhamkumar is interested in knowledge retrieval and decision support based on artificial intelligence greatly. He also puts an emphasis on transferring research into practice covering scalable and modifiable AI applications. He is determined to make contributions to interdisciplinary discussions that will define the future of the intelligent systems and its contributions in developing information accessibility and innovation in business.

Sonukumar Pandit is a full-stack developer and a machine learning enthusiast whose core skills lie in Java, Python and MERN stack. He has worked on the breakthrough projects such as AI-based healthcare providers and smart inventory management. Having secured a place among the top few at hackathons nationwide, Sonukumar is a combination of both technical skills and creativity through problem-solving. He has experience in backend development, deep learning and real-time web applications. He is an enthusiast of technology and the benefits it can bring to the world and as such, he works on the development of solutions that matter.

Kiruba Buri R. presently serves as a Teaching Fellow in the Department of Computer Science and Engineering at the University College of Engineering, Anna University, Pattukottai Campus. He earned his Bachelor of Engineering (B.E.) degree in Computer Science and Engineering from Anna University, Chennai, in 2010, followed by a Master of Engineering (M.E.) degree from the (from Anna University, Chennai) same institution in 2013. Additionally, he attained a Master of Business Administration (M.B.A.) with a specialization in Systems from Manonmaniam Sundaranar University in 2017. In 2024, he was conferred a Doctor of Philosophy (Ph.D.) in Information and Communication Engineering by Anna University, Chennai. He has served in multiple editorial roles, including positions in prominent SCOPUS-indexed journals and his dedication to education and innovation to expertise reflect his commitment to advancing interdisciplinary knowledge and fostering technological progress. His scholarly interests encompass the Internet of Things (IoT), Networking, Wireless Ad Hoc Networks, Big Data, Cloud Computing, and Machine Learning.

Ravikumar R. Natarajan is an experienced academician with over 12+ years of teaching and research expertise across institutions in India, the UAE, and the Maldives. He currently serves as an Assistant Professor at Marwadi University, Rajkot, specializing in Machine Learning, Recommender Systems, and AI-driven applications in healthcare and security. He holds a Ph.D. in Machine Learning and Recommender Systems from Amity University, Jaipur, a Master of Engineering in CSE from VMRF University, Salem, and a Bachelor of Engineering in CSE from Anna University. He also has a MiniMBA from the Swiss eLearning Institute. Dr. Ravikumar has authored more than 8+ multiple SCIE/Scopus-indexed journal articles, 25+ IEEE/Springer conference papers, and co-authored a book on Artificial Intelligence. He has also published 9 patents and contributed 25+ book chapters in emerging domains such as Federated Learning, Blockchain, and Digital Twins for IGI Global, Wiley and Springer publisher. He actively mentors students through initiatives like the Research Activities Club (RAC) and Student Research Internship Training (SRIT), promoting innovation and research culture.

Aedah Abd Rahman is a distinguished academic at the Faculty of Education, Universiti Kebangsaan Malaysia (UKM), where she serves as a Professor and an active researcher in the field of Educational Technology. She holds a PhD in Educational Technology from UKM, a Master's degree in the same field from the University of Malaya, and a Bachelor's degree in Mathematics from UKM. Her expertise lies in instructional design, digital learning environments, ICT integration in education, and mobile learning innovations. Prof. Aedah is particularly known for her work in gamification, e-learning strategies, and blended learning models, which have been widely applied in national educational reform initiatives. She has led several research projects, including the "Innovative Learning Design for Future Classrooms" funded by the Malaysian Ministry of Higher Education, and has co-led national evaluations of digital pedagogy strategies. Her scholarly contributions include numerous publications in peer-reviewed journals such as the Malaysian Journal of Learning and Instruction and the International Journal of Instructional Technology. In addition to her academic roles, she actively contributes as a journal editor, policy advisor, and trainer in MOOCs development and digital education transformation. Prof. Aedah continues to play a vital role in shaping Malaysia's future-ready education ecosystem through her research, teaching, and national-level engagements.

TItik Hawa Abd Rahman, received a B.Sc. in Electrical and Electronics Engineering from Loughborough University of Technology, UK. She holds a Ph.D in Power Systems from the University of Malaya. She is the current Deputy Vice Chancellor of Asia e University. Her expertise includes artificial intelligence, biologically inspired optimization, machine learning, power system analysis and operation, and smart grid.

Madhura Raut is a Principal Data Scientist at leading enterprise company, where she leads the design of large-scale machine learning systems. Her work integrates time series modeling, AI, and reinforcement learning to enable real-time workforce optimization. She is the lead inventor on two U.S. patents related to advanced time series techniques. With over nine years of experience spanning enterprise machine learning and applied research, Madhura is passionate about building intelligent, production-grade systems that address complex operational challenges. Besides her professional work, she also serves as an IEEE Senior Member, IOA Fellow and mentor, actively contributing to the broader AI and data science community. Madhura has been keynote speaker at many prestigious data science conferences including KDD 2025 and has served as judge and mentor to many codecrunch hackathons.

Tomy Ronaldi is Doctoral Student at Indonesia Defense University, Bogor Indonesia.

Aarthi S is a dedicated Assistant Professor in Computer Science with over six years of teaching experience. She is currently pursuing her Ph.D. at Marwadi University, Gujarat, and holds an M.Phil. and MCA from Bharathiar University, Tamil Nadu. Throughout her academic career, Aarthi has specialized in programming languages such as C++, Java, and C, and contributed to research in recommender systems and AI-related data privacy. She has published journal articles, contributed to book chapters, and actively participated in technical workshops. Known for her student mentorship and engagement in academic events, Aarthi remains committed to fostering innovation and excellence in computer science education.

Shreya Bhaveshbhai Sejani was born and raised in Gujarat, India, and is currently based in Ahmedabad. She has recently completed her Bachelor of Technology (B.Tech) in Computer Science and Engineering from Nirma University. With a strong foundation in full-stack development and experience in ERP systems, microservices, and process automation, she brings both technical proficiency and a passion for innovation to her work. Outside of academics, she is deeply curious about the intersection of software systems and real-world impact, and she actively seeks out opportunities to contribute to research-driven, forward-thinking projects. As an aspiring technologist and researcher, she aims to be part of shaping smarter, more sustainable digital futures.

Rajasrkaran Selvaraju Having 22+ years of teaching and research experience in the fields of Computer Science and Information Technology. Presently, he is working at the University of Technology and Applied Sciences - Ibri in the Sultanate of Oman. He has acted as a resource person for various international conferences, webinars and workshops. Also, serving as a reviewer for many reputed international journals and conferences. Received many awards from various institutions. His area of interest is not limited to Data Science, Artificial Intelligence and Machine Learning, Internet of Things and Software Engineering.

Bhavya Shah is a final-year undergraduate student in Computer Science and Engineering at the Institute of Technology, Nirma University. His research interests lie in the fields of Artificial Intelligence, Machine Learning, Healthcare technologies, and the Internet of Things (IoT).

Anshika Sharma received her B.Tech. degree in Computer Science and Engineering from the University Institute of Engineering and Technology (UIET), CSJM University, Kanpur, in 2025, where she was awarded the Gold Medal for academic excellence. She is currently pursuing an M.Tech. at the National Institute of Technology (NIT), Raipur. Her research interests include Machine Learning, Deep Learning, Computer Vision, Image Segmentation, Artificial Intelligence, and Pattern Recognition. She is particularly focused on applying intelligent computational techniques to solve real-world challenges.

Vikas Sharma is an active academician and researcher in the field of computer science with 8 years of experience. He earned M. Tech degree specialized in Computer Science. He is involved in academic development and scholarly activities. He is member of International Association of Engineers, The Society of Digital Information and Wireless Communications, Global Professors Welfare Association, International Association of Academic plus Corporate (IAAC), Computer Science Teachers Association and Internet Society. His research interest includes Cloud Computing, Artificial Intelligence, Machine Learning; published more than 55 research articles with 4 books and 90 book chapters in edited books. He is contributing as member of editorial and reviewers' boards in conferences and various computer journals and edited books published by Elsevier, CRC Press, Taylor and Francis, Springer, IGI global, Wiley Publication, IET Publication, De Gruyter, River Publication etc.

Hemant Soni is a seasoned technology and product leader driving AI-powered digital transformation across Telecommunications, Media, and Technology. With over two decades of global experience, he specializes in AI platform architecture and IoE-based customer experience optimization, delivering impactful solutions that modernize infrastructure, elevate engagement, and accelerate innovation. Hemant is recognized for his strategic vision, technical excellence, and thought leadership in intelligent automation and enterprise AI.

Viraj Soni is the AVP of Portfolio Management Strategy at Merrick Bank, where he leads enterprise-wide initiatives to optimize fraud strategy, risk controls, and the customer experience. With nearly a decade of hands-on experience spanning banking, fintechs, and start-ups, Viraj has delivered transformative results—by reimagining authorization strategy and implementing cutting-edge fraud decisioning engines. His leadership includes significant roles at PayPal, Zip Co, and Citi, consistently driving measurable improvements in fraud prevention, approval rates, and operational efficiency across diverse financial segments. Viraj holds a Master's in Information Management from Syracuse University, along with a certificate in Data Science. Renowned for blending data-driven innovation with practical strategy, he is passionate about secure, seamless finance in rapidly evolving digital environments. Viraj's research focuses on big data analytics and advanced fraud detection methodologies, contributing to data-driven solutions that enhance risk identification and prevention in financial services.

Sridevi, Professor, Department of Computer Science, Karnatak University, Dharwad. Obtained a PhD from Mangalore University, Mangalore in 2017. Research interest includes Advanced Computer Networks, Mobile and Wireless Communication, Network Security, Cloud Computing and the Internet of Things. At present guiding four research scholars and one M.Phil. awarded. And also, Reviewer for the Journal of Advances in Mathematics and Computer Science. Published more than 30 research papers in a national and international journal.

Musdalifa Thamrin, Born in Ujung Pandang on February 11, 1971, completed a Bachelor of Computer Science (S.Kom.) majoring in Informatics Engineering in 2009, and a Master of Computer Science (M.Kom.) majoring in Computer Systems in 2015. Currently teaching at STMIK Profesional Makassar. Active as a member of the Association of Informatics and Computer Colleges (APTIKOM) and the Indonesian Open Source Centre of Excellence in Information Systems (INDOCEIS).

Darshana Upadhyay is a Research Scientist at TeleAI Corporation and holds an adjunct faculty position at Dalhousie University, where she co-supervises both graduate and undergraduate students in Dr. Sampalli's MYTech lab. She received her Ph.D. degree from the Faculty of Computer Science at Dalhousie University, Canada. She holds a master's degree in computer science from Nirma University, India, where she also served as an assistant professor. During her graduate studies, she was awarded the Gold Medal for securing the first position. Darshana's primary research focuses on network and information security, algorithm conceptualization, hardware design in the field of embedded systems, vulnerability assessments, and intrusion detection and prevention techniques for IoT/SCADA-based systems. She is the co-recipient of the Indo-Canadian Shastri research grant in the field of secure mobile communication. Additionally, she was selected as one of Canada's 2020 Emerging Thought Leaders by the Women in International Security - CANADA. Furthermore, she has received the 2021 Citizenship Award and the 2022 Leadership Award and also featured her portfolio on Alma website for mentoring the students at Dalhousie University.

Tarun Kumar Vashishth is an active academician and researcher in the field of Computer Science with 22 years of experience. He earned Ph.D. Mathematics degree specialized in Operations Research; served several academic positions such as Dy. Director, HoD, NAAC Coordinator, Academic Coordinator, Member Secretary of Department Research Committee, Assistant Center Superintendent and Head Examiner in University Examinations. He is involved in academic development and scholarly activities. He is member of International Association of Engineers, The Society of Digital Information and Wireless Communications, Global Professors Welfare Association, International Association of Academic plus Corporate (IAAC), Computer Science Teachers Association and Internet Society . His research interest includes Cloud Computing, Artificial Intelligence, Machine Learning and Operations Research; published more than 50 research articles & conference papers with 2 book and 80 book chapters. He is contributing as member of editorial and reviewers boards in conferences and various computer journals and edited books published by Elsevier, CRC Press, Taylor and Francis, Springer, IGI global, Wiley Publication, IET Publication, De Gruyter, River Publication etc.

Yashwant Waykar is a dedicated educator and researcher at Dr. Babasaheb Ambedkar Marathwada University, specializing in Software Engineering, Machine Learning, Artificial Intelligence, and Learning Management Systems (LMS). With over 14 years of experience and a Ph.D., Dr. Waykar has authored 40+ publications in prestigious journals indexed in SCOPUS, UGC CARE, and Web of Science. Their research has earned accolades and includes several book chapters and four patents (3 granted, 1 filed) in LMS and AI. Dr. Waykar's teaching in AI, Web Development, and Software Testing is highly regarded, fostering a dynamic learning environment. They have completed three research projects funded by UGC and Dr. B.A.M.U. and have mentored 3 M.Phil and 2 Ph.D. students, guiding them in Machine Learning and LMS research.

Sucheta S. Yambal has been an Assistant Professor in the Department of Management Science of Dr. Babasaheb Ambedkar Marathwada University in Chhatrapati Sambhaji Nagar, Maharashtra. Her total teaching experience is 19 years, and she has about 16 years of research experience. She has taken on several significant duties as a member of the Board of Directors, organized multiple workshops, and given guest lectures at numerous universities. Dr. Yambal has supervised four M.Phil. and four Ph.D. research students at Dr. Babasaheb Ambedkar Marathwada University to date. She has also reviewed Ph.D. theses for Tilak Maharastra Vidyapeeth, Pune. More than thirty national and international research papers have been published by her at prestigious institutes' international conferences, such as IIT Delhi and IIM Raipur. Her research has also been published in prestigious journals, such as SCOPUS and UGC- Care Listed. In November 2019, Dr. Yambal scored highly in the e-business course on the Nptel platform during an exam administered by IIT Kharagpur. She is deemed a "Women of Substance" 2018 Award which was organized at MGM, Chhtrapati Sambhaji Nagar. She had a stellar academic record from elementary school to the Dr. Babasaheb Ambedkar Marathwada University Ph.D. admission exam, where she received the best possible mark. Her areas of interest include employment, AI/ML, automation in industries, database management systems and data mining, modern and inclusive education, and employability.

Index

Symbols

A

B

C

D

www.ingramcontent.com/pod-product-compliance
Ingram Content Group UK Ltd.
Pitfield, Milton Keynes, MK11 3LW, UK
UKHW050843101026
15624UKWH00018B/472

9 798337 335766